CHAPMAN & HALL/CRC FINANCIAL MATHEMATICS SERIES

Financial Modelling With Jump Processes

CHAPMAN & HALL/CRC
Financial Mathematics Series

Aims and scope:
The field of financial mathematics forms an ever-expanding slice of the financial sector. This series aims to capture new developments and summarize what is known over the whole spectrum of this field. It will include a broad range of textbooks, reference works and handbooks that are meant to appeal to both academics and practitioners. The inclusion of numerical code and concrete real-world examples is highly encouraged.

Proposals for the series should be submitted to one of the series editors above or directly to:

CRC Press UK
23 Blades Court
Deodar Road
London SW15 2NU
UK

CHAPMAN & HALL/CRC FINANCIAL MATHEMATICS SERIES

Financial Modelling With Jump Processes

Rama Cont
Peter Tankov

CRC Press
Taylor & Francis Group
Boca Raton London New York

CRC Press is an imprint of the
Taylor & Francis Group, an **informa** business
A CHAPMAN & HALL BOOK

Chapman & Hall/CRC
Taylor & Francis Group
6000 Broken Sound Parkway NW, Suite 300
Boca Raton, FL 33487-2742

ISBN 13: 978-1-58488-413-2 (hbk)

Visit the Taylor & Francis Web site at
http://www.taylorandfrancis.com

and the CRC Press Web site at
http://www.crcpress.com

Library of Congress Card Number 2003063470

Library of Congress Cataloging-in-Publication Data

Cont, Rama.
Financial modeling with jump processes / Rama Cont, Peter Tankov.
p. cm. — (Chapman & Hall/CRC financial mathematics series)
Includes bibliographical references and index.
ISBN 1-58488-413-4 (alk. paper)
1. Finance—Mathematical models. 2. Jump processes. I. Tankov, Peter. II. Title. III. Series.

HG106.C66 2004
332'.01'519233—dc22 2003063470

To Rokhsaneh, for her patience and constant encouragement

To Maria

Preface

During the last decade Lévy processes and other stochastic processes with jumps have become increasingly popular for modelling market fluctuations, both for risk management and option pricing purposes. More than a hundred research papers related to this topic have been published to this date in various finance and applied mathematics journals, leading to a considerable literature which is difficult to master for the nonspecialist. The time seems therefore ripe for a book which can give a self-contained overview of the important aspects of this body of research in a way that can be relevant for applications.

While there exists a considerable volume of mathematical literature related to processes with jumps and Lévy processes in particular, this literature is quite technical and difficult to access for readers not specialized in stochastic analysis. On the other hand many of the applications of jump processes in financial modelling use fairly sophisticated analytical and probabilistic tools which are only explained in advanced mathematical texts. As a result, the recent body of research on the use of jump processes in financial modelling has been difficult to access for end users, who are sometimes under the impression that jump processes and Lévy processes are complicated notions beyond their reach.

We believe that it is not so; the concepts and tools necessary for understanding and implementing these models can be explained in simple terms and in fact are sometimes much more simple and intuitive than the ones involved in the Black-Scholes model and diffusion models.

The motivation for our manuscript is precisely to provide a self-contained overview of theoretical, numerical and empirical research on the use of jump processes in financial modelling, understandable by students, researchers and quants familiar with quantitative methods in finance at the level of classical Black-Scholes option pricing theory.

Our goal has been to:

- explain the motivation for using Lévy processes in financial modelling in terms understandable for nonspecialists

- motivate, through intuitive explanations, the necessity of introducing the various mathematical tools used in the modelling process

- provide precise mathematical statements of results while trying to avoid unnecessary technicalities

- emphasize clarity of exposition over the generality of results in order to maximize their ease of use in applications

- illustrate the mathematical concepts by many numerical and empirical examples

- provide details of numerical implementation for pricing and calibration algorithms

- provide real examples of uses of jump processes in option pricing and risk management

- provide a pedagogical exposition which can be used as teaching material for a graduate course for students in applied mathematics or quantitative finance.

The goal of the present work is *not*:

- to be a comprehensive treatise on mathematical properties of Lévy processes: such treatises already exist [49, 215, 345]. Here we intend to focus on the mathematical tools necessary in the context of financial modelling, thus omitting other mathematically interesting topics related to Lévy processes.

- to give the most general statement of mathematical results: we have preferred to explain a concept on an example relevant for financial modelling instead of giving abstract theorems.

- to provide an exhaustive survey of the literature on Lévy processes in finance: rather than presenting a catalogue of models we have emphasized common aspects of models in order to provide *modelling tools* and tried to illustrate their use through examples.

The first part of the book (Chapters 2, 3, 4 and 5) presents a concise introduction to the mathematical theory of Lévy processes — processes with independent stationary increments which are building blocks for constructing models with jumps. Chapter 2 presents some preliminary notions on probability spaces, random variables and the Poisson process. Lévy processes are defined in Chapter 3 and their main properties are discussed: behavior of sample paths, distributional properties, the Markov property and their relation to martingales. Examples of one-dimensional Lévy processes frequently used in mathematical finance are presented and studied in Chapter 4. Chapter 5 presents some multidimensional models and tools for building them.

The second part (Chapters 6 and 7) deals with simulation and estimation of models with jumps. Chapter 6 presents various methods for Monte Carlo simulation of Lévy processes in one or several dimensions. Chapter 7 discusses statistical properties of Lévy processes, their advantages and their drawbacks for modelling financial time series.

The third and longest part of the book (Chapters 8 to 13) focuses on option pricing models based on jump processes. After a short introduction to stochastic calculus for jump processes in Chapter 8, we study in Chapter 9 the concept of equivalent change of measure, its relevance in arbitrage pricing theory and its application to Lévy processes. This allows us to show that the models with jumps we consider correspond to arbitrage-free, *incomplete* markets. These notions are further developed in Chapter 10, where we review different approaches to option pricing and hedging in incomplete markets. We then focus on a tractable class of models with jumps: exponential-Lévy models, in which the price of the underlying asset is modelled by the exponential of a Lévy process. Chapter 11 explores properties of option prices in these models, using Fourier-based pricing methods. Option prices in exponential-Lévy models can also be expressed as solutions of certain integro-differential equations: these equations are derived in Chapter 12 and numerical algorithms for solving them are presented. Chapter 13 discusses the problem of model calibration: retrieving parameters of an option pricing model from market prices of options. Several algorithms for solving this problem are presented in the context of exponential-Lévy models and their implementation and empirical performance is discussed.

The last part of the book deals with models with jumps which are not in the exponential-Lévy class. The simplest extensions of exponential-Lévy models are models in which log-returns are independent but not stationary. These time-inhomogeneous models are discussed in Chapter 14. Stochastic volatility models are another important class of models which have been the focus of a lot of recent research. Chapter 15 discusses stochastic volatility models based on Lévy processes.

Some of the results in the book are standard and well known in the literature. In this case our effort has focused on presenting them in a pedagogical way, avoiding unnecessary technicalities, giving appropriate references for further reading when necessary. Some other parts — in particular the material in Chapters 5, 12, 13, 14 — are based on research work done by the authors and collaborators.

One important question was the level of mathematical detail used to treat all these topics. Many research papers dealing with financial applications of jump processes are so technical that they are inaccessible to readers without a graduate degree in probability theory. While technical content is unavoidable, we believe that an alternative exposition is possible, provided that generality is sacrificed in order to gain in clarity. In particular we have chosen to explain the main ideas using Poisson processes and Lévy processes which are tractable examples of models with jumps, mentioning semimartingales briefly in Chapter 8. Accordingly, we have adopted the approach to stochastic integration proposed by P. Protter [324] , which is more amenable to the applications considered here. Mathematical definitions and proofs are given in detail when we believe that they are important in the context of financial modelling: this is the case, for example, for the construction of stochastic integrals in Chapter 8. For results of purely "technical" nature we have given appropriate references. Sections with higher technical content are signaled by a (*) and can be skipped at first reading.

Another issue was the level of generality. What classes of models should be considered? Here the approaches in the financial modelling literature tend to be extreme; while some books are entirely focused on diffusion models and Brownian motion, others consider a knowledge of semimartingale theory more or less as a prerequisite. While semimartingales provide the general framework for stochastic integration and theoretical developments in arbitrage theory, financial modelling is focused on *computing* quantities so model building in quantitative finance has almost *exclusively* focused on Markovian models and more precisely *tractable* Markovian models, which are the only ones used in option pricing and risk management. We have therefore chosen to develop the main ideas using Lévy processes, which form a tractable subclass of jump processes for which the theory can be explained in reasonably simple terms. Extensions beyond Lévy processes are considered in the last part of the book, where time-inhomogeneous models and stochastic volatility models with jumps are considered.

We have assumed that the reader is typically familiar with the Black-Scholes model and the machinery behind it — Brownian motion, the Itô formula for continuous processes — but have tried to explain in detail notions specific to jump processes. In particular the Poisson process seems not to be known to many students trained in Black-Scholes theory!

We have tried not to give a catalog of models: since research papers typically focus on specific models we have chosen here a complementary approach namely to provide tools for building, understanding and using models with jumps and referring the reader to appropriate references for details of a particular model. The reader will judge whether we have succeeded in attaining these objectives!

This book grew out of a graduate course on "Lévy processes and applications in finance" given by R. Cont at ENSAE in 2000. R. Cont thanks Griselda Deelstra and Jean-Michel Grandmont for the opportunity to teach this course and the graduate students of ENSAE and Université de Paris IX (DEA MASE) for their participation and interest in this topic which encouraged us to write this book. The material in Chapter 12 resulted from joint work with Ekaterina Voltchkova, who deserves special thanks for the numerous discussions we had on the subject and suggestions for improving this chapter. We also thank Yann Braouezec, Andreas Kyprianou, Cecilia Mancini and Olivier Pantz for their comments on a preliminary version of the manuscript. Finally we are grateful to Dilip Madan and the editor Sunil Nair for encouraging this project and to the CRC editorial staff who helped us during the final stages: Jasmin Naim, Helena Redshaw, Jamie Sigal and especially Andrea Demby for her careful reading of the manuscript.

Though we have done our best to avoid mistakes, they are unavoidable and we will be grateful to the readers who take their time to inform us of the errors and omissions they might remark. An updated list of corrections, as well as other additional material, will be made available on the website:

`http://www.cmap.polytechnique.fr/~rama/Jumps/`

We hope that this volume will stimulate the interest of students and researchers in applied mathematics and quantitative finance and make the realm of discontinuous stochastic models more accessible to those interested in using them.

Rama CONT and Peter TANKOV
Palaiseau (France), July 2003.

Contents

Chapter 1

Financial modelling beyond Brownian motion

In the end, a theory is accepted not because it is confirmed by conventional empirical tests, but because researchers persuade one another that the theory is correct and relevant.

Fischer Black (1986)

In the galaxy of stochastic processes used to model price fluctuations, Brownian motion is undoubtedly the brightest star. A Brownian motion is a random process W_t with independent, stationary increments that follow a Gaussian distribution. Brownian motion is the most widely studied stochastic process and the mother of the modern stochastic analysis. Brownian motion and financial modelling have been tied together from the very beginning of the latter, when Louis Bachelier [17] proposed to model the price S_t of an asset at the Paris Bourse as:

$$S_t = S_0 + \sigma W_t. \tag{1.1}$$

The multiplicative version of Bachelier's model led to the commonly used Black-Scholes model [60] where the log-price $\ln S_t$ follows a Brownian motion:

$$S_t = S_0 \exp[\mu t + \sigma W_t]$$

or, in local form:

$$\frac{dS_t}{S_t} = \sigma dW_t + (\mu + \frac{\sigma^2}{2})dt. \tag{1.2}$$

The process S is sometimes called a *geometric Brownian motion*. Figure 1.1 represents two curves: the evolution of (the logarithm of) the stock price for SLM Corporation (NYSE:SLM) between January 1993 and December 1996 and a sample path of Brownian motion, with the same average volatility as the stock over the three-year period considered. For the untrained eye, it may be difficult to tell which is which: the evolution of the stock does look like a sample path of Brownian motion and examples such as Figure 1.1 are given in many texts on quantitative finance to motivate the use of Brownian motion for modelling price movements.

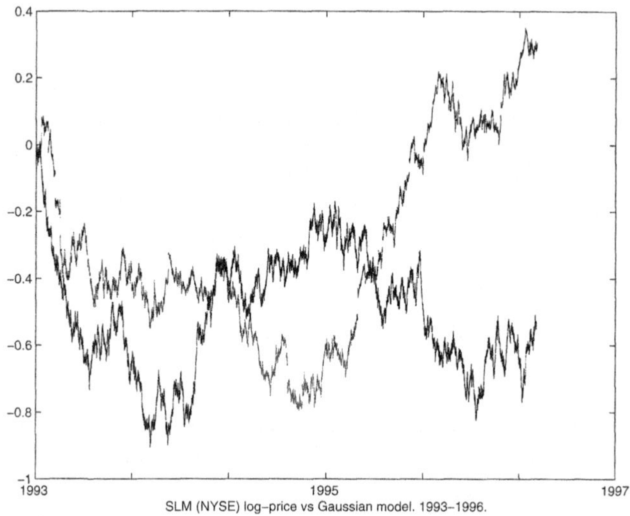

FIGURE 1.1: Evolution of the log price for SLM (NYSE), 1993-1996, compared with a sample path of Brownian motion with same annualized return and volatility. Which is which?

An important property of Brownian motion is the *continuity* of its sample paths: a typical path $t \mapsto B_t$ is a continuous function of time. This remark already allows to distinguish the two curves seen on Figure 1.1: a closer look shows that, unlike Brownian motion, the SLM stock price undergoes several abrupt downward *jumps* during this period, which appear as discontinuities in the price trajectory.

Another property of Brownian motion is its *scale invariance*: the statistical properties of Brownian motion are the same at all time resolutions. Figure 1.2 shows a zoom on the preceding figure, with only the first three months of the three-year period considered above. Clearly, the Brownian path in Figure 1.2 (left) resembles the one in Figure 1.1 and, if the scales were removed from the vertical axis one could not tell them apart. But the evolution of stock price (Figure 1.2, right) does not seem to verify this scale invariance property: the jumps become more visible and now account for more than half of the downward moves in the three-month period! The difference becomes more obvious when we zoom in closer on the price behavior: Figure 1.3 shows the evolution of SLM over a one-month period (February 1993), compared

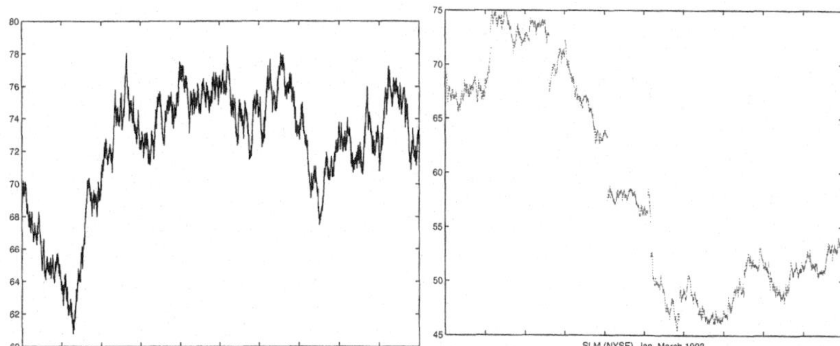

FIGURE 1.2: Evolution of SLM (NYSE), January-March 1993, compared with a scenario simulated from a Black-Scholes model with same annualized return and volatility.

to the simulated sample path of the Brownian motion over the same period. While the Brownian path looks the same as over three years or three months, the price behavior over this period is clearly dominated by a large downward jump, which accounts for half of the monthly return. Finally, if we go down to an *intraday* scale, shown in Figure 1.4, we see that the price moves essentially through jumps while the Brownian model retains the same continuous behavior as over the long horizons.

These examples show that while Brownian motion does not distinguish between time scales, price behavior does: prices move essentially by jumps at intraday scales, they still manifest discontinuous behavior at the scale of months and only after coarse-graining their behavior over longer time scales do we obtain something that resembles Brownian motion. Even though a Black-Scholes model can be chosen to give the right variance of returns at a given time horizon, it does not behave properly under *time aggregation*, i.e., across time scales. Since it is difficult to model the behavior of asset returns equally well across all time scales, ranging from several minutes to several years, it is crucial to specify from the onset which time scales are relevant for applications. The perspective of this book being oriented towards option pricing models, the relevant time scales for our purpose range between several days and several months. At these time scales, as seen in Figures 1.2 and 1.3, discontinuities cannot be ignored.

Of course, the Black-Scholes model is not the only continuous time model built on Brownian motion: nonlinear Markov diffusions where instantaneous volatility can depend on the price and time via a *local volatility* function have been proposed by Dupire [122], Derman and Kani [112]:

$$\frac{dS_t}{S_t} = \sigma(t, S_t)dW_t + \mu dt. \tag{1.3}$$

Another possibility is given by stochastic volatility models [196, 203] where

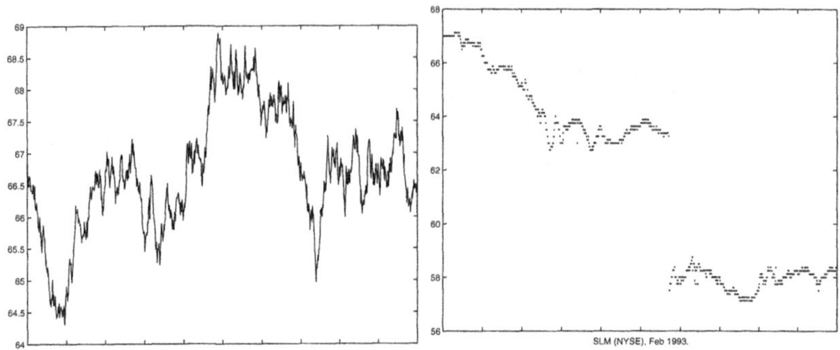

FIGURE 1.3: Price behavior of SLM (NYSE), February 1993, compared with a scenario simulated from a Black-Scholes model with same annualized return and volatility.

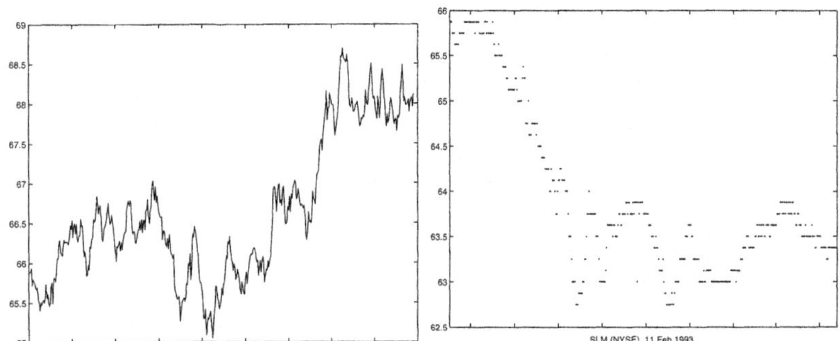

FIGURE 1.4: Price behavior of SLM (NYSE) on February 11, 1993, compared with a scenario simulated from a Black-Scholes model with same annualized return and volatility.

the price S_t is the component of a bivariate diffusion (S_t, σ_t) driven by a two-dimensional Brownian motion (W_t^1, W_t^2):

$$\frac{dS_t}{S_t} = \sigma_t dW_t^1 + \mu dt, \tag{1.4}$$

$$\sigma_t = f(Y_t) \quad dY_t = \alpha_t dt + \gamma_t dW_t^2. \tag{1.5}$$

While these models have more flexible statistical properties, they share with Brownian motion the property of *continuity*, which does not seem to be shared by real prices over the time scales of interest. Assuming that prices move in a continuous manner amounts to neglecting the abrupt movements in which most of the *risk* is concentrated.

Since the continuity of paths plays a crucial role in the properties of diffusion models, one would like to know whether results obtained in such models are robust to the removal of the continuity hypothesis. This book presents various stochastic models in which prices are allowed to display a *discontinuous* behavior, similar to that of market prices at the time scales of interest. By examining some of the main issues studied in quantitative finance in the framework of models with *jumps*, we will observe that many results obtained in diffusion models are actually *not* robust to the presence of jumps in prices and thus deserve to be reconsidered anew when jumps are taken into account.

A common approach to promote the use of models with jumps, has been to compare them systematically with the Black-Scholes model, given by Equation (1.2), and conclude that the alternative model is superior in describing empirical observations and its modelling flexibility. Since most classes of models with jumps include the Black-Scholes model as a particular instance, this approach is not serious and we shall not adopt it here. The universe of diffusion models extends far beyond the Black-Scholes framework and the full spectrum of diffusion models, including local volatility models and stochastic volatility models has to be considered in a comparative evaluation of modelling approaches. Our objective is not so much to promote the use of discontinuous models as to provide the reader with the necessary background to understand, explore and compare models with jumps with the more well-known diffusion models.

In the rest of this introductory chapter we will review some of the strengths and weaknesses of diffusion models in three contexts: capturing the empirical properties of asset returns, representing the main features of option prices and providing appropriate tools and insights for hedging and risk management. We will see that, while diffusion models offer a high flexibility and can be *fine-tuned* to obtain various properties, these properties appear as *generic* in models with jumps.

1.1 Models in the light of empirical facts

More striking than the comparison of price trajectories to those of Brownian paths is the comparison of *returns*, i.e., increments of the log-price, which are the relevant quantities for an investor. Figure 1.5 compares the five-minute returns on the Yen/Deutschemark exchange rate to increments of a Brownian motion with the same average volatility. While both return series have the same variance, the Brownian model achieves it by generating returns which always have roughly the same amplitude whereas the Yen/DM returns are widely dispersed in their amplitude and manifest frequent large peaks corresponding to "jumps" in the price. This high variability is a constantly observed feature of financial asset returns. In statistical terms this results in *heavy tails* in the empirical distribution of returns: the tail of the distribution decays slowly at infinity and very large moves have a significant probability of occurring. This well-known fact leads to a poor representation of the distribution of returns by a normal distribution. And no book on financial

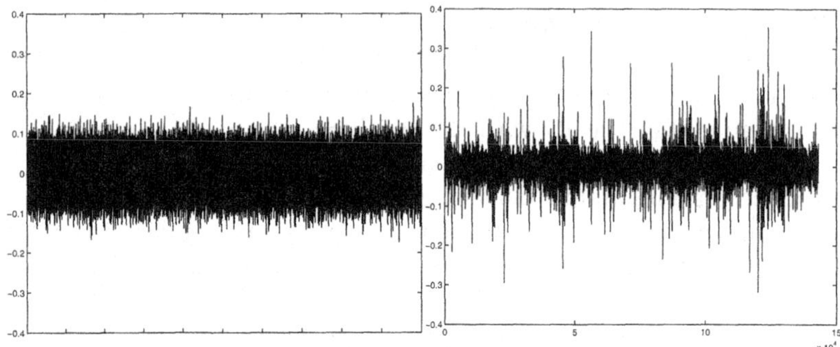

FIGURE 1.5: Five-minute log-returns for Yen/Deutschemark exchange rate, 1992–1995, compared with log-returns of a Black-Scholes model with same annualized mean and variance.

risk is nowadays complete without a reference to the traditional six-standard deviation market moves which are commonly observed on all markets, even the largest and the most liquid ones. Since for a normal random variable the probability of occurrence of a value six times the standard deviation is less than 10^{-8}, in a Gaussian model a daily return of such magnitude occurs less than once in a million years! Saying that such a model underestimates risk is a polite understatement. Isn't this an overwhelming argument against diffusion models based on Brownian motion?

Well, not really. Let us immediately dissipate a frequently encountered misconception: nonlinear diffusion processes such as (1.3) or (1.4) are *not* Gaussian processes, even though the driving noise is Gaussian. In fact, as pointed out by Bibby and Sorensen [367], an appropriate choice of a nonlinear diffusion coefficient (along with a linear drift) can generate diffusion processes with arbitrary heavy tails. This observation discards some casual arguments that attempt to dismiss diffusion models simply by pointing to the heavy tails of returns. But, since the only degree of freedom for tuning the local behavior of a diffusion process is the diffusion coefficient, these heavy tails are produced at the price of obtaining highly varying (nonstationary) diffusion coefficients in local volatility models or unrealistically high values of "volatility of volatility" in diffusion-based stochastic volatility models.

By contrast, we will observe that the simplest Markovian models with jumps — Lévy processes — generically lead to highly variable returns with realistic tail behavior without the need for introducing nonstationarity, choosing extreme parameter values or adding unobservable random factors.

But the strongest argument for using discontinuous models is not a statistical one: it is the presence of jumps in the price! While diffusion models can generate heavy tails in returns, they *cannot* generate sudden, discontinuous moves in prices. In a diffusion model tail events are the result of the accumulation of many small moves. Even in diffusion-based stochastic volatility models where market volatility can fluctuate autonomously, it cannot change suddenly. As a result, short-term market movements are approximately Gaussian and their size is predictable. A purist would argue that one cannot tell whether a given large price move is a true discontinuity since observations are made in discrete time. Though true, this remark misses a point: the question is not really to identify whether the price trajectory is *objectively* discontinuous (if this means anything at all), but rather to propose a model which reproduces the realistic properties of price behavior at the time scales of interest in a generic manner, i.e., without the need to fine-tune parameters to extreme values. While large sudden moves are *generic* properties of models with jumps, they are only obtainable in diffusion processes at the price of fine tuning parameters to extreme values. In a diffusion model the notion of a *sudden*, unpredictable market move, which corresponds to our perception of risk, is difficult to capture and this is where jumps are helpful. We will review the statistical properties of market prices in more detail in Chapter 7 but it should be clear from the onset that the question of using continuous or discontinuous models has important consequences for the representation of risk and is not a purely statistical issue.

1.2 Evidence from option markets

Although an outsider could imagine that the main objective of a stochastic model is to capture the empirical properties of prices, the driving force behind the introduction of continuous-time stochastic models in finance has been the development of *option pricing models*, which serve a somewhat different purpose. Here the logic is different from the traditional time series models in econometrics: an option pricing model is used as a device for capturing the features of option prices quoted on the market, relating prices of market instruments in an arbitrage-free manner (pricing of "vanilla" options consistently with the market) and extrapolating the notion of value to instruments not priced on the market (pricing of exotic options). In short, an option pricing model is an arbitrage-free interpolation and extrapolation tool. Option pricing models are also used to compute hedging strategies and to quantify the risk associated with a given position. Given these remarks, a particular class of models may do a good job in representing time series of returns, but a poor one as a model for pricing and hedging.

1.2.1 Implied volatility smiles and skews

A first requirement for an option pricing model is to capture the state of the options market at a given instant. To achieve this, the parameters of the model are chosen to "fit" the market prices of options or at least to reproduce the main features of these prices, a procedure known as the "calibration" of the model to the market prices. The need for models which can calibrate market prices has been one of the main thrusts behind the generalization of the Black-Scholes model.

The market prices of options are usually represented in terms of their Black-Scholes implied volatilities of the corresponding options. Recall that a European call option on an asset S_t paying no dividends, with maturity date T and strike price K is defined as a contingent claim with payoff $(S_T - K)^+$ at maturity. Denoting by $\tau = T - t$ the time remaining to maturity, the Black-Scholes formula for the value of this call option is:

$$C^{BS}(S_t, K, \tau, \sigma) = S_t N(d_1) - K e^{-r\tau} N(d_2), \tag{1.6}$$

$$d_1 = \frac{-\ln m + \tau(r + \frac{\sigma^2}{2})}{\sigma\sqrt{\tau}}, \quad d_2 = \frac{-\ln m + \tau(r - \frac{\sigma^2}{2})}{\sigma\sqrt{\tau}}. \tag{1.7}$$

where $m = K/S_t$ is the moneyness and $N(u) = (2\pi)^{-1/2} \int_{-\infty}^{u} \exp(-\frac{z^2}{2}) dz$. Let us now consider, in a market where the hypotheses of the Black-Scholes model do not necessarily hold, a call option whose (observed) market price is denoted by $C_t^*(T, K)$. Since the Black-Scholes value of a call option, as a function of the volatility parameter, is strictly increasing from $]0, +\infty[$ to $](S_t - Ke^{-r\tau})^+, S_0[$, given any observed market price within this range, one can find a value of the volatility parameter $\Sigma_t(T, K)$ such that the corresponding Black-Scholes price matches the market price:

$$\exists! \quad \Sigma_t(T, K) > 0, \qquad C^{BS}(S_t, K, \tau, \Sigma_t(T, K)) = C_t^*(T, K). \tag{1.8}$$

In Rebonato's terms [330] the implied volatility is thus a "wrong number which, plugged into the wrong formula, gives the right answer." Prices in option markets are commonly quoted in terms of Black-Scholes implied volatility. This does not mean that market participants believe in the hypotheses of the Black-Scholes model — they do not : the Black-Scholes formula is not used as a pricing model for vanilla options but as a tool for *translating* market prices into a *representation* in terms of implied volatility.

For fixed t, the implied volatility $\Sigma_t(T, K)$ depends on the characteristics of the option such as the maturity T and the strike level K: the function

$$\Sigma_t : (T, K) \to \Sigma_t(T, K) \tag{1.9}$$

is called the *implied volatility surface* at date t. A typical implied volatility surface is displayed in Figure 1.6. A large body of empirical and theoretical literature deals with the profile of the implied volatility surface for various markets as a function of (T,K) -or (m,τ) - at a given date, i.e., with (t,S_t) fixed. While the Black-Scholes model predicts a flat profile for the implied volatility surface:

$$\Sigma_t(T,K)=\sigma$$

it is a well documented empirical fact that the implied volatility is not constant as a function of strike nor as a function of time to maturity [95, 96, 121, 330]. This phenomenon can be seen in Figure 1.6 in the case of DAX index options and in Figure 1.7 for S&P 500 index options. The following properties of implied volatility surfaces have been empirically observed [95, 96, 330]:

1. Smiles and skews: for equity and foreign exchange options, implied volatilities $\Sigma_t(T,K)$ display a strong dependence with respect to the strike price: this dependence may be decreasing ("skew") or U-shaped ("smile") and has greatly increased since the 1987 crash.

2. Flattening of the smile: the dependence of $\Sigma_t(T,K)$ with respect to K decreases with maturity; the smile/ skew flattens out as maturity increases.

3. Floating smiles: if expressed in terms of relative strikes (moneyness $m=K/S_t$), implied volatility patterns vary less in time than when expressed as a function of the strike K.

Coming up with a pricing model which can reproduce these features has become known as the "smile problem" and, sure enough, a plethora of generalizations of the Black-Scholes model have been proposed to deal with it.

How do diffusion models fare with the smile problem? Well, at the level of "fitting" the shape of the implied volatility surface, they do fairly well: as shown by Dupire [122] for any arbitrage-free profile $C_0(T,K)$,$T\in[0,T^*]$,$K>0$ of call option prices observed at $t=0$, there is a unique "local volatility function" $\sigma(t,S)$ given by

$$\sigma(T,K)=\sqrt{2\frac{\frac{\partial C_0}{\partial T}(T,K)+Kr\frac{\partial C_0}{\partial K}(T,K)}{K^2\frac{\partial^2 C_0}{\partial K^2}(T,K)}}, \qquad (1.10)$$

which is consistent with these option prices, in the sense that the model (1.3) with $\sigma(.,.)$ given by (1.10) gives back the market prices $C_t(T,K)$ for the call options.

For long maturities, this leads to local volatilities which are roughly constant, predicting a future smile that is much flatter than current smiles which is, in the words of E. Derman, "an uncomfortable and unrealistic forecast that contradicts the omnipresent nature of the skew." More generally, though local volatility models can fit practically any cross section of prices they give rise to non-intuitive profiles of local volatility which, to this day, have received no interpretation in terms of market dynamics. This means that local volatility models, while providing an elegant solution to the "calibration" problem, do not give an *explanation* of the smile phenomenon.

Diffusion-based stochastic volatility models can also reproduce the profile of implied volatilities at a given maturity fairly well [196, 150]. However, they have more trouble across maturities, i.e., they cannot yield a realistic term structure of implied volatilities [330, 42]. In particular the "at-the-money skew", which is the slope of implied volatility when plotted against $\ln(K/S_t)$, decays as $1/T$ in most stochastic volatility models [150], at odds with market skews which decay more slowly. In addition, stochastic volatility models require a negative correlation between movements in stock and movements in volatility for the presence of a skew. While this can be reasonably interpreted in terms of a "leverage" effect, it does not explain why in some markets such as options on major foreign exchange rates the "skew" becomes a smile: does the nature of the leverage effect vary with the underlying asset? Nor does this interpretation explain why the smile/skew patterns increased right after the 1987 crash: did the "leverage" effect change its nature after the crash? Since the instantaneous volatility is unobservable, assertions about its (instantaneous) correlation with the returns are difficult to test but it should be clear from these remarks that the explanation of the implied volatility skew offered by stochastic volatility models is no more "structural" than the explanation offered by local volatility models.

Models with jumps, by contrast, not only lead to a variety of smile/ skew patterns but also propose a simple explanation in terms of market anticipations: the presence of a skew is attributed to the fear of large negative jumps by market participants. This is clearly consistent with the fact that the skew/smile features in implied volatility patterns have greatly increased since the 1987 crash; they reflect the "jump fear" of the market participants having experienced the crash [42, 40]. Jump processes also allow to explain the distinction between skew and smile in terms of asymmetry of jumps anticipated by the market: for index options, the fear of a large downward jump leads to a downward skew as in Figure 1.6 while in foreign exchange markets such as USD/EUR where the market moves are symmetric, jumps are expected to be symmetric thus giving rise to smiles.

1.2.2 Short-term options

The shortcomings discussed above are exacerbated when we look at options with short maturities. The very *existence* of a market for short-term options is evidence that jumps in the price are not only present but also recognized as being present by participants in the options market. How else could the underlying asset move 10% out of the money in a few days?

Not only are short-term options traded at significant prices but their market implied volatilities also exhibit a significant skew, as shown for S&P 500 options in Figure 1.7. This feature is unattainable in diffusion-based stochastic volatility models: in these models, the volatility and the price are both

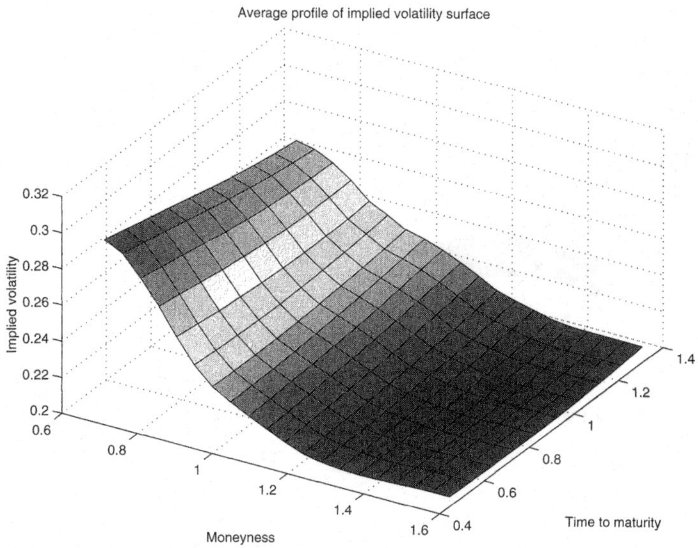

FIGURE 1.6: Implied volatilities of DAX index options, 2001.

continuous and their movements are conditionally Gaussian so one would require ridiculously high values of "volatility of volatility" to obtain realistic short-term skews. In a local volatility model, one can always obtain an arbitrary short-term skew but at the price of a very high variability in the local volatility surface, which is difficult to use or interpret. By contrast, we will see in Chapter 11 that models with jumps generically lead to significant skews for short maturities and this behavior can be used to test for the presence of jumps using short-term option prices [85, 314]. More generally, we will see in Chapter 15 that by adding jumps to returns in a stochastic volatility model as in [41] one can easily enhance the empirical performance for short maturities of stochastic volatility models which have an otherwise reasonable behavior for long maturities.

1.3 Hedging and risk management

In the language of financial theory, one-dimensional diffusion models ("local volatility" models) are examples of complete markets: any option can be perfectly replicated by a self-financing strategy involving the underlying and cash. In such markets, options are redundant; they are perfectly substitutable by trading in the underlying so the very existence of an options market becomes a mystery. Of course, this mystery is easily solved: in real markets, perfect

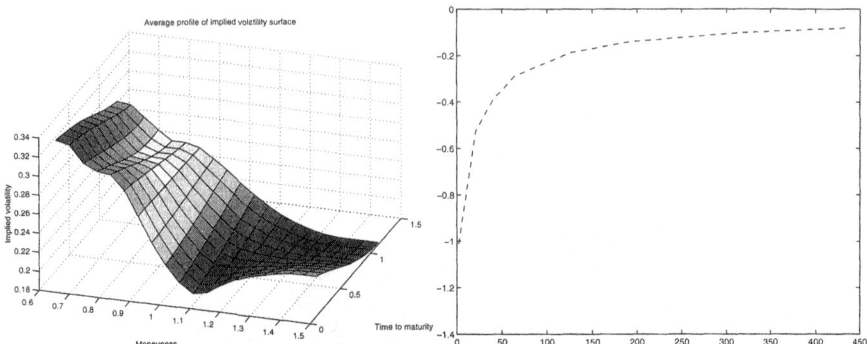

FIGURE 1.7: Left: The implied volatility surface for S&P 500 options. Right: At-the-money skew $\frac{\partial \Sigma}{\partial K}(K = S, T)$ as a function of the maturity T (in days).

hedging is not possible and options enable market participants to hedge risks that cannot be hedged by trading in the underlying only. Options thus allow a better allocation and transfer of risk among market participants, which was the purpose for the creation of derivatives markets in the first place [339].

While these facts are readily recognized by most users of option pricing models, the usage has been to twist the complete market framework of diffusion models to adapt it to market realities. On the practical side one complements delta hedging (hedging with the underlying) with gamma and vega hedging. These strategies — while clearly enhancing the performance of "replication" strategies proposed in such models — appear clearly at odds with the model: indeed, in a complete market diffusion model vega and gamma hedges are redundant with respect to delta hedging. On the theoretical side, it has been shown [129] that the Black-Scholes delta-hedging strategy is valid outside the lognormal framework if one uses upper bounds for volatility to price and hedge contingent claims: this property is known as the robustness of the Black-Scholes formula. However, as we will see in Chapter 10, the upper bound for "volatility" in a model with jumps is...infinity! In other words, the only way to perfectly hedge a call option against jumps is to buy and hold the underlying asset. This remark shows that, when moving from diffusion-based complete market models to more realistic models, the concept of "replication," which is central in diffusion models, does not provide the right framework for hedging and risk management.

Complete market models where every claim can be perfectly hedged by the underlying also fail to explain the common practice of quasi-static hedging of exotic options with vanilla options [4]. Again, this is a natural thing to do in a model with jumps since in such incomplete markets options are not redundant assets and such static (vega) hedges may be used to reduce the residual risk associated with the jumps. Also, as we will see in Chapter 10, the hedge ratio in models with jumps takes into account the possibility of a large move in the underlying asset and therefore partly captures the gamma risk.

Stochastic volatility models do recognize the impossibility of perfectly hedging options with the underlying. However in diffusion-based stochastic volatility models completeness can be restored by adding a single option to the set of available hedging instruments: stochastic volatility models then recommend setting up a perfect hedge by trading dynamically in the underlying and one option. While options are available and used for hedging, this is often done in a static framework for liquidity reasons: dynamic hedging with options remains a challenge both in theory and practice.

By contrast, from the point of view of the discontinuous price models considered in this book, the nonexistence of a perfect hedge is not a market imperfection but an imperfection of complete market models! We will see that in models with jumps, "riskless replication" is an exception rather than the rule: any hedging strategy has a residual risk which cannot be hedged away to zero and should be taken into account in the exposure of the portfolio. This offers a more realistic picture of risk management of option portfolios. Unlike what is suggested by complete market models, option trading is a risky business!

In the models that we will consider — exponential-Lévy models, jump-diffusion models, stochastic volatility models with jumps — one has to recognize from the onset the impossibility of perfect hedges and to distinguish the theoretical concept of replication from the practical concept of hedging: the hedging problem is concerned with *approximating* a target future payoff by a trading strategy and involves some risks which need to quantified and minimized by an appropriate choice of hedging strategy, instead of simply being ignored. These points will be discussed in more detail in Chapter 10.

1.4 Objectives

Table 1.1 lists some of the main messages coming out of more than three decades of financial modelling and risk management and compares them with the messages conveyed by diffusion models and models with jumps. This brief comparison shows that, aside from having various empirical, computational and statistical features that have motivated their use in the first place, discontinuous models deliver *qualitatively different* messages about the key issues of hedging, replication and risk.

Our point, which will be stressed again in Chapter 10, is not so much that diffusion models such as (1.3), (1.4) or even (1.2) do not give good "fits" of empirical data: in fact, they do quite well in some circumstances. The point is that they have the wrong *qualitative* properties and therefore can

TABLE 1.1: Modelling market moves: diffusion models vs. models with jumps.

Empirical facts	**Diffusion models**	**Models with jumps**
Large, sudden movements in prices.	Difficult: need very large volatilities.	Generic property.
Heavy tails.	Possible by choosing nonlinear volatility structures.	Generic property.
Options are risky investments.	Options can be hedged in a risk-free manner.	Perfect hedges do not exist: options are risky investments.
Markets are incomplete; some risks cannot be hedged.	Markets are complete.	Markets are incomplete.
Concentration: losses are concentrated in a few large downward moves.	Continuity: price movements are conditionally Gaussian; large sudden moves do not occur.	Discontinuity: jumps/ discontinuities in prices can give rise to large losses.
Some hedging strategies are better than others.	All hedging strategies lead to the zero residual risk, regardless of the risk measure used.	Hedging strategy is obtained by solving portfolio optimization problem.
Exotic options are hedged using vanilla (call/put) options.	Options are redundant: any payoff can be replicated by dynamic hedging with the underlying.	Options are not redundant: using vanilla options can allow to reduce hedging error.

convey erroneous intuitions about price fluctuations and the risk resulting from them. We will argue that, when viewed as a subset of the larger family of jump-diffusion models, which are the object of this book, diffusion models should be considered as singularities: while they should certainly be included in all finance textbooks as pedagogical examples, their conclusions for risk measurement and management cannot be taken seriously.

The points outlined above should have convinced the reader that in the models considered in this book we are not merely speaking about a generalization of the classical Black-Scholes model in view of "fitting" the distribution of returns or implied volatility curves with some additional parameters. In addition to matching empirical observations, these models will force us to critically reconsider some of the main concepts which are the backbone of the realm of diffusion models: arbitrage pricing, riskless hedges, market completeness and even the Itô formula!

Our goal has been to provide the reader with the necessary tools for understanding these models and the concepts behind them. Instead of heading for the full generality of "semimartingale" theory we have chosen to focus on tractable families of models with jumps — Lévy processes, additive processes and stochastic volatility models with jumps. The main ideas and modelling tools can be introduced in these models without falling into excessive abstraction.

Exponential Lévy models, introduced in Chapters 3 and 4, offer analytically tractable examples of positive jump processes and are the main focus of the book. They are simple enough to allow a detailed study both in terms of statistical properties (Chapter 7) and as models for risk-neutral dynamics, i.e., option pricing models (Chapter 11). The availability of closed-form expressions for characteristic function of Lévy processes (Chapter 3) enables us to use Fourier transform methods for option pricing. Also, the Markov property of the price will allow us to express option prices as solutions of partial integro-differential equations (Chapter 12). The flexibility of choice of the Lévy measure allows us to calibrate the model to market prices of options and reproduce implied volatility skews/smiles (Chapter 13).

We will see nevertheless that time-homogeneous models such as Lévy processes do not allow for a flexible representation of the term structure of implied volatility and imply empirically undesirable features for forward smiles/skews. In the last part of the book, we will introduce extensions of these models allowing to correct these shortcomings while preserving the mathematical tractability: additive processes (Chapter 14) and stochastic volatility models with jumps (Chapter 15).

Finally, let us stress that we are not striving to promote the systematic use of the models studied in this book. In the course of the exposition we will point our their shortcomings as well as their advantages. We simply aim at providing the necessary background so that jump processes and models built using them will, hopefully, hold no mystery for the reader by the time (s)he has gone through the material proposed here. Table 1.2 provides an

TABLE 1.2: Topics presented in this book

Concepts	**Mathematical tools**	**Chapter**
Constructing models with jumps	Poisson random measures Lévy processes	2 3,4
Multivariate models	Lévy copulas	5
Time series modelling	Statistical methods	7
Arbitrage pricing	Changes of measure	9
Hedging in incomplete markets	Stochastic calculus	8
Model calibration	Inverse problems and regularization methods	13
Numerical methods for option pricing	Monte Carlo simulation Fourier transform methods Finite difference methods	6 11 12
Time-inhomogeneous models	Additive processes	14
Stochastic volatility models with jumps	Ornstein-Uhlenbeck processes	15

outline of the different topics presented in this book and the chapters where they are discussed. The chapters have been designed to be as self-contained as possible. However, to learn the necessary mathematical tools, the reader should go through Chapters 2 and 3 before passing to the rest of the book. In addition, it is recommended to read Chapter 8 before passing to Chapters 10 and 12 and to read Chapter 9 before continuing with Chapter 13.

Part I

Mathematical tools

Chapter 2

Basic tools

> Lorsque l'on expose devant un public de mathématiciens [...] on peut supposer que chacun connait les variétés de Stein ou les nombres de Betti d'un espace topologique; mais si l'on a besoin d'une intégrale stochastique, on doit définir à partir de zéro les filtrations, les processus prévisibles, les martingales, etc. Il y a là quelque chose d'anormal. Les raisons en sont bien sûr nombreuses, à commencer par le vocabulaire esotérique des probabilistes...
>
> Laurent Schwartz

Modern probability theory has become the natural language for formulating quantitative models of financial markets. This chapter presents, in the form of a crash course, some of its tools and concepts that will be important in the sequel. Although the reader is assumed to have some background knowledge on random variables and probability, there are some tricky notions that we have found useful to recall. Instead of giving a catalog of definitions and theorems, which can be found elsewhere, we have tried to justify *why* these definitions are relevant and to promote an intuitive understanding of the main results.

The mathematical concept of *measure* is important in the study of stochastic processes in general and jump processes in particular. Basic definitions and notions from measure theory are recalled in Section 2.1. Section 2.2 recalls some facts about random variables, probability spaces and characteristic functions. Basic notions on stochastic processes are recalled in Section 2.4. A fundamental example of a stochastic process is the Poisson process, discussed in Section 2.5. The study of the Poisson process naturally leads to the notion of *Poisson random measures*, introduced in Section 2.6. Our presentation is concise and motivated by applications of these concepts in the sequel: references for further reading are provided at the end of the chapter.

2.1 Measure theory

2.1.1 σ-algebras and measures

The notion of measure is a straightforward generalization of the more familiar notions of length, area and volume to more abstract settings. Consider a set E, which we will usually take to be $\mathbb{R}^d$ or some space of $\mathbb{R}^d$-valued functions. Intuitively, a measure μ on E associates to certain subsets $A \subset E$, called *measurable* sets, a positive (possibly infinite) number $\mu(A) \in [0, \infty]$, called the measure of A. By analogy with the notions of area or volume, it is natural to say that the empty set $\emptyset$ has measure 0: $\mu(\emptyset) = 0$. Also, if A and B are disjoint measurable sets, $A \bigcup B$ should also be measurable and its measure is naturally defined to be $\mu(A \bigcup B) = \mu(A) + \mu(B)$. This is the *additivity* property. In order to consider limits it is useful to extend this property to infinite sequences: if $(A_n)_{n \in \mathbb{N}}$ is a sequence of disjoint measurable subsets then

$$\mu(\bigcup_{n \geq 1} A_n) = \sum_{n \geq 1} \mu(A_n). \tag{2.1}$$

This countable additivity property is sometimes known under the (obscure) name of σ-additivity.

Note that we have not excluded that $\mu(A) = \infty$ for some A: returning to the analogy with volume, the volume of a half space is infinite, for instance. In particular $\mu(E)$ may be finite or infinite. If $\mu(E) < \infty$ then for any measurable set A, since its complement A^c verifies $A \bigcup A^c = E$, the additivity property can be used to define the measure of the complement A^c by

$$\mu(A^c) = \mu(E) - \mu(A).$$

Therefore it is natural to require that for any measurable set A its complement A^c is also measurable.

These remarks can be summarized by saying that the domain of definition of a measure on E is a collection of subsets of E which

- contains the empty set : $\emptyset \in \mathcal{E}$.
- is stable under unions:

$$A_n \in \mathcal{E}, (A_n)_{n \geq 1} \text{ disjoint} \quad \Rightarrow \bigcup_{n \geq 1} A_n \in \mathcal{E}. \tag{2.2}$$

- contains the complementary of every element: $\forall A \in \mathcal{E}, A^c \in \mathcal{E}$.

Such a collection of subsets is called a σ-algebra. We will usually denote σ-algebras by curly letters like $\mathcal{E}, \mathcal{B}$ or $\mathcal{F}$. A measurable set will then be an element of the σ-algebra.

All this is fine, but how does one construct such σ-algebras? Usually we start with a collection of sets that we would like to be measurable (say, intervals if we are on the real line) and then keep adding new sets by taking reunions and complements until the collection forms a σ-algebra. The following result (see e.g., [153, Chapter 1]) shows that this operation is always possible:

PROPOSITION 2.1
Given a collection $\mathcal{A}$ of subsets of E, there exists a unique σ-algebra denoted $\sigma(\mathcal{A})$ with the following property: if any σ-algebra $\mathcal{F}'$ contains $\mathcal{A}$ then $\sigma(\mathcal{A}) \subset \mathcal{F}'$. $\sigma(\mathcal{A})$ is the smallest σ-algebra containing $\mathcal{A}$ and is called the σ-algebra generated by $\mathcal{A}$.

An important example is the case where E has a topology, i.e., the notion of open subset is defined on E. This is the case of $E = \mathbb{R}^d$ and more generally, of function spaces for which a notion of convergence is available. The σ-algebra generated by all open subsets is called the Borel σ-algebra [1] and we will denote it by $\mathcal{B}(E)$ or simply $\mathcal{B}$. An element $B \in \mathcal{B}$ is called a Borel set. Obviously any open or closed set is a Borel set but a Borel set can be horribly complicated. Defining measures on $\mathcal{B}$ will ensure that all open and closed sets are measurable. Unless otherwise specified, we will systematically consider the Borel σ-algebra in all the examples of measures encountered in the sequel.

Having defined σ-algebras, we are now ready to properly define a measure:

DEFINITION 2.1 Measure *Let $\mathcal{E}$ be a σ-algebra of subsets of E. $(E, \mathcal{E})$ is called a measurable space. A (positive) measure on $(E, \mathcal{E})$ is defined as a function*

$$\begin{aligned} \mu : \mathcal{E} &\to [0, \infty] \\ A &\mapsto \mu(A) \end{aligned}$$

such that

1. $\mu(\emptyset) = 0$.

2. *For any sequence of disjoint sets $A_n \in \mathcal{E}$*

$$\mu\Big(\bigcup_{n\geq 1} A_n\Big) = \sum_{n\geq 1} \mu(A_n). \tag{2.3}$$

An element $A \in \mathcal{E}$ is called a measurable set and $\mu(A)$ its measure.

[1] In French: *tribu borélienne*, literally "Borelian tribe"!

Here we have required that $\mu(A)$ be positive but one can also consider non-positive measures: if μ_+ and μ_- are two (positive) measures then $\mu = \mu_+ - \mu_-$ verifies the properties given in Definition (2.1) and is called a signed measure. It is not clear at this stage why such objects might be interesting to consider so we postpone their discussion until later. In the sequel the term "measure" will be used synonymously with "positive measure."

A well-known example of a measure is the Lebesgue measure on $\mathbb{R}^d$: it is defined on the Borel σ-algebra $\mathcal{B}(\mathbb{R}^d)$ and corresponds to the (d-dimensional) notion of volume:

$$\lambda(A) = \int_A dx.$$

More generally for any positive continuous function $\rho : \mathbb{R}^d \to \mathbb{R}^+$ one can define a measure on $\mathbb{R}^d$ as follows:

$$\begin{aligned} \mu : \mathcal{B}(\mathbb{R}^d) &\to [0, \infty] \\ A &\mapsto \int_A \rho(x)dx = \int 1_A \rho(x)dx. \end{aligned} \tag{2.4}$$

The function ρ is called the density of μ with respect to the Lebesgue measure λ. More generally, the Equation (2.4) defines a positive measure for every positive measurable function ρ (measurable functions are defined in the next section).

Dirac measures are other important examples of measures. The Dirac measure δ_x associated to a point $x \in E$ is defined as follows: $\delta_x(A) = 1$ if $x \in A$ and $\delta_x(A) = 0$ if $x \notin A$. More generally one can consider a sum of such Dirac measures. Given a countable set of points $X = \{x_i, i = 0, 1, 2, ...\} \subset E$ the counting measure $\mu_X = \sum_i \delta_{x_i}$ is defined in the following way: for any $A \subset E$, $\mu_X(A)$ counts the number of points x_i in A:

$$\mu(A) = \#\{i, x_i \in A\} = \sum_{i \geq 1} 1_{x_i \in A}. \tag{2.5}$$

A measurable set may have zero measure without being empty. Going back to the analogy with length and volume, we can note that the area of a line segment is zero, the "length" of a point is also defined to be zero. The existence of nontrivial sets of measure zero is the origin of many subtleties in measure theory. If A is a measurable set with $\mu(A) = 0$, it is then natural to set $\mu(B) = 0$ for any $B \subset A$. Such sets — subsets of sets of measure zero — are called *null sets*. If all null sets are not already included in $\mathcal{E}$, one can always include them by adding all null sets to $\mathcal{E}$: the new σ-algebra is then said to be *complete*.

A measure μ is said to be *integer valued* if for any measurable set A, $\mu(A)$ is a (positive) integer. An example of an integer valued measure is a Dirac measure. More generally, any counting measure is an integer valued measure.

If μ is a measure on E, $\mu(E)$ need not be finite. A measure μ on $(E, \mathcal{E})$ is said to be *finite* if $\mu(E) < +\infty$ (which entails that $\mu(A)$ is finite for any measurable set A). The quantity $\mu(E)$ is usually called the (total) mass of μ. For example, a Dirac measure is a finite measure with mass 1. The counting measure μ_X associated to the set $X = \{x_1, x_2, \ldots\}$ is finite if X is finite and its mass $\mu_X(E)$ is simply the number of elements in X.

A finite measure with mass 1 is called a probability measure. In this case $(E, \mathcal{E}, \mu)$ is called a probability space. Probability measures will be considered in more detail in Section 2.2.

Not all measures we will encounter will be finite measures. A well-known example is the Lebesgue measure on $\mathbb{R}^d$: the total mass is infinite. A more flexible notion is that of a Radon measure:

DEFINITION 2.2 Radon measure *Let $E \subset \mathbb{R}^d$. A* Radon measure *on $(E, \mathcal{B})$ is a measure μ such that for every* compact [2] *measurable set $B \in \mathcal{B}$, $\mu(B) < \infty$.*

For example the Lebesgue measure on $\mathbb{R}$ is a Radon measure: the length of any bounded interval is finite. Dirac measures and any finite linear combination of Dirac measures are also examples of Radon measures. We will encounter these fundamental examples recurrently later on. More generally, a measure μ on $E \subset \mathbb{R}^d$ is called σ-finite if

$$E = \bigcup_{i=1}^{\infty} E_i \quad \text{with } \mu(E_i) < \infty. \tag{2.6}$$

This condition holds for example if μ is a finite measure or a Radon measure.

The examples given above — the Lebesgue measure and Dirac measures — differ in a fundamental way: while Dirac measures are concentrated on a finite number of points, the Lebesgue measure assigns zero measure to any finite set. A measure μ_0 which gives zero mass to any point is said to be *diffuse* or atomless:

$$\forall x \in E, \ \mu_0(\{x\}) = 0. \tag{2.7}$$

The Lebesgue measure is an example of a diffuse measure. Measures with a continuous density with respect to the Lebesgue measure on $\mathbb{R}^d$ are other examples of diffuse measures.

The following result shows that any Radon measure can be decomposed into a diffuse part and a sum of Dirac measures:

[2] A compact subset of $\mathbb{R}^d$ is simply a bounded closed subset.

PROPOSITION 2.2 Decomposition of Radon measures
Any Radon measure μ can be decomposed into a sum of a diffuse measure μ_0 and a linear combination of Dirac measures [227]:

$$\mu = \mu_0 + \sum_{j\geq 1} b_j \delta_{x_j} \qquad x_j \in E, b_j > 0. \tag{2.8}$$

If $\mu_0 = 0$ then μ is said to be a (purely) atomic measure. Dirac measures and linear combinations of Dirac measures are atomic measures.

2.1.2 Measures meet functions: integration

Let us now consider two measurable spaces $(E, \mathcal{E})$ and $(F, \mathcal{F})$. In most applications we will be interested in evaluating a measure on a set of the form:

$$\{x \in E, f(x) \in A\}.$$

For a given function f, there is no reason for this set to be measurable. This motivates the following definition:

DEFINITION 2.3 Measurable function *A function $f : E \to F$ is called measurable if for any measurable set $A \in \mathcal{F}$, the set*

$$f^{-1}(A) = \{x \in E, f(x) \in A\}$$

is a measurable subset of E.

As noted above we will often consider measures on sets which already have a metric or topological structure such as $\mathbb{R}^d$ equipped with the Euclidean metric. On such spaces the notion of continuity for functions is well defined and one can then ask whether there is a relation between continuity and measurability for a function $f : E \to F$. In general, there is no relation between these notions. However it is desirable that the notion of measurability be defined such that all continuous functions be measurable. This is automatically true if the Borel σ-algebra is chosen (since f is continuous if and only if $f^{-1}(A)$ is open for every open set), which explains why we will choose it all the time. Hence in the following, whenever the notion of continuous function makes sense, all continuous functions will be measurable.

Simple examples of measurable functions $f : E \to \mathbb{R}$ are functions of the form

$$f = \sum_{j=1}^{n} c_j 1_{A_j}, \tag{2.9}$$

where (A_j) are measurable sets and $c_j \in \mathbb{R}$. These functions are sometimes called *simple functions*. The *integral* of such a simple function with respect to a measure μ is defined as

$$\mu(f) = \sum_{j=1}^{n} c_j \mu(A_j). \tag{2.10}$$

Having defined integrals for simple functions, we extend it to any positive measurable function $f : E \to \mathbb{R}$ by setting

$$\mu(f) = \sup\{\mu(\varphi),\ \varphi \text{ simple function},\ \varphi \leq f\}. \tag{2.11}$$

The integral $\mu(f)$ is allowed to be infinite. Since any measurable function $f : E \to \mathbb{R}$ can be written as the sum of its positive and negative parts $f = f^+ - f^-$, $f^+, f^- \geq 0$ we can define separately $\mu(f^+), \mu(f^-)$ as above. If $\mu(f^+), \mu(f^-)$ are not infinite we say that f is μ-integrable and we define $\mu(f) = \mu(f^+) - \mu(f^-)$, called the integral of μ with respect to f. When μ is the Lebesgue measure, $\mu(f)$ is simply the Lebesgue integral of f; by analogy with this case $\mu(f)$ is denoted using the "differential" notation

$$\mu(f) = \int_{x \in E} f(x) \mu(dx). \tag{2.12}$$

If the measure μ has the decomposition (2.8) then the integral in (2.12) can be interpreted as

$$\mu(f) = \int f(x) \mu_0(dx) + \sum_{j \geq 1} b_j f(x_j). \tag{2.13}$$

Two measurable functions f, g on E are said to be equal μ-*almost everywhere* if they differ only on a null set:

$$f = g \ \mu - a.e. \qquad \Longleftrightarrow \ \mu(\{x \in E,\ f(x) \neq g(x)\}) = 0.$$

If f, g are μ-integrable then

$$f = g \ \mu - a.e. \qquad \Rightarrow \quad \int_E f(x)\mu(dx) = \int_E g(x)\mu(dx).$$

2.1.3 Absolute continuity and densities

Consider now a measurable space $(E, \mathcal{E})$ with measures μ_1 and μ_2 defined on it. How can these two measures be compared?

A natural idea to compare μ_1 and μ_2 is to look at the ratio $\mu_2(A)/\mu_1(A)$ for various measurable sets A. Of course this is only possible if $\mu_2(A) = 0$ every time $\mu_1(A) = 0$. This remark motivates the following definition:

DEFINITION 2.4 Absolute continuity *A measure μ_2 is said to be* absolutely continuous *with respect to μ_1 if for any measurable set A*

$$\mu_1(A) = 0 \Rightarrow \mu_2(A) = 0. \tag{2.14}$$

Absolute continuity can be characterized in the following way:

PROPOSITION 2.3 Radon-Nikodym theorem
If μ_2 is absolutely continuous with respect to μ_1 then there exists a measurable function $Z : E \to [0, \infty[$ such that for any measurable set A

$$\mu_2(A) = \int_A Z d\mu_1 = \mu_1(Z 1_A). \tag{2.15}$$

The function Z is called the density *or* Radon-Nikodym derivative *of μ_2 with respect to μ_1 and denoted as $\frac{d\mu_2}{d\mu_1}$. For any μ_2-integrable function f*

$$\mu_2(f) = \int_E f d\mu_2 = \mu_1(fZ) = \int_E d\mu_1 Z f. \tag{2.16}$$

Therefore if μ_2 is absolutely continuous with respect to μ_1, an integral with respect to μ_2 is a weighted integral with respect to μ_1, the weight being given by the density Z.

If both μ_2 is absolutely continuous with respect to μ_1 and μ_1 is absolutely continuous with respect to μ_2 then μ_1 and μ_2 are said to be *equivalent* measures. This is equivalent to stating that $Z > 0$. The term "equivalent" is somewhat confusing: it would be more appropriate to say that μ_1 and μ_2 are comparable. However this is the usual terminology and we will continue to use it.

2.2 Random variables

2.2.1 Random variables and probability spaces

Consider a set Ω, called the set of scenarios, equipped with a σ-algebra $\mathcal{F}$. In a financial modelling context, Ω will represent the different scenarios which can occur in the market, each scenario $\omega \in \Omega$ being described in terms of the evolution of prices of different instruments. A *probability measure* on $(\Omega, \mathcal{F})$ is a positive finite measure $\mathbb{P}$ with total mass 1. $(\Omega, \mathcal{F}, \mathbb{P})$ is then called a probability space. A measurable set $A \in \mathcal{F}$, called an event, is therefore a set of scenarios to which a probability can be assigned. A probability measure assigns a *probability* between 0 and 1 to each event:

$$\begin{aligned} \mathbb{P} : \mathcal{F} &\to [0, 1] \\ A &\mapsto \mathbb{P}(A). \end{aligned}$$

Probability measures will be usually denoted by letters such as $\mathbb{P}, \mathbb{Q}$.

An event A with probability $\mathbb{P}(A) = 1$ is said to occur almost surely. If $\mathbb{P}(A) = 0$ this is interpreted by saying the event A is impossible. If we are dealing with several probability measures defined on the same set then one should be more specific: we will then replace "almost surely" or "impossible" by "$\mathbb{P}$-almost surely" or "impossible under $\mathbb{P}$." A $\mathbb{P}$-null set is a subset of an impossible event. As before, we can complete $\mathcal{F}$ to include all null sets. This means we assign probability zero to subsets of impossible events, which is intuitively reasonable. Unless stated otherwise, we shall consider complete versions of all σ-algebras. We will say that a property holds $\mathbb{P}$-almost surely ($\mathbb{P}$ a.s. for short) if the set of $\omega \in \Omega$ for which the property does not hold is a null set.

As in Section 2.1.3, one can speak of absolute continuity and equivalence for probability measures: two probability measures $\mathbb{P}$ and $\mathbb{Q}$ on $(\Omega, \mathcal{F})$ are equivalent (or comparable) if they define the same impossible events:

$$\mathbb{P} \sim \mathbb{Q} \iff [\forall A \in \mathcal{F},\ \mathbb{P}(A) = 0 \iff \mathbb{Q}(A) = 0]. \tag{2.17}$$

A *random variable* X taking values in E is a measurable function

$$X : \Omega \to E,$$

where $(\Omega, \mathcal{F}, \mathbb{P})$ is a probability space. An element $\omega \in \Omega$ is called a scenario of randomness. $X(\omega)$ represents the outcome of the random variable if the scenario ω happens and is called the realization of X in the scenario ω. If X and Y are two random variables, we write "$X = Y$ $\mathbb{P}$ a.s." (almost surely) if $\mathbb{P}\{\omega \in \Omega, X(\omega) = Y(\omega)\} = 1$. The law (or distribution) of X is the probability measure on E defined by:

$$\mu_X(A) = \mathbb{P}(X \in A). \tag{2.18}$$

If $\mu_X = \mu_Y$ then X and Y are said to be identical in law and we write $X \stackrel{d}{=} Y$. Obviously if $X = Y$ almost surely, they are identical in law.

$X : \Omega \mapsto E$ is called a real-valued random variable when $E \subset \mathbb{R}$. As in Section 2.1.2, one can define the integral of a positive random variable X with respect to $\mathbb{P}$: this quantity, called the *expectation* of X with respect to $\mathbb{P}$ and denoted by $E^{\mathbb{P}}[X] = \int_E X(\omega)d\mathbb{P}(\omega)$, is either a positive number or $+\infty$. If $E^{\mathbb{P}}[X] < \infty$ then X is said to be $\mathbb{P}$-integrable. By decomposing any real-valued random variable Y into its positive and negative parts $Y = Y_+ - Y_-$, one sees that if $E^{\mathbb{P}}[\ |Y|\ ] < \infty$ then Y_-, Y_+ are integrable and the expectation $E[Y] = E[Y_+] - E[Y_-]$ is well-defined. The set of (real-valued) random variables Y verifying $||Y||_1 = E^{\mathbb{P}}[\ |Y|\ ] < \infty$ is denoted by $L^1(\Omega, \mathbb{P})$.

It is sometimes useful to allow "infinite" values for positive random variables, i.e., choose $E = [0, \infty[\cup \{+\infty\}$. Of course if $Y \in L^1(\Omega, \mathbb{P})$ then Y is almost surely finite.

2.2.2 What is $(\Omega, \mathcal{F}, \mathbb{P})$ anyway?

While many discussions of stochastic models start with the magic sentence "let $(\Omega, \mathcal{F}, \mathbb{P})$ be a probability space" one can actually follow such discussions without having the slightest idea what Ω is and who lives inside. So here comes the question that many beginners are dying to ask without daring to: what is "$\Omega, \mathcal{F}, \mathbb{P}$" and why do we need it? Indeed, for many users of probability and statistics, a random variable X is synonymous with its probability distribution μ_X and all computations such as sums, expectations, etc., done on random variables amount to analytical operations such as integrations, Fourier transforms, convolutions, etc., done on their distributions. For defining such operations, you do not need a probability space. Isn't this all there is to it?

One *can* in fact compute quite a lot of things without using probability spaces in an essential way. However the notions of probability space and random variable are central in modern probability theory so it is important to understand why and *when* these concepts are relevant.

From a modelling perspective, the starting point is a set of observations taking values in some set E (think for instance of numerical measurement, $E = \mathbb{R}$) for which we would like to build a stochastic model. We would like to represent such observations $x_1, \ldots, x_n$ as samples drawn from a random variable X defined on some probability space $(\Omega, \mathcal{F}, \mathbb{P})$. It is important to see that the only natural ingredient here is the set E where the random variables will take their values: the set of events Ω is not given a priori and there are many different ways to construct a probability space $(\Omega, \mathcal{F}, \mathbb{P})$ for modelling the same set of observations.

Sometimes it is natural to identify Ω with E, i.e., to identify the randomness ω with its observed effect. For example if we consider the outcome of a dice rolling experiment as an integer-valued random variable X, we can define the set of events to be precisely the set of possible outcomes: $\Omega = \{1, 2, 3, 4, 5, 6\}$. In this case, $X(\omega) = \omega$: the outcome of the randomness is identified with the randomness itself. This choice of Ω is called the *canonical space* for the random variable X. In this case the random variable X is simply the identity map $X(\omega) = \omega$ and the probability measure $\mathbb{P}$ is formally the same as the distribution of X. Note that here X is a one-to-one map: given the outcome of X one knows which scenario has happened so any other random variable Y is completely determined by the observation of X. Therefore using the canonical construction for the random variable X, we cannot define, on the same probability space, another random variable which is independent of X: X will be the sole source of randomness for all other variables in the model. These remarks also show that, although the canonical construction is the simplest way to construct a probability space for representing a given random variable, it forces us to identify this particular random variable with the "source of randomness" in the model. Therefore when we want to deal with models with a sufficiently rich structure, we need to distinguish Ω — the set of scenarios of randomness — from E, the set of values of our random variables.

Let us give an example where it is natural to distinguish the source of randomness from the random variable itself. For instance, if one is modelling the market value of a stock at some date T in the future as a random variable S_1, one may consider that the stock value is affected by many factors such as external news, market supply and demand, economic indicators, etc., summed up in some abstract variable ω, which may not even have a numerical representation: it corresponds to a scenario for the future evolution of the market. $S_1(\omega)$ is then the stock value if the market scenario which occurs is given by ω. If the only interesting quantity in the model is the stock price then one can always label the scenario ω by the value of the stock price $S_1(\omega)$, which amounts to identifying all scenarios where the stock S_1 takes the same value and using the canonical construction described above. However if one considers a richer model where there are now *other* stocks $S_2, S_3, \ldots$ involved, it is more natural to distinguish the scenario ω from the random variables $S_1(\omega), S_2(\omega), \ldots$ whose values are observed in these scenarios but may not completely pin them down: knowing $S_1(\omega), S_2(\omega), \ldots$ one does not necessarily know which scenario has happened. In this way one reserves the possibility of adding more random variables later on without changing the probability space.

These comments, although a bit abstract at first sight, have the following important consequence: the probabilistic description of a random variable X can be reduced to the knowledge of its distribution μ_X *only* in the case where the random variable X is the only source of randomness. In this case, a stochastic model can be built using a canonical construction for X. In all other cases — in fact as soon as we are concerned with a second random variable which is not a deterministic function of X — the underlying probability measure $\mathbb{P}$ contains *more* information on X than just its distribution. In particular, it contains all the information about the *dependence* of the random variable X with respect to all other random variables in the model: specifying $\mathbb{P}$ means specifying the joint distributions of all random variables constructed on Ω. For instance, knowing the distributions μ_X, μ_Y of two variables X, Y does not allow to compute their covariance or joint moments. Only in the case where all random variables involved are mutually independent can one reduce all computations to operations on their distributions. This is the case covered in most introductory texts on probability, which explains why one can go quite far, for example in the study of random walks, without formalizing the notion of probability space.

2.2.3 Characteristic functions

The characteristic function of a random variable is the Fourier transform of its distribution. Many probabilistic properties of random variables correspond to analytical properties of their characteristic functions, making this concept very useful for studying random variables.

DEFINITION 2.5 Characteristic function *The* characteristic function *of the $\mathbb{R}^d$-valued random variable X is the function $\Phi_X : \mathbb{R}^d \to \mathbb{R}$ defined by*

$$\forall z \in \mathbb{R}^d,\ \Phi_X(z) = E[\exp(iz.X)] = \int_{\mathbb{R}^d} e^{iz.x} d\mu_X(x). \tag{2.19}$$

The characteristic function of a random variable completely characterizes its law: two variables with the same characteristic function are identically distributed. A characteristic function is always continuous and verifies $\Phi_X(0) = 1$. Additional smoothness properties of Φ_X depend on the existence of *moments* of the random variable X. The n-th moment of a random variable X on $\mathbb{R}$ is defined by $m_n(X) = E[X^n]$. The absolute moments of X are the quantities $m_n(|X|) = E[|X|^n]$. The n-th centered moment μ_n is defined as the n-th moment of $X - E[X]$:

$$\mu_n(X) = E[(X - E[X])^n]. \tag{2.20}$$

The moments of a random variable may or may not exist, depending on how fast the distribution μ_X decays at infinity. For example, for the exponential distribution all moments are well-defined while the Student t distribution with n degrees of freedom (see Table 2.1) only has moments of orders up to n. The moments of a random variable are related to the derivatives at 0 of its characteristic function:

PROPOSITION 2.4 Characteristic function and moments

1. *If $E[|X|^n] < \infty$ then Φ_X has n continuous derivatives at $z = 0$ and*

$$\forall k = 1 \dots n, \qquad m_k \equiv E[X^k] = \frac{1}{i^k}\frac{\partial^k \Phi_X}{\partial z^k}(0). \tag{2.21}$$

2. *If Φ_X has $2n$ continuous derivatives at $z = 0$ then $E[|X|^{2n}] < \infty$ and*

$$\forall k = 1 \dots 2n, \qquad m_k \equiv E[X^k] = \frac{1}{i^k}\frac{\partial^k \Phi_X}{\partial z^k}(0). \tag{2.22}$$

3. *X possesses finite moments of all orders iff $z \mapsto \Phi_X(z)$ is C^∞ at $z = 0$. Then the moments of X are related to the derivatives of Φ_X by:*

$$m_n \equiv E[X^n] = \frac{1}{i^n}\frac{\partial^n \Phi_X}{\partial z^n}(0). \tag{2.23}$$

If $(X_i, i = 1 \dots n)$ are independent random variables, the characteristic function of $S_n = X_1 + X_2 + \cdots + X_n$ is the *product* of characteristic functions of individual variables:

$$\Phi_{S_n}(z) = \prod_{i=1}^{n} \Phi_{X_i}(z). \tag{2.24}$$

2.2.4 Moment generating function

DEFINITION 2.6 Moment generating function *The* moment generating function *of* $\mathbb{R}^d$*-valued random variable* X *is the function* M_X *defined by*

$$\forall u \in \mathbb{R}^d, \;\; M_X(u) = E[\exp(u.X)]. \tag{2.25}$$

Contrarily to the characteristic function, which is always well-defined (as the Fourier transform of a probability measure), the moment generating function is not always defined: the integral in (2.25) may not converge for some (or all) values of u. When M_X is well-defined, it can be formally related to the characteristic function Φ_X by:

$$M_X(u) = \Phi_X(-iu). \tag{2.26}$$

If the moment generating function M_X of a random variable X on $\mathbb{R}$ is defined on a neighborhood $[-\epsilon, \epsilon]$ of zero then in particular all (polynomial) moments of X are finite and can be recovered from the derivatives of M in the following manner:

$$m_n = \frac{\partial^n M_X}{\partial u^n}(0). \tag{2.27}$$

2.2.5 Cumulant generating function

Let X be a random variable and Φ_X its characteristic functions. As mentioned above $\Phi_X(0) = 1$ and Φ_X is continuous at $z = 0$ so $\Phi_X(z) \neq 0$ in a neighborhood of $z = 0$. One can then define a continuous version of the logarithm of Φ_X: there exists a unique continuous function Ψ_X defined in a neighborhood of zero such that

$$\Psi_X(0) = 0 \quad \text{and} \quad \Phi_X(z) = \exp[\Psi_X(z)]. \tag{2.28}$$

The function Ψ_X is called the *cumulant generating function* or log-characteristic function of X. Note that if $\Phi_X(z) \neq 0$ for all z, the cumulant generating function can be extended to all $\mathbb{R}^d$. The *cumulants* or *semi-invariants* of X are defined by:

$$c_n(X) = \frac{1}{i^n}\frac{\partial^n \Psi_X}{\partial u^n}(0). \tag{2.29}$$

By expanding the exponential function at $z = 0$ and using (2.25), the n-th cumulant can be expressed as a polynomial function of the moments $m_k(X)$, $k =$

$1 \ldots n$ or the central moments $\mu_k(X)$ defined in (2.20). For instance

$$c_1(X) = m_1(X) = EX, \tag{2.30}$$

$$c_2(X) = \mu_2(X) = m_2(X) - m_1(X)^2 = \text{Var}(X), \tag{2.31}$$

$$c_3(X) = \mu_3(X) = m_3(X) - 3m_2(X)m_1(X) + 2m_1(X)^3, \tag{2.32}$$

$$c_4(X) = \mu_4(X) - 3\mu_2(X). \tag{2.33}$$

Scale-free versions of cumulants can be obtained by normalizing c_n by the n-th power of the standard deviation:

$$s(X) = \frac{c_3(X)}{c_2(X)^{3/2}}, \qquad \kappa(X) = \frac{c_4(X)}{c_2(X)^2}. \tag{2.34}$$

The quantity $s(X)$ is called the *skewness* coefficient of X: if $s(X) > 0$, X is said to be positively skewed. The quantity $\kappa(X)$ is called the (excess) kurtosis of X. X is said to be leptokurtic or "fat-tailed" if $\kappa(X) > 0$. Note that if X follows a normal distribution, Ψ_X is a second degree polynomial so $\forall n \geq 3, c_n(X) = 0$. This allows to view $s(X), \kappa(X)$ and higher cumulants as measures of deviation from normality. By construction the skewness and the kurtosis are invariant to a change of scale:

$$\forall \lambda > 0, \qquad s(\lambda X) = s(X) \quad \kappa(\lambda X) = \kappa(X). \tag{2.35}$$

Cumulant generating functions of independent variables add up when the variables are added: from (2.24) one easily deduces that, if $(X_i)_{i=1\ldots n}$ are independent random variables then

$$\Psi_{X_1+\ldots+X_n}(z) = \sum_{i=1}^{n} \Psi_{X_i}(z). \tag{2.36}$$

2.3 Convergence of random variables

When considering sequences or families of random variables, one can give several different meanings to the notion of convergence. In this section we define and compare these different notions of convergence which will be useful in the sequel. While we define convergence in terms of sequences (indexed by integers), our definitions also hold for continuously indexed families.

2.3.1 Almost-sure convergence

Consider a sequence of random variables $(X_n)_{n\geq 1}$ taking values in some normed vector space E, for instance $E = \mathbb{R}^d$. Recall that a random variable

TABLE 2.1: Some probability distributions and their characteristic functions.

Distribution	**Density**	**Characteristic function**
Exponential	$\mu(x) = \alpha e^{-\alpha x} 1_{x \geq 0}$	$\Phi(z) = \frac{\alpha}{\alpha - iz}$
Two-sided exponential	$\mu(x) = \frac{\alpha}{2} e^{-\alpha\|x\|}$	$\Phi(z) = \frac{\alpha}{\alpha^2 + \|z\|^2}$
Gamma	$\mu(x) = \frac{\lambda^c}{\Gamma(c)} x^{c-1} e^{-\lambda x} 1_{x \geq 0}$	$\Phi(z) = \frac{1}{(1 - i\lambda^{-1} z)^c}$
Gaussian	$\mu(x) = \frac{\exp(\frac{-(x-\gamma)^2}{2\sigma^2})}{\sqrt{2\pi\sigma}}$	$\Phi(z) = \exp(-\frac{\sigma^2 z^2}{2} + i\gamma z)$
Cauchy	$\mu(x) = \frac{c}{\pi[(x-\gamma)^2 + c^2]}$	$\Phi(z) = \exp(-c\|z\| + i\gamma z)$
Symmetric α-stable	Not known in closed form	$\Phi(z) = \exp(-c\|z\|^\alpha)$
Student t	$\frac{\Gamma((n+1)/2)}{\sqrt{n\pi}\Gamma(n/2)} (1 + \frac{x^2}{n})^{-(n+1)/2}$	Not known in closed form

is defined as a function $X : \Omega \to E$ of the "randomness" $\omega \in \Omega$. One could then consider applying to random variables various notions of convergence which exist for sequences of functions. The simplest notion of convergence for functions is that of *pointwise convergence* which requires that for *each* $\omega \in \Omega$, the sequence $(X_n(\omega))_{n\geq 1}$ converge to $X(\omega)$ in E. This notion turns out to be too strong in many cases since we are asking convergence for *all* samples $\omega \in \Omega$ without taking into account that many events may in fact be negligible, i.e., of probability zero. The notion of almost-sure convergence takes this fact into account and requires pointwise convergence only for realizations which have a nonzero probability of occurrence:

DEFINITION 2.7 Almost-sure convergence *A sequence (X_n) of random variables on $(\Omega, \mathcal{F}, \mathbb{P})$ is said to converge almost surely to a random variable X if*

$$\mathbb{P}(\lim_{n\to\infty} X_n = X) = 1. \tag{2.37}$$

For the above definition to make sense the variables $(X_n)_{n\geq 1}$ have to be defined on the *same* probability space $(\Omega, \mathcal{F}, \mathbb{P})$. Note that almost sure convergence does not imply convergence of moments: if $X_n \to X$ almost surely, $E[X_n^k]$ may be defined for all $n \geq 1$ but have no limit as $n \to \infty$.

2.3.2 Convergence in probability

While the almost-sure convergence of $(X_n)_{n\geq 1}$ deals with the behavior of typical samples $(X_n(\omega))_{n\geq 1}$, the notion of convergence in probability only puts a condition on the probability of events when $n \to \infty$:

DEFINITION 2.8 Convergence in probability *A sequence (X_n) of random variables on $(\Omega, \mathcal{F}, \mathbb{P})$ is said to converge in probability to a random variable X if for each $\epsilon > 0$*

$$\lim_{n\to\infty} \mathbb{P}(|X_n - X| > \epsilon) = 0. \tag{2.38}$$

We denote convergence in probability by

$$X_n \underset{n\to\infty}{\overset{\mathbb{P}}{\to}} X. \tag{2.39}$$

Almost sure convergence implies convergence in probability but the two notions are not equivalent. Also note that convergence in probability requires that the variables $(X_n)_{n\geq 1}$ be defined on the *same* probability space $(\Omega, \mathcal{F}, \mathbb{P})$.

2.3.3 Convergence in distribution

In many situations, especially in a modelling context, the random variable is not a directly observable quantity itself and the only observable quantities are expectations of various functions of this random variable: $E[f(X)]$. These quantities will in fact be the same for two random variables having the same distribution. In this context, a meaningful notion of convergence from the point of view of observation is asking that $E[f(X_n)]$ converge to $E[f(X)]$ for a given set of "observables" or test functions $f : E \to \mathbb{R}$. If the set of test functions is rich enough, this will ensure the uniqueness of the law of X but will not distinguish between two limits X and X' with the same distribution μ. A commonly used choice of test functions is the set of bounded continuous functions $f : E \to \mathbb{R}$, which we denote by $C^b(E, \mathbb{R})$.

DEFINITION 2.9 Convergence in distribution *A sequence (X_n) of random variables with values in E is said to converge in distribution to a random variable X if, for any bounded continuous function $f : E \to \mathbb{R}$*

$$E[f(X_n)] \underset{n\to\infty}{\to} E[f(X)]. \tag{2.40}$$

Denote by μ_n the distribution of X_n: μ_n is a probability measure on E. Then Equation (2.40) is equivalent to:

$$\forall f \in C^b(E, \mathbb{R}), \qquad \int_E d\mu_n(x) f(x) \underset{n\to\infty}{\to} \int_x d\mu(x) f(x). \tag{2.41}$$

In this form it is clear that, unlike almost-sure convergence, convergence in distribution is defined not in terms of the random variables themselves but in terms of their distributions. Therefore sometimes convergence in distribution of $(X_n)_{n\geq 1}$ is also called *weak convergence* of the measures $(\mu_n)_{n\geq 1}$ on E. We write:

$$\mu_n \Rightarrow \mu \quad \text{or} \quad X_n \overset{d}{\to} X. \tag{2.42}$$

An important feature of convergence in distribution is that, unlike the other notions of convergence mentioned above, it does not require the variables to be defined on a common probability space. Nevertheless, in the case where the variables (X_n) are defined on the same space, convergence in distribution leads to convergence of probabilities of (regular) events in the following sense:

PROPOSITION 2.5
If (X_n) converges in distribution to X, then for any set A with boundary ∂A,

$$\mu(\partial A) = 0 \Rightarrow [\mathbb{P}(X_n \in A) = \mu_n(A) \underset{n\to\infty}{\to} \mathbb{P}(X \in A) = \mu(A)]. \tag{2.43}$$

Note also that, unlike other notions of convergence, convergence in distribution is not "stable under sums": if (X_n) and (Y_n) converge in distribution to X and Y, it is not true in general that $(X_n + Y_n)$ converges in distribution to $(X + Y)$. However from the definition it is readily observed that if (X_n) converge in distribution to X then for any continuous function g, $g(X_n)$ converges in distribution to $g(X)$.

The notion of weak convergence is relevant in studying numerical approximations obtained by discretizing continuous time models [243, 193]. As we will see in Chapter 10, option prices can be expressed as (risk-neutral) expectations of their payoffs, so weak convergence of discrete time models to their continuous time counterparts will imply convergence of the option prices in the discretized model to the option prices in the continuous time model [193].

The following characterization of convergence in distribution in terms of pointwise convergence of characteristic functions is useful in practice:

PROPOSITION 2.6
$(X_n)_{n\geq 1}$ converges in distribution to X, if and only if for every $z \in \mathbb{R}^d$

$$\Phi_{X_n}(z) \to \Phi_X(z). \tag{2.44}$$

Note however that pointwise convergence of (Φ_{X_n}) does not imply the existence of a weak limit for (X_n) since the pointwise limit of (Φ_{X_n}) is not necessarily a characteristic function.

Convergence in distribution does not necessarily entail convergence of moments: one cannot choose f to be a polynomial in (2.41) since polynomials are not bounded. In fact the moments of X_n, X need not even exist.

When (X_n) and X are defined on the same probability space, convergence in distribution is the weakest notion of convergence among the above; almost sure convergence entails convergence in probability, which entails convergence in distribution.

2.4 Stochastic processes

A stochastic[3] process is a family $(X_t)_{t\in[0,T]}$ of random variables indexed by time. The time parameter t may be either discrete or continuous, but in this book we will consider continuous-time stochastic processes. For each realization of the randomness ω, the trajectory $X_.(\omega) : t \to X_t(\omega)$ defines a function of time, called the *sample path* of the process. Thus stochastic processes can also be viewed as *random functions*: random variables taking values in function spaces. Some of these function spaces will be defined in Section 2.4.1.

[3]Stochastic is just a fancy word for random!

A further step must be taken if one interprets the index t as *time*: one needs to take into account the fact that events become less uncertain as more information becomes available through time. In order to precisely formulate these intuitively important notions one needs to describe how *information* is revealed progressively. This is done by introducing the important (and delicate) notion of *filtration*, discussed in Section 2.4.2, which will allow us to define the important notions of past information, predictability and non-anticipativeness and to classify processes and random times according to these properties.

Finally, a stochastic process can also be seen as a function $X : [0,T]\times\Omega \mapsto E$ of both time t and the randomness ω. This point of view leads us to define notions of joint measurability and the concepts of optional and predictable processes, discussed in Section 2.4.5.

2.4.1 Stochastic processes as random functions

In order to define stochastic processes as function-valued random variables, one needs to define measures on function spaces. The simplest choice of function space for a stochastic process taking values in $\mathbb{R}^d$ is the set of all functions $f : [0,T] \to \mathbb{R}^d$ but this space happens to be too large: it contains many "pathological" functions and it is not easy to define measures on this space. Furthermore we expect the stochastic processes we work with to have sample paths with some more specific properties.

Random processes with continuous sample paths can be constructed as random variables defined on the space of continuous functions $C([0,T],\mathbb{R}^d)$. The usual topology on this space is defined by the sup norm

$$||f||_\infty = \sup_{t\in[0,T]} ||f(t)||, \tag{2.45}$$

which in turn can be used to construct a Borel σ-algebra, on which measures can be defined. The most well-known example is the Wiener measure, a Gaussian measure on $C([0,T],\mathbb{R}^d)$ describing the Wiener process. However most of the processes encountered in this book will not have continuous sample paths. We need therefore a space that allows for discontinuous functions. The class of *cadlag*[4] functions happens to be a convenient class of discontinuous functions:

DEFINITION 2.10 Cadlag function *A function $f : [0,T] \to \mathbb{R}^d$ is said to be* cadlag *if it is right-continuous with left limits: for each $t \in [0,T]$*

[4]The obscure word "cadlag" is a French acronym for "continu à droite, limite à gauche" which simply means "right-continuous with left limits." While most books use this terminology some authors use the (unpronounceable) English acronym "rcll."

the limits

$$f(t-) = \lim_{s \to t, s<t} f(s) \qquad f(t+) = \lim_{s \to t, s>t} f(s) \tag{2.46}$$

exist and $f(t) = f(t+)$.

Of course, any continuous function is cadlag but cadlag functions can have discontinuities. If t is a discontinuity point we denote by

$$\Delta f(t) = f(t) - f(t-) \tag{2.47}$$

the "jump" of f at t. However, cadlag functions cannot jump around too wildly. A cadlag function f can have at most a countable number of discontinuities: $\{t \in [0,T], f(t) \neq f(t-)\}$ is finite or countable [153]. Also, for any $\epsilon > 0$, the number of discontinuities ("jumps") on $[0,T]$ larger than ϵ should be finite. So a cadlag function on $[0,T]$ has a finite number of "large jumps" (larger than ϵ) and a possibly infinite, but countable number of small jumps.

An example of cadlag function is a step function having a jump at some point T_0, whose value at T_0 is defined to be the value *after* the jump: $f = 1_{[T_0,T[}(t)$. In this case $f(T_0-) = 0$, $f(T_0+) = f(T_0) = 1$ and $\Delta f(T_0) = 1$. More generally, given a continuous function $g : [0,T] \to \mathbb{R}$ and constants $f_i, i = 0 \ldots n-1$ and $t_0 = 0 < t_1 < \cdots < t_n = T$, the following function is cadlag:

$$f(t) = g(t) + \sum_{i=0}^{n-1} f_i 1_{[t_i,t_{i+1}[}(t). \tag{2.48}$$

The function g can be interpreted as the continuous component of f to which the jumps have been added: the jumps of f occur at $t_i, i \geq 1$ with $\Delta f(t_i) = f_i - f_{i-1}$. Not every cadlag function has such a neat decomposition into a continuous and a jump part but this example is typical. Cadlag functions are therefore natural models for the trajectories of processes with jumps.

REMARK 2.1 Cadlag or caglad? In the above example the function is right-continuous at jump times t_i simply because we have defined its value at t_i to be the value *after* the jump: $f(t_i) := f(t_i+)$. If we had defined $f(t_i)$ to be the value before the jump $f(t_i) := f(t_i-)$ we would have obtained a *left-continuous* function with right limits ("caglad"). The reader might wonder whether it makes any difference to interchange left and right. Yes, it does make a difference: since t is interpreted as a time variable, right means "after" and left means "before"! If a right continuous function has a jump at time t, then the value $f(t)$ is not foreseeable by following the trajectory up to time t: the discontinuity is seen as a sudden event. By contrast, if the sample paths were *left*-continuous, an observer approaching t along the path could predict the value at t. In the context of financial modelling, jumps represent sudden, unforeseeable events so the choice of right-continuity is natural. On the other hand, if we want to model a discontinuous process whose values are *predictable* we should use a caglad process. This will be the case when we model trading strategies in Chapter 8. □

It is possible to define a topology and a notion of convergence on the space of cadlag functions. For details we refer the reader to [215] or [364]. Equipped with this topology and the corresponding Borel σ-algebra (see Section 2.1.2), the space of cadlag functions is known as the Skorokhod space and denoted by $D([0,T],\mathbb{R}^d)$ or simply $D([0,T])$ if the context is clear. Obviously $C([0,T]) \subset D([0,T])$. A random variable with values in $D([0,T])$ is called a cadlag process. In all the models considered in this book, prices will be modelled as cadlag processes so when we will speak of "path space," we will refer to $D([0,T])$.

2.4.2 Filtrations and histories

The interpretation of the index t as a time variable introduces a dynamic aspect which needs to be taken into account by properly defining the notions of information, causality and predictability in the context of a stochastic model.

In a dynamic context, as time goes on, more information is progressively revealed to the observer. The result is that many quantities which are viewed as "random" at $t = 0$ may change status at a later time $t > 0$ if their value is revealed by the information available at time t. We must add some time-dependent ingredient to the structure of our probability space $(\Omega, \mathcal{F}, \mathbb{P})$ to accommodate this additional feature. This is usually done using the concept of *filtration*:

DEFINITION 2.11 Information flow *A filtration or information flow on $(\Omega, \mathcal{F}, \mathbb{P})$ is an increasing family of σ-algebras $(\mathcal{F}_t)_{t\in[0,T]}$: $\forall t \geq s \geq 0, \mathcal{F}_s \subseteq \mathcal{F}_t \subseteq \mathcal{F}$.*

$\mathcal{F}_t$ is then interpreted as the information known at time t, which increases with time. Naturally if we start with a set of events $\mathcal{F}$ then $\mathcal{F}_t \subseteq \mathcal{F}$. A probability space $(\Omega, \mathcal{F}, \mathbb{P})$ equipped with a filtration is called a *filtered probability space*. From an intuitive point of view, the probability of occurrence of a random event will change with time as more information is revealed. However, instead of changing the probability measure $\mathbb{P}$ with time, we will keep $\mathbb{P}$ fixed and model the impact of information by conditioning on the information $\mathcal{F}_t$.

The information flow being described by the filtration $\mathcal{F}_t$, we can now distinguish quantities which are known given the current information from those which are still viewed as random at time t. An event $A \in \mathcal{F}_t$ is an event such that given the information $\mathcal{F}_t$ at time t the observer can decide whether A has occurred or not. Similarly, an $\mathcal{F}_t$-measurable random variable is nothing else but a random variable whose value will be *revealed* at time t. A process whose value at time t is revealed by the information $\mathcal{F}_t$ is said to be nonanticipating:

DEFINITION 2.12 Nonanticipating process *A stochastic process $(X_t)_{t\in[0,T]}$ is said to be* nonanticipating *with respect to the information structure $(\mathcal{F}_t)_{t\in[0,T]}$ or $\mathcal{F}_t$*-adapted *if, for each $t \in [0,T]$, the value of X_t is revealed at time t: the random variable X_t is $\mathcal{F}_t$-measurable.*

A nonanticipating process is also called an adapted process: $(X_t)_{t\in[0,T]}$ is said to be $(\mathcal{F}_t)_{t\in[0,T]}$ -adapted. While the term "adapted" is certainly less self-explanatory than "nonanticipating," it is commonly used in the literature. If the only observation available is the past values of a stochastic process X, then the information is represented by the *history* (also called the natural filtration) of X defined as follows:

DEFINITION 2.13 History of a process *The history of a process X is the information flow $(\mathcal{F}_t^X)_{t\in[0,T]}$ where $\mathcal{F}_t^X$ is the σ-algebra generated by the past values of the process, completed by the null sets:*

$$\mathcal{F}_t^X = \sigma(X_s, s \in [0,t]) \bigvee \mathcal{N}. \tag{2.49}$$

One can think of $\mathcal{F}_t^X$ as containing all the information one can extract from having observed the path of X between 0 and t. All explicit examples of information flow we will use will correspond to the history of a set of asset prices. Notice that $\mathcal{F}_t^X$ has been completed by adding the null sets; all the null sets are stuffed into $\mathcal{F}_0$. This means that if a certain evolution for X between 0 and T is deemed impossible, its impossibility is already known at $t = 0$.

2.4.3 Random times

We will often have to deal with events happening at random times. A random time is nothing else than a positive random variable $T \geq 0$ which represents the time at which some event is going to take place. Given an information flow $(\mathcal{F}_t)$, a natural question is whether given the information in $\mathcal{F}_t$ one can determine whether the event has happened $(\tau \leq t)$ or not $(\tau > t)$. If the answer is yes, the random time τ is called a nonanticipating random time or *stopping time.* In other words, T is a nonanticipating random time ($(\mathcal{F}_t)$-stopping time) if

$$\forall t \geq 0, \quad \{T \leq t\} \in \mathcal{F}_t.$$

If T_1 and T_2 are stopping times then $T_1 \wedge T_2 = \inf\{T_1, T_2\}$ is also a stopping time. The term "stopping time" seems to imply that something is going to stop at the τ: given a stopping time τ and a nonanticipating process (X_t) one can define a new process $X_{\tau\wedge t}$, the process X stopped at τ, by:

$$X_{\tau\wedge t} = X_t \text{ if } t < \tau \quad X_{\tau\wedge t} = X_\tau \text{ if } t \geq \tau. \tag{2.50}$$

Examples of stopping times are *hitting times*: given a nonanticipating cadlag process X, the hitting time of an open set A is defined by the first time when X reaches A:

$$T_A = \inf\{t \geq 0, X_t \in A\}. \tag{2.51}$$

At any given time t, it is enough to know the past positions of X_t to see whether the set A has been reached ($T_A \leq t$) or not ($T_A > t$). A frequently encountered example is the following: if X is a real-valued process starting from 0 and $a > 0$, the *exit time* of X from the interval $]-\infty, a]$ is defined as

$$T_a = \inf\{t > 0, X_t > a\}. \tag{2.52}$$

It is the hitting time T_A associated with the open interval $A =]a, \infty[$.

An example of a random time which is not a stopping time is the first instant $t \in [0, T]$ when X reaches its maximum:

$$T_{\max} = \inf\{t \in [0,T], X_t = \sup_{s\in[0,T]} X_s\}. \tag{2.53}$$

Obviously in order to know the value of the maximum one must first wait until T to observe the whole path on $[0, T]$. Therefore given the information $\mathcal{F}_t$ at time $t < T$ one cannot decide whether $T_{\max}$ has occurred or not.

Given an information flow $\mathcal{F}_t$ and a nonanticipating random time τ, the information set $\mathcal{F}_\tau$ can be defined as the information obtained by observing all nonanticipating (cadlag) processes at τ, i.e., the σ-algebra generated by these observations: $\mathcal{F}_\tau = \sigma(X_\tau, X$ nonanticipating cadlag process). It can be shown that this definition is equivalent to the following (see [110, p. 186] or [324, Chapter 1]):

$$\mathcal{F}_\tau = \{A \in \mathcal{F},\ \forall t \in [0,T], A \cap \{t \leq \tau\} \in \mathcal{F}_t\}. \tag{2.54}$$

2.4.4 Martingales

Consider now a probability space $(\Omega, \mathcal{F}, \mathbb{P})$ equipped with an information flow $\mathcal{F}_t$.

DEFINITION 2.14 Martingale *A cadlag process $(X_t)_{t\in[0,T]}$ is said to be a martingale if X is nonanticipating (adapted to $\mathcal{F}_t$), $E[|X_t|]$ is finite for any $t \in [0,T]$ and*

$$\forall s > t, \qquad E[X_s|\mathcal{F}_t] = X_t. \tag{2.55}$$

In other words, the best prediction of a martingale's future value is its present value. A familiar example of a martingale is the Wiener process (W_t).

Notice that the definition of martingale makes sense only when the underlying information flow $(\mathcal{F}_t)_{t\in[0,T]}$ and the probability measure $\mathbb{P}$ have been specified. To avoid confusion one should speak of $(\mathbb{P},\mathcal{F}_t)$-martingales. When several probability measures are involved, we shall use the term $\mathbb{P}$-*martingale* to emphasize the fact the notion of martingale depends on the probability measure $\mathbb{P}$.

A typical way to construct a martingale is the following: given a random variable H revealed at T (i.e., $\mathcal{F}_T$-measurable) with $E|H| < \infty$, the process $(M_t)_{t\in[0,T]}$ defined by $M_t = E[H|\mathcal{F}_t]$ is a martingale. Conversely any martingale $(M_t)_{t\in[0,T]}$ can be written in this form, by choosing as $H = M_T$ the terminal value.[5]

An obvious consequence of (2.55) is that a martingale has constant expectation: $\forall t \in [0,T], E[X_t] = E[X_0]$. One can wonder whether any "driftless" process is a martingale. The answer is no; if (W_t) is a scalar Wiener process, W_t^3 has constant expectation $E[W_t^3] = 0$ but is not a martingale: indeed, if $s > t$

$$\begin{aligned} E[W_s^3|\mathcal{F}_t] &= E[(W_s - W_t + W_t)^3|\mathcal{F}_t] \\ &= E[(W_s - W_t)^3 + W_t^3 + 3(W_t - W_s)W_t^2 + 3(W_t - W_s)^2 W_t|\mathcal{F}_t] \\ &= E[(W_s - W_t)^3] + W_t^3 + 3W_t^2 E[W_t - W_s] + 3E[(W_t - W_s)^2]W_t \\ &= 0 + W_t^3 + 0 + 3(t-s)W_t \neq W_t^3. \end{aligned}$$

However, if one asks that the process be driftless when computed at random times, then this property actually characterizes martingales [325]: if $E[X_\tau] = E[X_0]$ for any stopping time τ then X is a martingale.

A fundamental property of martingales is the "sampling property": the martingale property (2.55) is verified when t, s are replaced by nonanticipating random times.

PROPOSITION 2.7 Sampling theorem

If $(M_t)_{t\in[0,T]}$ is a martingale and T_1, T_2 are nonanticipating random times (stopping times) with $T \geq T_2 \geq T_1 \geq 0$ a.s. then

$$E[M_{T_2}|\mathcal{F}_{T_1}] = M_{T_1}. \tag{2.56}$$

For a proof see [116] or [324, Section I.2]. In particular, a martingale stopped at a nonanticipating random time is still a martingale.

A process $(X_t)_{t\in[0,T]}$ is a called a local martingale if there exists a sequence of stopping times (T_n) with $T_n \to \infty$ a.s. such that $(X_{t\wedge T_n})_{t\in[0,T]}$ is a martingale. Thus a local martingale behaves like a martingale up to some stopping

[5] Note that here we define processes and martingales in particular on a finite time interval $[0,T]$; these results do not hold if the time parameter $t \in [0,\infty[$.

time T_n, which can be chosen as large as one wants. Obviously any martingale is a local martingale but there exist local martingales which are not martingales; this is the origin of many subtleties in martingale theory.

2.4.5 Predictable processes (*)

So far we have defined a stochastic process $(X_t)_{t\in[0,T]}$ either by considering it as a function of time for fixed ω — the sample path $t \mapsto X_t(\omega)$ — or as a function of ω for a fixed t — the random variable X_t. Of course it is natural to consider these two aspects jointly by considering X as a function defined on $[0,T] \times \Omega$. This requires defining σ-algebras and measurable functions [6] on $[0,T] \times \Omega$. At first sight we could consider the σ-algebra generated by products $B \times A$ where $A \in \mathcal{F}$ and $B \subset [0,T]$ is measurable. But if we are given an information flow $(\mathcal{F}_t)_{t\in[0,T]}$ we would like the previously defined class of nonanticipating cadlag processes to be measurable functions on $[0,T]$. The simplest choice is then to take the σ-algebra generated by these processes:

DEFINITION 2.15 Optional processes *The* optional σ-algebra *is the σ-algebra $\mathcal{O}$ generated on $[0,T] \times \Omega$ by all nonanticipating (adapted) cadlag processes. A process $X : [0,T] \times \Omega \mapsto \mathbb{R}^d$ which is measurable with respect to $\mathcal{O}$ is called an* optional process.[7]

With this definition, any nonanticipating cadlag process is optional but the sample paths of an optional process need not be cadlag in general: nonanticipating cadlag processes "generate" optional processes the same way that continuous functions "generate" measurable functions.

The distinction made in Section 2.4.1 between left and right continuity and its interpretation in terms of sudden vs. predictable jumps motivates the definition of another σ-algebra on $[0,T] \times \Omega$:

DEFINITION 2.16 Predictable processes *The* predictable σ-algebra *is the σ-algebra $\mathcal{P}$ generated on $[0,T]\times\Omega$ by all nonanticipating (adapted) left-continuous processes. A mapping $X : [0,T] \times \Omega \mapsto \mathbb{R}^d$ which is measurable with respect to $\mathcal{P}$ is called a* predictable process.

While the name "predictable" is justified given the discussion above, the name "optional" is less transparent. We will not use this notion very often so we will stick to this terminology. Any left-continuous process is therefore predictable (by definition): this is intuitive if $\lim_{s\to t, s<t} X_s = X_t$ then the value of X_t is "announced" by the values at preceding instants. In the sense of

[6] See Section 2.1.2.

[7] The term "optional" has nothing to do with options!

Definition 2.16, all predictable processes are "generated" from left-continuous processes (in the same way that all Borel sets are "generated" from open intervals). However there are predictable processes which are not left continuous (which is less intuitive). In the same way, any nonanticipating process with cadlag trajectories is "optional," but the notion of optional process allows sample paths to be more irregular.

$$\text{Cadlag / Right-continuous} + \text{nonanticipating} \Rightarrow \text{Optional}$$

$$\text{Caglad / Left-continuous} + \text{nonanticipating} \Rightarrow \text{Predictable}$$

The distinction between optional and predictable process will become clear in Chapter 10 when we will use these notions in a financial modelling context: state variables such as market prices will be modelled as optional processes while the decisions of an investor — hedging strategies, portfolios — will be represented by predictable processes.

The practical-minded reader might wonder why we bother to consider processes more general than cadlag, if cadlag or piecewise continuous processes are already rich enough to model discontinuities. As we shall observe in the next chapters, in all models that we will encounter, prices will follow cadlag processes and all explicit examples of predictable processes will be left-continuous. Also, on the theoretical side we will see that it is possible to discuss stochastic integration, Itô's formula, stochastic differential equations and changes of measure only using cadlag processes. However when using results of the type "there exists a predictable process such that..." we will need to consider more general predictable processes. Typical examples are martingale representation theorems, discussed in Chapter 10.

2.5 The Poisson process

The Poisson process is a fundamental example of a stochastic process with discontinuous trajectories and will be used as a building block for constructing more complex jump processes.

2.5.1 Exponential random variables

Properties of exponential random variables and their sums will play an important role in defining Markov processes with jumps. We review some of these important properties in this section.

A positive random variable Y is said to follow an exponential distribution with parameter $\lambda > 0$ if it has a probability density function of the form

$$\lambda e^{-\lambda y} 1_{y \geq 0}. \tag{2.57}$$

The distribution function of Y is then given by

$$\forall y \in [0, \infty[\ , \quad F_Y(y) = \mathbb{P}(Y \leq y) = 1 - \exp(-\lambda y). \tag{2.58}$$

F_Y is invertible and its inverse is given by:

$$\forall y \in [0, 1],\ F_Y^{-1}(y) = -\frac{1}{\lambda} \ln(1 - y). \tag{2.59}$$

A simple consequence is that if U is uniformly distributed on $[0, 1]$ then $-\frac{1}{\lambda} \ln U$ is exponentially distributed with parameter λ. This result will be used in Chapter 6 to simulate random variables with exponential distribution.

The exponential distribution has the following important property: if T is an exponential random variable,

$$\forall t, s > 0,\ \mathbb{P}(T > t + s | T > t) = \frac{\int_{t+s}^{\infty} \lambda e^{-\lambda y} dy}{\int_t^{\infty} \lambda e^{-\lambda y} dy} = \mathbb{P}(T > s). \tag{2.60}$$

In other words, if one interprets T as a random time, the distribution of $T - t$ knowing $T > t$ is the same as the distribution of T itself: this property is usually called the "absence of memory." In fact the exponential distribution is the only distribution with this property:

PROPOSITION 2.8 Absence of memory
Let $T \geq 0$ be a (nonzero) random variable such that

$$\forall t, s > 0,\ \mathbb{P}(T > t + s | T > t) = \mathbb{P}(T > s). \tag{2.61}$$

Then T has an exponential distribution.

PROOF Let $g(t) = \mathbb{P}(T > t)$. Using Bayes rule we obtain that g is a multiplicative function:

$$\forall t, s > 0\ g(t + s) = \mathbb{P}(T > t + s | T > t) \mathbb{P}(T > t) = g(s) g(t).$$

Since $1 - g$ is a distribution function, g is decreasing and right-continuous; together with the multiplicative property this implies that $g(t) = \exp(-\lambda t)$ for some $\lambda > 0$. ∎

Let $(\tau_i, i = 1 \dots n)$ be independent exponential random variables with parameter λ and define $T_k = \tau_1 + \dots + \tau_k$. Then $(T_1, T_2, \dots, T_n)$ has a probability density on $\mathbb{R}^n$ given by

$$\lambda^n e^{-\lambda t_n} 1_{0 < t_1 < \dots < t_n}(t_1, \dots, t_n), \tag{2.62}$$

and the random variable $T_n = \tau_1 + \cdots + \tau_n$ has probability density

$$p_n(t) = \lambda e^{-\lambda t} \frac{(\lambda t)^{n-1}}{(n-1)!} 1_{[0,\infty[}(t). \tag{2.63}$$

Sums of i.i.d. exponential variables bear a close relationship with the order statistics of uniformly distributed random variables. Let $U_1, \ldots, U_n$ be independent random variables, uniformly distributed on $[a, b]$. Arrange them in increasing order and call them $(Y_1, \ldots, Y_n)$: $Y_n = \max\{U_i, i = 1 \ldots n\}$, $Y_{n-1} = \max(\{U_i, i = 1 \ldots n\} \setminus \{Y_n\})$, etc. $Y_1 \leq \cdots \leq Y_n$ are called the *order statistics* of $U_1, \ldots, U_n$ and their density is given by

$$\frac{n!}{(b-a)^n} 1_{\{a<y_1<y_2<\cdots<y_n<b\}}(x). \tag{2.64}$$

The law with density given by (2.64) is sometimes called the Dirichlet distribution and denoted by $D_n([a, b])$. Starting with Equation (2.62), one can see that the expression (2.62) can be rearranged in order to make (2.63) appear:

$$\lambda^n e^{-\lambda t_n} 1_{0<t_1<\cdots<t_n}(t_1, \ldots, t_n)$$
$$= \lambda e^{-\lambda t_n} \frac{(\lambda t_n)^{n-1}}{(n-1)!} 1_{t_n \geq 0} \times \frac{(n-1)!}{t_n^{n-1}} 1_{0<t_1<\cdots<t_n}(t_1, \ldots, t_n),$$

which is simply the product of (2.63) with a $D_{n-1}([0, t_n])$ density. This leads to the following result:

PROPOSITION 2.9
Let $(\tau_i, i = 1 \ldots n+1)$ be independent exponential random variables with parameter λ and define $T_k = \tau_1 + \cdots + \tau_k$. Then

1. *The law of $(T_1, \ldots, T_n)$ knowing $T_{n+1} = t$ is $D_n([0, t])$.*
2. *The vector $(\frac{T_1}{T_{n+1}}, \ldots, \frac{T_n}{T_{n+1}})$ is independent from T_{n+1} and has the law $D_n([0, 1])$.*

The above proposition is interesting in practice since it gives us a way to simulate from the joint law of $(T_1, \ldots, T_n)$ by using independent uniformly distributed random variables. We will put this property to use in Chapter 6.

PROPOSITION 2.10
Let $U_1, \ldots, U_n$ be independent random variables, uniformly distributed on $[0, 1]$, $U_{(1)} \leq \cdots \leq U_{(n)}$ be the corresponding order statistics and V be an independent random variable with a density given by (2.63). Then the variables

$$VU_{(1)}, V(U_{(2)} - U_{(1)}), V(U_{(3)} - U_{(2)}), \ldots, V(U_{(n)} - U_{(n-1)})$$

form a sequence of independent exponential variables with parameter λ.

2.5.2 The Poisson distribution

An integer valued random variable N is said to follow a Poisson distribution with parameter λ if

$$\forall n \in \mathbb{N},\ \mathbb{P}(N = n) = e^{-\lambda}\frac{\lambda^n}{n!}. \tag{2.65}$$

The Poisson distribution has a well-defined moment generating function given by:

$$M(u) = \exp[\lambda(e^u - 1)]. \tag{2.66}$$

There is an intimate connection between the Poisson distribution and sums of independent exponential random variables:

PROPOSITION 2.11
If $(\tau_i)_{i\geq 1}$ are independent exponential random variables with parameter λ then, for any $t > 0$ the random variable

$$N_t = \inf\{n \geq 1, \sum_{i=1}^{n} \tau_i > t\} \tag{2.67}$$

follows a Poisson distribution with parameter λt:

$$\forall n \in \mathbb{N},\ \mathbb{P}(N_t = n) = e^{-\lambda t}\frac{(\lambda t)^n}{n!}. \tag{2.68}$$

PROOF Let $T_k = \sum_{i=1}^{k} \tau_i\ \forall k$. The density of $(T_1, ..., T_k)$ is given by

$$\lambda^k 1_{0<t_1<...<t_k} e^{-\lambda t_k} dt_1...dt_k.$$

Since $\mathbb{P}(N_t = n) = \mathbb{P}(T_n \leq t < T_{n+1})$, it can be computed as:

$$\begin{aligned}\mathbb{P}(N_t = n) &= \int_{0<t_1<...t_n<t<t_{n+1}} \lambda^n e^{-\lambda t_{n+1}} dt_1...dt_n dt_{n+1}\\ &= \lambda^n e^{-\lambda t}\int_{0<t_1<...t_n<t} dt_1...dt_n = e^{-\lambda t}\frac{(\lambda t)^n}{n!}.\end{aligned}$$

□

One interesting property of the Poisson distribution is the stability under convolution: if Y_1 and Y_2 are *independent* Poisson variables with parameters λ_1 and λ_2, then $Y_1 + Y_2$ also follows a Poisson law with parameter $\lambda_1 + \lambda_2$. This can be readily deduced from the form of the moment generating function, noting that $M_{Y_1+Y_2} = M_{Y_1}M_{Y_2}$.

In particular this leads to the following consequence: for any integer n, a Poisson random variable Y with parameter λ can be expressed as a sum of n independent (Poisson) random variables Y_i, with parameter λ/n. This property, called *infinite divisibility*, can be interpreted as saying that a Poisson random variable can be "divided" into an arbitrary number of i.i.d. random variables. We will encounter this important property again in Chapter 3.

2.5.3 The Poisson process: definition and properties

DEFINITION 2.17 Poisson process *Let $(\tau_i)_{i\geq 1}$ be a sequence of independent exponential random variables with parameter λ and $T_n = \sum_{i=1}^{n} \tau_i$. The process $(N_t, t \geq 0)$ defined by*

$$N_t = \sum_{n\geq 1} 1_{t\geq T_n} \tag{2.69}$$

is called a Poisson process *with intensity λ.*

The Poisson process is therefore defined as a *counting* process: it counts the number of random times (T_n) which occur between 0 and t, where $(T_n - T_{n-1})_{n\geq 1}$ is an i.i.d. sequence of exponential variables. The following properties of the Poisson process can be easily deduced:

PROPOSITION 2.12
Let $(N_t)_{t\geq 0}$ be a Poisson process.

1. *For any $t > 0$, N_t is almost surely finite.*
2. *For any ω, the sample path $t \to N_t(\omega)$ is piecewise constant and increases by jumps of size 1.*
3. *The sample paths $t \mapsto N_t$ are right continuous with left limite (cadlag).*
4. *For any $t > 0$, $N_{t-} = N_t$ with probability 1.*
5. *(N_t) is continuous in probability:*

$$\forall t > 0,\ N_s \underset{s\to t}{\overset{\mathbb{P}}{\to}} N_t. \tag{2.70}$$

6. *For any $t > 0$, N_t follows a Poisson distribution with parameter λt:*

$$\forall n \in \mathbb{N},\ \mathbb{P}(N_t = n) = e^{-\lambda t}\frac{(\lambda t)^n}{n!}. \tag{2.71}$$

7. *The characteristic function of N_t is given by*

$$E[e^{iuN_t}] = \exp\{\lambda t(e^{iu} - 1)\},\ \forall u \in \mathbb{R}. \tag{2.72}$$

8. *(N_t) has independent increments: for any $t_1 < \cdots < t_n$, $N_{t_n} - N_{t_{n-1}}, \ldots,$ $N_{t_2} - N_{t_1}, N_{t_1}$ are independent random variables.*

9. *The increments of N are homogeneous: for any $t > s$, $N_t - N_s$ has the same distribution as N_{t-s}.*

10. *(N_t) has the Markov property:*

$$\forall t > s, \quad E[f(N_t)|N_u, u \leq s] = E[f(N_t)|N_s].$$

PROOF

1. Let $\Omega_1 = \{\omega \in \Omega, \frac{T_n}{n}(\omega) \to \frac{1}{\lambda}\}$. By the law of large numbers, $\frac{T_n}{n} \to \frac{1}{\lambda}$ with probability 1 so $\mathbb{P}(\Omega_1) = 1$. For any $\omega \in \Omega_1, T_n(\omega) \to \infty$ so

$$\forall \omega \in \Omega_1, \exists n_0(\omega) \geq 1, \ \forall n \geq n_0(\omega), T_n(\omega) > t.$$

   So $\mathbb{P}(N_t < \infty) = \mathbb{P}(\Omega_1) = 1$: the number of terms in the sum (2.17) is almost surely finite.

2. From the expression (2.17) it is obvious that N_t is constant on each interval $]T_n, T_{n+1}[$ and increases by one at each T_n. Since the number of jump points in each interval $[0, t]$ is almost surely finite, property (2) follows.

3. The cadlag property can be seen by comparing the definition of the Poisson process to the example (2.48).

4. For a given $\omega \in \Omega$, the points of discontinuity of $N_t(\omega)$ are $\{T_n(\omega), n \geq 1\}$. But for a given t, $\mathbb{P}(t \in \{T_n(\omega), n \geq 1\}) = 0$. So, with probability 1, t is not a discontinuity point: $N_{t-} = N_t$ with probability 1.

5. Consequence of the above, since almost sure convergence entails convergence in probability.

6. This point was shown in proposition (2.11).

7. This is a easy consequence of the previous one.

8. Let $0 < t_1 < \cdots < t_n$ and compute:

$$\mathbb{P}(N_{t_1} = k_1, N_{t_2} - N_{t_1} = k_2, \ldots, N_{t_n} - N_{t_{n-1}} = k_n). \qquad (2.73)$$

   Define $j_n = \sum_{i \leq n} k_i$ for $i \geq 1$. Then the above probability can be rewritten as:

$$\mathbb{P}(T_{j_1} \leq t_1 < T_{j_1+1}, T_{j_2} \leq t_2 < T_{j_2+1}, \ldots, T_{j_n} \leq t_n < T_{j_n+1}).$$

Conditionally on $T_{j_n} < t_n < T_{j_n+1}$, $(T_1, T_2, \ldots, T_{j_n})$ are distributed as the order statistics of j_n uniform random variables on $[0, t_n]$. The conditional probability

$$\mathbb{P}(T_{j_1} \leq t_1 < T_{j_1+1}, T_{j_2} \leq t_2 < T_{j_2+1}, \ldots \big| T_{j_n} \leq t_n < T_{j_n+1}) \quad (2.74)$$

is then equal to the probability that, given j_n independent random variables $U_1, \ldots, U_{j_n}$, uniformly distributed on $[0, t_n]$, k_1 of them fall into the interval $[0, t_1]$, k_2 of them fall into the interval $]t_1, t_2]$, etc. The probability that U_k belongs to $[t_{i-1}, t_i]$ being $(t_i - t_{i-1})/t_n$, the conditional probability (2.74) is given by

$$\frac{j_n!}{t_n^{j_n}} \frac{t_1^{k_1}}{k_1!} \prod_{i=2}^{n} \frac{(t_i - t_{i-1})^{k_i}}{k_i!}.$$

To obtain the unconditional probability (2.73), we multiply this expression by the density of N_{t_n}, which is a Poisson distribution with parameter λt_n. After simplification this gives

$$\lambda^{j_n} e^{-\lambda t_n} \frac{t_1^{k_1}}{k_1!} \prod_{i=2}^{n} \frac{(t_i - t_{i-1})^{k_i}}{k_i!}.$$

Finally, substituting $j_n = k_1 + k_2 + \cdots + k_n$ we see that the probability of interest factorizes into a product of n terms:

$$\begin{aligned} \mathbb{P}(N_{t_1} = k_1, N_{t_2} - N_{t_1} = k_2, \ldots, N_{t_n} - N_{t_{n-1}} = k_n) \\ = \frac{\{\lambda t_1\}^{k_1} e^{-\lambda t_1}}{k_1!} \prod_{i=2}^{n} \frac{\{\lambda(t_i - t_{i-1})\}^{k_i} e^{-\lambda(t_i - t_{i-1})}}{k_i!}. \end{aligned}$$

The joint law of the increments has a product form, which shows their independence. Moreover, each term in the product is recognized to be the density of a Poisson law with parameter $\lambda(t_i - t_{i-1})$ which shows the homogeneity of the increments.

9. The Markov property follows from the independence of increments:

$$\begin{aligned} E[f(N_t)|N_u, u \leq s] &= E[f(N_t - N_s + N_s)|N_u, u \leq s] \\ &= E[f(N_t - N_s + N_s)|N_s]. \end{aligned}$$

since $N_t - N_s$ is independent of $N_u, u \leq s$.

□

Let us now comment on the properties (2), (3) and (4) which seem somewhat paradoxical: on one hand we assert that with probability one, any sample path of the Poisson process is discontinuous (in fact it only moves by jumps) and on the other hand at any given point t, the sample function is continuous with probability 1! We will encounter this feature for all the jump processes considered later: it is due to the fact that the points of discontinuity of the process form a set of zero measure. A typical path of a Poisson process is shown in Figure 2.1.

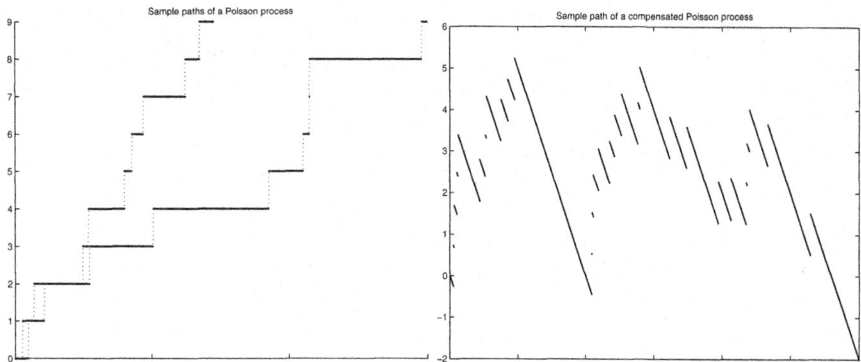

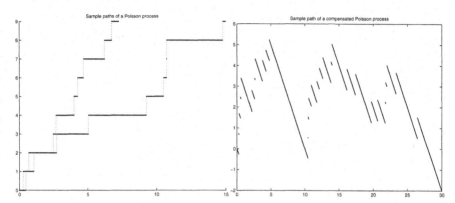

FIGURE 2.1: Left: Sample paths of a Poisson process with intensity $\lambda = 1$. Right: Sample path of a compensated Poisson process with intensity $\lambda = 1$.

The right continuity (cadlag property) of the Poisson process is not really a "property": we have *defined* N_t in such a way that at a discontinuity point $N_t = N_{t+}$, i.e., we have chosen the right-continuous (cadlag) version of the Poisson process. Another choice would have been the left-continuous one:

$$N'_t = \sum_{n \geq 0} 1_{t > T_n}. \tag{2.75}$$

The difference between N and N' is that the values of N' in the near future are foreseeable since $N'_s \to N_t$ as $s \to t, s < t$ while the values of N are "unpredictable" given the past. The jumps of N are interpreted as sudden events[8], which corresponds to our initial motivation for including jumps in models of price dynamics. Hence we will always use the cadlag version $(N_t)_{t \geq 0}$ of the Poisson process. Following these remarks, for other examples of jump process that we will encounter, we will always choose a right-continuous (cadlag) version.

Poisson processes have some other useful properties which we mention here. First, a superposition of independent Poisson processes is again a Poisson process:

[8]See the discussion in Section 2.4.5.

PROPOSITION 2.13 Sum of independent Poisson processes
If $(N_t^1)_{t\geq 0}$ and $(N_t^2)_{t\geq 0}$ are independent Poisson processes with intensities λ_1, λ_2 then $(N_t^1 + N_t^2)_{t\geq 0}$ is a Poisson process with intensity $\lambda_1 + \lambda_2$.

A second useful property of Poisson processes is the so-called thinning property. Let $(N_t)_{t\geq 0}$ be a Poisson process with intensity λ and define a new process X_t by "thinning" N_t: take all the jump events $(T_n, n \geq 1)$ corresponding to N, keep them with probability $0 < p < 1$ or delete them with probability $1 - p$, independently from each other. Now order the points which have not been deleted: $T'_1, \ldots, T'_n, \ldots$ and define

$$X_t = \sum_{n\geq 1} 1_{T'_n \geq t}. \tag{2.76}$$

Then the process X is a Poisson process with intensity $p\lambda$. Another way to see this result is the following: if the arrival T_n of each event in the Poisson process N is *marked* with probability p, independently from event to event, then the process of marked events thus obtained is again a Poisson process whose intensity is equal to the intensity of N, decreased by the marking probability: $\lambda_X = p\lambda$.

2.5.4 Compensated Poisson processes

Define the "centered" version of the Poisson process N_t by

$$\tilde{N}_t = N_t - \lambda t. \tag{2.77}$$

$\tilde{N}_t$ follows a centered version of the Poisson law with characteristic function:

$$\Phi_{\tilde{N}_t}(z) = \exp[\lambda t(e^{iz} - 1 - iz)]. \tag{2.78}$$

As the Poisson process, $\tilde{N}$ also has independent increments and it is easy to show that:

$$\begin{aligned} E[N_t|N_s, s \leq t] &= E[N_t - N_s + N_s|N_s] \\ &= E[N_t - N_s] + N_s = \lambda(t-s) + N_s, \end{aligned}$$

so $(\tilde{N}_t)$ has the *martingale property*:

$$\forall t > s, E[\tilde{N}_t|\tilde{N}_s] = \tilde{N}_s. \tag{2.79}$$

$(\tilde{N}_t)_{t\geq 0}$ is called a *compensated* Poisson process and (the deterministic expression) $(\lambda t)_{t\geq 0}$ is called the compensator of $(N_t)_{t\geq 0}$: it is the quantity which has to be subtracted from N in order to obtain a martingale. Sample paths of the Poisson and compensated Poisson process are shown in Figure 2.1. Note that the compensated Poisson process is no longer integer valued: unlike the

Poisson process, it is not a counting process. Its rescaled version $\tilde{N}_t/\lambda$ has the same first two moments as a standard Wiener process:

$$E\left[\frac{\tilde{N}_t}{\lambda}\right] = 0 \quad \text{Var}\left[\frac{\tilde{N}_t}{\lambda}\right] = t. \tag{2.80}$$

Figure 2.2 compares $\tilde{N}_t/\lambda$ with a standard Wiener process. The two graphs look similar and this is not a coincidence: when the intensity of its jumps increases, the (interpolated) compensated Poisson process converges in distribution to a Wiener process [153]:

$$\left(\frac{\tilde{N}_t}{\lambda}\right)_{t\in[0,T]} \overset{\lambda\to\infty}{\Rightarrow} (W_t)_{t\in[0,T]}. \tag{2.81}$$

This result is a consequence of the Donsker invariance principle [228, Theorem 14.9] and may be seen as a "functional" central limit theorem on $\Omega = D([0,T])$.

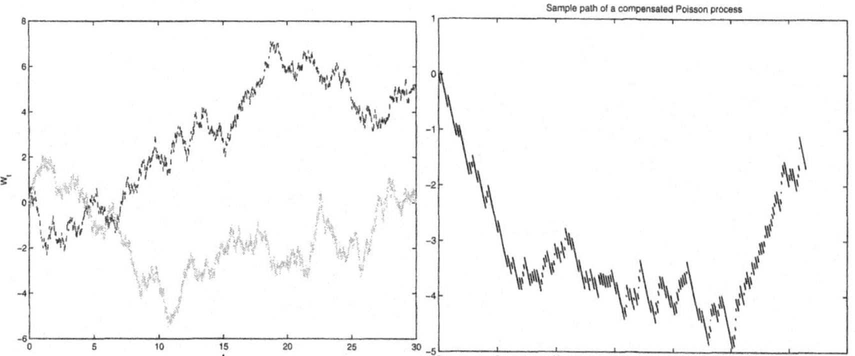

FIGURE 2.2: Left: Sample paths of a Wiener process with $\sigma = 1$. Right: Sample path of a compensated Poisson process with intensity $\lambda = 5$, rescaled to have the same variance as the Wiener process.

2.5.5 Counting processes

The Poisson process N_t counts the number of random times $\{T_n, n \geq 1\}$ occurring in $[0,t]$, where the random times T_n are partial sums of a sequence of i.i.d. exponential random variables. More generally, given an increasing sequence of random times $\{T_n, n \geq 1\}$ with $\mathbb{P}(T_n \to \infty) = 1$, one can define the associated counting process $(X_t)_{t\geq 0}$ by

$$X_t = \sum_{n\geq 1} 1_{t\geq T_n} = \#\{n \geq 1, T_n \geq t\}.$$

X_t is simply the number of random times $\{T_n, n \geq 1\}$ occurring in $[0, t]$. The condition $\mathbb{P}(T_n \to \infty) = 1$ guarantees that, with probability 1, X_t is well-defined (finite) for any $t \geq 0$. Like the Poisson process, $(X_t)_{t\geq 0}$ is a cadlag process with piecewise constant trajectories: its sample paths move by jumps of size $+1$.

If the random times (T_n) are constructed as partial sums of a sequence of i.i.d. exponential random variables, then X is a Poisson process. For a general counting process, the sequence of random times (T_n) can have any distribution and dependence structure. The following lemma shows that the only counting processes with independent stationary increments are Poisson processes:

LEMMA 2.1
Let (X_t) be a counting process with stationary independent increments. Then (X_t) is a Poisson process.

PROOF This is an elementary result, but its combinatorial proof (see for example [324, Chapter 1]) is rather long, so we prefer to give a shorter but less elementary one, which has an additional benefit of illustrating Doob's sampling theorem (Proposition 2.7).

Let $T_k = \inf\{t \geq 0 : X_t \geq k\}$. Define for $n \geq 1$, $Y_n = T_n - T_{n-1}$. The first step is to prove that Y_1 has exponential distribution.

$$\begin{aligned}\mathbb{P}\{Y_1 > t + s \mid Y_1 > t\} &= \mathbb{P}\{X_{t+s} = 0 \mid X_t = 0\} \\ &= \mathbb{P}\{X_{t+s} - X_t = 0 \mid X_t = 0\} = \mathbb{P}\{X_{t+s} - X_t = 0\} \\ &= \mathbb{P}\{X_s = 0\} = \mathbb{P}\{Y_1 > s\}.\end{aligned}$$

We have shown Y_1 to have the memoryless property. Thus, by Proposition 2.8 it is an exponential random variable.

The second step is to show that $(X_{t+Y_1} - X_{Y_1})_{t\geq 0}$ is independent from Y_1 and has the same law as $(X_t)_{t\geq 0}$. To see this observe that Y_1 is a nonanticipating (stopping) time for the process X_t because the event $\{Y_1 \leq t\} = \{X_t \geq 1\}$ does not depend on the trajectory of X after t. Let $f(t) = E[e^{iuX_t}]$ for u fixed. Independence and stationarity of increments of (X_t) entail that for all $s, t > 0$, $f(s + t) = f(s)f(t)$ and that

$$M_t = \frac{e^{iuX_t}}{f(t)}$$

is a martingale. Let $Y_1^n = n \wedge Y_1$. As Y_1^n is a bounded stopping time, using Doob's optional sampling theorem we obtain:

$$E[e^{iu(X_{Y_1^n+t}-X_{Y_1^n})+ivY_1^n}|\mathcal{F}_{Y_1^n}] = \frac{f(t+Y_1^n)}{f(Y_1^n)}e^{ivY_1^n} = f(t)e^{ivY_1^n}.$$

Therefore

$$E[e^{iu(X_{Y_1^n+t}-X_{Y_1^n})+ivY_1^n}] = E[e^{ivY_1^n}]E[e^{iuX_t}].$$

By the dominated convergence theorem we can compute the limit $n \to \infty$:

$$E[e^{iu(X_{Y_1+t}-X_{Y_1})+ivY_1}] = E[e^{ivY_1}]E[e^{iuX_t}].$$

This result shows both that the *process* $(X_{Y_1+t} - X_{Y_1})_{t\geq 0}$ is independent from Y_1 (because we could have taken any number of increments) and that it has the same distribution as $(X_t)_{t\geq 0}$. This property allows to conclude that Y_2 is independent from Y_1 and has an exponential distribution, Y_3 is independent from Y_2 and Y_1 and has exponential distribution so the result is obtained by induction. □

2.6 Random measures and point processes

The Poisson process $(N_t)_{t\geq 0}$ was defined in Section 2.5 as a counting process: if $T_1, T_2, \ldots$ is the sequence of jump times of N, then N_t is simply the number of jumps between 0 and t:

$$N_t = \#\{i \geq 1,\ T_i \in [0,t]\}. \tag{2.82}$$

Similarly, if $t > s$ then

$$N_t - N_s = \#\{i \geq 1,\ T_i \in]s,t]\}.$$

The jump times $T_1, T_2, \ldots$ form a random configuration of points on $[0, \infty[$ and the Poisson process N_t counts the number of such points in the interval $[0,t]$. This counting procedure defines a *measure* M on $[0, \infty[$: for any measurable set $A \subset \mathbb{R}^+$ let

$$M(\omega, A) = \#\{i \geq 1,\ T_i(\omega) \in A\}. \tag{2.83}$$

Then $M(\omega,.)$ is a positive, integer valued measure and $M(A)$ is finite with probability 1 for any bounded set A. Note that the measure $M(\omega,.)$ depends on ω: it is thus a *random measure*. The intensity λ of the Poisson process determines the *average* value of the random measure M: $E[M(A)] = \lambda|A|$ where $|A|$ is the Lebesgue measure of A.

M is called the random jump measure[9] associated to the Poisson process N. The Poisson process may be expressed in terms of the random measure M in the following way:

$$N_t(\omega) = M(\omega, [0,t]) = \int_{[0,t]} M(\omega, ds). \tag{2.84}$$

The properties of the Poisson process translate into the following properties for the measure M: for disjoint intervals $[t_1, t_1'], \ldots, [t_n, t_n']$

1. $M([t_k, t_k'])$ is the number of jumps of the Poisson process in $[t_k, t_k']$: it is a Poisson random variable with parameter $\lambda(t_k' - t_k)$.

2. For two disjoint intervals $j \neq k$, $M([t_j, t_j'])$ and $M([t_k, t_k'])$ are independent random variables.

3. More generally for any (measurable) set A, $M(A)$ follows a Poisson distribution with parameter $\lambda|A|$ where $|A| = \int_A dx$ is the Lebesgue measure of A.

The random measure M can also be viewed as the "derivative" of the Poisson process. Recall that each trajectory $t \mapsto N_t(\omega)$ of a Poisson process is an increasing step function. Hence its derivative (in the sense of distributions) is a positive measure: in fact it is simply the superposition of Dirac masses located at the jump times:

$$\frac{d}{dt} N_t(\omega) = M(\omega, [0,t]) \text{ where } M = \sum_{i \geq 1} \delta_{T_i(\omega)}. \tag{2.85}$$

In the same way, one can associate a random measure to the *compensated* Poisson process $\tilde{N}_t$, defined in Section 2.5.4, by:

$$\tilde{M}(\omega, A) = M(\omega, A) - \int_A \lambda dt = M(\omega, A) - \lambda|A|. \tag{2.86}$$

$\tilde{M}(A)$ then verifies $E[\tilde{M}(A)] = 0$ and $\text{Var}[\tilde{M}(A)] = \lambda|A|$. Note that unlike M, $\tilde{M}$ is neither integer valued (counting measure) nor positive: it is a signed measure. $\tilde{M}$ is an example of a *compensated* random measure and the measure $A \mapsto \lambda|A|$ is called the *compensator* of M. Note that here the compensator is none other than λ times the Lebesgue measure: $\lambda|A| = E[M(A)]$ and $\tilde{M}$ is the "centered" version of M.

This construction can be generalized in various ways, leading to the notions of Poisson random measure, point process, and marked point process.

[9]Note that in most texts M and N are denoted by the same letter, which can be quite confusing for the beginner.

2.6.1 Poisson random measures

The measure M defined in (2.83) specifies a random counting measure on $\mathbb{R}^+$ such that for any measurable set $A \subset \mathbb{R}^+$, $E[M(A)]$ is given by λ times the Lebesgue measure of A. One can extend this construction to more general settings, replacing $\mathbb{R}^+$ by a $E \subset \mathbb{R}^d$ and the Lebesgue measure by any Radon[10] measure μ on E:

DEFINITION 2.18 Poisson random measure *Let $(\Omega, \mathcal{F}, \mathbb{P})$ be a probability space, $E \subset \mathbb{R}^d$ and μ a given (positive) Radon measure μ on $(E, \mathcal{E})$. A Poisson random measure on E with intensity measure μ is an integer valued random measure:*

$$\begin{aligned} M : \Omega \times \mathcal{E} &\to \mathbb{N} \\ (\omega, A) &\mapsto M(\omega, A), \end{aligned}$$

such that

1. *For (almost all) $\omega \in \Omega$, $M(\omega, .)$ is an integer-valued Radon measure on E: for any bounded measurable $A \subset E$, $M(A) < \infty$ is an integer valued random variable.*

2. *For each measurable set $A \subset E$, $M(., A) = M(A)$ is a Poisson random variable with parameter $\mu(A)$:*

$$\forall k \in \mathbb{N}, \quad \mathbb{P}(M(A) = k) = e^{-\mu(A)} \frac{(\mu(A))^k}{k!}. \tag{2.87}$$

3. *For disjoint measurable sets $A_1, \ldots, A_n \in \mathcal{E}$, the variables $M(A_1), \ldots,, M(A_n)$ are independent.*

The following result allows to construct, given any Radon measure μ, a Poisson random measure with intensity μ:

PROPOSITION 2.14 Construction of Poisson random measures *For any Radon measure μ on $E \subset \mathbb{R}^d$, there exists a Poisson random measure M on E with intensity μ.*

PROOF We give an explicit construction of M from a sequence of independent random variables. We begin by considering the case $\mu(E) < \infty$.

1. Take $X_1, X_2, \ldots$ to be i.i.d. random variables so that $\mathbb{P}(X_i \in A) = \frac{\mu(A)}{\mu(E)}$.

[10] See Definition 2.2.

2. Take $M(E)$ to be a Poisson random variable on $(\Omega, \mathcal{F}, \mathbb{P})$ with mean $\mu(E)$, independent of the X_i.

3. Define $M(A) = \sum_{i=1}^{M(E)} 1_A(X_i)$, for all $A \in \mathcal{E}$.

It is then easily verified that this M is a Poisson random measure with intensity μ. If $\mu(E) = \infty$, since μ is a Radon measure we can represent $E \subset \mathbb{R}^d$ as $E = \bigcup_{i=1}^{\infty} E_i$ where $\mu(E_i) < \infty$ and construct Poisson random measures $M_i(\cdot)$, where the intensity of M_i is the restriction of μ to E_i. Make the $M_i(\cdot)$ independent and define $M(A) = \sum_{i=1}^{\infty} M_i(A)$ for all $A \in \mathcal{E}$. The superposition and thinning properties of Poisson random variables (see Section 2.5) imply that M has the desired properties. □

The construction given in this proof shows that in fact any Poisson random measure on E can be represented as a counting measure associated to a random sequence of points[11] in E: there exists $\{X_n(\omega), n \geq 1\}$ such that:

$$\forall A \in \mathcal{E}, \qquad M(\omega, A) = \sum_{n \geq 1} 1_A(X_n(\omega)). \tag{2.88}$$

M is thus a sum of Dirac masses located at the random points $(X_n)_{n \geq 1}$:

$$M = \sum_{n \geq 1} \delta_{X_n}.$$

Since we require $M(A)$ to be finite for any compact $A \subset E$, this puts a constraint on the sequence $(X_n)_{n \geq 1}$: $A \cap \{X_n, n \geq 1\}$ should be a.s. finite for any compact subset $A \subset E$, i.e., the sequence should not accumulate at a point in E.

A Poisson random measure on E can also be considered as as a random variable taking values in the $\mathcal{M}(E)$, the set of Radon measures on E (see Definition 2.2), on which a topology is defined as follows: a sequence μ_n of Radon measures on $E \subset \mathbb{R}^d$ is said to converge to a Radon measure μ if for any $f : E \to \mathbb{R}$ with compact support $\int f d\mu_n \to \int f d\mu$. Therefore the notions of convergence defined for random variables in Section 2.2 also apply to Poisson random measures. The following criterion [228] is useful for studying their convergence:

PROPOSITION 2.15 Convergence of Poisson random measures
Let $(M_n)_{n \geq 1}$ be a sequence of Poisson random measures on $E \subset \mathbb{R}^d$ with intensities $(\mu_n)_{n \geq 1}$. Then $(M_n)_{n \geq 1}$ converges in distribution if and only if the intensities (μ_n) converge to a Radon measure μ. Then $M_n \Rightarrow M$ where M is a Poisson random measure with intensity μ.

[11]For this reason Poisson random measures are also called "Poisson point processes", although the word "process" is confusing since there is no time variable involved yet.

2.6.2 Compensated Poisson random measure

In the same way as we defined the compensated Poisson process from the Poisson process in Section 2.5.4, one can construct the compensated Poisson random measure $\tilde{M}$ by subtracting from M its intensity measure:

$$\tilde{M}(A) = M(A) - \mu(A). \tag{2.89}$$

From the definition of Poisson random measures, one easily deduces that for disjoint compact sets $A_1, \ldots, A_n \in \mathcal{E}$, the variables $\tilde{M}(A_1), \ldots, \tilde{M}(A_n)$ are independent and verify

$$E[\tilde{M}(A_i)] = 0 \qquad \mathrm{Var}[\tilde{M}(A_i)] = \mu(A_i).$$

2.6.3 Building jump processes from Poisson random measures

Consider now a Poisson random measure M on $E = [0,T] \times \mathbb{R}^d \backslash \{0\}$: as above, it can be described as the counting measure associated to a random configuration of points $(T_n, Y_n) \in E$:

$$M = \sum_{n \geq 1} \delta_{(T_n, Y_n)}. \tag{2.90}$$

Intuitively, each point $(T_n(\omega), Y_n(\omega)) \in [0,T] \times \mathbb{R}^d$ corresponds to an observation made at time T_n and described by a (nonzero) random variable $Y_n(\omega) \in \mathbb{R}^d$.

Since we want to interpret the first coordinate t as time, we introduce as in Section 2.4.2 an information flow $\mathcal{F}_t$ on $(\Omega, \mathcal{F}, \mathbb{P})$. We will say that M is a nonanticipating Poisson random measure (or a Poisson random measure adapted to $\mathcal{F}_t$) if:

- $(T_n)_{n \geq 1}$ are nonanticipating random times.
- Y_n is revealed at T_n: Y_n is $\mathcal{F}_{T_n}$-measurable.

For each ω, $M(\omega, .)$ is a measure on $E = [0,T] \times \mathbb{R}^d \backslash \{0\}$ and we can define, as in Section 2.1.2, integrals with respect to this measure. First, for simple functions $f = \sum_{i=1}^n c_i 1_{A_i}$ where $c_i \geq 0$ and $A_i \subset E$ are disjoint measurable sets, we define $M(f) = \sum_{i=1}^n c_i M(A_i)$: $M(f)$ is a *random variable* with expectation $E[M(f)] = \sum_{i=1}^n c_i \mu(A_i)$. Next, for positive measurable function $f : E \to [0, \infty[$ we define $M(f) = \lim_{n \to \infty} M(f_n)$ where $f_n \uparrow f$ is an increasing sequence of simple functions. By the monotone convergence theorem, $M(f)$ is a random variable with values in $[0, \infty[\cup \{\infty\}$ and (possibly infinite) expectation $E[M(f)] = \mu(f)$. For a real-valued (measurable) function $f : E \mapsto \mathbb{R}$, one can decompose f into its positive and negative part: $f = f_+ - f_-$. If f verifies

$$\mu(|f|) = \int_{[0,T]} \int_{\mathbb{R}^d \backslash \{0\}} |f(s,y)| \mu(ds \times dy) < \infty \tag{2.91}$$

then the positive random variables $M(f_+), M(f_-)$ have finite expectation: $E[M(f_+)] = \mu(f_+) \leq \mu(|f|) < \infty$. In particular, $M(f_+), M(f_-)$ are almost surely finite: we can therefore define $M(f) := M(f_+) - M(f_-)$ and $M(f)$ thus defined is a random variable with expectation

$$E[M(f)] = \mu(f) = \int_{[0,T]} \int_{\mathbb{R}^d} f(s,y)\mu(ds \times dy).$$

Integrating f with respect to M up to time t, i.e., restricting the integral to $[0,t] \times \mathbb{R}^d \setminus \{0\}$ yields a nonanticipating stochastic process:

$$X_t = \int_0^t \int \mathbb{R}^d \setminus \{0\} f(s,y)M(ds\ dy) = \sum_{\{n, T_n \in [0,t]\}} f(T_n, Y_n). \qquad (2.92)$$

The second sum runs over the events (T_n, Y_n) which have occurred before t, i.e., $T_n \leq t$. $(X_t(f))_{t\in[0,T]}$ is thus a jump process whose jumps happen at the random times T_n and have amplitudes given by $f(T_n, Y_n)$. As remarked above, this construction makes sense if f verifies (2.91).

Similarly, under condition (2.91), one can define the integral of f with respect to the compensated Poisson measure $\tilde{M}$. The resulting process, called the compensated integral, is in fact a martingale:

PROPOSITION 2.16 Compensated Poisson integrals
Let M be a nonanticipating Poisson random measure on $E = [0,T] \times \mathbb{R}^d \setminus \{0\}$ with intensity μ with compensated random measure $\tilde{M} = M - \mu$ and $f : E \to \mathbb{R}^d$ verifying (2.91). Then the process

$$\begin{aligned} X_t &= \int_0^t \int_{\mathbb{R}^d \setminus \{0\}} f(s,y)\tilde{M}(ds\ dy) \\ &= \int_0^t \int_{\mathbb{R}^d \setminus \{0\}} f(s,y)M(ds\ dy) - \int_0^t \int_{\mathbb{R}^d \setminus \{0\}} f(s,y) \end{aligned} \qquad (2.93)$$

is a martingale.

Note that, even though the integrals defined in this section are random variables, they are defined pathwise i.e., for any $\omega \in \Omega$.

2.6.4 Marked point processes (*)

As observed above, a Poisson random measure on $[0,T] \times \mathbb{R}^d$ can be represented as a counting measure:

$$M(\omega, .) = \sum_{n\geq 1} \delta_{(T_n(\omega), Y_n(\omega))} \qquad (2.94)$$

for some random sequence $(T_n, Y_n)_{n\geq 1}$ of points in $[0,T]\times\mathbb{R}^d$. Using this representation one can define integer valued random measures with more complex dependence properties: given a random sequence $(T_n, Y_n) \in [0,T] \times E$ where $(T_n)_{n\geq 1}$ is a sequence of nonanticipating random times describing the occurrence of some events and $Y_n \in E \subseteq \mathbb{R}^d$ a quantity observed at time T_n (Y_n is $\mathcal{F}_{T_n}$ measurable), we can define a counting measure M by (2.94). M is called an integer valued random measure on $[0,T] \times E$ and the random sequence $(T_n, Y_n) \in [0,T] \times E$ is then called a marked point process. The random variables $(Y_n)_{n\geq 1}$ are called "marks".

DEFINITION 2.19 Marked point process *A marked point process on $(\Omega, \mathcal{F}, (\mathcal{F}_t), \mathbb{P})$ is a sequence $(T_n, Y_n)_{n\geq 1}$ where*

- *$(T_n)_{n\geq 1}$ is an increasing sequence of nonanticipating random times with $T_n \to \infty$ almost surely as $n \to \infty$.*
- *$(Y_n)_{n\geq 1}$ is a sequence of random variables taking values in E.*
- *The value of Y_n is revealed at T_n: Y_n is $\mathcal{F}_{T_n}$ measurable.*

The condition $T_n \to \infty$ guarantees that the number of events occurring on $[0,T]$ is a.s. finite. For each $\omega \in \Omega$, $M(\omega, .)$ is a counting measure on $[0,T]\times E$. If μ is a diffuse measure[12], i.e., $\mu(\{(t,y)\}) = 0$ for all points $(t,y) \in E$ then, with probability 1, each point occurs at most once: $M(\{(t,y)\}) = 0$ or 1.

Marked point processes do not have the independence properties of Poisson random measures: if $A_1 \cap A_2 = \emptyset$ then $M([0,t] \times A_1), M([0,t] \times A_2)$ are not independent anymore, nor are they Poisson random variables: they allow for arbitrary distributions and dependence structures. Any Poisson random measure on $[0,T]\times\mathbb{R}$ can be represented as in (2.94) but (T_n) does not necessarily verify $T_n \to \infty$ so all Poisson random measures cannot be represented by marked point processes: only those with $\mu([0,T] \times \mathbb{R}^d) < \infty$.

For a function $f : [0,T] \times E \to \mathbb{R}^d$ verifying $\int_{[0,T]\times E} |f(t,y)|\mu(dt\, dy)$ one can construct the integral with respect to the random measure M: it is given by the random variable

$$M(f) = \int_{[0,T]\times E} f(t,y)M(dt\ dy) = \sum_{n\geq 1} f(T_n, Y_n). \tag{2.95}$$

One can then construct a jump process from f as in Section 2.6.1:

$$X_t(f) = \int_{[0,t]\times\mathbb{R}^d\setminus\{0\}} f(s,y)M(ds\ dy) = \sum_{\{n, T_n\in[0,t]\}} f(T_n, Y_n). \tag{2.96}$$

[12]See Section 2.1.

$(X_t(f))_{t\in[0,T]}$ is a nonanticipating jump process with cadlag trajectories whose jumps are described by the marked point process M: the jumps occur at $(T_n)_{n\geq 1}$ and their amplitudes are given by $f(T_n, Y_n)$. This construction gives a systematic way of generating jump processes from marked point processes.

Conversely, to each cadlag process $(X_t)_{t\in[0,T]}$ with values in $\mathbb{R}^d$ one can associate a random measure J_X on $[0,T]\times\mathbb{R}^d$ called the jump measure, in the following manner. As mentioned in Section 2.4, X has at most a countable number of jumps: $\{t \in [0,T],\ \Delta X_t = X_t - X_{t-} \neq 0\}$ is countable: its elements can be arranged in a sequence $(T_n)_{n\geq 1}$ (not necessarily increasing), which are the (random) jump times of X. At time T_n, the process X has a discontinuity of size $Y_n = X_{T_n} - X_{T_n-} \in \mathbb{R}^d \setminus \{0\}$. $(T_n, Y_n)_{n\geq 1}$ defines a marked point process on $[0,T] \times \mathbb{R}^d \setminus \{0\}$ which contains all information about the jumps of the process X: the jump times T_n and the jump sizes Y_n. The associated random measure, which we denote by J_X, is called the jump measure of the process X:

$$J_X(\omega,.) = \sum_{n\geq 1} \delta_{(T_n(\omega),Y_n(\omega))} = \sum_{t\in[0,T]}^{\Delta X_t\neq 0} \delta_{(t,\Delta X_t)}. \tag{2.97}$$

In intuitive terms, for any measurable subset $A \subset \mathbb{R}^d$:

$$J_X([0,t]\times A) := \text{number of jumps of } X \text{ occurring between } 0 \text{ and } t$$
$$\text{whose amplitude belongs to } A.$$

The random measure J_X contains all information about the discontinuities (jumps) of the process X: it tells us *when* the jumps occur and how big they are. J_X does not tell us anything about the continuous component of X. It is easy to see that X has continuous sample paths if and only if $J_X = 0$ almost surely (which simply means that there are no jumps!).

All quantities involving the jumps of X can be computed by integrating various functions against J_X. For example if $f(t,y) = y^2$ then one obtains the sum of the squares of the jumps of X:

$$\int_{[0,T]\times\mathbb{R}} y^2 J_X(dtdy) = \sum_{t\in[0,T]} (\Delta X_t)^2. \tag{2.98}$$

Such expressions may involve infinite sums and we will see, in Chapters 3 and 8, when and in what sense they converge.

Example 2.1 Jump measure of a Poisson process
The jump measure of the Poisson process (2.69) is given by $J_N = \sum_{n\geq 1} \delta_{(T_n,1)}$:

$$J_N([0,t]\times A) = \#\{i\geq 1,\ T_i\in[0,t]\} \quad \text{if } 1\in A$$
$$= 0 \qquad \text{if } 1 \notin A.$$

□

Example 2.2
Consider a sequence of nonanticipating random times $T_1, T_2, \ldots$ and let $(X_n)_{n\geq 1}$ be a sequence of random variables such that X_n is revealed at T_n. Then

$$X_t = \sum_{n\geq 1} X_n 1_{[T_n, T_{n+1}[}$$

defines a cadlag process with jump times T_n and jump sizes $\Delta X_{T_n} = X_n - X_{n-1}$. Its jump measure is given by:

$$J_X = \sum_{n\geq 1} \delta_{(T_n, X_n - X_{n-1})}.$$

If $T_n \to \infty$ a.s. then $(T_n, X_n - X_{n-1})_{n\geq 1}$ defines a marked point process. □

Finally, let us note that if X_t is a jump process built from a Poisson random measure M as in (2.96):

$$M = \sum_{n\geq 1} \delta_{(T_n, Y_n)} \qquad X_t = \int_{[0,t]\times\mathbb{R}^d\setminus\{0\}} f(s,y) M(ds\ dy)$$

then the jump measure J_X can be expressed in terms of the Poisson random measure M by

$$J_X = \sum_{n\geq 1} \delta_{(T_n, f(T_n, Y_n))}. \tag{2.99}$$

Further reading

General references on topics covered in this chapter are [228] or [153]. For general notions on probability spaces, random variables, characteristic functions and various notions of convergence we refer the reader to the excellent introductory text by Jacod and Protter [214]. Poisson processes are treated in most texts on stochastic processes; the link with point processes is treated, in increasing level of difficulty by Kingman [235], Bouleau [68], Resnick [333], Fristedt & Gray [153], Neveu [306]. A short introduction to Poisson random measures is given in [205]. Point processes are considered in more detail in [227] using a measure theory viewpoint. A modern treatment using martingales is given in [306, 73] see also [295].

Simeon-Denis Poisson

The Poisson distribution and the Poisson process were named after the French mathematician Simeon-Denis Poisson (1781–1840). Poisson studied mathematics at Ecole Polytechnique in Paris, under Joseph Louis Lagrange, Pierre Simon Laplace and Jean Baptiste Fourier. Elected to the French Academy of Sciences at age 30, he held academic positions at Ecole Polytechnique and Sorbonne. Poisson made major contributions to the theories of electricity and magnetism, the motion of the moon, the calculus of variations, differential geometry and, of course, probability theory. Poisson found the limiting form of the binomial distribution that is now named after him: the Poisson distribution. The importance of this discovery was only recognized years later, by Chebyshev in Russia. His works on probability appeared in the book *Recherches sur la probabilité des jugements en matière criminelle et en matière civile, précédées des règles générales du calcul des probabilités*, published in 1837, where he was the first to use the notion of cumulative distribution function, to define the density as its derivative and to develop an asymptotic theory of central limit type with remainder for hypergeometric sampling [362]. The expression "law of large numbers" is due to Poisson: "La loi universelle des grands nombres est à la base de toutes les applications du calcul des probabilités" [13]. Poisson was also the first to compute the Fourier transform of $x \to \exp(-|x|)$ and thus discovered what is now (erroneously)

[13] The universal law of large numbers lies at the foundation of all applications of probability theory.

called the "Cauchy distribution" (also called the Lorentzian curve by physicists). He observed in particular that the law of large numbers did not apply to this distribution, which does not possess a first moment, thus finding the first counterexample to the law of large numbers [173, 120].

Poisson was also interested in the use of statistics in the study of social phenomena. In his book *Recherches sur la probabilité des jugements*, Poisson advocated the use of probability as the natural way of describing socio-economic phenomena. Herbert Solomon writes in [173]: "While Poisson's name is commonplace to us, the breadth and variety of Poisson's work is neglected in formal courses undertaken by even the most advanced students."

Chapter 3

Lévy processes: definitions and properties

> Dans la galaxie des mathématiques, le cas du calcul des probabilités est un peu spécial car tout le monde n'est pas d'accord pour savoir si cette discipline appartient aux mathématiques... Voilà qui, en un certain sens, nous redonne de l'espoir car si les probabilités ne sont pas des mathématiques peut-être gardons nous une chance d'y comprendre quelque chose?
>
> Marc Petit *L'équation de Kolmogoroff*, Ramsay: Paris, p 221.

Just as random walks — sums of independent identically distributed random variables — provide the simplest examples of stochastic processes in discrete time, their continuous-time relatives — processes with independent stationary increments, called Lévy processes in honor of the French mathematician Paul Lévy — provide key examples of stochastic processes in continuous time and provide ingredients for building continuous-time stochastic models. The Poisson process and the Wiener process, discussed in Chapter 2, are fundamental examples of Lévy processes. We will see later in this chapter that they can be thought of as building blocks of Lévy processes because every Lévy process is a superposition of a Wiener process and a (possibly infinite) number of independent Poisson processes.

In this chapter, we introduce Lévy processes and discuss some of their general properties. The next one will be devoted to several parametric examples of Lévy processes. First, in Sections 3.2 and 3.3, we discuss compound Poisson processes, which are the simplest examples of Lévy processes and can be considered as Poisson processes with random jump sizes. The class of compound Poisson processes is both simple to study and rich enough to introduce two important theoretical tools: the Lévy-Khinchin formula that allows to study distributional properties of Lévy processes and the Lévy-Itô decomposition, that describes the structure of their sample paths. For compound Poisson processes the proofs are elementary and we give them in full detail. In Section 3.4 we show how the fundamental results obtained in compound Poisson case can be extended to a more general setting. Here we prefer to explain the principles underlying the proofs rather than give full mathematical details. Finally, in last sections of the chapter we use the Lévy-Khinchin formula and the Lévy-Itô decomposition to derive the fundamental properties of Lévy processes.

3.1 From random walks to Lévy processes

DEFINITION 3.1 Lévy process *A cadlag*[1] *stochastic process* $(X_t)_{t\geq 0}$ *on* $(\Omega, \mathcal{F}, \mathbb{P})$ *with values in* $\mathbb{R}^d$ *such that* $X_0 = 0$ *is called a* Lévy process *if it possesses the following properties:*

1. *Independent increments: for every increasing sequence of times* $t_0 \ldots t_n$*, the random variables* $X_{t_0}, X_{t_1} - X_{t_0}, \ldots, X_{t_n} - X_{t_{n-1}}$ *are independent.*

2. *Stationary increments: the law of* $X_{t+h} - X_t$ *does not depend on* t*.*

3. *Stochastic continuity:* $\forall \varepsilon > 0, \ \lim_{h\to 0} \mathbb{P}(|X_{t+h} - X_t| \geq \varepsilon) = 0.$

The third condition does not imply in any way that the *sample paths* are continuous: as noted in Proposition 2.12, it is verified by the Poisson process. It serves to exclude processes with jumps at fixed (nonrandom) times, which can be regarded as "calendar effects" and are not interesting for our purpose. It means that for given time t, the probability of seeing a jump at t is zero: discontinuities occur at random times.

If we sample a Lévy process at regular time intervals $0, \Delta, 2\Delta, \ldots$, we obtain a random walk: defining $S_n(\Delta) \equiv X_{n\Delta}$, we can write $S_n(\Delta) = \sum_{k=0}^{n-1} Y_k$ where $Y_k = X_{(k+1)\Delta} - X_{k\Delta}$ are i.i.d random variables whose distribution is the same as the distribution of X_Δ. Since this can be done for any sampling interval Δ we see that by specifying a Lévy process we have specified a whole family of random walks $S_n(\Delta)$: these models simply correspond to sampling the Lévy process X at different frequencies.

Choosing $n\Delta = t$, we see that for any $t > 0$ and any $n \geq 1$, $X_t = S_n(\Delta)$ can be represented as a sum of n i.i.d. random variables whose distribution is that of $X_{t/n}$: X_t can be "divided" into n i.i.d. parts. A distribution having this property is said to be infinitely divisible:

---

[1]Some authors do not impose the cadlag (right-continuity and left limits) property in the definition of a Lévy process but it can be shown (see [324, Theorem 30] or [345, Chapter 1]) that every Lévy process (defined without the cadlag property) has a unique modification that is cadlag, therefore the cadlag property can be assumed without loss of generality.

DEFINITION 3.2 Infinite divisibility *A probability distribution F on $\mathbb{R}^d$ is said to be infinitely divisible if for any integer $n \geq 2$, there exists n i.i.d. random variables $Y_1, ...Y_n$ such that $Y_1 + ... + Y_n$ has distribution F.*

Since the distribution of i.i.d. sums is given by convolution of the distribution of the summands, if we denote by μ the distribution of the Y_k-s in the definition above, then $F = \mu * \mu * \cdots * \mu$ is the n-th convolution of μ. So an infinitely divisible distribution can also be defined as a distribution F for which the n-th convolution root is still a probability distribution, for any $n \geq 2$.

Thus, if X is a Lévy process, for any $t > 0$ the distribution of X_t is infinitely divisible. This puts a constraint on the possible choices of distributions for X_t: whereas the increments of a discrete-time random walk can have arbitrary distribution, the distribution of increments of a Lévy process has to be infinitely divisible.

The most common examples of infinitely divisible laws are: the Gaussian distribution, the gamma distribution, α-stable distributions and the Poisson distribution: a random variable having any of these distributions can be decomposed into a sum of n i.i.d. parts having the same distribution but with modified parameters. For example, if $X \sim N(\mu, \sigma^2)$ then one can write $X = \sum_{k=0}^{n-1} Y_k$ where Y_k are i.i.d. with law $N(\mu/n, \sigma^2/n)$. Less trivial examples are the log-normal, Pareto, Student distributions. Finally, an example of distribution which is not infinitely divisible is the uniform law on an interval.

Conversely, given an infinitely divisible distribution F, it is easy to see that for any $n \geq 1$ by chopping it into n i.i.d. components we can construct a random walk model on a time grid with step size $1/n$ such that the law of the position at $t = 1$ is given by F. In the limit, this procedure can be used to construct a continuous time Lévy process $(X_t)_{t \geq 0}$ such that the law of X_1 if given by F:

PROPOSITION 3.1 Infinite divisibility and Lévy processes
Let $(X_t)_{t \geq 0}$ be a Lévy process. Then for every t, X_t has an infinitely divisible distribution. Conversely, if F is an infinitely divisible distribution then there exists a Lévy process (X_t) such that the distribution of X_1 is given by F.

The direct implication was shown above. For the proof of the converse statement see [345, Corollary 11.6].

Define the characteristic function of X_t:

$$\Phi_t(z) \equiv \Phi_{X_t}(z) \equiv E[e^{iz.X_t}], \quad z \in \mathbb{R}^d.$$

For $t > s$, by writing $X_{t+s} = X_s + (X_{t+s} - X_s)$ and using the fact that $X_{t+s} - X_s$ is independent of X_s, we obtain that $t \mapsto \Phi_t(z)$ is a multiplicative function:

$$\Phi_{t+s}(z) = \Phi_{X_{t+s}}(z) = \Phi_{X_s}(z)\Phi_{X_{t+s}-X_s}(z)$$
$$= \Phi_{X_s}(z)\Phi_{X_t}(z) = \Phi_s\Phi_t.$$

The stochastic continuity of $t \mapsto X_t$ implies in particular that $X_t \to X_s$ in distribution when $s \to t$. Therefore, from Proposition 2.6, $\Phi_{X_s}(z) \to \Phi_{X_t}(z)$ when $s \to t$ so $t \mapsto \Phi_t(z)$ is a continuous function of t. Together with the multiplicative property $\Phi_{s+t}(z) = \Phi_s(z).\Phi_t(z)$ this implies that $t \mapsto \Phi_t(z)$ is an exponential function:

PROPOSITION 3.2 Characteristic function of a Lévy process
Let $(X_t)_{t\geq 0}$ be a Lévy process on $\mathbb{R}^d$. There exists a continuous function $\psi : \mathbb{R}^d \mapsto \mathbb{R}$ called the characteristic exponent *of X, such that:*

$$E[e^{iz.X_t}] = e^{t\psi(z)}, \quad z \in \mathbb{R}^d. \tag{3.1}$$

Recalling the definition of the cumulant generating function of a random variable (2.29), we see that ψ is the cumulant generating function of X_1: $\psi = \Psi_{X_1}$ and that the cumulant generating function of X_t varies linearly in t: $\Psi_{X_t} = t\Psi_{X_1} = t\psi$. The law of X_t is therefore determined by the knowledge of the law of X_1 : the only degree of freedom we have in specifying a Lévy process is to specify the distribution of X_t for a single time (say, $t = 1$).

3.2 Compound Poisson processes

DEFINITION 3.3 Compound Poisson process A compound Poisson process *with intensity $\lambda > 0$ and jump size distribution f is a stochastic process X_t defined as*

$$X_t = \sum_{i=1}^{N_t} Y_i, \tag{3.2}$$

where jumps sizes Y_i are i.i.d. with distribution f and (N_t) is a Poisson process with intensity λ, independent from $(Y_i)_{i\geq 1}$.

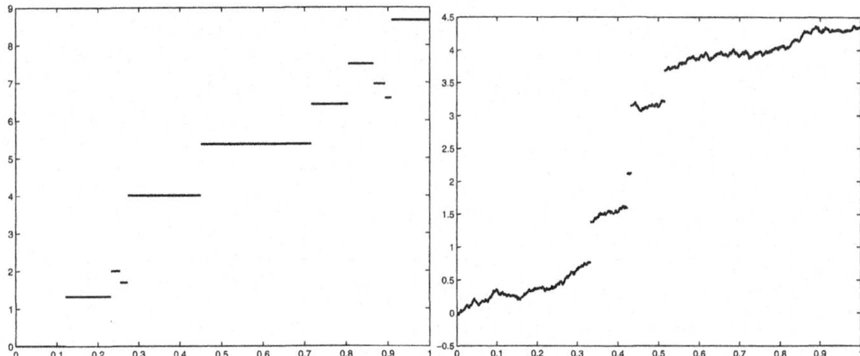

FIGURE 3.1: Left: A compound Poisson process with a Gaussian distribution of jump sizes. Right: A jump diffusion: Lévy process with Gaussian component and finite jump intensity.

The following properties of a compound Poisson process are easily deduced from the definition:

1. The sample paths of X are cadlag piecewise constant functions.
2. The jump times $(T_i)_{i\geq 1}$ have the same law as the jump times of the Poisson process N_t: they can be expressed as partial sums of independent exponential random variables with parameter λ.
3. The jump sizes $(Y_i)_{i\geq 1}$ are independent and identically distributed with law f.

The Poisson process itself can be seen as a compound Poisson process on $\mathbb{R}$ such that $Y_i \equiv 1$. This explains the origin of term "compound Poisson" in the definition.

Let $R(n), n \geq 0$ be a random walk with step size distribution f: $R(n) = \sum_{i=0}^{n} Y_i$. The compound Poisson process X_t can be obtained by changing the time of R with an independent Poisson process N_t: $X_t = R(N_t)$. X_t thus describes the position of a random walk after a random number of time steps, given by N_t. This operation is similar to the *subordination* of Lévy processes, to be discussed in Chapter 4.

The left graph in Figure 3.1 depicts a typical trajectory of a compound Poisson process — note the piecewise constant path. Compound Poisson processes are Lévy processes and they are the only Lévy processes with piecewise constant sample paths, as shown by the following proposition.

PROPOSITION 3.3

$(X_t)_{t\geq 0}$ is compound Poisson process if and only if it is a Lévy process and its sample paths are piecewise constant functions.

PROOF of the "if" part Let $(X_t)_{t\geq 0}$ be a Lévy process with piecewise constant paths. To show that it is a compound Poisson process we need to prove properties 2 and 3 of Definition 3.3. Let us first prove property 2. We can construct, path by path, a process $(N_t, t \geq 0)$ which counts the jumps of X:

$$N_t = \#\{0 < s \leq t : X_{s-} \neq X_s\}. \tag{3.3}$$

Since the trajectories of X are piecewise constant, X has a finite number of jumps in any finite interval which entails that N_t is finite for all finite t. Hence, it is a counting process. Let $h < t$. Then

$$N_t - N_h = \#\{h < s \leq t : X_{s-} \neq X_s\} = \#\{h < s \leq t : X_{s-} - X_h \neq X_s - X_h\}$$

Hence, $N_t - N_h$ depends only on $(X_s - X_h)$, $h \leq s \leq t$. Therefore, from the independence and stationarity of increments of (X_t) it follows that (N_t) also has independent and stationary increments. Lemma 2.1 entails that (N_t) is a Poisson process, which proves property 2 of Definition 3.3.

Using the process N, we can compute the jump sizes of X: $Y_n = X_{S_n} - X_{S_n-}$ where $S_n = \inf\{t : N_t \geq n\}$. To check property 3 of Definition 3.3 and complete the proof of the "if" part, it remains to show that these jump sizes are i.i.d. Let us prove that they are independent. First we would like to show that the increments of X conditionally on the trajectory of N are independent. Let $t > s$ and consider the following four sets:

$$\begin{aligned} A_1 &\in \sigma(X_s) \qquad & B_1 &\in \sigma(N_r, r \leq s) \\ A_2 &\in \sigma(X_t - X_s) \qquad & B_2 &\in \sigma(N_r - N_s, r > s) \end{aligned}$$

such that $P(B_1) > 0$ and $P(B_2) > 0$. The independence of increments of X implies that processes $(X_r - X_s, r > s)$ and $(X_r, r \leq s)$ are independent. Hence,

$$\mathbb{P}[A_1 \cap B_1 \cap A_2 \cap B_2] = \mathbb{P}[A_1 \cap B_1]\mathbb{P}[A_2 \cap B_2].$$

Moreover,
- A_1 and B_1 are independent from B_2.
- A_2 and B_2 are independent from B_1.
- B_1 and B_2 are independent from each other.

Therefore, the conditional probability of interest can be expressed as follows:

$$\begin{aligned} \mathbb{P}[A_1 \cap A_2 | B_1 \cap B_2] &= \frac{\mathbb{P}[A_1 \cap B_1]\mathbb{P}[A_2 \cap B_2]}{\mathbb{P}[B_1]\mathbb{P}[B_2]} \\ &= \frac{\mathbb{P}[A_1 \cap B_1 \cap B_2]\mathbb{P}[A_2 \cap B_1 \cap B_2]}{\mathbb{P}[B_1]^2\mathbb{P}[B_2]^2} = \mathbb{P}[A_1 | B_1 \cap B_2]\mathbb{P}[A_2 | B_1 \cap B_2]. \end{aligned}$$

This proves that $X_t - X_s$ and X_s are independent conditionally on the trajectory of N. In particular, choosing $B_1 = \{N_s = 1\}$ and $B_2 = \{N_t - N_s = 1\}$

we obtain that Y_1 and Y_2 are independent. Since we could have taken any number of increments of X and not just two of them, this proves that $(Y_i)_{i\geq 1}$ are independent.

Finally, to prove that the jump sizes have the same law, observe that the two-dimensional process (X_t, N_t) has stationary increments. Therefore, for every $n \geq 0$ and for every $s > h > 0$,

$$E[f(X_h)\big|N_h = 1, N_s - N_h = n] \\ = E[f(X_{s+h} - X_s)\big|N_{s+h} - N_s = 1, N_s - N_h = n],$$

where f is any bounded Borel function. This entails that for every $n \geq 0$, Y_1 and Y_{n+2} have the same law which completes the proof. □

PROOF of the "only if" part Let $(X_t)_{t\geq 0}$ be a compound Poisson process.

• Independence of increments. Let $0 < r < s$ and let f and g be bounded Borel functions on $\mathbb{R}^d$. To ease the notation, we prove only that X_r is independent from $X_s - X_r$, but the same reasoning applies to any finite number of increments. We must show that

$$E[f(X_r)g(X_s - X_r)] = E[f(X_r)]E[g(X_s - X_r)].$$

From the representation $X_r = \sum_{i=1}^{N_r} Y_i$ and $X_s - X_r = \sum_{i=N_r+1}^{N_s} Y_i$ the following observations can be made:

- Conditionally on the trajectory of N_t for $t \in [0, s]$, X_r and $X_s - X_r$ are independent because the first expression only depends on Y_i for $i \leq N_r$ and the second expression only depends on Y_i for $i > N_r$.

- The expectation $E[f(X_r)\big|N_t, t \leq s]$ depends only on N_r and the expectation $E[g(X_s - X_r)\big|N_t, t \leq s]$ depends only on $N_s - N_r$.

Now, using the independence of increments of the Poisson process, we can write:

$$\begin{aligned} E[f(X_r)g(X_s - X_r)] &= E[E[f(X_r)g(X_s - X_r)\big|N_t, t \leq s]] \\ &= E[E[f(X_r)\big|N_t, t \leq s]E[g(X_s - X_r)\big|N_t, t \leq s]] \\ &= E[E[f(X_r)\big|N_t, t \leq s]]E[E[g(X_s - X_r)\big|N_t, t \leq s]] \\ &= E[f(X_r)]E[g(X_s - X_r)]. \end{aligned}$$

• Stationarity of increments. Let $0 < r < s$ as above and let f be a bounded Borel function. Using the observations made above, we can write:

$$E[f(X_s - X_r)] = E[E[\sum_{i=N_r+1}^{N_s} Y_i\big|N_t, t \leq s]]$$

$$= E[E[\sum_{i=1}^{N_s - N_r} Y_i\big|N_t, t \leq s]] = E[E[\sum_{i=1}^{N_{s-r}} Y_i\big|N_t, t \leq s]] = E[f(X_{s-r})].$$

• Stochastic continuity. X_t only jumps if N_t does. By Proposition 2.12, for every $t > 0$,

$$\mathbb{P}(N_s \overset{s<t}{\underset{s\to t}{\to}} N_t) = 1.$$

Hence, for every $t > 0$,

$$\mathbb{P}(X_s \overset{s<t}{\underset{s\to t}{\to}} X_t) = 1.$$

Since almost sure convergence entails convergence in probability, this implies stochastic continuity. □

Since any cadlag function may be approximated by a piecewise constant function, one may expect that general Lévy processes can be well approximated by compound Poisson ones and that by studying compound Poisson processes one can gain an insight into the properties of Lévy processes.

PROPOSITION 3.4 Characteristic function of a compound Poisson process
Let $(X_t)_{t\geq 0}$ be a compound Poisson process on $\mathbb{R}^d$. Its characteristic function has the following representation:

$$E[\exp iu.X_t] = \exp\left\{t\lambda \int_{\mathbb{R}^d} (e^{iu.x} - 1) f(dx)\right\}, \quad \forall\, u \in \mathbb{R}^d, \tag{3.4}$$

where λ denotes the jump intensity and f the jump size distribution.

Comparing (3.4) with the characteristic function of the Poisson process (2.72) we see that a compound Poisson random variable can be represented as a superposition of independent Poisson processes with different jump sizes. The total intensity of Poisson processes with jump sizes in the interval $[x, x + dx]$ is determined by the density $\lambda f(dx)$.

PROOF Conditioning the expectation on N_t and denoting the characteristic function of f by $\hat{f}$, we find

$$\begin{aligned}
E[\exp iu.X_t] &= E[E[\exp iu.X_t]|N_t] = E[(\hat{f}(u))^{N_t}] \\
&= \sum_{n=0}^{\infty} \frac{e^{-\lambda t}(\lambda t)^n (\hat{f}(u))^n}{n!} = \exp\{\lambda t(\hat{f}(u) - 1)\} \\
&= \exp\left\{t\lambda \int_{\mathbb{R}^d} (e^{iu.x} - 1) f(dx)\right\}.
\end{aligned}$$

□

For one-dimensional compound Poisson processes the characteristic function has a simpler form:

$$E[\exp\{iuX_t\}] = \exp\left\{t\lambda \int_{-\infty}^{\infty} (e^{iux} - 1) f(dx)\right\}, \quad \forall u \in \mathbb{R}.$$

Introducing a new measure $\nu(A) = \lambda f(A)$, we can rewrite the formula (3.4) as follows:

$$E[\exp iu.X_t] = \exp\left\{t \int_{\mathbb{R}^d} (e^{iu.x} - 1)\nu(dx)\right\}, \quad \forall u \in \mathbb{R}^d. \tag{3.5}$$

ν is called the *Lévy measure* of process $(X_t)_{t\geq 0}$. ν is a positive measure on $\mathbb{R}$ but not a probability measure since $\int \nu(dx) = \lambda \neq 1$. Formula (3.5) is a particular case of the so called Lévy-Khinchin representation (see Theorem 3.1).

3.3 Jump measures of compound Poisson processes

We will now use the notion of random measure, introduced in Section 2.6, to study the behavior of jumps of a compound Poisson process. As shown in Section 2.6.4, to every cadlag process and in particular to every compound Poisson process $(X_t)_{t\geq 0}$ on $\mathbb{R}^d$ one can associate a random measure on $\mathbb{R}^d \times [0, \infty[$ describing the jumps of X: for any measurable set $B \subset R^d \times [0, \infty[$

$$J_X(B) = \#\{(t, X_t - X_{t-}) \in B\}. \tag{3.6}$$

For every measurable set $A \subset \mathbb{R}^d$, $J_X([t_1, t_2] \times A)$ counts the number of jump times of X between t_1 and t_2 such that their jump sizes are in A. The following proposition shows that J_X is a Poisson random measure in the sense of Definition 2.18.

PROPOSITION 3.5 Jump measure of a compound Poisson process

Let $(X_t)_{t\geq 0}$ be a compound Poisson process with intensity λ and jump size distribution f. Its jump measure J_X is a Poisson random measure on $\mathbb{R}^d \times [0, \infty[$ with intensity measure $\mu(dx \times dt) = \nu(dx)dt = \lambda f(dx)dt$.

This proposition suggests an alternative interpretation of the Lévy measure of a compound Poisson process as the average number of jumps per unit of time. In fact, we will see in the sequel that this interpretation is much more general than the one that uses the jump size distribution. It can be used to define the Lévy measure for *all* Lévy processes and not only compound Poisson ones as follows:

DEFINITION 3.4 Lévy measure *Let $(X_t)_{t\geq 0}$ be a Lévy process on $\mathbb{R}^d$. The measure ν on $\mathbb{R}^d$ defined by:*

$$\nu(A) = E[\#\{t \in [0,1] : \Delta X_t \neq 0,\ \Delta X_t \in A\}], \quad A \in \mathcal{B}(\mathbb{R}^d) \tag{3.7}$$

is called the Lévy measure *of X: $\nu(A)$ is the expected number, per unit time, of jumps whose size belongs to A.*

PROOF of Proposition 3.5 From the Definition (3.6) it is clear that J_X is an integer valued measure. Let us first check that $J_X(B)$ is Poisson distributed. It is sufficient to prove this property for a set of the form $B = A \times [t_1, t_2]$ with $A \in \mathcal{B}(\mathbb{R}^d)$. Let $(N_t)_{t\geq 0}$ be the Poisson process, counting the jumps of X. Conditionally on the trajectory of N, the jump sizes Y_i are i.i.d. and $J_X([t_1, t_2] \times A)$ is a sum of $N(t_2) - N(t_1)$ i.i.d. Bernoulli variables taking value 1 with probability $f(A)$. Therefore,

$$\begin{aligned} &E[e^{iuJ_X([t_1,t_2]\times A)}] = E[E[e^{iuJ_X([t_1,t_2]\times A)}|N_t, t \geq 0]] \\ &\quad = E[\{e^{iu} f(A) + 1 - f(A)\}^{N(t_2)-N(t_1)}] = \exp\{\lambda(t_2 - t_1) f(A)(e^{iu} - 1)\} \end{aligned}$$

because $N(t_2) - N(t_1)$ is Poisson distributed with parameter $\lambda(t_2 - t_1)$. Thus, $J_X([t_1, t_2] \times A)$ is a Poisson random variable with parameter $f(A)\lambda(t_2 - t_1)$ which was to be shown.

Now let us check the independence of measures of disjoint sets. First, let us show that if A and B are two disjoint Borel sets in $\mathbb{R}^d$ then $J_X([t_1, t_2] \times A)$ and $J_X([t_1, t_2] \times B)$ are independent. Conditionally on the trajectory of N, the expression $iuJ_X([t_1, t_2] \times A) + ivJ_X([t_1, t_2] \times B)$ is a sum of $N(t_2) - N(t_1)$ i.i.d. random variables taking values:

$$\begin{aligned} &iu \quad \text{with probability } f(A); \\ &iv \quad \text{with probability } f(B); \\ &0 \quad \text{with probability } 1 - f(A) - f(B). \end{aligned}$$

Proceeding like in the first part of the proof, we factorize the characteristic function as follows

$$
\begin{aligned}
&E[e^{iuJ_X([t_1,t_2]\times A)+ivJ_X([t_1,t_2]\times B)}] \\
&\quad = E[\{(e^{iu}-1)f(A)+(e^{iv}-1)f(B)+1\}^{N(t_2)-N(t_1)}] \\
&\quad = \exp\{\lambda(t_2-t_1)(f(A)(e^{iu}-1)+f(B)(e^{iv}-1))\} \\
&\quad = E[e^{iuJ_X([t_1,t_2]\times A)}]E[e^{ivJ_X([t_1,t_2]\times B)}].
\end{aligned}
$$

Second, let $[t_1, t_2]$ and $[s_1, s_2]$ be two disjoint intervals. The independence of $J_X([t_1,t_2]\times A)$ and $J_X([s_1,s_2]\times B)$ follows directly from the independence of increments of the process X.

Now the independence of jump measures of any finite number of disjoint sets of $[0,\infty[\times\mathbb{R}^d$ follows from the additivity of J_X and from the fact that the methods used in this proof work for any finite number of sets. □

Proposition 3.5 implies that every compound Poisson process can be represented in the following form:

$$X_t = \sum_{s\in[0,t]} \Delta X_s = \int_{[0,t]\times\mathbb{R}^d} x J_X(ds\times dx), \tag{3.8}$$

where J_X is a Poisson random measure with intensity measure $\nu(dx)dt$. This is a special case of the Lévy-Itô decomposition for Lévy processes. Here we have only rewritten the process X as the sum of its jumps. Since a compound Poisson process has almost surely a finite number of jumps in interval $[0,t]$, the stochastic integral appearing in (3.8) is a finite sum, so there are no convergence problems.

The following lemma is a useful tool for computing various quantities, related to Poisson random measures. It is somewhat analogous to Proposition 2.9.

LEMMA 3.1

Let M be a Poisson random measure with intensity measure μ and let A be a measurable set such that $0<\mu(A)<\infty$. Then the following two random measures on the subsets of A have the same distribution conditionally on $M(A)$:

- *$M\big|_A$, the restriction of M to A.*

- *$\widehat{M}_A$ defined by $\widehat{M}_A(B)=\#\{X_i\in B\}$ for all measurable subsets B of A, where $X_i, i=1,\dots,M(A)$ are independent and distributed on A with the law $\frac{\mu(dx)}{\mu(A)}$. In other words, $\widehat{M}_A$ is the counting measure of $M(A)$ independent random points, identically distributed on A.*

PROOF Let $B_1, \cdots, B_k$ be measurable disjoint subsets of A and denote $\tilde{B} = A \setminus (B_1 \cup \cdots \cup B_k)$ and $\tilde{n} = n - \sum_i n_i$. Then

$$\begin{aligned}
\mathbb{P}\{M(B_1) = n_1, \ldots, M(B_k) = n_k | M(A) = n\} \\
&= \frac{\mathbb{P}\{M(B_1) = n_1, \ldots, M(B_k) = n_k, M(\tilde{B}) = \tilde{n}\}}{\mathbb{P}\{M(A) = n\}} \\
&= \frac{n!}{n_1! \ldots n_k! \tilde{n}!} \left(\frac{\mu(B_1)}{\mu(A)}\right)^{n_1} \cdots \left(\frac{\mu(B_k)}{\mu(A)}\right)^{n_k} \left(\frac{\mu(\tilde{B})}{\mu(A)}\right)^{\tilde{n}},
\end{aligned}$$

which is clearly equal to the distribution of $\widehat{M}_A(B_1), \ldots, \widehat{M}_A(B_k)$. Using the additivity of measures, we can show that the sets $B_1, \ldots, B_k$ must not necessarily be disjoint. The statement of the lemma follows. □

As an application of the preceding lemma, consider the following result.

***PROPOSITION 3.6* Exponential formula for Poisson random measures**

Let M be a Poisson random measure with intensity measure μ. Then the following formula holds for every measurable set B such that $\mu(B) < \infty$ and for all functions f such that $\int_B e^{f(x)} \mu(dx) < \infty$:

$$E \exp\left\{\int_B f(x) M(dx)\right\} = \exp\left\{\int_B (e^{f(x)} - 1)\mu(dx)\right\}. \tag{3.9}$$

PROOF Condition the expectation on $\mu(B)$ and use Lemma 3.1. □

In the sequel we will see that to obtain this formula we do not need the assumption that both $\mu(B)$ and $\int_B e^{f(x)}\mu(dx)$ be finite, it suffices only to require $\int_B |e^{f(x)} - 1| \mu(dx) < \infty$.

Proposition 3.6 allows to establish a one-to-one correspondence between compound Poisson processes and Poisson random measures with intensity measures of the form $\nu(dx)dt$ with ν finite. Indeed, let ν be a finite measure on $\mathbb{R}^d$ and let M be a Poisson random measure on $\mathbb{R}^d \times [0, \infty[$ with intensity measure $\nu(dx)dt$. Then one can show using Proposition 3.6 that Equation (3.8) defines a compound Poisson process with Lévy measure ν.

3.4 Infinite activity Lévy processes

In the preceding section, we saw that every piecewise constant Lévy process X_t^0 can be represented in the form (3.8) for some Poisson random measure with intensity measure of the form $\nu(dx)dt$ where ν is a finite measure, defined by

$$\nu(A) = E[\#\{t \in [0,1] : \Delta X_t^0 \neq 0, \ \Delta X_t^0 \in A\}], \quad A \in \mathcal{B}(\mathbb{R}^d). \tag{3.10}$$

Given a Brownian motion with drift $\gamma t + W_t$, independent from X^0, the sum $X_t = X_t^0 + \gamma t + W_t$ defines another Lévy process, which can be decomposed as:

$$X_t = \gamma t + W_t + \sum_{s \in [0,t]} \Delta X_s = \gamma t + W_t + \int_{[0,t] \times \mathbb{R}^d} x J_X(ds \times dx),$$

where J_X is a Poisson random measure on $[0, \infty[\times \mathbb{R}^d$ with intensity $\nu(dx)dt$.

Can every Lévy process be represented in this form? Given a Lévy process X_t, we can still define its Lévy measure ν as above. $\nu(A)$ is still finite for any compact set A such that $0 \notin A$: if this were not true, the process would have an infinite number of jumps of finite size on $[0,T]$, which contradicts the cadlag property. So ν defines a Radon measure on $\mathbb{R}^d \setminus \{0\}$. But ν is not necessarily a finite measure: the above restriction still allows it to blow up at zero and X may have an infinite number of small jumps on $[0,T]$. In this case the sum of the jumps becomes an infinite series and its convergence imposes some conditions on the measure ν, under which we obtain a decomposition of X similar to the one above:

PROPOSITION 3.7 Lévy-Itô decomposition

Let $(X_t)_{t \geq 0}$ be a Lévy process on $\mathbb{R}^d$ and ν its Lévy measure, given by Definition 3.4.

- *ν is a Radon measure on $\mathbb{R}^d \setminus \{0\}$ and verifies:*

$$\int_{|x| \leq 1} |x|^2 \nu(dx) < \infty \qquad \int_{|x| \geq 1} \nu(dx) < \infty.$$

- *The jump measure of X, denoted by J_X, is a Poisson random measure on $[0, \infty[\times \mathbb{R}^d$ with intensity measure $\nu(dx)dt$.*

- *There exist a vector γ and a d-dimensional Brownian motion[2] $(B_t)_{t \geq 0}$*

with covariance matrix A *such that*

$$X_t = \gamma t + B_t + X_t^l + \lim_{\varepsilon \downarrow 0} \tilde{X}_t^\varepsilon, \quad \text{where} \tag{3.11}$$

$$X_t^l = \int_{|x|\geq 1, s\in[0,t]} x J_X(ds \times dx) \quad \text{and}$$

$$\tilde{X}_t^\varepsilon = \int_{\varepsilon\leq |x|<1, s\in[0,t]} x\{J_X(ds \times dx) - \nu(dx)ds\}$$

$$\equiv \int_{\varepsilon\leq |x|<1, s\in[0,t]} x\tilde{J}_X(ds \times dx).$$

The terms in (3.11) are independent and the convergence in the last term is almost sure and uniform in t *on* $[0,T]$.

The Lévy-Itô decomposition entails that for every Lévy process there exist a vector γ, a positive definite matrix A and a positive measure ν that uniquely determine its distribution. The triplet (A, ν, γ) is called *characteristic triplet* or *Lévy triplet* of the process X_t.

Given the importance of this result, let us comment a bit on the meaning of the terms in (3.11). First, $\gamma t + B_t$ is a continuous Gaussian Lévy process and *every Gaussian Lévy process is continuous and can be written in this form* and can be described by two parameters: the drift γ and the covariance matrix of Brownian motion, denoted by A.

The other two terms are discontinuous processes incorporating the jumps of X_t and are described by the Lévy measure ν. The condition $\int_{|y|\geq 1} \nu(dy) < \infty$ means that X has a finite number of jumps with absolute value larger than 1. So the sum

$$X_t^l = \sum_{0\leq s\leq t}^{|\Delta X_s|\geq 1} \Delta X_s$$

contains almost surely a finite number of terms and X_t^l is a compound Poisson process. There is nothing special about the threshold $\Delta X = 1$: for any $\varepsilon > 0$, the sum of jumps with amplitude between ε and 1:

$$X_t^\varepsilon = \sum_{0\leq s\leq t}^{1>|\Delta X_s|\geq \epsilon} \Delta X_s = \int_{\varepsilon\leq |x|<1, s\in[0,t]} x J_X(ds \times dx) \tag{3.12}$$

[2]In terminology used here a Brownian motion may have arbitrary covariance matrix whereas the term Wiener process is used for a standard Brownian motion, with unit covariance matrix.

is again a well-defined compound Poisson process. However, contrarily to the compound Poisson case, ν can have a singularity at zero: there can be infinitely many small jumps and their sum does not necessarily converge. This prevents us from making ε go to 0 directly in Expression 3.12. In order to obtain convergence we have to center the remainder term, i.e., replace the jump integral by its *compensated* version, defined in Section 2.6.2:

$$\tilde{X}_t^\varepsilon = \int_{\varepsilon \leq |x| < 1, s \in [0,t]} x \tilde{J}_X(ds \times dx) \tag{3.13}$$

which, as seen in Proposition 2.16, is a *martingale*. While X^ε can be interpreted as an infinite superposition of independent Poisson processes, $\tilde{X}_t^\varepsilon$ should be seen as an infinite superposition of independent *compensated*, i.e., centered Poisson processes to which a "central-limit" type argument can now be applied to show convergence (see below).

An important implication of the Lévy-Itô decomposition is that every Lévy process is a combination of a Brownian motion with drift and a possibly infinite sum of independent compound Poisson processes. This also means that every Lévy process can be approximated with arbitrary precision by a jump-diffusion process, that is by the sum of Brownian motion with drift and a compound Poisson process, a point which is useful both in theory and in practice. The right graph of Figure 3.1 shows a typical trajectory of a jump-diffusion process.

The Lévy-Itô decomposition was originally found by Lévy [251] using a direct analysis of the paths of Lévy processes and completed by Itô [206]. There are many proofs available in the literature. A probabilistic approach close to the original proof of Lévy is given in [164]. We will not give a detailed proof but sketch the main ideas of this approach:

PROOF of the Lévy-Itô decomposition (outline)
First, we construct a Poisson random measure J_X on $[0,t] \times \mathbb{R}^d$ from the jumps of (X_t). Since (X_t) is cadlag, for any positive ε the set $\{t : |X_t - X_{t-}| \geq \varepsilon\}$ is finite and the Poisson random measure (of any closed set not containing 0) can be constructed using Proposition 3.5. The intensity measure of J_X is homogeneous and equal to $\nu(dx)dt$. Throughout the rest of the proof we can suppose without loss of generality that all jumps of (X_t) are smaller than 1 in absolute value. The next step is to prove the following lemma.

LEMMA 3.2
Let (X_t, Y_t) be a Lévy process. If (Y_t) is compound Poisson and (X_t) and (Y_t) never jump together, then they are independent.

For a proof see for example [228, Lemma 13.6].

This lemma together with the exponential formula (3.6) allows to prove that the Lévy measure ν satisfies the integrability condition

$$\int (|x|^2 \wedge 1)\nu(dx) < \infty. \tag{3.14}$$

We give this part of the proof in full detail to explain the origin of this condition. Since the Lévy measure of any closed set not containing zero is finite, it is sufficient to prove that for some $\delta > 0$, $\int_{|x|\leq\delta} |x|^2\nu(dx) < \infty$.

Let X_t^ε be as above in (3.13) and let $R_t^\varepsilon = X_t - X_t^\varepsilon$. Then $(X_t^\varepsilon, R_t^\varepsilon)$ is a Lévy process because (X_t) is. Clearly for some u and some t we have $|E\exp\{iuX_t\} > 0|$. Let us fix this u and this t. Since by Lemma 3.2, (X_t^ε) and (R_t^ε) are independent,

$$E\exp\{iuX_t\} = E\exp\{iuR_t^\varepsilon\}E\exp\{iuX_t^\varepsilon\},$$

and this means that $|E\exp\{iuX_t^\varepsilon\}|$ is bounded from below by a positive number which does not depend on ε. By the exponential formula (Proposition 3.6) this is equivalent to

$$\left|\exp\left\{t\int_{|x|\geq\varepsilon}(e^{iux} - 1)\nu(dx)\right\}\right| \geq C > 0,$$

which implies that $\int_{|x|\geq\varepsilon}(1-\cos(ux))\nu(dx) \leq \tilde{C} < \infty$. Making ε tend to zero, we obtain the desired result (3.14).

Now we can use it to show the convergence of $\tilde{X}_t^\varepsilon$. Consider a sequence $\{\varepsilon_n\} \downarrow 0$ and let $Y_n = \tilde{X}_t^{\varepsilon_{n+1}} - \tilde{X}_t^{\varepsilon_n}$. All the variables Y_i have zero mean and (3.14) entails that $\sum \operatorname{Var} Y_i < \infty$. Hence, by Kolmogorov's three series Theorem [228, Theorem 3.18], $\sum Y_i$ converges almost surely, which means that $\tilde{X}_t^\varepsilon$ converges almost surely as $\varepsilon \to 0$. Using Kolmogorov's maximum inequality [228, Lemma 3.15], one can show that the convergence is uniform in t.

To complete the proof, consider the process $X_t^c = X_t - \lim \tilde{X}_t^\varepsilon$. It is a Lévy process which is independent from $\lim \tilde{X}_t^\varepsilon$ by Lemma 3.2. It is continuous, because $\tilde{X}_t^\varepsilon$ converges uniformly in t and therefore one can interchange the limits. Finally, the Feller-Lévy central limit Theorem [228, Theorem 4.15] implies that it is also Gaussian. □

Our knowledge of the structure of paths of a Lévy process allows to obtain almost without additional work the second fundamental result of the theory: the expression of the characteristic function of a Lévy process in terms of its characteristic triplet (A, ν, γ).

THEOREM 3.1 Lévy-Khinchin representation
Let $(X_t)_{t\geq 0}$ be a Lévy process on $\mathbb{R}^d$ with characteristic triplet (A, ν, γ). Then

$$E[e^{iz.X_t}] = e^{t\psi(z)}, \; z \in \mathbb{R}^d \tag{3.15}$$
$$\textit{with} \quad \psi(z) = -\frac{1}{2} z.Az + i\gamma.z + \int_{\mathbb{R}^d} (e^{iz.x} - 1 - iz.x 1_{|x|\leq 1})\nu(dx).$$

For real-valued Lévy processes, the formula (3.15) takes the form

$$E[e^{izX_t}] = e^{t\psi(z)}, \; z \in \mathbb{R}$$
$$\text{with} \quad \psi(z) = -\frac{1}{2} Az^2 + i\gamma z + \int_{-\infty}^{\infty} (e^{izx} - 1 - izx 1_{|x|\leq 1})\nu(dx).$$

An equivalent version of the Lévy-Khinchin representation may be obtained by truncating the jumps larger than an arbitrary number ε:

$$\psi(z) = -\frac{1}{2} z.Az + i\gamma^{\varepsilon}.z + \int_{\mathbb{R}^d} (e^{iz.x} - 1 - iz.x 1_{|x|\leq \varepsilon})\nu(dx),$$
$$\text{where} \qquad \gamma^{\varepsilon} = \gamma + \int_{\mathbb{R}^d} x(1_{|x|\leq\varepsilon} - 1_{|x|\leq 1})\nu(dx).$$

More generally, for every bounded measurable function $g : \mathbb{R}^d \to \mathbb{R}$ satisfying $g(x) = 1 + o(|x|)$ as $x \to 0$ and $g(x) = O(1/|x|)$ as $x \to \infty$, one can write:

$$\psi(z) = -\frac{1}{2} z.Az + i\gamma^{g}.z + \int_{\mathbb{R}^d} (e^{iz.x} - 1 - iz.x g(x))\nu(dx).$$

Such a function g is called the *truncation function* and the characteristic triplet (A, ν, γ^g) is called the characteristic triplet of X with respect to the truncation function g. Different choices of g do not affect A and ν which are intrinsic parameters of the Lévy process, but γ depends on the choice of truncation function so one should avoid calling it "the drift" of the process. Various choices of the truncation function have been used in the literature. Paul Lévy used the truncation function $g(x) = \frac{1}{1+|x|^2}$ while most recent texts use $g(x) = 1_{|x|\leq 1}$. In the sequel, when we refer to the Lévy triplet of a Lévy process we implicitly refer to the truncation function $g(x) = 1_{|x|\leq 1}$.

If the Lévy measure satisfies the additional condition $\int_{|x|\geq 1} |x|\nu(dx) < \infty$ there is no need to truncate large jumps and one can use the simpler form

$$\psi(z) = -\frac{1}{2} z.Az + i\gamma_c.z + \int_{\mathbb{R}^d} (e^{iz.x} - 1 - iz.x)\nu(dx).$$

In this case it can be shown that $E[X_t] = \gamma_c t$ and γ_c is called the *center* of process (X_t). It is linked to γ by the relation $\gamma_c = \gamma + \int_{|x|\geq 1} x\nu(dx)$.

PROOF of Theorem 3.1 The Lévy-Itô decomposition (Theorem 3.7) shows that for every t, the random variable $X_t^c + X_t^l + X_t^\varepsilon$ converges almost surely to X_t when ε tends to 0. Since almost sure convergence implies convergence in distribution, the characteristic function of $X_t^c + X_t^l + X_t^\varepsilon$ converges to the characteristic function of X_t. Since X_t^c, X_t^l and X_t^ε are independent,

$$E[\exp\{iz.(X_t^c + X_t^l + X_t^\varepsilon)\}] = \exp\{-\frac{1}{2}tz.Az + it\gamma.z\}$$
$$\times \exp\{t\int_{|x|\geq 1}(e^{iz.x}-1)\nu(dx)\}\exp\{t\int_{\varepsilon\leq|x|<1}(e^{iz.x}-1-iz.x)\nu(dx)\}$$

and this expression converges to (3.15) for every z when ε tends to 0. □

When $\nu(\mathbb{R}^d) = \infty$ (infinite activity case), the set of jump times of every trajectory of the Lévy process is countably infinite and dense in $[0,\infty[$. The countability follows directly from the fact that the paths are cadlag. To prove that the set of jump times is dense in $[0,\infty[$, consider a time interval $[a,b]$ and let

$$\varepsilon(n) = \sup\{r : \int_{|x|\geq r}\nu(dx) \geq n\}$$
$$\text{and} \quad Y_n = \int_{\varepsilon(n)\leq|x|<\varepsilon(n-1), t\in[a,b]} J_X(dx\times dt).$$

Then, if the Lévy measure has no atoms, Y_i are independent and identically Poisson distributed random variables. The total number of jumps in the interval $[a,b]$ is equal to $\sum_{i=1}^\infty Y_i$, hence, by the law of large numbers, it is almost surely infinite. Since this is true for every nonempty time interval $[a,b]$, this means that the set of jump times is dense in $[0,\infty[$. The proof can be easily modified to include the case when the Lévy measure has atoms.

Since any infinitely divisible distribution is the distribution at time $t = 1$ of some Lévy process, the Lévy-Khinchin formula also gives the a general representation for the characteristic function of any infinitely divisible distribution:

***THEOREM 3.2* Characteristic function of infinitely divisible distributions**
Let F be an infinitely divisible distribution on $\mathbb{R}^d$. Its characteristic function can be represented as:

$$\Phi_F(z) = e^{\psi(z)}, \quad z \in \mathbb{R}^d$$
$$\psi(z) = -\frac{1}{2}z.Az + i\gamma.z + \int_{\mathbb{R}^d} (e^{iz.x} - 1 - iz.x1_{|x|\leq 1})\nu(dx),$$

where A is a symmetric positive $n \times n$ matrix, $\gamma \in \mathbb{R}^d$ and ν is a positive Radon measure on $\mathbb{R}^d \setminus \{0\}$ verifying:

$$\int_{|x|\leq 1} |x|^2\nu(dx) < \infty \qquad \int_{|x|\geq 1} \nu(dx) < \infty.$$

ν is called the Lévy measure *of the distribution F.*

3.5 Pathwise properties of Lévy processes

In this section, using the Lévy-Itô decomposition, we deduce some properties of typical sample paths of a Lévy process from analytical properties of its characteristic triplet (A, ν, γ).

Piecewise constant trajectories We saw in the preceding section (Proposition 3.3) that almost all trajectories of a Lévy process are piecewise constant if and only if it is of compound Poisson type. Combining this with the Formula (3.4), which gives the characteristic function of a compound Poisson processes, we obtain the following criterion:

PROPOSITION 3.8
A Lévy process has piecewise constant trajectories if and only if its characteristic triplet satisfies the following conditions: $A = 0$, $\int_{\mathbb{R}^d} \nu(dx) < \infty$ and $\gamma = \int_{|x|\leq 1} x\nu(dx)$ or equivalently, if its characteristic exponent is of the form:

$$\psi(z) = \int_{-\infty}^{\infty} (e^{iux} - 1)\nu(dx) \quad \textit{with } \nu(\mathbb{R}) < \infty.$$

Lévy processes of finite variation We recall that the total variation of a function $f : [a, b] \to \mathbb{R}^d$ is defined by

$$TV(f) = \sup \sum_{i=1}^{n} |f(t_i) - f(t_{i-1})|,$$

where the supremum is taken over all finite partitions $a = t_0 < t_1 < \cdots < t_{n-1} < t_n = b$ of the interval $[a, b]$. In particular, in one dimension every increasing or decreasing function is of finite variation and every function of finite variation is a difference of two increasing functions.

A Lévy process is said to be of finite variation if its trajectories are functions of finite variation with probability 1.

PROPOSITION 3.9 Finite variation Lévy processes

A Lévy process is of finite variation if and only if its characteristic triplet (A, ν, γ) *satisfies:*

$$A = 0 \quad \textit{and} \quad \int_{|x|\leq 1} |x|\nu(dx) < \infty. \tag{3.16}$$

PROOF The *if* part. Under the stated conditions, X_t can be represented in the following form:

$$X_t = bt + \int_{|x|\geq 1, s\in[0,t]} x J_X(ds \times dx) + \lim_{\varepsilon \downarrow 0} \tilde{X}_t^\varepsilon,$$
$$\text{where} \quad \tilde{X}_t^\varepsilon = \int_{\varepsilon \leq |x| < 1, s\in[0,t]} x J_X(ds \times dx).$$

The first two terms are of finite variation, therefore we only need to consider the third term. Its variation on the interval $[0, t]$ is

$$TV(\tilde{X}_t^\varepsilon) = \int_{\varepsilon \leq |x| < 1, s\in[0,t]} |x| J_X(ds \times dx).$$

Since the integrand in the right-hand side is positive, we obtain, using Fubini's theorem

$$E[TV(\tilde{X}_t^\varepsilon)] = t \int_{\varepsilon \leq |x| < 1} |x|\nu(dx),$$

which converges to a finite value when $\varepsilon \to 0$. Therefore $E[TV(\lim_{\varepsilon\downarrow 0} \tilde{X}_t^\varepsilon)] < \infty$, which implies that the variation of X_t is almost surely finite.

The *only if* part. Consider the Lévy-Itô decomposition (3.11) of X_t. Since the variation of any cadlag function is greater or equal to the sum of its jumps, we have for every $\varepsilon > 0$:

$$TV(X_t) \geq \int_{\varepsilon \leq |x| < 1, s \in [0,t]} |x| J_X(ds \times dx)$$
$$= t \int_{\varepsilon \leq |x| < 1} |x| \nu(dx) + \int_{\varepsilon \leq |x| < 1, s \in [0,t]} |x| (J_X(ds \times dx) - \nu(dx) ds).$$

Using the exponential formula (Proposition 3.6), one can show that the variance of the second term in the last line is equal to $t \int_{\varepsilon \leq |x| < 1, s \in [0,t]} |x|^2 \nu(dx)$. Hence, by the same argument that was used in the proof of Lévy-Itô decomposition, the second term converges almost surely to something finite. Therefore, if the condition $\int (|x| \wedge 1) \nu(dx) < \infty$ is not satisfied, the first term in the last line will diverge and the variation of X_t will be infinite. Suppose now that this condition is satisfied. This means that X_t may be written as

$$X_t = X_t^c + \int_{[0,t] \times \mathbb{R}^d} x J_X(ds \times dx),$$

where the second term is of finite variation. Since trajectories of Brownian motion are almost surely of infinite variation (see [335]), if A is nonzero, X_t will also have infinite variation. Therefore we must have $A = 0$. □

The preceding proposition shows that in the finite variation case Lévy-Itô decomposition and Lévy-Khinchin representation can be simplified:

COROLLARY 3.1 Lévy-Itô decomposition and Lévy-Khinchin representation in the finite-variation case

Let $(X_t)_{t \geq 0}$ be a Lévy process of finite variation with Lévy triplet given by $(\nu, 0, \gamma)$. Then X can be expressed as the sum of its jumps between 0 and t and a linear drift term:

$$X_t = bt + \int_{[0,t] \times \mathbb{R}^d} x J_X(ds \times dx) = bt + \sum_{s \in [0,t]}^{\Delta X_s \neq 0} \Delta X_s \tag{3.17}$$

and its characteristic function can be expressed as:

$$E[e^{iz.X_t}] = \exp t \left\{ ib.z + \int_{\mathbb{R}^d} (e^{iz.x} - 1) \nu(dx) \right\}, \tag{3.18}$$

where $b = \gamma - \int_{|x| \leq 1} x \nu(dx)$.

Note that the Lévy triplet of X is not given by $(b, 0, \nu)$ but by $(\gamma, 0, \nu)$. In fact, as mentioned before γ is not an intrinsic quantity and depends on the truncation function used in the Lévy-Khinchin representation while bt has an intrinsic interpretation as the continuous part of X.

Increasing Lévy processes (subordinators) Increasing Lévy processes are also called subordinators because they can be used as time changes for other Lévy process (see Chapter 4). They are very important ingredients for building Lévy-based models in finance.

PROPOSITION 3.10

Let $(X_t)_{t\geq 0}$ be a Lévy process on $\mathbb{R}$. The following conditions are equivalent:

i. $X_t \geq 0$ *a.s. for some* $t > 0$.

ii. $X_t \geq 0$ *a.s. for every* $t > 0$.

iii. *Sample paths of* (X_t) *are almost surely nondecreasing:* $t \geq s \Rightarrow X_t \geq X_s$ *a.s.*

iv. *The characteristic triplet of* (X_t) *satisfies* $A = 0$, $\nu((-\infty, 0]) = 0$, $\int_0^\infty (x \wedge 1)\nu(dx) < \infty$ *and* $b \geq 0$*, that is,* (X_t) *has no diffusion component, only positive jumps of finite variation and positive drift.*

PROOF $i \Rightarrow iii$. For every n, X_t is a sum of n i.i.d. random variables $X_{t/n}, X_{2t/n} - X_{t/n}, \ldots, X_t - X_{(n-1)t/n}$. This means that all these variables are almost surely nonnegative. With the same logic we can prove that for any two rationals p and q such that $0 < p < q$, $X_{qt} - X_{pt} \geq 0$ a.s. Since the trajectories are right-continuous, this entails that they are nondecreasing.

The implications $iii \Rightarrow ii$ and $ii \Rightarrow i$ are trivial.

$iv \Rightarrow iii$. Under the conditions of (iv) the process is of finite variation, therefore equal to the sum of its jumps plus an increasing linear function. For every trajectory the number of negative jumps on any fixed interval is a Poisson random variable with intensity 0, hence, almost surely zero. This means that almost every trajectory is nondecreasing.

$iii \Rightarrow iv$. Since the trajectories are nondecreasing, they are of finite variation. Therefore, $A = 0$ and $\int_{-\infty}^{\infty} (x \wedge 1)\nu(dx) < \infty$. For trajectories to be nonincreasing, there must be no negative jumps, hence $\nu(]-\infty, 0]) = 0$. If a function is nondecreasing then after removing some of its jumps, we obtain another nondecreasing function. When we remove all jumps from a trajectory of X_t, we obtain a deterministic function bt which must therefore be nondecreasing. This allows to conclude that $b \geq 0$. □

An important example of subordinator is introduced by the following proposition.

PROPOSITION 3.11
Let $(X_t)_{t\geq 0}$ be a Lévy process on $\mathbb{R}^d$ and let $f : \mathbb{R}^d \to [0,\infty[$ be a positive function such that $f(x) = O(|x|^2)$ when $x \to 0$. Then the process $(S_t)_{t\geq 0}$ defined by

$$S_t = \sum_{\substack{s\leq t\\ \Delta X_s\neq 0}} f(\Delta X_s) \tag{3.19}$$

is a subordinator.

PROOF Let us first show that the sum in (3.19) converges to something finite. By truncating large jumps we can suppose that for each s, $\Delta X_s \leq \varepsilon$ for some $\varepsilon > 0$ and $f(\Delta X_s) \leq C\Delta X_s^2$ for some $C > 0$. But then

$$E[S_t] = \int_{[0,t]\times\mathbb{R}} f(x)ds\nu(dx) < \infty. \tag{3.20}$$

Since all the terms in the sum are positive, this means that it always converges and S_t is almost surely finite for all t. The fact that S has independent and stationary increments follows directly from independence and stationarity of increments of X. To prove that it is continuous in probability one can once again suppose that jumps of X_t are bounded (because the compound Poisson part is always continuous in probability). But then $E[|S_t - S_s|] \to 0$ as $s \to t$. Therefore, S is continuous in probability. □

The choice $f(x) = x^2$ yields the sum of squared jumps

$$S_t = \sum_{\substack{s\leq t\\ \Delta X_s\neq 0}} |\Delta X_s|^2. \tag{3.21}$$

This process which by the above proposition is a subordinator, is usually denoted $[X,X]^d$ and called the "discontinuous quadratic variation" of X. We will encounter it again in Section 8.2.

REMARK 3.1 There exist Lévy processes without diffusion component, having no negative jumps, but satisfying $\int_0^1 |x|\nu(dx) = \infty$. The above proposition entails that these processes cannot have increasing trajectories, whatever drift coefficient they may have. The explanation of this "mysterious" behavior is that in this case the process is not equal to the sum of its jumps, because the jumps must be compensated. These compensation terms add up to an "infinitely strong" negative drift between the jumps, which cannot be made positive by changing the drift coefficient. □

3.6 Distributional properties

If $(X_t)_{t\geq 0}$ is a Lévy process then for any $t > 0$, the distribution of X_t is infinitely divisible and has a characteristic function of the form (3.15). However, X_t does not always have a density: indeed, if X_t is a compound Poisson process we have

$$\mathbb{P}(X_t = 0) = e^{-\lambda t} > 0 \tag{3.22}$$

so the probability distribution has an atom at zero for all t. But if X is not a compound Poisson process, then X_t has a continuous density; we give the following result for $d = 1$ from [312]:

PROPOSITION 3.12 Existence of a smooth density
Let X be a real-valued Lévy process with Lévy triplet (σ^2, ν, γ).

- *If $\sigma > 0$ or $\nu(\mathbb{R}) = \infty$ then X_t has a continuous density $p_t(.)$ on $\mathbb{R}^d$.*
- *If the Lévy measure ν verifies*

$$\exists \beta \in]0,2[, \quad \liminf_{\epsilon \downarrow 0} \epsilon^{-\beta} \int_{-\epsilon}^{\epsilon} |x|^2 d\nu(x) > 0 \tag{3.23}$$

then for each $t > 0$, X_t has a smooth density $p_t(.)$ such that

$$p_t(.) \in C^\infty(\mathbb{R}) \;\; \forall n \geq 1, \frac{\partial^n p_t}{\partial x^n}(t,x) \underset{|x|\to\infty}{\to} 0. \tag{3.24}$$

These and other properties of the density may be obtained using the Lévy-Khinchin representation and the properties of the Fourier transform, see [345, Chapter 5].

Relation between probability density and Lévy density In the compound Poisson case there is a simple relation between probability distribution at time t and the jump size distribution/Lévy measure. Let $(X_t)_{t\geq 0}$ be a compound Poisson process with intensity λ and jump size distribution f and $(N_t)_{t\geq 0}$ be the number of jumps of X on $[0,t]$. Then

$$\mathbb{P}\{X_t \in A\} = \sum_{n=0}^{\infty} \mathbb{P}\{X_t \in A \big| N_t = n\} \frac{e^{-\lambda t}(\lambda t)^n}{n!}$$

$$= e^{-\lambda t}\delta_0 + \sum_{n=1}^{\infty} f^{*n}(A) \frac{e^{-\lambda t}(\lambda t)^n}{n!}, \tag{3.25}$$

where f^{*n} denotes the n-th convolution power of f, and δ_0 is the Dirac measure concentrated at 0. As noted above, this probability measure does not have a density because $P\{X_t = 0\} > 0$. However, if the jump size distribution has a density with respect to Lebesgue measure, then the law of X_t is absolutely continuous everywhere except at zero (because convolution of absolutely continuous distributions is absolutely continuous), i.e., the law of X_t can be decomposed as

$$\mathbb{P}\{X_t \in A\} = e^{-\lambda t} 1_{0\in A} + \int_A p_t^{ac}(x)dx \qquad \text{where}$$
$$p_t^{ac}(x) = \sum_{n=1}^{\infty} f^{*n}(x)\frac{e^{-\lambda t}(\lambda t)^n}{n!} \qquad \forall x \neq 0.$$

where we denote the jump size density by $f(x)$. p_t^{ac} is the density conditional on the fact that the process has jumped at least once. This implies in particular the following asymptotic relation:

$$\lim_{t\downarrow 0} \frac{1}{t} p_t^{ac}(x) = \lambda f(x) = \nu(x) \quad \forall x \neq 0,$$

where $\nu(x)$ is the Lévy density. This means that the Lévy density describes the small time behavior of the probability density.

This relation also gives the small time behavior for expectations of functions of X_t: given any bounded measurable function f such that $f(0) = 0$,

$$\lim_{t\downarrow 0} \frac{1}{t} E[f(X_t)] = \lim_{t\downarrow 0} \frac{1}{t} \int_{\mathbb{R}^d} f(x) p_t(dx) = \int_{\mathbb{R}^d} f(x)\nu(dx). \tag{3.26}$$

In the infinite activity setting the classical asymptotic result for expectations (see [345, Corollary 8.9]) is weaker: it states that the formula (3.26) holds for any bounded continuous function f vanishing in the neighborhood of zero. More results on the relation between probability density of X_t and the Lévy measure for infinite-activity processes may be found in [33] and [341].

Moments and cumulants The tail behavior of the distribution of a Lévy process and its moments are determined by the Lévy measure, as shown by the following proposition, which is a consequence of [345, Theorem 25.3].

PROPOSITION 3.13 Moments and cumulants of a Lévy process
Let $(X_t)_{t\geq 0}$ be a Lévy process on $\mathbb{R}$ with characteristic triplet (A, ν, γ). The n-th absolute moment of X_t, $E[|X_t|^n]$ is finite for some t or, equivalently, for every $t > 0$ if and only if $\int_{|x|\geq 1} |x|^n \nu(dx) < \infty$. In this case moments of X_t can be computed from the its characteristic function by differentiation. In particular, the form of cumulants (defined in Section 2.2.5) of X_t is especially

simple:

$$E[X_t] = t(\gamma + \int_{|x|\geq 1} x\nu(dx)),$$

$$c_2(X_t) = \text{Var}\, X_t = t(A + \int_{-\infty}^{\infty} x^2\nu(dx)),$$

$$c_n(X_t) = t\int_{-\infty}^{\infty} x^n\nu(dx) \quad \textit{for } n \geq 3.$$

This entails in particular that all infinitely divisible distributions are leptokurtic since $c_4(X_t) > 0$. Also, the cumulants of the distribution of X_t increase linearly with t. In particular the kurtosis and skewness[3] of X_Δ (or, equivalently, of the increments $X_{t+\Delta} - X_t$ are given by:

$$s(X_\Delta) = \frac{c_3(X)}{c_2(X)^{3/2}} = \frac{s(X_1)}{\sqrt{\Delta}}, \qquad \kappa(X_\Delta) = \frac{c_4(X_\Delta)}{c_2(X_\Delta)^2} = \frac{\kappa(X_1)}{\Delta}. \quad (3.27)$$

Therefore the increments of a Lévy process or, equivalently, all infinitely divisible distributions are always leptokurtic but the kurtosis (and skewness, if there is any) decreases with the time scale over which increments are computed: the skewness falls as $\Delta^{-1/2}$ while the kurtosis decays as $1/\Delta$.

PROPOSITION 3.14 Exponential moments
Let $(X_t)_{t\geq 0}$ be a Lévy process on $\mathbb{R}$ with characteristic triplet (A, ν, γ) and let $u \in \mathbb{R}$. The exponential moment $E[e^{uX_t}]$ is finite for some t or, equivalently, for all $t > 0$ if and only if $\int_{|x|\geq 1} e^{ux}\nu(dx) < \infty$. In this case

$$E[e^{uX_t}] = e^{t\psi(-iu)}.$$

where ψ is the characteristic exponent of the Lévy process defined by (3.15).

For a proof see [345, Theorem 25.17].

3.7 Stable laws and processes

A remarkable property of Brownian motion is its selfsimilarity property: if W is a Wiener process on $\mathbb{R}$ then

$$\forall a > 0, \quad \left(\frac{W_{at}}{\sqrt{a}}\right)_{t\geq 0} \stackrel{d}{=} (W_t)_{t\geq 0}.$$

[3] See Equation (2.34) for a definition of skewness and kurtosis.

If we consider a Brownian motion with drift $B_t = W_t + \gamma t$ then this property is only verified up to a translation:

$$\forall a > 0, \quad \left(\frac{B_{at}}{\sqrt{a}}\right)_{t\geq 0} \stackrel{d}{=} (B_t + \sqrt{a}\gamma t)_{t\geq 0}.$$

A natural question is whether there exist other real valued Lévy processes that share this selfsimilarity property: a Lévy process X_t is said to be *selfsimilar* if

$$\forall a > 0, \; \exists b(a) > 0 : \quad \left(\frac{X_{at}}{b(a)}\right)_{t\geq 0} \stackrel{d}{=} (X_t)_{t\geq 0}.$$

Since the characteristic function of X_t has the form $\Phi_{X_t}(z) = \exp[-t\psi(z)]$, this property is equivalent to the following property of the characteristic function:

$$\forall a > 0, \; \exists b(a) > 0 : \quad \Phi_{X_t}(z)^a = \Phi_{X_t}(zb(a)) \quad \forall z.$$

The distributions that verify this property are called *strictly stable distributions*. More precisely, we have the following definition.

DEFINITION 3.5 *A random variable $X \in \mathbb{R}^d$ is said to have* stable distribution *if for every $a > 0$ there exist $b(a) > 0$ and $c(a) \in \mathbb{R}^d$ such that*

$$\Phi_X(z)^a = \Phi_X(zb(a))e^{ic.z}, \quad \forall z \in \mathbb{R}^d. \tag{3.28}$$

It is said to have a strictly stable distribution *if*

$$\Phi_X(z)^a = \Phi_X(zb(a)), \quad \forall z \in \mathbb{R}^d. \tag{3.29}$$

The name *stable* comes from the following *stability under addition* property: if X has stable distribution and $X^{(1)}, \ldots, X^{(n)}$ are independent copies of X then there exist a positive number c_n and a vector $\mathbf{d}$ such that

$$X^{(1)} + \cdots + X^{(n)} \stackrel{d}{=} c_n X + \mathbf{d}. \tag{3.30}$$

This property is clearly verified if the distribution of X is that of a selfsimilar Lévy process at a given time t.

It can be shown (see [344, corollary 2.1.3]) that for every stable distribution there exists a constant $\alpha \in (0, 2]$ such that in Equation (3.28), $b(a) = a^{1/\alpha}$. This constant is called the *index of stability* and stable distributions with index α are also referred to as α-stable distributions. The only 2-stable distributions are Gaussian.

A selfsimilar Lévy process therefore has strictly stable distribution at all times. For this reason, such processes are also called strictly stable Lévy processes. A strictly α-stable Lévy process satisfies:

$$\forall a > 0, \quad \left(\frac{X_{at}}{a^{1/\alpha}}\right)_{t\geq 0} \stackrel{d}{=} (X_t)_{t\geq 0}. \tag{3.31}$$

In the case of the Wiener process $\alpha = 2$. More generally, an α-stable Lévy process satisfies this relation up to a translation:

$$\forall a > 0,\ \exists c \in \mathbb{R}^d : \quad (X_{at})_{t\geq 0} \stackrel{d}{=} (a^{1/\alpha} X_t + ct)_{t\geq 0}.$$

A stable Lévy process defines a family of stable distributions and the converse is also true: every stable distribution is infinitely divisible and can be seen as the distribution at a given time of a stable Lévy process. The following result gives the form of characteristic triplet of all stable distributions (and, therefore, Lévy processes).

PROPOSITION 3.15 Stable distributions and Lévy processes
A distribution on $\mathbb{R}^d$ is α-stable with $0 < \alpha < 2$ if and only if it is infinitely divisible with characteristic triplet $(0, \nu, \gamma)$ and there exists a finite measure λ on S, a unit sphere of $\mathbb{R}^d$, such that

$$\nu(B) = \int_S \lambda(d\xi) \int_0^\infty 1_B(r\xi) \frac{dr}{r^{1+\alpha}}. \tag{3.32}$$

A distribution on $\mathbb{R}^d$ is α-stable with $\alpha = 2$ if and only if it is Gaussian.

A proof is given in [345, Theorem 14.3], see also [344].

For real-valued stable variables and Lévy processes ($d = 1$) the above representation can be made explicit: if X is a real-valued α-stable variable with $0 < \alpha < 2$ then its Lévy measure is of the form

$$\nu(x) = \frac{A}{x^{\alpha+1}} 1_{x>0} + \frac{B}{|x|^{\alpha+1}} 1_{x<0} \tag{3.33}$$

for some positive constants A and B. The characteristic function at time 1 of a real-valued stable random variable X has the form

$$\Phi_X(z) = \exp\left\{-\sigma^\alpha |z|^\alpha (1 - i\beta \operatorname{sgn} z \tan\frac{\pi\alpha}{2}) + i\mu z\right\}, \quad \text{if } \alpha \neq 1,$$
$$\Phi_X(z) = \exp\left\{-\sigma |z| (1 + i\beta\frac{2}{\pi} \operatorname{sgn} z \log|z|) + i\mu z\right\}, \quad \text{if } \alpha = 1, \tag{3.34}$$

where $\alpha \in (0, 2]$, $\sigma \geq 0$, $\beta \in [-1, 1]$ and $\mu \in \mathbb{R}$. In the sequel, a stable distribution on $\mathbb{R}$ in this parameterization is denoted by $S_\alpha(\sigma, \beta, \nu)$. In this representation, σ is the scale parameter (note that is has nothing to do with the Gaussian component if $\alpha < 2$), μ is the shift parameter (when $\alpha \neq 1$ this is not true: see [344, Section 1.2]), α determines the shape of the distribution and β its skewness. When $\beta = 0$ and $\mu = 0$, X is said to have a symmetric stable distribution and the characteristic function is given by

$$\Phi_X(z) = \exp(-\sigma^\alpha |z|^\alpha).$$

The explicit form of the Lévy measure (3.33) shows that α-stable distributions on $\mathbb{R}$ never admit a second moment, and they only admit a first moment if $\alpha > 1$. The probability density of an α-stable law is not known in closed form except in the following three cases (plus the degenerate case of a constant random variable)

- The Gaussian distribution $S_2(\sigma, 0, \mu)$ with density (note the nonstandard parameterization)

$$\frac{1}{2\sigma\sqrt{\pi}} e^{-(x-\mu)^2/4\sigma^2}. \tag{3.35}$$

- The Cauchy distribution $S_1(\sigma, 0, \mu)$ with density

$$\frac{\sigma}{\pi((x-\mu)^2+\sigma^2)}. \tag{3.36}$$

- The Lévy distribution $S_{1/2}(\sigma, 1, \mu)$ with density

$$\left(\frac{\sigma}{2\pi}\right)^{1/2} \frac{1}{(x-\mu)^{3/2}} \exp\left\{-\frac{\sigma}{2(x-\mu)}\right\} 1_{x>\mu}. \tag{3.37}$$

While the first two distributions are symmetric around their mean, the last one is concentrated on (μ, ∞). Despite the fact that closed formulae for probability density are only available in these three cases, closed-form algorithms for simulating stable random variables on $\mathbb{R}$ exist for all values of parameters (see Chapter 6).

3.8 Lévy processes as Markov processes

An important property of Lévy processes is the Markov property, which states that conditionally on X_t, the evolution of the process after time t is independent on its past before this moment. In other words, for every random variable Y depending on the history $\mathcal{F}_t$ of X_t one must have

$$E[Y|\mathcal{F}_t] = E[Y|X_t].$$

The *transition kernel* of process X_t is defined as follows:

$$P_{s,t}(x, B) = \mathbb{P}\{X_t \in B | X_s = x\}, \quad \forall B \in \mathcal{B}. \tag{3.38}$$

The Markov property implies the following relation between transition kernels (known as the Chapman-Kolmogorov equations):

$$P_{s,u}(x, B) = \int_{\mathbb{R}^d} P_{s,t}(x, dy) P_{t,u}(y, B).$$

It is easily seen from the Definition (3.38) that the transition kernels of Lévy processes are homogeneous in space and time, that is,

$$P_{s,t}(x,B) = P_{0,t-s}(0, B-x).$$

Lévy processes are completely characterized by this condition (see [345, Theorem 10.5]): they are the only Markov processes which are homogeneous in space and time.

Lévy processes satisfy a stronger version of the Markov property, namely, for all t, the process $(X_{t+s} - X_t)_{s\geq 0}$ has the same law as the process $(X_s)_{s\geq 0}$ and is independent from $(X_s)_{0\leq s\leq t}$.

Finally, the *strong Markov property* of Lévy processes allows to replace the nonrandom time t by any random time which is nonanticipating with respect to the history of X (see Section 2.4.2): if τ is a nonanticipating random time, then the process $Y_t = X_{t+\tau} - X_t$ is again a Lévy process, independent from $\mathcal{F}_\tau$ and with same law as $(X_t)_{t\geq 0}$.

The transition operator for Markov processes is defined as follows:

$$P_t f(x) = E[f(x + X_t)].$$

Chapman-Kolmogorov equations and the time homogeneity of transition kernels imply the following *semigroup* relation between transition operators:

$$P_t P_s = P_{t+s}.$$

Let $\mathcal{C}_0$ be the set of continuous functions vanishing at infinity. Then for any $t > 0, P_t f \in \mathcal{C}_0$ and

$$\forall x \quad \lim_{t\downarrow 0} P_t f(x) = f(x). \tag{3.39}$$

where the convergence is in the sense of supremum norm on $\mathcal{C}_0$. This property is called the Feller property. A semigroup P_t verifying the Feller property (3.39) can be described by means of its infinitesimal generator L which is a linear operator defined by

$$Lf = \lim_{t\downarrow 0} t^{-1}(P_t f - f), \tag{3.40}$$

where the convergence is in the sense of supremum norm on $\mathcal{C}_0$ and f should be such that the right-hand side of (3.40) exists. The infinitesimal generator of a Lévy process can be expressed in terms of its characteristic triplet:

PROPOSITION 3.16 Infinitesimal generator of a Lévy process
Let $(X_t)_{t\geq 0}$ be a Lévy process on $\mathbb{R}^d$ with characteristic triplet (A, ν, γ). Then

the infinitesimal generator of X is defined for any $f \in C_0^2(\mathbb{R})$ as

$$Lf(x) = \frac{1}{2}\sum_{j,k=1}^{d} A_{jk}\frac{\partial^2 f}{\partial x_j \partial x_k}(x) + \sum_{j=1}^{d}\gamma_j \frac{\partial f}{\partial x_j}(x) + \int_{\mathbb{R}^d}\left(f(x+y) - f(x) - \sum_{j=1}^{d} y_j \frac{\partial f}{\partial x_j}(x) 1_{|y|\leq 1}\right)\nu(dy), \quad (3.41)$$

where $C_0^2(\mathbb{R}^d)$ is the set of twice continuously differentiable functions, vanishing at infinity.

We will give a proof of this result in a slightly different setting in Chapter 12. For a classical proof see [345, Theorem 31.5]. We will see in Chapters 8 and 12 that the computation of expectations of various functionals of Lévy processes can be transformed into partial integro-differential equations involving the infinitesimal generator. Due to this fact, infinitesimal generators are important tools in option pricing.

3.9 Lévy processes and martingales

The notion of martingale (see Section 2.4.4) is crucial for probability theory and mathematical finance. Different martingales can be constructed from Lévy processes using their independent increments property.

PROPOSITION 3.17
Let $(X_t)_{t\geq 0}$ be a real-valued process with independent increments. Then

1. $\left(\frac{e^{iuX_t}}{E[e^{iuX_t}]}\right)_{t\geq 0}$ *is a martingale $\forall u \in \mathbb{R}$.*
2. *If for some $u \in \mathbb{R}$, $E[e^{uX_t}] < \infty$ $\forall t \geq 0$ then $\left(\frac{e^{uX_t}}{E[e^{uX_t}]}\right)_{t\geq 0}$ is a martingale.*
3. *If $E[X_t] < \infty$ $\forall t \geq 0$ then $M_t = X_t - E[X_t]$ is a martingale (and also a process with independent increments).*
4. *If $\mathrm{Var}[X_t] < \infty$ $\forall t \geq 0$ then $(M_t)^2 - E[(M_t)^2]$ is a martingale, where M is the martingale defined above.*

If (X_t) is a Lévy processes, for all of the processes of this proposition to be martingales it suffices that the corresponding moments be finite for one value of t (see Theorems 25.17 and 25.3 in [345]).

These statements follow from the independent increments property. Details of the proof are left to the reader as an exercise.

Sometimes, in particular in financial applications, it is important to check whether a given Lévy process or its exponential is a martingale. We will now obtain the necessary and sufficient conditions.

PROPOSITION 3.18

Let $(X_t)_{t\geq 0}$ be a Lévy process on $\mathbb{R}$ with characteristic triplet (A, ν, γ).

1. *(X_t) is a martingale if and only if $\int_{|x|\geq 1} |x|\nu(dx) < \infty$ and*

$$\gamma + \int_{|x|\geq 1} x\nu(dx) = 0.$$

2. *$\exp(X_t)$ is a martingale if and only if $\int_{|x|\geq 1} e^x \nu(dx) < \infty$ and*

$$\frac{A}{2} + \gamma + \int_{-\infty}^{\infty} (e^x - 1 - x1_{|x|\leq 1})\nu(dx) = 0. \tag{3.42}$$

This proposition is a consequence of Proposition 3.17 and the Lévy-Khinchin formula.

Further reading

A summary of properties of Lévy processes can be found in [51]. The monograph by Sato [345] is a detailed study of Lévy processes and their properties using an analytic viewpoint. The original probabilistic approach introduced by Paul Lévy was to analyze the sample paths of a Lévy process directly. An easy to read reference where details on this approach can be found is [164]. Bertoin's book [49] treats more advanced topics on Lévy processes, not discussed in the first two references, including local times and excursion theory. A detailed treatment of subordinators is given in the Saint Flour lectures of Bertoin [50]. Stroock [370] discusses in detail the Lévy-Khinchin representation and the construction of Lévy processes as Markov processes using Itô's original approach. Stable laws and processes are discussed in detail in [344] and in the monograph by Zolotarev [394].

REMARK 3.2 Lévy flights α-stable Lévy processes (see Section 3.7) are known in the physics literature under the name of *Lévy flights* or *anomalous diffusions* and have been extensively used for modelling physical phenomena; see, e.g., [26]. Lévy flights are Lévy processes with infinite variance and possess scale invariance and self-similarity properties (see Section 7.4 for a discussion of self-similarity). Some authors have used the names "Lévy process" and "Lévy flights" interchangeably, giving the wrong impression that all Lévy processes have infinite variance, have scaling properties, etc.

It should be clear to the reader that Lévy processes are much more general than Lévy flights and do not share most of their properties (except, of course, independence and stationarity of increments). In fact, all the examples of Lévy processes given in Chapter 4 have finite variance and are neither self-similar nor self-affine.

Non-Gaussian α-stable distributions are also referred to as "Lévy distributions" in the physics literature. Finally, to add to the confusion, some authors in the mathematics literature call the α-stable distribution with $\alpha = 1/2$ the "Lévy distribution." □

Paul Lévy

Lévy processes were named after the French mathematician Paul Lévy, one of the founding fathers of the modern theory of stochastic processes. Lévy was born into a family counting several mathematicians. His grandfather was a professor of mathematics and Paul's father, Lucien Lévy, was an examiner at the Ecole Polytechnique and wrote papers on geometry. Paul attended the Lycée Saint Louis in Paris where he achieved brilliant results, winning prizes not only in mathematics but also in Greek, chemistry and physics. He ranked first in the entrance examination to the Ecole Normale Supérieure and second for entry to the Ecole Polytechnique. He chose to attend the Ecole Polytechnique and in 1905, while still an undergraduate there, he published his first paper on semi-convergent series. In 1919 Lévy was asked to give three lectures at the École Polytechnique on "... notions of calculus of probabilities and the role of Gaussian law in the theory of errors." Taylor writes in [377]: "At that time there was no mathematical theory of probability — only a collection of small computational problems. Now it is a fully-fledged branch of mathematics using techniques from all branches of modern analysis and making its own contribution of ideas, problems, results and useful machinery to be applied elsewhere. If there is one person who has influenced the establishment and growth of probability theory more than any other, that person must be Paul Lévy."

This was the beginning of his lifelong interest in probability theory, which lead to the discovery of a wealth of results, many of which have become today standard material for undergraduate and graduate courses in probability theory. He made major contributions to the study of Gaussian variables and processes, the law of large numbers, the central limit theorem, stable laws, infinitely divisible laws and pioneered the study of processes with independent and stationary increments, now known as Lévy processes. The book he wrote on this topic, *Théorie de l'addition des variables aléatoires*, has served as an inspiration to many researchers in probability and physics, where stable processes with independent increments have become known as Lévy flights. He pioneered the study of the properties of Brownian paths, in which he introduced the notion of local time. These studies culminated in his classic book *Processus stochastiques et mouvement Brownien* [260].

Michel Loève, in [264], gives a vivid description of Lévy's contributions: "Paul Lévy was a painter in the probabilistic world. Like the very great painting geniuses, his palette was his own and his paintings transmuted forever our vision of reality... His three main, somewhat overlapping, periods were: the limit laws period, the great period of additive processes and of martingales painted in pathtime colours, and the Brownian pathfinder period."

Although he was a contemporary of Kolmogorov, Lévy did not adopt the axiomatic approach to probability. Joseph Doob writes of Lévy: "[Paul Lévy] is not a formalist. It is typical of his approach to mathematics that he defines the random variables of a stochastic process successively rather than postulating a measure space and a family of functions on it with stated properties, that he is not sympathetic with the delicate formalism that discriminates between the Markov and strong Markov properties, and that he rejects the idea that the axiom of choice is a separate axiom which need not be accepted. He has always traveled an independent path, partly because he found it painful to follow the ideas of others."

This attitude was in strong contrast to the mathematicians of his time, especially in France where the Bourbaki movement dominated the academic scene. Adding this to the fact that probability theory was not regarded as a branch of mathematics by many of his contemporary mathematicians, one can see why his ideas did not receive in France the attention they deserved at the time of their publication. P.A. Meyer writes: "Malgré son titre de professeur, malgré son élection à l'Institut [...], Paul Lévy a été méconnu en France. Son oeuvre y était considérée avec condéscendance, et on entendait fréquemment dire que 'ce n'était pas un mathématicien."'[4]

[4]Authors translation: Although he was a professor and a member of the Institut [i.e., the Academy of Sciences], Paul Lévy was not well recognized in France. His work was not highly considered and one frequently heard that 'he was not a mathematician'.

However, Paul Lévy's work was progressively recognized at an international level. The first issue of *Annals of Probability*, an international journal of probability theory, was dedicated to his memory in 1973, two years after his death.

An excellent mathematical biography is given by Loève in [264]. For those who read French, the Web site:

`http://www.annales.org/archives/x/paullevy.html`

contains interesting biographical notes by Benoit Mandelbrot, Paul-André Meyer and Jacques Neveu. Paul Lévy has also written a scientific autobiography [253]. Lévy's role in the discovery of the central limit theorem is discussed by LeCam in [247]. Other biographies include [233, 377, 74, 354]. The collected papers of Paul Lévy have been published in six volumes by Ecole Polytechnique [254, 255, 256, 257, 258, 259].

Chapter 4

Building Lévy processes

Having discussed general properties of Lévy processes in Chapter 13, we will now give some tractable examples of Lévy processes which can be used in model building. Rather than giving an exhaustive inventory of all Lévy processes that can be found in the financial modelling literature, we will discuss various transformations which allow to build new Lévy processes from known ones, thus showing the relations between these models.

4.1 Model building with Lévy processes

Let us start by some general considerations on Lévy processes, their use as models for price dynamics and different ways to specify such models.

4.1.1 "Jump-diffusions" vs. infinite activity Lévy processes

Financial models with jumps fall into two categories. In the first category, called *jump-diffusion* models, the "normal" evolution of prices is given by a diffusion process, punctuated by jumps at random intervals. Here the jumps represent rare events — crashes and large drawdowns. Such an evolution can be represented by modelling the (log-)price as a Lévy process with a nonzero Gaussian component and a jump part, which is a compound Poisson process with finitely many jumps in every time interval. Examples of such models are the Merton jump-diffusion model with Gaussian jumps [291] and the Kou model with double exponential jumps [238]. In Chapter 15, we will see a model combining compound Poisson jumps and stochastic volatility: the Bates model [41]. In these models, the dynamical structure of the process is easy to understand and describe, since the distribution of jump sizes is known. They are easy to simulate and efficient Monte Carlo methods for pricing path-dependent options can be used. Models of this type also perform quite well for the purposes of implied volatility smile interpolation (see Chapter 13). However they rarely lead to closed-form densities: statistical estimation and computation of moments or quantiles may be quite difficult.

The second category consists of models with infinite number of jumps in

TABLE 4.1: Compound Poisson or infinite activity: a comparison of two modelling approaches

Jump-diffusion models	**Infinite activity models**
Must contain a Brownian component.	Do not necessarily contain a Brownian component.
Jumps are rare events.	The process moves essentially by jumps.
Distribution of jump sizes is known.	"Distribution of jump sizes" does not exist: jumps arrive infinitely often.
Perform well for implied volatility smile interpolation.	Give a realistic description of the historical price process.
Densities not known in closed form.	Closed form densities available in some cases.
Easy to simulate.	In some cases can be represented via Brownian subordination, which gives additional tractability.

every interval, which we will call *infinite activity* models. In these models, one does not need to introduce a Brownian component since the dynamics of jumps is already rich enough to generate nontrivial small time behavior [80] and it has been argued [270, 80, 160] that such models give a more realistic description of the price process at various time scales. In addition, many models from this class can be constructed via Brownian subordination (this point will be addressed in detail below), which gives them additional analytical tractability compared to jump-diffusion models.

One could also consider pure jump processes of finite activity without diffusion component [320] but these models do not lead to a realistic description of price dynamics.

Table 4.1 compares the advantages and drawbacks of these two categories. It should be kept in mind that since the price process is observed on a discrete grid, it is difficult if not impossible to see empirically to which category the price process belongs. The choice is more a question of modelling convenience than an empirical one.

There are three convenient ways to define a parametric Lévy process, summarized in Table 4.2.

The first approach is to obtain a Lévy process by subordinating a Brownian motion with an independent increasing Lévy process. Here the characteristic function of the resulting process can be obtained immediately, but we do not always have an explicit formula for the Lévy measure. Due to the conditionally Gaussian structure of the process, simulation and some computations can be considerably simplified (for instance, call option price can be expressed as an integral involving Black-Scholes prices). The interpretation of the subordinator as a "business time" [162] makes models of this type easier to understand and interpret. Multidimensional extensions are also possible: one can take a multidimensional Brownian motion and change the time scale of all components with the same subordinator. However, we will see in Chapter 5 that dependence structures obtained using this method are rather limited and sometimes have dependence properties with undesirable features.

The second approach is to specify the Lévy measure directly. We will illustrate it later in this chapter using the example of tempered stable process. This approach provides a dynamic vision of the Lévy process because we model directly the jump structure and we know, via the Lévy-Khinchin formula, the distribution of the process at any time, although sometimes it is not very explicit.

The third approach is to specify an infinitely divisible density as the density of increments at a given time scale, say Δ. Generalized hyperbolic processes (see Section 4.6) can be constructed in this way. In this approach it is easy to simulate the increments of the process at the same time scale and to estimate parameters of the distribution if data are sampled with the same period Δ, but in general the Lévy measure is not known. Therefore, unless this distribution belongs to some parametric class closed under convolution, we do not know the law of the increments at other time scales. In particular, given an infinitely divisible distribution it may not be easy to see whether the corresponding Lévy process has a Gaussian component or not, has finite or infinite activity, etc.

4.2 Building new Lévy processes from known ones

To construct new Lévy processes, we use three basic types of transformations, under which the class of Lévy processes is invariant: linear transformations, subordination (time changing a Lévy process with another increasing Lévy process) and exponential tilting of the Lévy measure. We start with a rather intuitive result about linear transformation of Lévy processes.

4.2.1 Linear transformations

THEOREM 4.1

Let $(X_t)_{t\geq 0}$ be a Lévy process on $\mathbb{R}^d$ with characteristic triplet (A, ν, γ) and let M be an $n \times d$ matrix. Then $Y_t = MX_t$ is a Lévy process on $\mathbb{R}^n$ with

TABLE 4.2: Three approaches to building Lévy processes

Brownian subordination	**Specifying the Lévy measure**	**Specifying probability density for $t = \Delta$**
Interpretation as "Brownian motion in business time".	Clear vision of the pathwise properties.	Structure of jumps is not known.
Simulation is easy if we know how to simulate the subordinator.	Simulation is quite involved.	Simulation is easy on a grid of size Δ.
Estimation via maximum likelihood may be difficult.	Estimation can be done by approximating the transition density.	Estimation is easy for data with sampling interval Δ.
Multivariate generalizations possible using multidimensional Brownian motion.	Rich variety of models.	The infinite divisibility of a given model may be diffucult to prove.

characteristic triplet (A_Y, ν_Y, γ_Y) *where*

$$A_Y = MAM^t, \tag{4.1}$$

$$\nu_Y(B) = \nu(\{x : Mx \in B\}), \quad \forall B \in \mathcal{B}(R^n), \tag{4.2}$$

$$\gamma_Y = M\gamma + \int_{\mathbb{R}^n} y(1_{\{|y|\leq 1\}}(y) - 1_{S_1}(y))\nu_Y(dy). \tag{4.3}$$

S_1 *is the image by* M *of a unit ball in* $\mathbb{R}^d$: $S_1 = \{Mx : |x| \leq 1\}$.

PROOF $(Y_t)_{t\geq 0}$ is clearly a Lévy process on $\mathbb{R}^n$, so we only need to prove that (4.2) defines a Lévy measure, that the integral in (4.3) is finite and that (A_Y, ν_Y, γ_Y) is the characteristic triplet of Y. The measure ν_Y is a positive measure on $\mathbb{R}^n$. It satisfies

$$\int_{\mathbb{R}^n} (|y|^2 \wedge 1)\nu_Y(dy) = \int_{\mathbb{R}^d} (|Mx|^2 \wedge 1)\nu(dx) < \infty$$

because the norm of M is finite. Hence, ν_Y is a Lévy measure on $\mathbb{R}^n$.

Let us now turn to the integral in (4.3). It is sufficient to integrate not over the whole $\mathbb{R}^n$ but over its subspace $S \subseteq \mathbb{R}^n$ defined by $S = \{Mx : x \in \mathbb{R}^d\}$ because the measure ν_Y is concentrated on this subspace. Therefore we can multiply the integrand in (4.3) by the indicator function of this subspace, obtaining

$$f(y) = y(1_{\{|y|\leq 1\}}(y) - 1_{S_1}(y))1_S(y).$$

We will show that the integral is finite by proving that there exist two constants $0 < c < C < \infty$ such that $f(y) = 0$ for all y such that $|y| \geq C$ and $f(y) = 0$ for all y such that $|y| \leq c$.

C can be taken equal to any constant greater than $\max(|M|, 1)$. Indeed, if $|y| > \max(|M|, 1)$ then one cannot find an $x \in \mathbb{R}^d$ satisfying at the same time $|x| \leq 1$ and $Mx = y$, therefore, $1_{\{|y|\leq 1\}}(y)$ and $1_{S_1}(y)$ are both equal to zero and $f(y) = 0$.

On the other hand, the mapping $\tilde{M} : \mathbb{R}^d \to S$ which to every $x \in \mathbb{R}^d$ associates Mx is a continuous linear surjection between Banach spaces, hence, by the open mapping theorem, it maps open sets into open sets. Therefore, the set $\tilde{S}_1 = \{Mx : x \in \mathbb{R}^d, |x| < 1\}$ is open in S. This means that the set $\tilde{S}_1 \cap \{y \in S : |y| < 1\}$ is also open in S. Since this set contains zero, we have found a neighborhood of zero in S such that $f(y) = 0$ for every y in this neighborhood. This means that there exists $c > 0$ such that $f(y) = 0$ for all y such that $|y| \leq c$ and the finiteness of the integral is shown.

To complete the proof, we use the Lévy-Khinchin formula for process X_t:

$$\begin{aligned}
&E[e^{iu.MX_t}] = E[e^{iM^t u.X_t}] \\
&= \exp t\{-\frac{1}{2}M^t u.AM^t u + i\gamma.M^t u + \int_{\mathbb{R}^d}(e^{iM^t u.x} - 1 - iM^t u.x1_{|x|\leq 1})\nu(dx)\} \\
&\qquad = \exp t\{-\frac{1}{2}u.A_Y u + i\gamma_Y.u + \int_{\mathbb{R}^n}(e^{iu.x} - 1 - iu.x1_{|x|\leq 1})\nu_Y(dx)\}.
\end{aligned}$$

□

Example 4.1 Sums of independent Lévy processes
Let $(X_t)_{t\geq 0}$ and $(Y_t)_{t\geq 0}$ be two independent Lévy processes with characteristic triplets (A_1, ν_1, γ_1) and (A_2, ν_2, γ_2). Using Theorem 4.1 with $M = \begin{pmatrix} 1 \\ 1 \end{pmatrix}$ and Proposition 5.3 from Chapter 5 we obtain that $X_t + Y_t$ is a Lévy process with characteristic triplet (A, ν, γ) where

$$\begin{aligned}
&A = A_1 + A_2, \\
&\nu(B) = \nu_1(B) + \nu_2(B) \quad \forall B \in \mathcal{B}(\mathbb{R}), \\
&\gamma = \gamma_1 + \gamma_2 - \int_{[-\sqrt{2},-1]\cup[1,\sqrt{2}]} y\nu(dy).
\end{aligned}$$

□

For Lévy processes of finite variation, Theorem 4.1 can be simplified. Namely, let $(X_t)_{t\geq 0}$ be a Lévy process of finite variation on $\mathbb{R}^d$ with characteristic function

$$E[e^{iz.X_t}] = \exp t\left\{ib.z + \int_{\mathbb{R}^d}(e^{iz.x} - 1)\nu(dx)\right\} \tag{4.4}$$

and let M be an $n \times d$ matrix. Then $Y_t = MX_t$ is a Lévy process on $\mathbb{R}^n$ with Lévy measure $\nu_Y(B) = \nu(\{x : Mx \in B\})$ and $b_Y = Mb$.

4.2.2 Subordination

Let $(S_t)_{t\geq 0}$ be a subordinator, that is, a Lévy process satisfying one of the equivalent conditions of Proposition 3.10, which means in particular that its trajectories are almost surely increasing. Since S_t is a positive random variable for all t, we describe it using Laplace transform rather than Fourier transform. Let the characteristic triplet of S be $(0, \rho, b)$. Then the moment generating function of S_t is

$$E[e^{uS_t}] = e^{t\ l(u)} \quad \forall u \leq 0, \quad \text{where} \quad l(u) = bu + \int_0^\infty (e^{ux} - 1)\rho(dx). \tag{4.5}$$

We call $l(u)$ the Laplace exponent of S. Since process S is increasing it can be interpreted as a "time deformation" and used to "time change" other Lévy processes as shown by the following theorem.

THEOREM 4.2 Subordination of a Lévy process
Fix a probability space $(\Omega, \mathcal{F}, \mathbb{P})$. Let $(X_t)_{t\geq 0}$ be a Lévy process on $\mathbb{R}^d$ with characteristic exponent $\Psi(u)$ and triplet (A, ν, γ) and let $(S_t)_{t\geq 0}$ be a subordinator with Laplace exponent $l(u)$ and triplet $(0, \rho, b)$. Then the process $(Y_t)_{t\geq 0}$ defined for each $\omega \in \Omega$ by $Y(t, \omega) = X(S(t, \omega), \omega)$ is a Lévy process. Its characteristic function is

$$E[e^{iuY_t}] = e^{tl(\Psi(u))}, \tag{4.6}$$

i.e., the characteristic exponent of Y is obtained by composition of the Laplace exponent of S with the characteristic exponent of X. The triplet (A^Y, ν^Y, γ^Y) of Y is given by

$$A^Y = bA,$$

$$\nu^Y(B) = b\nu(B) + \int_0^\infty p_s^X(B)\rho(ds), \quad \forall B \in \mathcal{B}(\mathbb{R}^d), \tag{4.7}$$

$$\gamma^Y = b\gamma + \int_0^\infty \rho(ds) \int_{|x|\leq 1} x p_s^X(dx), \tag{4.8}$$

where p_t^X is the probability distribution of X_t.

$(Y_t)_{t\geq 0}$ is said to be *subordinate* to the process $(X_t)_{t\geq 0}$.

PROOF Let us first prove that Y is a Lévy process. Denote by $\mathcal{F}_t^S$ the filtration of $(S_t)_{t\geq 0}$ with $\mathcal{F}^S \equiv \mathcal{F}_\infty^S$. For every sequence of times $t_0 <$

$t_1 < \ldots < t_n$ we obtain, using the independent increments property of X, Lévy-Khinchin formula for X and the independent increments property of S:

$$E\left[\prod_{i=1}^{n} e^{iu_i(X(S_{t_i})-X(S_{t_{i-1}}))}\right] = E\left\{E\left[\prod_{i=1}^{n} e^{iu_i(X(S_{t_i})-X(S_{t_{i-1}}))}\Big|\mathcal{F}^S\right]\right\}$$

$$= E\left\{\prod_{i=1}^{n} E\left[e^{iu_i(X(S_{t_i})-X(S_{t_{i-1}}))}\Big|\mathcal{F}^S\right]\right\} = E\left\{\prod_{i=1}^{n} e^{(S_{t_i}-S_{t_{i-1}})\Psi(u_i)}\right\}$$

$$= \prod_{i=1}^{n} E\left\{e^{(S_{t_i}-S_{t_{i-1}})\Psi(u_i)}\right\} = \prod_{i=1}^{n} E\left\{e^{iu_i(X(S_{t_i})-X(S_{t_{i-1}}))}\right\}.$$

Therefore, Y has independent increments. The stationarity of increments can be shown in the same way. To show that Y is continuous in probability, first observe that every Lévy process is *uniformly* continuous in probability, due to the stationarity of its increments. Further, for every $\varepsilon > 0$ and $\delta > 0$, one can write:

$$P\left\{|X(S_s) - X(S_t)| > \varepsilon\right\}$$
$$\leq P\left\{|X(S_s) - X(S_t)| > \varepsilon \Big| |S_s - S_t| < \delta\right\} + P\left\{|S_s - S_t| \geq \delta\right\}.$$

The first term can be made arbitrarily small simultaneously for all values of s and t by changing δ, because X is uniformly continuous in probability. As for the second term, its limit as $s \to t$ is always zero, because S is continuous in probability. Hence, $P\{|X(S_s) - X(S_t)| > \varepsilon\} \to 0$ as $s \to t$.

The formula (4.6) is easily obtained by conditioning on $\mathcal{F}^S$:

$$E\left[e^{iuX(S_t)}\right] = E\left\{E\left[e^{iuX(S_t)}\Big|\mathcal{F}^S\right]\right\} = E\left\{e^{S_t\Psi(u)}\right\} = e^{tl(\Psi(u))}.$$

For the detailed proof of the expression for the characteristic triplet of Y we refer the reader to [345, Theorem 30.1]. Here we will instead explain what is going on on a simple example. Suppose that S is a compound Poisson subordinator with characteristic triplet $(0, \rho, 0)$. Then Y is again a compound Poisson process with the same intensity, because it moves only by jumps and its jumps occur at the same times as those of S. Therefore its drift and Gaussian component are equal to zero. To compute its jump measure, suppose that S has a jump at t. Conditionally on $S_t - S_{t-} = s$, the size of jump in Y has the distribution p_s^X. Integrating with respect to jump measure of S, we obtain formula (4.7). Finally, rewriting the characteristic triplet of Y with respect to the truncation function $1_{|x|\leq 1}$, we obtain formula (4.8) for γ^Y. □

Example 4.2

A stable subordinator is an α-stable process with $\alpha \in (0, 1)$, Lévy measure concentrated on the positive half-axis and a nonnegative drift (such a process

is a subordinator because it satisfies the last condition of Proposition 3.10).
Let $(S_t)_{t\geq 0}$ be a stable subordinator with zero drift. Its Laplace exponent is

$$l(u) = c_1 \int_0^\infty \frac{e^{ux}-1}{x^{1+\alpha}}dx = -\frac{c_1\Gamma(1-\alpha)}{\alpha}(-u)^\alpha \tag{4.9}$$

for some constant $c_1 > 0$. Let $(X_t)_{t\geq 0}$ be a symmetric β-stable process on $\mathbb{R}$ with characteristic exponent $\Psi(u) = -c_2|u|^\beta$ for some constant $c_2 > 0$. Then the process $(Y_t)_{t\geq 0}$ subordinate to X by S has characteristic exponent $l(\Psi(u)) = -c|u|^{\beta\alpha}$, where $c = c_1 c_2 \Gamma(1-\alpha)/\alpha$, that is, Y is a symmetric stable process with index of stability $\beta\alpha$. In particular, when X is a Brownian motion, the subordinate process is 2α-stable. □

4.2.3 Tilting and tempering the Lévy measure

As we noted before, one way to specify a Lévy process is by giving an admissible Lévy triplet: in particular, the Lévy measure must verify the constraints:

$$\int_{|x|\leq 1} |x|^2 \nu(dx) < \infty \qquad \int_{|x|\geq 1} \nu(dx) < \infty.$$

Any transformation of the Lévy measure, respecting the integrability constraint above, will lead to a new Lévy process. Examples of such transformations are obtained by multiplying $\nu(.)$ by an exponential function. If there exists $\theta \in \mathbb{R}^d$ such that $\int_{|x|\geq 1} e^{\theta.x}\nu(dx) < \infty$ then the measure $\tilde{\nu}$ defined by

$$\tilde{\nu}(dx) := e^{\theta.x}\nu(dx) \tag{4.10}$$

is a Lévy measure. Then for any Lévy process $(X_t)_{t\geq 0}$ on $\mathbb{R}^d$ with characteristic triplet (A, ν, γ), the process with characteristic triplet $(A, \tilde{\nu}, \gamma)$ is also a Lévy process, called the Esscher transform of X. The transform given by (4.10) is called *exponential tilting* of the Lévy measure. Esscher transforms will be discussed in more detail in Chapter 9.

When $d = 1$, we can consider asymmetric version of this transformation: if ν is a Lévy measure on $\mathbb{R}$ then

$$\tilde{\nu}(dx) = \nu(dx)\left(1_{x>0}e^{-\lambda_+ x} + 1_{x<0}e^{-\lambda_-|x|}\right),$$

where λ_+ and λ_- are positive parameters, is also a Lévy measure and defines a Lévy process whose large jumps are "tempered," i.e., the tails of the Lévy measure are exponentially damped.

4.3 Models of jump-diffusion type

A Lévy process of jump-diffusion type has the following form:

$$X_t = \gamma t + \sigma W_t + \sum_{i=1}^{N_t} Y_i, \tag{4.11}$$

where $(N_t)_{t\geq 0}$ is the Poisson process counting the jumps of X and Y_i are jump sizes (i.i.d. variables). To define the parametric model completely, we must now specify the distribution of jump sizes $\nu_0(x)$. It is especially important to specify the tail behavior of ν_0 correctly depending on one's beliefs about behavior of extremal events, because as we have seen in Chapter 3, the tail behavior of the jump measure determines to a large extent the tail behavior of probability density of the process (cf. Propositions 3.13 and 3.14).

In the *Merton model* [291], jumps in the log-price X_t are assumed to have a Gaussian distribution: $Y_i \sim N(\mu, \delta^2)$. This allows to obtain the probability density of X_t as a quickly converging series. Indeed,

$$\mathbb{P}\{X_t \in A\} = \sum_{k=0}^{\infty} \mathbb{P}\{X_t \in A \big| N_t = k\}\mathbb{P}\{N_t = k\},$$

which entails that the probability density of X_t satisfies

$$p_t(x) = e^{-\lambda t} \sum_{k=0}^{\infty} \frac{(\lambda t)^k \exp\left\{-\frac{(x-\gamma t - k\mu)^2}{2(\sigma^2 t + k\delta^2)}\right\}}{k!\sqrt{2\pi(\sigma^2 t + k\delta^2)}}. \tag{4.12}$$

In a similar way, prices of European options in the Merton model can be obtained as a series where each term involves a Black-Scholes formula.

In the *Kou model* [238], the distribution of jump sizes is an asymmetric exponential with a density of the form

$$\nu_0(dx) = [p\lambda_+ e^{-\lambda_+ x} 1_{x>0} + (1-p)\lambda_- e^{-\lambda_-|x|} 1_{x<0}]dx \tag{4.13}$$

with $\lambda_+ > 0$, $\lambda_- > 0$ governing the decay of the tails for the distribution of positive and negative jump sizes and $p \in [0,1]$ representing the probability of an upward jump. The probability distribution of returns in this model has semi-heavy (exponential) tails. The advantage of this model compared to the previous one is that due to the memoryless property of exponential random variables, analytical expressions for expectations involving first passage times may be obtained [239]. Key properties of Merton and Kou models are summarized in Table 4.3.

TABLE 4.3: Two jump-diffusion models: the Merton model and the Kou model

	Merton model	**Kou model**
Model type	Compound Poisson jumps + Brownian motion	
Parameters (excluding drift)	4 parameters: σ — diffusion volatility, λ — jump intensity, μ — mean jump size and δ — standard deviation of jump size	5 parameters: σ — diffusion volatility, λ — jump intensity, λ_+, λ_-, p — parameters of jump size distribution
Lévy density	$\nu(x) = \frac{\lambda}{\delta\sqrt{2\pi}} \exp\{-\frac{(x-\mu)^2}{2\delta^2}\}$	$\nu(x) = p\lambda\lambda_+ e^{-\lambda_+ x} 1_{x>0} + (1-p)\lambda\lambda_- e^{-\lambda_-\lvert x\rvert} 1_{x<0}$
Characteristic exponent	$\Psi(u) = -\frac{\sigma^2 u^2}{2} + ibu + \lambda\{e^{-\delta^2 u^2/2 + i\mu u} - 1\}$	$\Psi(u) = -\frac{\sigma^2 u^2}{2} + ibu + iu\lambda\{\frac{p}{\lambda_+ - iu} - \frac{1-p}{\lambda_- + iu}\}$
Probability density	Admits a series expansion (4.12)	Not available in closed form
Cumulants:		
$E[X_t]$	$t(b + \lambda\mu)$	$t(b + \lambda p/\lambda_+ - \lambda(1-p)/\lambda_-)$
$\mathrm{Var}\, X_t$	$t(\sigma^2 + \lambda\delta^2 + \lambda\mu^2)$	$t(\sigma^2 + \lambda p/\lambda_+^2 + \lambda(1-p)/\lambda_-^2)$
c_3	$t\lambda(3\delta^2\mu + \mu^3)$	$t\lambda(p/\lambda_+^3 - (1-p)/\lambda_-^3)$
c_4	$t\lambda\{3\delta^3 + 6\mu^2\delta^2 + \mu^4\}$	$t\lambda(p/\lambda_+^4 + (1-p)/\lambda_-^4)$
Tail behavior of probability density	Tails are heavier than Gaussian but all exponential moments are finite	Semi-heavy (exponential) tails: $p(x) \sim e^{-\lambda_+ x}$ when $x \to +\infty$ and $p(x) \sim e^{-\lambda_-\lvert x\rvert}$ when $x \to -\infty$

4.4 Building Lévy processes by Brownian subordination

4.4.1 General results

Let $(S_t)_{t\geq 0}$ be a subordinator with Laplace exponent $l(u)$ and let $(W_t)_{t\geq 0}$ be a Brownian motion independent from S. Subordinating Brownian motion with drift μ by the process S, we obtain a new Lévy process $X_t = \sigma W(S_t) + \mu S_t$. This process is a Brownian motion if it is observed on a new time scale, that is, the stochastic time scale given by S_t. This time scale has the financial interpretation of business time (see [161]), that is, the integrated rate of information arrival. This interpretation makes models based on subordinated Brownian motion easier to understand than general Lévy models. Formula (4.6) entails that X has characteristic exponent $\Psi(u) = l(-u^2\sigma^2/2 + i\mu u)$. This allows in particular to compute cumulants of X_t from those of S_t. Consider the symmetric case $\mu = 0$. Then X_t is symmetric, therefore has zero mean and skewness, and one can easily compute its variance and excess kurtosis:

$$\begin{aligned} \operatorname{Var} X_t &= \sigma^2 E[S_t], \\ \kappa(X_t) &= \frac{3 \operatorname{Var} S_t}{E[S_t]^2}. \end{aligned}$$

Therefore, X_t is leptocurtic if the subordinator is not a deterministic process.

Although the representation via Brownian subordination is a nice property, which makes the model easier to understand and adds tractability, it imposes some important limitations on the form of the Lévy measure. The following theorem characterizes Lévy measures of processes that can be represented as subordinated Brownian motion with drift. We recall that a function $f : [a, b] \to \mathbb{R}$ is called completely monotonic if all its derivatives exist and $(-1)^k \frac{d^k f(u)}{du^k} > 0$ for all $k \geq 1$.

THEOREM 4.3

Let ν be a Lévy measure on $\mathbb{R}$ and $\mu \in \mathbb{R}$. There exists a Lévy process $(X_t)_{t\geq 0}$ with Lévy measure ν such that $X_t = W(Z_t) + \mu Z_t$ for some subordinator $(Z_t)_{t\geq 0}$ and some Brownian motion $(W_t)_{t\geq 0}$ independent from Z if and only if the following conditions are satisfied:

1. *ν is absolutely continuous with density $\nu(x)$.*
2. *$\nu(x)e^{-\mu x} = \nu(-x)e^{\mu x}$ for all x.*
3. *$\nu(\sqrt{u})e^{-\mu\sqrt{u}}$ is a completely monotonic function on $(0, \infty)$.*

This theorem allows to describe the jump structure of a process, that can be represented as time changed Brownian motion with drift. For example, the Lévy measure of such a process has an exponentially tilted version that is symmetric on $\mathbb{R}$. Since the exponential tilting mainly affects the big jumps, this means that small jumps of such a process will always be symmetric.

Let ν be a Lévy measure on $\mathbb{R}^d$. It can be the Lévy measure of a subordinated Brownian motion (without drift) if and only if it is symmetric and $\nu(\sqrt{u})$ is a completely monotonic function on $(0,\infty)$. Furthermore, consider a subordinator with zero drift and Lévy measure ρ. Formula (4.7) entails that a Brownian motion with drift μ time changed by this subordinator will have Lévy density $\nu(x)$ given by

$$\nu(x) = \int_0^\infty e^{-\frac{(x-\mu t)^2}{2t}} \frac{\rho(dt)}{\sqrt{2\pi t}}. \tag{4.14}$$

We can symbolically denote this operation by $BS_\mu(\rho) = \nu$, where BS stands for Brownian subordination. The inverse transform is denoted by $BS_\mu^{-1}(\nu) = \rho$. Then (4.14) allows to write

$$BS_\mu^{-1}(\nu) = e^{\mu^2 t/2} BS_0^{-1}(\nu e^{-\mu x}). \tag{4.15}$$

Hence, we can deduce the time changed Brownian motion representation for an exponentially tilted Lévy measure from the representation for its symmetric modification.

PROOF of Theorem 4.3 The *only if* part. The absolute continuity of ν is a direct consequence of (4.7), because the Gaussian probability distribution is absolutely continuous. Omitting the constant factor, the formula (4.14) can be rewritten as

$$\nu(x)e^{-\mu x} = \int_0^\infty e^{-\frac{x^2}{2t}} e^{-\frac{\mu^2 t}{2}} t^{-1/2} \rho(dt),$$

which shows that $\nu(x)e^{-\mu x}$ must be symmetric. Further, by making the variable change $u = x^2/2$ and $s = 1/t$ we obtain (to simplify the notation we suppose that ρ has a density):

$$\int_0^\infty e^{-us} e^{-\frac{\mu^2}{2s}} s^{-3/2} \rho(1/s) ds = \nu(\sqrt{2u}) e^{-\mu\sqrt{2u}}, \tag{4.16}$$

which shows that $\nu(\sqrt{2u})e^{-\mu\sqrt{2u}}$ is the Laplace transform of a positive measure and therefore, by Bernstein's theorem (see [141, Volume 2, page 439]) it is a completely monotonic function.

The *if* part. Using absolute continuity of ν and since Bernstein's theorem is a necessary and sufficient result, we can conclude, using the same reasoning as in the first part of the proof, that there exists a positive measure ρ on $(0,\infty)$ such that (4.14) holds. Therefore it remains to prove that $\int_0^\infty (t \wedge 1)\rho(dt) < \infty$.

We suppose without loss of generality that $\mu \geq 0$. The subordinated Lévy measure ν satisfies $\int_{\mathbb{R}}(x^2 \wedge 1)\nu(dx) < \infty$. Using Fubini's theorem we obtain

$$\int_0^\infty \rho(dt)t^{-1/2} \int_{\mathbb{R}} (x^2 \wedge 1)e^{-\frac{(x-\mu t)^2}{2t}} dx = \int_0^\infty \rho(dt)g(t) < \infty.$$

To complete the proof of theorem we must show that $g(t) \geq c(t \wedge 1)$ for some constant $c > 0$. After variable change we get:

$$\begin{aligned} g(t) = \int_{\mathbb{R}} ((u\sqrt{t} + \mu t)^2 \wedge 1)e^{-u^2/2}du &\geq \int_0^\infty ((u\sqrt{t} + \mu t)^2 \wedge 1)e^{-u^2/2}du \\ &\geq \int_0^\infty (u^2 t \wedge 1)e^{-u^2/2}du. \end{aligned}$$

If $t \leq 1$ then

$$g(t) \geq \int_0^\infty (u^2 t \wedge 1)e^{-u^2/2}du \geq \int_0^1 (u^2 t \wedge 1)e^{-u^2/2}du = \int_0^1 u^2 t e^{-u^2/2}du = tc_1$$

with $c_1 > 0$. On the other hand, if $t > 1$ then

$$\begin{aligned} g(t) \geq \int_0^\infty (u^2 t \wedge 1)e^{-u^2/2}du &\geq \int_1^\infty (u^2 t \wedge 1)e^{-u^2/2}du \\ &= \int_1^\infty e^{-u^2/2}du = c_2 > 0. \end{aligned}$$

Therefore, there exists a constant $c > 0$ such that $g(t) \geq c(t \wedge 1)$ and the proof is completed. □

4.4.2 Subordinating processes

Let us consider the tempered stable subordinator, that is, an exponentially tempered version of the stable subordinator, discussed in Example 4.2. It is a three-parameter process with Lévy measure

$$\rho(x) = \frac{ce^{-\lambda x}}{x^{\alpha+1}} 1_{x>0}, \tag{4.17}$$

where c and λ are positive constants and $1 > \alpha \geq 0$. For greater generality we include the case $\alpha = 0$ (the gamma process) although it cannot be obtained from a stable subordinator via exponential tilting. A tempered stable subordinator is, of course, a tempered stable process in the sense of formula (4.26). The parameter c alters the intensity of jumps of all sizes simultaneously; in other words, it changes the time scale of the process, λ fixes the decay rate of big jumps and α determines the relative importance of small jumps in the path of the process. The Laplace exponent of tempered stable subordinator in the general case ($\alpha \neq 0$) is

$$l(u) = c\Gamma(-\alpha)\{(\lambda - u)^\alpha - \lambda^\alpha\} \tag{4.18}$$

TABLE 4.4: Gamma subordinator and inverse Gaussian subordinator

Subordinator	**Gamma process**	**Inverse Gaussian process**
Lévy density	$\rho(x) = \frac{ce^{-\lambda x}}{x} 1_{x>0}$	$\rho(x) = \frac{ce^{-\lambda x}}{x^{3/2}} 1_{x>0}$
Laplace transform	$E[e^{uS_t}] = (1 - u/\lambda)^{-ct}$	$E[e^{uS_t}] = e^{-2ct\sqrt{\pi}(\sqrt{\lambda-u}-\sqrt{\lambda})}$
Probability density	$p_t(x) = \frac{\lambda^{ct}}{\Gamma(ct)} x^{ct-1} e^{-\lambda x}$ for $x > 0$	$p_t(x) = \frac{ct}{x^{3/2}} e^{2ct\sqrt{\pi\lambda}} e^{-\lambda x - \pi c^2 t^2/x}$ for $x > 0$

and $l(u) = -c\log(1 - u/\lambda)$ if $\alpha = 0$. The probability density of tempered stable subordinator is only known in explicit form for $\alpha = 1/2$ (inverse Gaussian subordinator) and $\alpha = 0$ (gamma subordinator). These two cases are compared in Table 4.4.

The tempered stable subordinator possesses the following scaling property. Let $(S_t(\alpha, \lambda, c))_{t\geq 0}$ be a tempered stable subordinator with parameters α, λ and c. Then for every $r > 0$, $rS_t(\alpha, \lambda, c)$ has the same law as $S_{r^\alpha t}(\alpha, \lambda/r, c)$. Because of this scaling property and the scaling property of Brownian motion (rW_t has the same law as $W_{r^2 t}$), in subordinated models it is sufficient to consider only tempered stable subordinators with $E[S_t] = t$, which in this case form a two-parameter family. For all computations related to characteristic function, moments and cumulants it is convenient to use the parameterization

$$\rho(x) = \frac{1}{\Gamma(1-\alpha)} \left(\frac{1-\alpha}{\kappa}\right)^{1-\alpha} \frac{e^{-(1-\alpha)x/\kappa}}{x^{1+\alpha}}, \tag{4.19}$$

where α is the index of stability. In this new parameterization κ is equal to the variance of subordinator at time 1. Since the expectation of the subordinator at time 1 is equal to 1, κ actually determines how random the time change is, and the case $\kappa = 0$ corresponds to a deterministic function. In the variance gamma case this formula simplifies to

$$\rho(x) = \frac{1}{\kappa} \frac{e^{-x/\kappa}}{x} \tag{4.20}$$

and in the inverse Gaussian case

$$\rho(x) = \frac{1}{\sqrt{2\pi\kappa}} \frac{e^{-\frac{x}{2\kappa}}}{x^{3/2}}. \tag{4.21}$$

4.4.3 Models based on subordinated Brownian motion

By time changing an independent Brownian motion (with volatility σ and drift θ) by a tempered stable subordinator, we obtain the so-called *normal tempered stable* process (it is customary in the literature to name processes

TABLE 4.5: Two models based on Brownian subordination: variance gamma process and normal inverse Gaussian process

Model name	**Variance gamma**	**Normal inverse Gaussian**
Model type	Finite variation process with infinite but relatively low activity of small jumps	Infinite variation process with stable-like ($\alpha = 1$) behavior of small jumps
Parameters (excluding drift)	3 parameters: σ and θ — volatility and drift of Brownian motion and κ — variance of the subordinator	
Lévy measure	$\nu(x) = \frac{1}{\kappa\|x\|}e^{Ax-B\|x\|}$ with $A = \frac{\theta}{\sigma^2}$ and $B = \frac{\sqrt{\theta^2+2\sigma^2/\kappa}}{\sigma^2}$	$\nu(x) = \frac{C}{\|x\|}e^{Ax}K_1(B\|x\|)$ with $C = \frac{\sqrt{\theta^2+\sigma^2/\kappa}}{2\pi\sigma\sqrt{\kappa}}$ and denoting $A = \frac{\theta}{\sigma^2}$ and $B = \frac{\sqrt{\theta^2+\sigma^2/\kappa}}{\sigma^2}$
Characteristic exponent	$\Psi(u) = -\frac{1}{\kappa}\log(1 + \frac{u^2\sigma^2\kappa}{2} - i\theta\kappa u)$	$\Psi(u) = \frac{1}{\kappa} - \frac{1}{\kappa}\sqrt{1 + u^2\sigma^2\kappa - 2i\theta u\kappa}$
Probability density	$p_t(x) = C\|x\|^{\frac{t}{\kappa}-\frac{1}{2}}e^{Ax}K_{\frac{t}{\kappa}-\frac{1}{2}}(B\|x\|)$ with $C = \sqrt{\frac{\sigma^2\kappa}{2\pi}}\frac{(\theta^2\kappa+2\sigma^2)^{\frac{1}{4}-\frac{\theta}{2\kappa}}}{\Gamma(t/\kappa)}$	$p_t(x) = Ce^{Ax}\frac{K_1(B\sqrt{x^2+t^2\sigma^2/\kappa})}{\sqrt{x^2+t^2\sigma^2/\kappa}}$ with $C = \frac{t}{\pi}e^{t/\kappa}\sqrt{\frac{\theta^2}{\kappa\sigma^2} + \frac{1}{\kappa^2}}$
Cumulants:		
$E[X_t]$	θt	θt
$\mathrm{Var}\, X_t$	$\sigma^2 t + \theta^2\kappa t$	$\sigma^2 t + \theta^2\kappa t$
c_3	$3\sigma^2\theta\kappa t + 2\theta^3\kappa^2 t$	$3\sigma^2\theta\kappa t + 3\theta^3\kappa^2 t$
c_4	$3\sigma^4\kappa t + 6\theta^4\kappa^3 t + 12\sigma^2\theta^2\kappa^2 t$	$3\sigma^4\kappa t + 15\theta^4\kappa^3 t + 18\sigma^2\theta^2\kappa^2$
Tail behavior	Both Lévy density and probability density have exponential tails with decay rates $\lambda_+ = B - A$ and $\lambda_- = B + A$.	

resulting from Brownian subordination by adding the word "normal" to the name of subordinator). Its characteristic exponent is

$$\Psi(u) = \frac{1-\alpha}{\kappa\alpha}\left\{1 - \left(1 + \frac{\kappa(u^2\sigma^2/2 - i\theta u)}{1-\alpha}\right)^\alpha\right\} \tag{4.22}$$

in the general case and

$$\Psi(u) = -\frac{1}{\kappa}\log\{1 + \frac{u^2\sigma^2\kappa}{2} - i\theta\kappa u\} \tag{4.23}$$

in the variance gamma case ($\alpha = 0$).

The Lévy measure of a normal tempered stable process can be computed using Equation (4.7). It has a density $\nu(x)$ given by

$$\begin{aligned}
\nu(x) &= \frac{2c}{\sigma\sqrt{2\pi}} \int_0^\infty e^{-\frac{(x-\theta t)^2}{2t\sigma^2} - \lambda t} \frac{dt}{t^{\alpha+3/2}} \\
&= \frac{2c}{\sigma\sqrt{2\pi}} \left(\frac{\sqrt{\theta^2+2\lambda\sigma^2}}{|x|}\right)^{\alpha+1/2} e^{\theta x/\sigma^2} K_{\alpha+1/2}\left(\frac{|x|\sqrt{\theta^2+2\lambda\sigma^2}}{\sigma^2}\right) \\
&= \frac{C(\alpha,\kappa,\sigma,\theta)}{|x|^{\alpha+1/2}} e^{\theta x/\sigma^2} K_{\alpha+1/2}\left(\frac{|x|\sqrt{\theta^2+\frac{2}{\kappa}\sigma^2(1-\alpha)}}{\sigma^2}\right),
\end{aligned} \tag{4.24}$$

where $C(\alpha,\kappa,\sigma,\theta) = \frac{2}{\Gamma(1-\alpha)\sigma\sqrt{2\pi}}\left(\frac{1-\alpha}{\kappa}\right)^{1-\alpha}(\theta^2+\frac{2}{\kappa}\sigma^2(1-\alpha))^{\alpha/2+1/4}$ and K is the modified Bessel function of the second kind (see Appendix A.1). In accordance with Theorem 4.3, this measure is an exponential tilt of a symmetric measure. Introducing tail decay rates $\lambda_+ = \frac{1}{\sigma^2}\left(\sqrt{\theta^2+\frac{2}{\kappa}\sigma^2(1-\alpha)}-\theta\right)$ and $\lambda_- = \frac{1}{\sigma^2}\left(\sqrt{\theta^2+\frac{2}{\kappa}\sigma^2(1-\alpha)}+\theta\right)$, we can rewrite this measure in the following form:

$$\nu(x) = \frac{C}{|x|^{\alpha+1/2}} e^{x(\lambda_- - \lambda_+)/2} K_{\alpha+1/2}(|x|(\lambda_- + \lambda_+)/2). \tag{4.25}$$

From the asymptotic behavior formulae for K, we deduce that

$$\begin{aligned}
\nu(x) &\sim \frac{1}{|x|^{2\alpha+1}}, \quad \text{when } x \to 0, \\
\nu(x) &\sim \frac{1}{|x|^{\alpha+1}} e^{-\lambda_+ x}, \quad \text{when } x \to \infty, \\
\nu(x) &\sim \frac{1}{|x|^{\alpha+1}} e^{-\lambda_- |x|}, \quad \text{when } x \to -\infty,
\end{aligned}$$

that is, the Lévy measure has stable-like behavior near zero and exponential decay with decay rates λ_+ and λ_- at the tails.

Because the probability density of tempered stable subordinator is known in closed form for $\alpha = 1/2$ and $\alpha = 0$, the corresponding subordinated processes are also more mathematically tractable and easier to simulate and therefore they have been widely used in the literature. Namely, the variance gamma process has been used as a model for the logarithm of stock prices in [83, 271] and the normal inverse Gaussian process (NIG) has been used for financial modelling in [342, 32, 30]. Properties of these two important models are summed up in Table 4.5.

4.5 Tempered stable process

The tempered stable process is obtained by taking a one-dimensional stable process and multiplying the Lévy measure with a decreasing exponential on each half of the real axis. After this exponential softening, the small jumps keep their initial stable-like behavior whereas the large jumps become much less violent. A tempered stable process is thus a Lévy process on $\mathbb{R}$ with no Gaussian component and a Lévy density of the form

$$\nu(x) = \frac{c_-}{|x|^{1+\alpha}} e^{-\lambda_-|x|} 1_{x<0} + \frac{c_+}{x^{1+\alpha}} e^{-\lambda_+ x} 1_{x>0}, \tag{4.26}$$

where the parameters satisfy $c_- > 0$, $c_+ > 0$, $\lambda_- > 0$, $\lambda_+ > 0$ and $\alpha < 2$. This model was introduced by Koponen in [237] and used in financial modelling in [94, 67]. A version of this model is also used in [81].

REMARK 4.1 Generalized tempered stable processes Unlike the case of stable processes, which can only be defined for $\alpha > 0$, in the tempered stable case there is no natural lower bound on α and the expression (4.26) yields a Lévy measure for all $\alpha < 2$. In fact, taking negative values of α we obtain compound Poisson models with rich structure. It may also be interesting to allow for different values of α on the two sides of real axis. To include these cases into our treatment, we use the name "tempered stable" for the process with Lévy measure of the form (4.26) with $\alpha > 0$ (because only in this case do the small jumps have stable-like behavior), and we use the term *generalized tempered stable model* for the process with Lévy measure

$$\nu(x) = \frac{c_-}{|x|^{1+\alpha_-}} e^{-\lambda_-|x|} 1_{x<0} + \frac{c_+}{x^{1+\alpha_+}} e^{-\lambda_+ x} 1_{x>0} \tag{4.27}$$

with $\alpha_+ < 2$ and $\alpha_- < 2$. All the formulae of this section will be given for generalized tempered stable model, but they are of course valid in the tempered stable case. □

The following proposition shows that the tempered stable model allows for richer structures than the subordinated Brownian motion models that we have treated in the preceding section.

PROPOSITION 4.1 Time changed Brownian motion representation for tempered stable process

A generalized tempered stable process (4.27) can be represented as a time changed Brownian motion (with drift) if and only if $c_- = c_+$ *and* $\alpha_- = \alpha_+ = \alpha \geq -1$.

REMARK 4.2 The subordinator in this representation can be expressed via special functions: the representation is given by $\mu = (\lambda_- - \lambda_+)/2$ and

$$\rho(t) = \frac{c}{t^{\alpha/2+1}} e^{t\mu^2/2 - t\lambda^2/4} D_{-\alpha}(\lambda\sqrt{t}), \tag{4.28}$$

where $\lambda = (\lambda_- + \lambda_+)/2$, c is a constant and $D_{-\alpha}(z)$ denotes the Whittaker's parabolic cylinder function (see [1, page 686]). □

REMARK 4.3 The condition on the coefficients means that small jumps must be symmetric whereas decay rates for big jumps may be different. In other words the class of tempered stable processes which are representable as time changed Brownian motion coincides with the models discussed by Carr et al. [81] under the name "CGMY." Hence, in the class of tempered stable processes one has a greater modelling freedom than with models based on Brownian subordination because tempered stable models allow for asymmetry of small jumps. However, since the main impact on option prices is due to large jumps, the CGMY subclass is probably as flexible as the whole class of tempered stable processes. □

PROOF In order for the first condition of Theorem 4.3 to be satisfied, we must clearly have $c_- = c_+ = c$ and $\alpha_- = \alpha_+$. In this case the Lévy measure has a symmetric exponentially tilted modification (with $\mu = (\lambda_- - \lambda_+)/2$) that is given by $\tilde{\nu}(x) = c\frac{e^{-\lambda|x|}}{|x|^{\alpha+1}}$. To finish the proof we must therefore show that $\frac{e^{-\lambda\sqrt{u}}}{u^{(\alpha+1)/2}}$ is completely monotonic on $(0,\infty)$ if and only if $\alpha \geq -1$. If $\alpha < -1$ this function cannot be completely monotonic (because it is not monotonic). When $\alpha \geq -1$, the function $\frac{1}{u^{(1+\alpha)/2}}$ is completely monotonic (this can be verified directly). Further, a product of two completely monotonic functions is completely monotonic, therefore it remains to prove the complete monotonicity of $e^{-\lambda\sqrt{u}}$. To see this, observe that $(-1)^n e^{\lambda\sqrt{u}} \frac{d^n e^{-\lambda\sqrt{u}}}{du^n}$ is a polynomial in $\frac{1}{\sqrt{u}}$, all coefficients of which are positive, because this is evident for $n = 1$ and at each successive differentiation all the coefficients of $e^{\lambda\sqrt{u}} \frac{d^n e^{-\lambda\sqrt{u}}}{du^n}$ change sign. Hence, $e^{-\lambda\sqrt{u}}$ is completely monotonic on $(0,\infty)$, which completes the proof of the first part.

The second part follows from (4.15) and the corresponding Laplace transform inversion formula. □

From Propositions 3.8, 3.9 and 3.10 we deduce that a generalized tempered stable process

- is of compound Poisson type if $\alpha_+ < 0$ and $\alpha_- < 0$,
- has trajectories of finite variation if $\alpha_+ < 1$ and $\alpha_- < 1$,

- is a subordinator (positive Lévy process) if $c_- = 0$, $\alpha_+ < 1$ and the drift parameter is positive.

The limiting case $\alpha_- = \alpha_+ = 0$ corresponds to an infinite activity process. If in addition $c_+ = c_-$, we recognize the variance gamma model of the previous section.

Excluding the deterministic drift parameter we see that the generalized tempered stable process is a parametric model with six parameters. We will discuss the role of the parameters a little later, after computing the characteristic function of the process.

Working with tempered stable process becomes more convenient if we use the version of Lévy-Khinchin formula without truncation of big jumps, namely we write

$$E[e^{iuX_t}] = \exp t\{iu\gamma_c + \int_{-\infty}^{\infty}(e^{iux} - 1 - iux)\nu(x)dx\}. \tag{4.29}$$

This form can be used because of exponential decay of the tails of Lévy measure. In this case $E[X_t] = \gamma_c t$. To compute the characteristic function, we first consider the positive half of the Lévy measure and suppose that $\alpha_\pm \neq 1$ and $\alpha_\pm \neq 0$.

$$\int_0^\infty (e^{iux} - 1 - iux)\frac{e^{-\lambda x}}{x^{1+\alpha}}dx$$

$$= \sum_{n=2}^{\infty} \frac{(iu)^n}{n!} \int_0^\infty x^{n-1-\alpha} e^{-\lambda x} dx = \sum_{n=2}^{\infty} \frac{(iu)^n}{n!} \lambda^{\alpha-n}\Gamma(n-\alpha)$$

$$= \lambda^\alpha \Gamma(2-\alpha)\left\{\frac{1}{2!}\left(\frac{iu}{\lambda}\right)^2 + \frac{2-\alpha}{3!}\left(\frac{iu}{\lambda}\right)^3 + \frac{(2-\alpha)(3-\alpha)}{4!}\left(\frac{iu}{\lambda}\right)^4 + \dots\right\}.$$

The expression in braces resembles to the well-known power series

$$(1+x)^\mu = 1 + \mu x + \mu(\mu-1)\frac{x^2}{2!} + \dots$$

Comparing the two series we conclude that

$$\int_0^\infty (e^{iux} - 1 - iux)\frac{e^{-\lambda x}}{x^{1+\alpha}}dx = \lambda^\alpha\Gamma(-\alpha)\left\{\left(1 - \frac{iu}{\lambda}\right)^\alpha - 1 + \frac{iu\alpha}{\lambda}\right\}. \tag{4.30}$$

The interchange of sum and integral and the convergence of power series that we used to obtain this expression can be formally justified if $|u| < \lambda$ but the resulting formula can be extended via analytic continuation to other values of u such that $\Im u > -\lambda$. To compute the power in (4.30) we choose a branch of z^α that is continuous in the upper half plane and maps positive half-line into positive half-line.

A similar computation in the case $\alpha = 1$, that is left to the reader as an exercise, yields that

$$\int_0^\infty (e^{iux} - 1 - iux)\frac{e^{-\lambda x}}{x^2}dx = (\lambda - iu)\log\left(1 - \frac{iu}{\lambda}\right) + iu \tag{4.31}$$

and if $\alpha = 0$,

$$\int_0^\infty (e^{iux} - 1 - iux)\frac{e^{-\lambda x}}{x}dx = \frac{u}{i\lambda} + \log\frac{i\lambda}{u + i\lambda}. \tag{4.32}$$

Assembling together both parts of the Lévy measure, we obtain the characteristic function of the generalized tempered stable process.

PROPOSITION 4.2

Let $(X_t)_{t\geq 0}$ be a generalized tempered stable process. In the general case ($\alpha_\pm \neq 1$ and $\alpha_\pm \neq 0$) its characteristic exponent $\Psi(u) = t^{-1}\log E[e^{iuX_t}]$ is

$$\Psi(u) = iu\gamma_c + \Gamma(-\alpha_+)\lambda_+^{\alpha_+}c_+\left\{\left(1 - \frac{iu}{\lambda_+}\right)^{\alpha_+} - 1 + \frac{iu\alpha_+}{\lambda_+}\right\}$$
$$+ \Gamma(-\alpha_-)\lambda_-^{\alpha_-}c_-\left\{\left(1 + \frac{iu}{\lambda_-}\right)^{\alpha_-} - 1 - \frac{iu\alpha_-}{\lambda_-}\right\}. \tag{4.33}$$

If $\alpha_+ = \alpha_- = 1$,

$$\Psi(u) = iu(\gamma_c + c_+ - c_-) + c_+(\lambda_+ - iu)\log\left(1 - \frac{iu}{\lambda_+}\right)$$
$$+ c_-(\lambda_- + iu)\log\left(1 + \frac{iu}{\lambda_-}\right), \tag{4.34}$$

and if $\alpha_+ = \alpha_- = 0$,

$$\Psi(u) = iu\gamma_c - c_+\left\{\frac{iu}{\lambda_+} + \log\left(1 - \frac{iu}{\lambda_+}\right)\right\}$$
$$- c_-\left\{-\frac{iu}{\lambda_-} + \log\left(1 + \frac{iu}{\lambda_-}\right)\right\}. \tag{4.35}$$

The other cases (when only one of the α-s is equal to 0 or 1) can be obtained in a similar fashion.

Proposition 3.13 allows to compute the first cumulants of the tempered stable process. Taking derivatives of the characteristic exponent, we find in the general case:

$$K_1 = E[X_t] = t\gamma_c,$$
$$K_2 = \operatorname{Var} X_t = t\Gamma(2 - \alpha_+)c_+\lambda_+^{\alpha_+ - 2} + t\Gamma(2 - \alpha_-)c_-\lambda_-^{\alpha_- - 2},$$
$$K_3 = t\Gamma(3 - \alpha_+)c_+\lambda_+^{\alpha_+ - 3} - t\Gamma(3 - \alpha_-)c_-\lambda_-^{\alpha_- - 3},$$
$$K_4 = t\Gamma(4 - \alpha_+)c_+\lambda_+^{\alpha_+ - 4} + t\Gamma(4 - \alpha_-)c_-\lambda_-^{\alpha_- - 4}.$$

These expressions do not completely clarify the role of different parameters. For example, suppose that the process is symmetric. Then the excess kurtosis of the distribution of X_t is

$$k = \frac{K_4}{(\mathrm{Var}\, X_t)^2} = \frac{(2-\alpha)(3-\alpha)}{ct\lambda^\alpha},$$

which shows that we can decrease the excess kurtosis by either increasing λ (the jumps become smaller and the process becomes closer to a continuous one) or increasing c (the jumps become more frequent and therefore the central limit theorem works better). However, this expression does not allow us to distinguish the effect of c and λ. To fully understand the role of different parameters we must pass to the dynamical viewpoint and look at the Lévy measure rather than at the moments of the distribution. Then it becomes clear that λ_- and λ_+ determine the tail behavior of the Lévy measure, they tell us how far the process may jump, and from the point of view of a risk manager this corresponds to the amount of money that we can lose (or earn) during a short period of time. c_+ and c_- determine the overall and relative frequency of upward and downward jumps; of course, the total frequency of jumps is infinite, but if we are interested only in jumps larger than a given value, then these two parameters tell us, how often we should expect such events. Finally, α_+ and α_- determine the local behavior of the process (how the price evolves between big jumps). When α_+ and α_- are close to 2, the process behaves much like a Brownian motion, with many small oscillations between big jumps. On the other hand, if α_+ and α_- is small, most of price changes are due to big jumps with periods of relative tranquillity between them.

Key properties of the tempered stable model are summed up in Table 4.6.

4.6 Generalized hyperbolic model

This last example of this chapter illustrates the modelling approach by specifying the probability density directly. Let us first come back to the inverse Gaussian subordinator with probability density at some fixed time

$$p(x) = c(\chi, \zeta)x^{-3/2}e^{-\frac{1}{2}(\chi x - \zeta/x)}1_{x>0}, \tag{4.36}$$

where we used a different parameterization than in Table 4.4. Introducing an additional parameter into this distribution, we obtain the so-called *generalized inverse Gaussian* law (GIG):

$$p(x) = c(\lambda, \chi, \zeta)x^{\lambda-1}e^{-\frac{1}{2}(\chi x - \zeta/x)}1_{x>0}. \tag{4.37}$$

TABLE 4.6: Tempered stable process

Model name	**Tempered stable process**
Model type	Infinite activity but finite variation if $\alpha_1 < 1$, $\alpha_2 < 1$ and at least one of them is non-negative; infinite variation if at least one of them is greater or equal to 1. Small jumps have stable-like behavior if α_- and α_+ are positive.
Parameters	6: three for each side of the Lévy measure. λ_- and λ_+ are tail decay rates, α_- and α_+ are describe the Lévy measure at 0 negative and positive jumps, and c_- and c_+ determine the arrival rate of jumps of given size.
Lévy measure	$\nu(x) = \frac{c_-}{\lvert x\rvert^{1+\alpha_-}} e^{-\lambda_- \lvert x\rvert} 1_{x<0} + \frac{c_+}{x^{1+\alpha_+}} e^{-\lambda_+ x} 1_{x>0}$
Characteristic exponent	$\Psi(u) = \Gamma(-\alpha_+)\lambda_+^{\alpha_+} c_+ \left\{\left(1 - \frac{iu}{\lambda_+}\right)^{\alpha_+} - 1 + \frac{iu\alpha_+}{\lambda_+}\right\} + \Gamma(-\alpha_-)\lambda_-^{\alpha_-} c_- \left\{\left(1 + \frac{iu}{\lambda_-}\right)^{\alpha_-} - 1 - \frac{iu\alpha_-}{\lambda_-}\right\}$ if $\alpha_\pm \neq 1$ and $\alpha_\pm \neq 0$. Expressions for these particular cases are given in Proposition 4.2.
Probability density	Not available in closed form.
Cumulants:	
$E[X_t]$	0 because we use the representation (4.29)
$\operatorname{Var} X_t$	$t\Gamma(2-\alpha_+)c_+\lambda_+^{\alpha_+-2} + t\Gamma(2-\alpha_-)c_-\lambda_-^{\alpha_--2}$
c_3	$t\Gamma(3-\alpha_+)c_+\lambda_+^{\alpha_+-3} - t\Gamma(3-\alpha_-)c_-\lambda_-^{\alpha_--3}$
c_4	$t\Gamma(4-\alpha_+)c_+\lambda_+^{\alpha_+-4} + t\Gamma(4-\alpha_-)c_-\lambda_-^{\alpha_--4}$
Tail behavior	Both Lévy density and probability density have exponential tails with decay rates λ_+ and λ_-.

This distribution was proven to be infinitely divisible [183] and can generate a Lévy process (a subordinator). However, GIG laws do not form a convolution closed class, which means that distributions of this process at other times will not be GIG. Let S be a GIG random variable and W be an independent standard normal random variable. Then the law of $\sqrt{S}W + \mu S$, where μ is a constant, is called normal variance-mean mixture with mixing distribution GIG. This can be seen as a static analog of Brownian subordination (indeed, the distribution of the subordinated process at time t is a variance-mean mixture with the mixing distribution being that of the subordinator at time t). Normal variance-mean mixtures of GIG laws are called *generalized hyperbolic* distributions (GH) and they are also infinitely divisible. GH laws have been used for financial modelling by a number of authors (see [123, 128, 125]). The one-dimensional GH law is a five-parameter family that is usually defined via its Lebesgue density:

$$p(x; \lambda, \alpha, \beta, \delta, \mu) = C(\delta^2 + (x-\mu)^2)^{\frac{\lambda}{2}-\frac{1}{4}} K_{\lambda-\frac{1}{2}}(\alpha\sqrt{\delta^2 + (x-\mu)^2}) e^{\beta(x-\mu)}$$

$$C = \frac{(\alpha^2-\beta^2)^{\lambda/2}}{\sqrt{2\pi}\alpha^{\lambda-1/2}\delta^\lambda K_\lambda(\delta\sqrt{\alpha^2-\beta^2})}, \tag{4.38}$$

where K is the modified Bessel function of the second kind. The characteristic function of this law has the following form:

$$\Phi(u) = e^{i\mu u}\left(\frac{\alpha^2-\beta^2}{\alpha^2-(\beta+iu)^2}\right)^{\lambda/2} \frac{K_\lambda(\delta\sqrt{\lambda^2-(\beta+iu)^2})}{K_\lambda(\delta\sqrt{\alpha^2-\beta^2})}. \tag{4.39}$$

The main disadvantage of GH laws is that they are not closed under convolution: the sum of two independent GH random variables is not a GH random variable. This fact makes GH laws inconvenient for working with data on different time scales. For example, it is difficult to calibrate a GH model to a price sheet with options of several maturities. In this case one has to choose one maturity at which the stock price distribution is supposed to be generalized hyperbolic, and distributions at other maturities must be computed as convolution powers of this one. On the contrary, it is relatively easy to sample from a GH distribution and to estimate its parameters when all data are on the same time scale (e.g., when one disposes of an equally spaced price series).

Because the GH law is infinitely divisible, one can construct a generalized hyperbolic Lévy process whose distributions at fixed times have characteristic functions $\Phi(u)^t$. The Lévy measure of this process is difficult to compute and work with but some of its properties can be read directly from the characteristic function, using the following lemma.

LEMMA 4.1

Let (X_t) be a Lévy process with characteristic exponent $\Psi(u) = \frac{1}{t}\log E[e^{iuX_t}]$ and let $\tilde{\Psi}(u) = \frac{\Psi(u)+\Psi(-u)}{2}$ be the symmetrized characteristic exponent.

1. *If X is a compound Poisson process then $\tilde{\Psi}(u)$ is bounded.*

2. *If X is a finite variation process then $\frac{\tilde{\Psi}(u)}{u} \to 0$ as $u \to \infty$.*

3. *If X has no Gaussian part then $\frac{\tilde{\Psi}(u)}{u^2} \to 0$ as $u \to \infty$.*

PROOF From the Lévy-Khinchin formula we have:

$$\tilde{\Psi}(u) = -\frac{1}{2}\sigma^2 u^2 - \int_{-\infty}^{\infty} (1 - \cos(ux))\nu(dx).$$

Suppose that X is a compound Poisson process. Then by Proposition 3.8, $A = 0$ and ν is a finite measure on $\mathbb{R}$. Since $|\tilde{\Psi}(u)| \leq 2\int_{-\infty}^{\infty} \nu(dx)$, it follows that $\tilde{\Psi}(u)$ is bounded.

Suppose now that X is a finite variation process. Then by Proposition 3.9, $A = 0$ and $\int_{-1}^{1} |x|\nu(dx) < \infty$. The latter property and the inequality $|1 - \cos(z)| \leq |z|$, $z \in \mathbb{R}$ allow to use the dominated convergence theorem to compute the limit:

$$\lim_{u\to\infty} \int_{-\infty}^{\infty} \frac{1 - \cos(ux)}{u} \nu(dx) = \int_{-\infty}^{\infty} \lim_{u\to\infty} \frac{1 - \cos(ux)}{u} \nu(dx) = 0.$$

Finally suppose that X has no Gaussian part, that is $A = 0$. Then using the fact that $|1 - \cos(z)| \leq z^2$, $z \in \mathbb{R}$ and once again the dominated convergence theorem, we conclude that

$$\lim_{u\to\infty} \frac{\tilde{\Psi}(u)}{u^2} = -\lim_{u\to\infty} \int_{-\infty}^{\infty} \frac{1 - \cos(ux)}{u^2} \nu(dx) = 0.$$

□

For the GH distribution we find, using the asymptotic properties of Bessel functions (see Appendix A.1), that $\tilde{\Psi}(u) \sim -\delta|u|$ when $u \to \infty$. This relation is valid in the general case but it does not hold in some particular cases (e.g., when $\delta = 0$). This means that except in some particular cases generalized hyperbolic Lévy process is an infinite variation process without Gaussian part.

The principal advantage of the GH law is its rich structure and great variety of shapes. Indeed, many of the well-known probability distributions are subclasses of GH family. For example,

- The normal distribution is the limiting case of GH distribution when $\delta \to \infty$ and $\delta/\alpha \to \sigma^2$.

- The case $\lambda = 1$ corresponds to the hyperbolic distribution with density

  $$p(x; \alpha, \beta, \delta, \mu) = a(\alpha, \beta, \delta) \exp(-\alpha\sqrt{\delta^2 + (x - \mu)^2} + \beta(x - \mu)).$$

  This distribution (and the GH distribution itself) owes its name to the fact that the logarithm of its density is a hyperbola.

TABLE 4.7: Generalized hyperbolic model

Model name	**Generalized hyperbolic model**
Model type	Infinite variation process without Gaussian part (in the general case).
Parameters	5: δ is the scale parameter, μ is the shift parameter, λ, α and β determine the shape of the distribution.
Lévy measure	The Lévy measure is known, but the expression is rather complicated as it involves integrals of special functions.
Characteristic function	$\Phi(u) = e^{i\mu u}\left(\frac{\alpha^2-\beta^2}{\alpha^2-(\beta+iu)^2}\right)^{\lambda/2}\frac{K_\lambda(\delta\sqrt{\lambda^2-(\beta+iu)^2})}{K_\lambda(\delta\sqrt{\alpha^2-\beta^2})}$.
Probability density	$p(x) = C(\delta^2+(x-\mu)^2)^{\frac{\lambda}{2}-\frac{1}{4}}K_{\lambda-\frac{1}{2}}(\alpha\sqrt{\delta^2+(x-\mu)^2})e^{\beta(x-\mu)}$ where $C = \frac{(\alpha^2-\beta^2)^{\lambda/2}}{\sqrt{2\pi}\alpha^{\lambda-1/2}\delta^\lambda K_\lambda(\delta\sqrt{\alpha^2-\beta^2})}$.
Cumulants:	
$E[X_t]$	$\mu + \frac{\beta\delta}{\sqrt{\alpha^2-\beta^2}}\frac{K_{\lambda+1}(\zeta)}{K_\lambda(\zeta)}$ where $\zeta = \delta\sqrt{\alpha^2-\beta^2}$.
Var X_t	$\delta^2\left(\frac{K_{\lambda+1}(\zeta)}{\zeta K_\lambda(\zeta)} + \frac{\beta^2}{\alpha^2-\beta^2}\left\{\frac{K_{\lambda+2}(\zeta)}{K_\lambda(\zeta)} - \left(\frac{K_{\lambda+1}(\zeta)}{K_\lambda(\zeta)}\right)^2\right\}\right)$.
Tail behavior	In the general case both Lévy density and probability density have exponential tails with decay rates $\lambda_+ = \alpha-\beta$ and $\lambda_- = \alpha+\beta$.

- The case $\lambda = -1/2$ corresponds to the normal inverse Gaussian law that we have already discussed.
- The case $\delta = 0$ and $\mu = 0$ corresponds to variance gamma process.
- Student t distributions are obtained for $\lambda < 0$ and $\alpha = \beta = \mu = 0$.

A summary of analytical facts about generalized hyperbolic distribution and examples of their application in finance can be found in [128]. Key properties of this model are summed up in Table 4.7.

Further reading

More details on the Merton jump-diffusion model, including pricing formulae for European options, are given in [291]. Kou's double exponential jump-diffusion model is described in [238], this paper also derives formulae for pricing European and barrier options. Bates [41] discusses a model with compound Poisson jumps and stochastic volatility.

The variance gamma model was introduced in the symmetric case by Madan and Seneta in [273]. The general case is discussed in [271]. Normal inverse Gaussian process was introduced by Barndorff-Nielsen (see [32, 30]) and extensively studied by Rydberg [342].

Tempered stable laws have appeared in the literature under many different names (see remarks below). First introduced by Koponen [237] under the name *truncated Lévy flights*, they have been applied in financial modelling in [94, 67, 71, 80]. Boyarchenko and Levendorskiĭ [71] study it under the name of *KoBoL* process. Carr et al. [80] discuss a four-parameter subclass of tempered stable models under the name *CGMY process*.

Generalized hyperbolic distributions were introduced by Barndorff-Nielsen [31]. They were proven to be infinitely divisible by Halgreen [183], which made it possible to construct generalized hyperbolic Lévy processes. Financial applications of this process were studied by Eberlein [123] and by Raible in his PhD thesis [329].

REMARK 4.4 Tempered stable processes and truncated Lévy flights The tempered stable process was defined in Section 4.5 as a Lévy process whose Lévy measure has a density of the form:

$$\nu(x) = \frac{c_-}{|x|^{1+\alpha_+}} e^{-\lambda_-|x|} 1_{x<0} + \frac{c_+}{x^{1+\alpha_-}} e^{-\lambda_+ x} 1_{x>0}.$$

In the case $\alpha_- = \alpha_+ = \alpha$ this process is obtained by multiplying the Lévy density of an α-stable process by a (two-sided) exponential factor which can be asymmetric. This exponentially decreasing factor has the effect of tempering the large jumps of a Lévy process and giving rise to finite moments, while retaining the same behavior for the small jumps hence the name of *tempered stable process.* Incidentally, this exponential tempering also destroys the self-similarity and scaling properties of the α-stable process, but one can argue that these scaling properties hold approximately for short times [94, 285].

First proposed in this form by Koponen [237] under the name of "truncated Lévy flight[1]" (and not truncated Lévy *process*), tempered stable processes were used in financial modelling in [94, 288, 67]. Variants with $\lambda_+ \neq \lambda_-$ have been studied by Boyarchenko and Levendorskiĭ [71] under the name of "KoBoL" processes and in [81] under the name of "CGMY" processes. We have used here the more explanatory denomination of "tempered stable".

In [71], Boyarchenko & Levendorskiĭ study a generalization of tempered stable processes, called regular Lévy processes of exponential type (RLPE), where the Lévy density does not exactly have the form above, but behaves similarly both when $x \to \infty$ and $x \to 0$: a Lévy process is said be a RLPE[2] of type $[-\lambda_-, \lambda_+]$ and order $\alpha \in]0, 2[$ if the Lévy measure has exponentially decaying tails with rates $\lambda_\pm$:

$$\int_{-\infty}^{-1} e^{\lambda_-|y|}\nu(dy) < \infty, \qquad \int_{1}^{-\infty} e^{\lambda_+ y}\nu(dy) < \infty$$

and behaves near zero as $|y|^{-(1+\alpha)}$: there exists $\alpha' < \alpha$ such that

$$\left| \int_{|y|>\epsilon} \nu(dy) - \frac{C}{|\epsilon|^{\alpha}} \right| \leq \frac{c}{|\epsilon|^{\alpha'}}.$$

The class of RLPEs includes hyperbolic, normal inverse Gaussian and tempered stable processes but does not include the variance gamma process. Since these processes are essentially the only analytical examples of RLPE, this class should be seen more as a unified framework for studying properties of such processes rather than a new class of models. □

[1]The term "Truncated Lévy flight" was actually coined by Mantegna and Stanley in [284] but for a discrete time random walk which does not have an infinitely divisible distribution and therefore does not correspond to a Lévy process. Note that these authors do not speak of a Lévy processes but of "Lévy flight" which is a discrete time process.

[2]Note that these authors give several different definitions of RLPE in their different papers.

Chapter 5

Multidimensional models with jumps

Apart from the pricing and hedging of options on a single asset, practically all financial applications require a multivariate model with dependence between components: examples are basket option pricing, portfolio optimization, simulation of risk scenarios for portfolios. In most of these applications, jumps in the price process must be taken into account. However, multidimensional models with jumps are more difficult to construct than one-dimensional ones. This has led to an important imbalance between the range of possible applications and the number of available models in the multidimensional and one-dimensional cases: a wide variety of one-dimensional models have been developed for relatively few applications, while multidimensional applications continue to be dominated by Brownian motion.

One reason for the omnipresence of Gaussian models is that dependence in the Gaussian world can be parameterized in a simple manner, in terms of correlation matrices. In particular, in this case marginal properties — given by volatilities — are easy to separate from dependence properties — described by correlations. A second reason is that it is easy to simulate Gaussian time series with arbitrary correlation matrices.

In this chapter, we will provide tools for building multivariate models with jumps and propose a framework for parameterizing these models that allows the same flexibility as the Gaussian multivariate case: separating dependence from marginal properties and easy simulation.

A simple method to introduce jumps into a multidimensional model is to take a multivariate Brownian motion and time change it with a univariate subordinator (see [123, 319]). This approach, that we summarize in Section 5.1, allows to construct multidimensional versions of the models we discussed in the preceding chapter, including variance gamma, normal inverse Gaussian and generalized hyperbolic processes. The principal advantage of this method is its simplicity and analytic tractability; in particular, processes of this type are easy to simulate. However, this approach to dependence modelling lacks flexibility; the range of possible dependence patterns is quite limited (for instance, independence is not included), only components of the same type can be used (e.g., either all of the components are variance gamma or all of the components are normal inverse Gaussian) and no quantitative measure of dependence is available.

In finite activity models, a more accurate modelling of dependence may be achieved by specifying directly the dependence of individual jumps in one-dimensional compound Poisson processes (see [263]). This approach is discussed in Section 5.2. If the jumps are Gaussian, their dependence can be described via correlation coefficients. If they are not Gaussian (e.g., if they have double exponential distribution like in Kou's model) then one has to model their dependence with copulas (described in Section 5.3). This approach is useful in presence of few sources of jump risk (e.g., when all components jump at the same time) because in this case it allows to achieve a precise description of dependence within a simple model. On the other hand, when there are several sources of jump risk, that is, the compound Poisson processes do not all jump at the same time, one has to introduce a separate copula for each jump risk source and the model quickly becomes very complicated. Another inconvenience of this modelling approach is that it does not allow to couple components of different types (all of them must be compound Poisson).

Comparison of advantages and drawbacks of these two methods leads to an understanding of the required properties of a multidimensional modelling approach for Lévy processes. In general, such an approach must satisfy the following conditions:

- The method should allow to specify any one-dimensional model for each of the components.
- The range of possible dependence structures should include complete dependence and independence with a "smooth" transition between these two extremes.
- The dependence should be modelled in a parametric fashion and the number of parameters must be small.

This program can be implemented if the dependence is separated from the marginal behavior of the components. In a more restrictive setting of random vectors this idea has long existed in the statistical literature: the dependence structure of a random vector can be disentangled from its margins via the notion of copula.[1] We discuss this notion in Section 5.3 and show how it can be carried over from the static setting of random variables to the dynamic framework of Lévy processes. The dependence among components of a multidimensional Lévy processes can be completely characterized with a Lévy copula — a function that has the same properties as ordinary copula but is defined on a different domain. This allows us to give a systematic method to construct multidimensional Lévy processes with specified dependence. We suggest several parametric families of Lévy copulas, which can be used to construct an n-dimensional Lévy processes by taking any set of n one-dimensional Lévy processes and coupling them with a Lévy copula from

[1]For an introduction to copulas see [305] — this monograph treats mostly bivariate case — and [223] for multivariate case.

a parametric family. This method is described in Section 5.5 in the setting of Lévy processes with positive jumps and generalized in Section 5.6 to all Lévy processes. The models constructed using the Lévy copula approach may be less mathematically tractable and somewhat harder to implement than the simple models of Sections 5.1 and 5.2 but their main advantage is the ability to describe every possible dependence pattern for a multidimensional Lévy process with arbitrary components.

In the following sections, for ease of presentation, examples are given in the two-dimensional case. A summary of important definitions and results in higher dimensions is provided at the end of sections.

5.1 Multivariate modelling via Brownian subordination

A possible generalization of the Black-Scholes model is to suppose that the stock prices do follow a multidimensional Brownian motion but that they do it on a different (stochastic) time scale. In other words, the clock of the market is not the one that we are used to: sometimes it may run faster and sometimes slower than ours. Suppose that this market time scale is modelled by a subordinator (Z_t) with Laplace exponent $l(u)$ and let $(B(t))_{t\geq 0}$ be a d-dimensional Brownian motion with covariance matrix Σ and $\mu \in \mathbb{R}^d$. Time changing the Brownian motion with drift μ with (Z_t), we obtain a new d-dimensional Lévy process $X_t = B(Z_t) + \mu Z_t$. This process can be used to model d dependent stocks: $S_t^i = \exp(X_t^i)$ for $i = 1 \dots d$. The characteristic function of X_t can be computed using Theorem 4.2: for every $u \in \mathbb{R}^d$

$$E\{e^{iu.X_t}\} = e^{tl(-\frac{1}{2}u.\Sigma u + i\mu.u)}.$$

Formulae (4.7) and (4.8) also allow to compute the characteristic triplet of (X_t). The conditionally Gaussian nature of this model makes it easy to simulate the increments of X_t. This can be done by first simulating the increment of the subordinator and then the increment of d-dimensional Brownian motion at the time scale given by the increment of subordinator.

Let us take a closer look at some properties of this model. To ease the notation, we switch to the two-dimensional case. Two stock price processes (S_t^1) and (S_t^2) are modelled as follows:

$$S_t^1 = \exp(X_t^1), \quad X_t^1 = B^1(Z_t) + \mu_1 Z_t,$$
$$S_t^2 = \exp(X_t^2), \quad X_t^2 = B^2(Z_t) + \mu_2 Z_t,$$

where B^1 and B^2 are two components of a planar Brownian motion, with variances σ_1^2 and σ_2^2 and correlation coefficient ρ. The correlation of returns,

$\rho(X_t^1, X_t^2)$, can be computed by conditioning on Z_t:

$$\rho(X_t^1, X_t^2) = \frac{\sigma_1\sigma_2\rho E[Z_t] + \mu_1\mu_2 \operatorname{Var} Z_t}{(\sigma_1^2 E[Z_t] + \mu_1^2 \operatorname{Var} Z_t)^{1/2}(\sigma_2^2 E[Z_t] + \mu_2^2 \operatorname{Var} Z_t)^{1/2}}.$$

In the completely symmetric case ($\mu_1 = \mu_2 = 0$) and in this case only $\rho(X_t^1, X_t^2) = \rho$: the correlation of returns equals the correlation of Brownian motions that are being subordinated. However, the distributions of real stocks are skewed and in the skewed case the correlation of returns will be different from the correlation of Brownian motions that we put into the model. Even if the Brownian motions are independent, the covariance of returns is equal to $\mu_1\mu_2 \operatorname{Var} Z_t$ and if the distributions of stocks are not symmetric, they are correlated.

In the symmetric case, if Brownian motions are independent, the two stocks are decorrelated but not independent. Since the components of the Brownian motion are time changed with the same subordinator, large jumps in the two stocks (that correspond to large jumps of the subordinator) will tend to arrive together, which means that absolute values of returns will be correlated. If $\mu_1 = \mu_2 = 0$ and $\rho = 0$ then the covariance of squares of returns is

$$\operatorname{Cov}((X_t^1)^2, (X_t^2)^2) = \sigma_1\sigma_2 \operatorname{Cov}(W_1^2(Z_t), W_2^2(Z_t)) = \sigma_1\sigma_2 \operatorname{Var} Z_t,$$

therefore squares of returns are correlated if the subordinator Z_t is not deterministic. This phenomenon can lead to mispricing and errors in evaluation of risk measures. The following example illustrates this point.

Example 5.1

Let (X_t) and (Y_t) be two symmetric identically distributed variance gamma Lévy processes and suppose that we are interested in the distribution of $X_t + Y_t$ for a horizon of 10 business days, which corresponds roughly to $t = 1/25$. The parameters of the distribution of X_t and Y_t have been chosen to have an annualized volatility of 30% and an excess kurtosis at time t equal to 3. In the first test the two processes, X_t and Y_t, were assumed completely independent, whereas in the second test they result from the subordination of independent Brownian motions by the same gamma subordinator. Various statistics of the distributions of $X_t + Y_t$ in the two cases were computed using the Monte Carlo method with 50,000 trials and are summed up in Table 5.1. They show that in the case when X_t and Y_t are obtained by subordination with the same subordinator, the distribution of $X_t + Y_t$ is more leptokurtic than in the completely independent case, and its small quantiles are higher in absolute value. This means in particular that the use for risk management of an incorrect model with dependence between assets when they are really independent will result in higher requirements in regulatory capital and therefore in unjustified extra costs for the bank. □

TABLE 5.1: Decorrelation vs. independence: independent processes cannot be modelled via independent Brownian motions time changed with the same subordinator

Statistics of $X_t + Y_t$	X_t **and** Y_t **independent**	X_t **and** Y_t **decorrelated**
Volatility	41.9%	41.8%
Excess kurtosis	1.54	2.64
5% quantile	−0.136	−0.133
1% quantile	−0.213	−0.221
0.1% quantile	−0.341	−0.359

Thus, subordination of a multidimensional Brownian motion allows to construct simple and tractable multivariate parametric models based on Lévy processes. However, this construction has two important drawbacks: first, one cannot specify arbitrary one-dimensional models for the components and second, the range of available dependence structures is too narrow, in particular, it does not include complete independence — the components obtained are always partially dependent.

5.2 Building multivariate models from common Poisson shocks

Suppose that we want to improve a d-dimensional Black-Scholes model by allowing for "market crashes." The dates of market crashes can be modelled as arrival times of a standard Poisson process (N_t). This leads us to the following model for the log-price processes of d assets:

$$X_t^i = \mu^i t + B_t^i + \sum_{j=1}^{N_t} Y_j^i, \quad i = 1 \ldots d,$$

where (B_t) is a d-dimensional Brownian motion with covariance matrix Σ, and $\{Y_j\}_{j=1}^{\infty}$ are i.i.d. d-dimensional random vectors which determine the sizes of jumps in individual assets during a market crash. This model contains only one driving Poisson shock (N_t) because we only account for jump risk of one type (global market crash affecting all assets). To construct a parametric model, we need to specify the distribution of jumps in individual assets during a crash (distribution of Y^i for all i) and the dependence between jumps in assets. If we make a simplifying assumption that Y_j are Gaussian random vectors, then we need to specify their covariance matrix Σ' and the mean vector m. We thus obtain a multivariate version of Merton's model (see

Chapter 4 and the numerical example at the end of Chapter 11). If the jumps are not Gaussian, we must specify the distribution of jumps in each component and the copula function describing their dependence.

In the case of only one driving Poisson shock, everything is relatively easy and simple, but sometimes it is necessary to have several independent shocks to account for events that affect individual companies or individual sectors rather than the entire market. In this case we need to introduce several driving Poisson processes into the model, which now takes the following form:

$$X_t^i = \mu^i t + B_t^i + \sum_{k=1}^{M} \sum_{j=1}^{N_t^k} Y_{j,k}^i, \quad i = 1 \dots d,$$

where $N_t^1, \dots, N_t^M$ are Poisson processes driving M independent shocks and $Y_{j,k}^i$ is the size of jump in i-th component after j-th shock of type k. The vectors $\{Y_{j,k}^i\}_{i=1}^d$ for different j and/or k are independent. To define a parametric model completely, one must specify a one-dimensional distribution for each component for each shock type — because different shocks influence the same stock in different ways — and one d-dimensional copula for each shock type. This adds up to $M \times d$ one-dimensional distributions and M one-dimensional copulas. How many different shocks do we need to describe sufficiently rich dependence structures? The answer depends on the particular problem, but to describe *all* possible dependence structures, such that the d-dimensional process remains a Lévy process of compound Poisson type, one needs a total of $2^M - 1$ shocks (M shocks that affect only one stock, $\frac{M(M-1)}{2}$ shocks that affect two stocks etc., adding up to $2^M - 1$). It is clear that as the dimension of the problem grows, this kind of modelling quickly becomes infeasible. Not only the number of parameters grows exponentially, but also, when the number of shocks is greater than one, one cannot specify directly the laws of components because the laws of jumps must be given separately for each shock.

Building multivariate models from compound Poisson shock is therefore a feasible approach in low-dimensional problems and/or when few sources of jump risk need to be taken into account. For a detailed discussion of this modelling approach including examples from insurance and credit risk modelling, the reader is referred to [263].

5.3 Copulas for random variables

The law of a two-dimensional random vector (X, Y) is typically described via its cumulative distribution function

$$F(x, y) = P[X \leq x, Y \leq y]. \tag{5.1}$$

We call *marginal laws* or *margins* of (X, Y) the laws of X and Y taken separately. These marginal laws can be described via their respective distribution functions $F_1(x) = P[X \leq x]$ and $F_2(y) = P[Y \leq y]$ which can, of course, be obtained from the two-dimensional distribution function:

$$F_1(x) = F(x, \infty), \qquad F_2(y) = F(\infty, y).$$

To know the distribution of a two-dimensional random vector, it is not enough to know its marginal distributions; in addition to the margins the distribution function $F(x, y)$ contains the information about dependence. Two vectors with the same margins but different dependence structures have different distribution functions. For example, for a random vector with marginal distributions $F_1(x)$ and $F_2(y)$, $F(x, y) = F_1(x)F_2(y)$ if the components are independent and $F(x, y) = \min(F_1(x), F_2(y))$ if they are completely dependent, that is, one of the components is a strictly increasing deterministic function of the other.

The *copula* of a two-dimensional random vector (or, more precisely, of its distribution) is a two-place function that characterizes its dependence structure and does not depend on the margins. The pair copula + margins gives an alternative description of the law of a random vector. The distribution function can be computed from the copula $C(x, y)$ and the margins: $F(x, y) = C(F_1(x), F_2(y))$. If $F_1(x)$ and $F_2(y)$ are continuous then the copula is unique and can be computed from the distribution function and the margins: $C(x, y) = F(F_1^{-1}(x), F_2^{-1}(y))$.

Saying that the copula of a distribution does not depend on the margins means that it is invariant under strictly increasing transformations of the components of the random vector. That is, for every two strictly increasing functions f and g the copula of X and Y is the same as the copula of $f(X)$ and $g(Y)$. The term "dependence structure" refers to all characteristics of a distribution that do not depend on the margins and that, together with the margins, allow to reconstruct the entire distribution.

To proceed, we must add some mathematics to these intuitive explanations. Let us come back to the distribution function $F(x, y)$ of a random vector. This function has the following properties:

1. For any rectangle $B = [x_1, x_2] \times [y_1, y_2]$,

$$F(x_2, y_2) - F(x_2, y_1) - F(x_1, y_2) + F(x_1, y_1) = P[(X, Y) \in B] \geq 0.$$

2. For every x, $F(-\infty, x) = F(x, -\infty) = 0$.

The first property can be seen as a generalization of the notion of increasing function to multiple dimensions, leading to the following definitions:

DEFINITION 5.1 *Let S_1 and S_2 be two possible infinite closed intervals of $\bar{\mathbb{R}} = \mathbb{R} \cup \{\infty\} \cup \{-\infty\}$ and consider a function $F : S_1 \times S_2 \to \bar{\mathbb{R}}$.*

- *F-volume of a rectangle* $B = [x_1, x_2] \times [y_1, y_2] \subset S_1 \times S_2$ *is defined by*

$$V_F(B) = F(x_2, y_2) - F(x_2, y_1) - F(x_1, y_2) + F(x_1, y_1).$$

- *F is 2-increasing if for every rectangle B in its domain,* $V_F(B) \geq 0$.

- *F is grounded if for every* $x \in S_1$, $F(x, \min S_2) = 0$ *and for every* $y \in S_2$, $F(\min S_1, y) = 0$.

REMARK 5.1 If a two-place function is increasing in each of its arguments, it is not in general 2-increasing (the example of such function that is not 2-increasing is $F(x, y) = -\frac{1}{xy}$ for $x, y > 0$). If a function is 2-increasing, it is not necessarily increasing in each of its arguments (for example, $F(x, y) = xy$ is 2-increasing but not increasing in its arguments if $x, y < 0$). However, if a function is increasing and grounded, then it is increasing in each of its arguments. □

In the sequel, we will need the following two properties of increasing functions (both of which the reader can easily prove).

LEMMA 5.1
Let F be a 2-increasing function.

1. *Increasing variable changes: if the functions* f_1 *and* f_2 *are increasing then* $F(f_1(x), f_2(y))$ *is 2-increasing.*

2. *Transformation by a function with positive derivatives: if F is grounded and f is increasing, convex and satisfies* $f(0) = 0$ *then the superposition* $f \circ F$ *is a 2-increasing grounded function.*

In the theory of copulas the definition of a distribution function is slightly non-standard.

DEFINITION 5.2 *An* abstract *distribution function is a function from* $\bar{\mathbb{R}}^2$ *to* $[0, 1]$ *that is grounded, 2-increasing and satisfies* $F(\infty, \infty) = 1$.

In addition to the above properties, a *probability* distribution function, defined by Equation (5.1) must possess some continuity properties: for every $x_0, y_0 \in \mathbb{R}$,

$$\lim_{x \downarrow x_0, y \downarrow y_0} F(x, y) = F(x_0, y_0).$$

However, it can be shown, that for every abstract distribution function F there exists a unique probability distribution function $\tilde{F}$ such that $V_F(B) = V_{\tilde{F}}(B)$ for every *continuity rectangle* B of F, where a continuity rectangle is defined as a rectangle B such that for every sequence of rectangles whose vertices converge to B, the sequence of F-volumes of these rectangles converges to $V_F(B)$. Having said that, we are not going to use the word "abstract" in the sequel, the meaning of the term "distribution function" being clear from the context.

Generalizing the notion of margins of a distribution function, for any function $F : S_1 \times S_2 \to \bar{\mathbb{R}}$ the one-place functions $F_1(x) = F(x, \max S_2)$ and $F_2(y) = F(\max S_1, y)$ are called margins of F.

DEFINITION 5.3 (Copula) *A two-dimensional copula is a function C with domain $[0,1]^2$ such that*

1. *C is grounded and 2-increasing.*
2. *C has margins $C_k, k = 1, 2$, which satisfy $C_k(u) = u$ for all u in $[0,1]$.*

From the probabilistic point of view this definition means that a two-dimensional copula is a distribution function on $[0,1]^2$ with uniform margins.

The next theorem is fundamental for the theory of copulas.

THEOREM 5.1 (Sklar)
Let F be a two-dimensional distribution function with margins F_1, F_2. Then there exists a two-dimensional copula C such that for all $x \in \bar{\mathbb{R}}^2$,

$$F(x_1, x_2) = C(F_1(x_1), F_2(x_2)). \tag{5.2}$$

if F_1, F_2 are continuous then C is unique, otherwise C is uniquely determined on $\operatorname{Ran} F_1 \times \operatorname{Ran} F_2$. *Conversely, if C is a copula and F_1, F_2 are distribution functions, then the function F defined by (5.2) is an two-dimensional distribution function with margins F_1 and F_2.*

PROOF We suppose that F_1 and F_2 are continuous (a complete proof for the general case can be found in [364]).

The direct statement. Since F_1 is continuous, one can find an increasing function $F_1^{-1}(u)$ satisfying $F_1^{-1}(0) = -\infty$, $F_1^{-1}(1) = +\infty$ and $F_1(F^{-1}(u)) = u$ for all $u \in \bar{\mathbb{R}}$. The inverse function of F_2 can be chosen in the same way. Let $\tilde{C}(u,v) = F(F_1^{-1}(u), F_2^{-1}(v))$. Then by property 1 of Lemma 5.1, C is increasing. Moreover, $\tilde{C}(0,v) = F(F_1^{-1}(0), F_2^{-1}(v)) = F(-\infty, F_2^{-1}(v)) = 0$ and $\tilde{C}(1,v) = F(F_1^{-1}(1), F_2^{-1}(v)) = F(\infty, F_2^{-1}(v)) = F_2(F_2^{-1}(v)) = v$. Hence, $\tilde{C}$ is a copula. To show the uniqueness, suppose that there exists another copula $\tilde{\tilde{C}}$ that corresponds to the same distribution. Then for every $x \in \bar{\mathbb{R}}$ and $y \in \bar{\mathbb{R}}$, $\tilde{C}(F_1(x), F_2(y)) = \tilde{\tilde{C}}(F_1(x), F_2(y))$. Therefore, from the continuity of F_1 and F_2, we conclude that for every $u \in [0,1]$ and $v \in [0,1]$, $\tilde{C}(u,v) = \tilde{\tilde{C}}(u,v)$.

The converse statement. By property 1 of Lemma 5.1, the function defined in (5.2) is 2-increasing. One can then check by direct substitution that it is grounded, has margins F_1 and F_2 and that $F(1,1) = 1$. Therefore, it is a distribution function. □

Examples of copulas

- If X and Y are independent, the copula of (X,Y) is $C_\perp(u,v) = uv$. If the distributions of X and/or Y are not continuous, the copula is not unique, so $C_\perp$ is only one of the possible copulas. This remark also applies to the next two examples.

- If $X = f(Y)$ for some strictly increasing deterministic function f, then $C_\parallel(u,v) = \min(u,v)$. This copula is called the complete dependence copula or the upper Fréchet bound, because for every copula C one has $C(u,v) \leq C_\parallel(u,v),\ \forall u,v$.

- If $X = f(Y)$ with f strictly decreasing then $C(u,v) = \max(x+y-1,0)$. This is the complete negative dependence copula or the lower Fréchet bound.

- The Clayton family of copulas is a one-parameter family given by

$$C_\theta(x,y) = (x^{-\theta} + y^{-\theta} - 1)^{-1/\theta},\ \theta > 0. \tag{5.3}$$

  The Clayton family includes the complete dependence copula and the independence copula as limiting cases:

$$\begin{aligned} C_\theta &\to C_\parallel \quad \text{as } \theta \to \infty, \\ C_\theta &\to C_\perp \quad \text{as } \theta \to 0. \end{aligned}$$

- The Gaussian copula family corresponds to the dependence structure of a Gaussian random vector:

$$C_\rho(x,y) = N_{2,\rho}(N^{-1}(x), N^{-1}(y))$$

  where $N_{2,\rho}$ is the distribution function of a bivariate standard normal distribution with correlation ρ and N is the univariate standard normal distribution function. Here the case $\rho = 1$ corresponds to complete dependence and $\rho = 0$ corresponds to independence.

Figure 5.1 illustrates the difference between the Gaussian copula and the Clayton copula. Both graphs depict 1000 realizations of a two-dimensional random vector with standard normal margins and a correlation of 80%. However, the dependence of the left graph is given by the Gaussian copula and the dependence of the right graph is parameterized by the Clayton copula. The Clayton copula is asymmetric: the structure of the lower tail is different

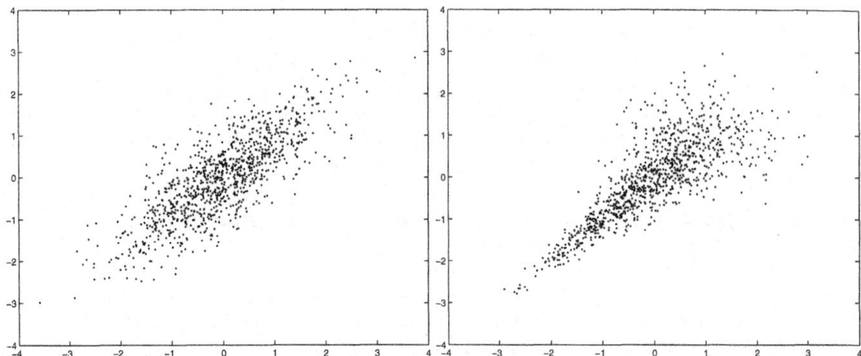

FIGURE 5.1: 1000 realizations of a random vector with standard normal margins and correlation $\rho = 0.8$. Left: Gaussian copula. Right: Clayton copula.

from the upper one. There is more tail dependence in the lower tail: there is a relatively high probability of having a large negative jump in one component given that there is a large negative jump in the other. This is not the case for the Gaussian copula on the left.

The Clayton copula is a representative of the class of Archimedean copulas defined by the following proposition.

PROPOSITION 5.1
Let $\phi : [0,1] \to [0,\infty]$ be a continuous strictly decreasing function. Then

$$C(u,v) = \phi^{-1}(\phi(u) + \phi(v))$$

is a copula if and only if ϕ is convex.

A proof of this proposition and multivariate generalizations can be found in [305]. The function ϕ is called the generator of an Archimedean copula. $\phi(t) = (t^{-\theta} - 1)/\theta$ produces the Clayton family and $\phi(t) = -\log(t)$ produces the independence copula $C_\perp$.

Summary of definitions and results for multivariate case

DEFINITION 5.4 F-volume *Let $S_1, \cdots, S_n$ be nonempty subsets of $\bar{\mathbb{R}}$, let F be a real function of n variables such that $\operatorname{Dom} F = S_1 \times \cdots \times S_n$ and for $\mathbf{a} \le \mathbf{b}$ ($a_k \le b_k$ for all k) let $B = [\mathbf{a}, \mathbf{b}]$ be an n-box whose vertices are in $\operatorname{Dom} F$. Then the F-volume of B is defined by*

$$V_F(B) = \sum \operatorname{sgn}(\mathbf{c}) F(\mathbf{c}),$$

where the sum is taken over all vertices $\mathbf{c}$ *of* B*, and* $\operatorname{sgn}(\mathbf{c})$ *is*

$$\operatorname{sgn}(\mathbf{c}) = \begin{cases} 1, & \text{if } c_k = a_k \text{ for an even number of vertices,} \\ -1, & \text{if } c_k = a_k \text{ for an odd number of vertices.} \end{cases}$$

DEFINITION 5.5 n-increasing function, grounded function, margins

- *A real function* F *of* n *variables is called* n*-increasing if* $V_F(B) \geq 0$ *for all* n*-boxes* B *whose vertices lie in* $\operatorname{Dom} F$.
- *Suppose that the domain of* F *is* $S_1 \times \cdots \times S_n$ *where each* S_k *has a smallest element* a_k*.* F *is said to be* grounded *if* $F(\mathbf{t}) = 0$ *for all* $\mathbf{t}$ *in* $\operatorname{Dom} F$ *such that* $t_k = a_k$ *for at least one* k.
- *If each* S_k *is nonempty and has a greatest element* b_k*, then (one-dimensional) margins of* F *are functions* F_k *with* $\operatorname{Dom} F_k = S_k$ *defined by* $F_k(x) = F(b_1, \cdots, b_{k-1}, x, b_{k+1}, \cdots, b_n)$ *for all* x *in* S_k.

DEFINITION 5.6 Copula *An* n*-dimensional copula is a function* C *with domain* $[0,1]^n$ *such that*

1. C *is grounded and* n*-increasing.*
2. C *has margins* $C_k, k = 1, 2, \cdots, n$*, which satisfy* $C_k(u) = u$ *for all* u *in* $[0,1]$.

An n-dimensional distribution function F is an n-increasing grounded function with domain $\bar{\mathbb{R}}^n$ such that $F(\infty, \cdots, \infty) = 1$. Copulas can be seen as distribution functions with uniform margins.

THEOREM 5.2 Sklar

Let F *be an* n*-dimensional distribution function with margins* $F_1, \cdots, F_n$*. Then there exists an* n*-dimensional copula* C *such that for all* $\mathbf{x} \in \bar{\mathbb{R}}^n$,

$$F(x_1, x_2, \cdots, x_n) = C(F_1(x_1), F_2(x_2), \cdots, F_n(x_n)). \tag{5.4}$$

If $F_1, \cdots, F_n$ *are all continuous then* C *is unique; otherwise,* C *is uniquely determined on* $\operatorname{Ran} F_1 \times \cdots \times \operatorname{Ran} F_n$*. Conversely, if* C *is an* n*-copula and* $F_1, \cdots, F_n$ *are distribution functions, then the function* F *defined by (5.4) is a* n*-dimensional distribution function with margins* $F_1, \cdots, F_n$.

5.4 Dependence concepts for Lévy processes

In Chapter 3, we saw that the law of a Lévy process (X_t) is completely determined by the law of X_t for some $t > 0$. Therefore, the dependence structure of a two-dimensional Lévy process (X_t, Y_t) can be parameterized by the copula C_t of X_t and Y_t for some $t > 0$. However, this approach has a number of drawbacks.

- Except in the case of stable processes, the copula C_t may depend on t (an explicit example of a Lévy process with nonconstant copula is given in [372]). C_s for some $s \neq t$ cannot in general be computed from C_t because it also depends on the margins.

- For given infinitely divisible distributions of X_t and Y_t, it is unclear, which copulas C_t will yield a two-dimensional infinitely divisible distribution.

- The laws of components of a multidimensional Lévy process are usually specified via their Lévy measures (see Chapter 4). In this case it would be inconvenient to model dependence using the copula of probability distribution.

The traditional concept of dependence for random variables, meaning all characteristics of a distribution that are invariant under strictly increasing transformations of the margins, is not suitable in the framework of Lévy processes. The property of infinite divisibility of a random variable is destroyed under strictly increasing transformations. We need therefore to redefine the notion of dependence with respect to some other type of margin transformation, the one that preserves the Lévy property and reflects the dynamic structure of Lévy processes. The following example clarifies this point.

Example 5.2 Dynamic complete dependence for Lévy processes
Let (X_t) be a pure jump Lévy process and (Y_t) be a Lévy process, constructed from the jumps of (X_t): $Y_t = \sum_{s \leq t} \Delta X_s^3$. From the dynamic point of view (X_t) and (Y_t) are completely dependent, because the trajectory of one of them can be reconstructed from the trajectory of the other. However, the copula of X_t and Y_t is not that of complete dependence because Y_t is not a deterministic function of X_t. □

This example shows that the important dependence concept for Lévy processes is the dependence of jumps, that should be studied using the Lévy measure. Knowledge of jump dependence gives a "microscopic" vision of a Lévy process and allows to characterize its dynamic structure, which is very important for risk management and other financial applications.

Here we will not dwell on the dependence structure of the continuous Gaussian part of Lévy processes, because on one hand it is very simple and well studied (it is completely characterized by the covariance matrix) and on the other hand it can be treated separately from jump dependence because the Gaussian part of every Lévy process is independent from its jump part. In the rest of this chapter, all Lévy processes are supposed to have no Gaussian part.

Before introducing the equivalent of copulas to describe dependence of Lévy processes, we show how independence of Lévy processes and their marginal laws are expressed in terms of the Lévy measure.

PROPOSITION 5.2 Margins of Lévy measure
Let (X_t, Y_t) be a Lévy process with characteristic triplet (A, ν, γ). Then (X_t) has characteristic triplet (A_X, ν_X, γ_X) where

$$\begin{aligned} A_X &= A_{11}, \\ \nu_X(B) &= \nu(B \times]-\infty, \infty[), \ \forall B \in \mathcal{B}(\mathbb{R}), \\ \gamma_X &= \gamma_1 + \int_{\mathbb{R}^2} x(1_{x^2 \leq 1} - 1_{x^2+y^2 \leq 1})\nu(dx \times dy). \end{aligned} \tag{5.5}$$

The most important fact in this proposition is that the margins of a Lévy measure can be computed in exactly the same way as the margins of a probability measure.

PROOF This result follows directly from Theorem 4.1. □

PROPOSITION 5.3 Independence of Lévy processes
Let (X_t, Y_t) be a Lévy process with Lévy measure ν and without Gaussian part. Its components are independent if and only if the support of ν is contained in the set $\{(x, y) : xy = 0\}$, that is, if and only if they never jump together. In this case

$$\nu(A) = \nu_X(A_X) + \nu_Y(A_Y), \tag{5.6}$$

where $A_X = \{x : (x, 0) \in A\}$ and $A_Y = \{y : (0, y) \in A\}$, and ν_X and ν_Y are Lévy measures of (X_t) and (Y_t).

PROOF The only if part. Suppose that (X_t) and (Y_t) are independent and have characteristic triplets $(0, \nu_X, \gamma_X)$ and $(0, \nu_Y, \gamma_Y)$. Then, using the Lévy-Khinchin formula for X_t and Y_t, we obtain

$$\begin{aligned} &E[e^{iuX_t + ivY_t}] = E[e^{iuX_t}]E[e^{ivY_t}] = \exp t\{i\gamma_X u + i\gamma_Y v \\ &+ \int_{\mathbb{R}} (e^{iux} - 1 - iux1_{|x|\leq 1})\nu_X(dx) + \int_{\mathbb{R}} (e^{ivy} - 1 - ivy1_{|y|\leq 1})\nu_Y(dy)\}. \end{aligned} \tag{5.7}$$

One-dimensional integrals in this expression can be rewritten in two-dimensional form:

$$\int_{\mathbb{R}} (e^{iux} - 1 - iux 1_{|x|\le 1})\nu_X(dx)$$
$$= \int_{\mathbb{R}^2} (e^{iux+ivy} - 1 - (iux + ivy)1_{x^2+y^2\le 1})\tilde{\nu}_X(dx \times dy), \quad (5.8)$$

where $\tilde{\nu}_X(A) = \nu_X(A_X)\ \forall A \in \mathcal{B}(\mathbb{R})$. The equality in (5.8) holds because $\tilde{\nu}_X$ is supported by the set $\{(x, 0), x \in \mathbb{R}\}$ which means that in the integrand y is in reality always zero. Rewriting the second integral in (5.7) in the same manner, we complete the proof of the first part.

The if part. Since the Lévy measure is supported by the set $\{(x, y) : xy = 0\}$, it can be represented in the form (5.6) for some positive measures ν_X and ν_Y. Proposition (5.2) entails that these measures coincide with the Lévy measures of (X_t) and (Y_t). To conclude it remains to apply the Lévy-Khinchin formula for process (X_t, Y_t). □

5.5 Copulas for Lévy processes with positive jumps

The key idea of our approach is that for parametrizing the dependence between jumps of Lévy processes, the Lévy measure plays the same role as the probability measure for random variables. Hence, to model the dependence, we must construct copulas for Lévy measures. The principal difference from the random variable case and the main difficulty is that Lévy measures are not necessarily finite: they may have a nonintegrable singularity at zero. Due to this fact, Lévy copulas that we are about to introduce are defined on infinite intervals rather than on $[0, 1]^2$. Lévy copulas for processes with positive jumps are functions from $[0, \infty]^2$ to $[0, \infty]$ and Lévy copulas for general Lévy processes are functions from $[-\infty, \infty]^2$ to $[-\infty, \infty]$.

In this section, we discuss the conceptually simpler case of Lévy processes with positive jumps, that is, processes with Lévy measure concentrated on $]0, \infty[^2$. The next section shows how to generalize this approach to other types of Lévy processes.

The role of distribution function is now played by the tail integral. At this stage we do not impose any integrability or continuity conditions.

DEFINITION 5.7 *A d-dimensional* tail integral *is a function* $U : [0, \infty] \to [0, \infty]$ *such that*

1. $(-1)^d U$ *is a d-increasing function.*

2. U *is equal to zero if one of its arguments is equal to* ∞.

3. U is finite everywhere except at zero and $U(0,\dots,0)=\infty$.

Later, U will be interpreted as the tail integral of a Lévy measure:

$$U(x_1,\dots,x_d)=\nu([x_1,\infty[\times\dots\times[x_d,\infty[),\ x_1,\dots,x_d\in[0,\infty[^d\setminus\{0\}, \quad (5.9)$$

but at this stage we only need this abstract tail integral, analogous to abstract distribution function (Definition 5.2). The *margins* of a tail integral are defined similarly to the margins of a distribution function:

$$U(0,\dots,0,x_k,0,\dots,0)=U_k(x_k).$$

DEFINITION 5.8 Lévy copula for processes with positive jumps
A two-dimensional Lévy copula for Lévy processes with positive jumps, or, for short, a positive Lévy copula, is a 2-increasing grounded function F : $[0,\infty]^2\to[0,\infty]$ with uniform margins, that is, $F(x,\infty)=F(\infty,x)=x$.

REMARK 5.2 Copulas for general Lévy processes that we will define in the next section, are functions from $[-\infty,\infty]^2$ to $[-\infty,\infty]$. However, positive Lévy copulas can be extended to copulas for general Lévy processes, by setting $F(x,y)=0$ if $x<0$ or $y<0$. □

The next theorem is a reformulation of Sklar's theorem for tail integrals and Lévy copulas. It shows that Lévy copulas link multidimensional tail integrals to their margins in the same way as the copulas link the distribution functions to their margins.

THEOREM 5.3
Let U be a two-dimensional tail integral with margins U_1 and U_2. There exists a positive Lévy copula F such that

$$U(x_1,x_2)=F(U_1(x_1),U_2(x_2)). \quad (5.10)$$

If U_1 and U_2 are continuous, this Lévy copula is unique. Otherwise it is unique on $\operatorname{Ran}U_1\times\operatorname{Ran}U_2$, the product of ranges of one-dimensional tail integrals.

Conversely, if F is a positive Lévy copula and U_1,U_2 are one-dimensional tail integrals then (5.10) defines a two-dimensional tail integral.

PROOF We suppose for simplicity that U_1 and U_2 are continuous. The proof in the general case can be carried out along the lines of the proof of Sklar's theorem [364] but it is rather lengthy.

The direct statement. Choose the inverses U_1^{-1} and U_2^{-1} such that $U_1^{-1}(0)=U_2^{-1}(0)=\infty$ and $U_1^{-1}(\infty)=U_2^{-1}(\infty)=0$ and let

$$\tilde F(y_1,y_2)=U(U_1^{-1}(y_1),U_2^{-1}(y_2)). \quad (5.11)$$

It is now easy to check directly that $\tilde{F}$ is a Lévy copula and that it satisfies (5.10). Suppose that there exists another such Lévy copula $\tilde{\tilde{F}}$. Then for every $x_1 \in [0,\infty]$ and $x_2 \in [0,\infty]$, $\tilde{F}(U_1(x_1), U_2(x_2)) = \tilde{\tilde{F}}(U_1(x_1), U_2(x_2))$. Therefore, from the continuity of U_1 and U_2, we conclude that for every $y_1 \in [0,\infty]$ and $y_2 \in [0,\infty]$, $\tilde{F}(y_1, y_2) = \tilde{\tilde{F}}(y_1, y_2)$.

The converse statement is a matter of straightforward verification. □

Tail integrals and Lévy measures

For every Lévy measure ν on $[0,\infty[^2$, one can define its tail integral as follows:

$$\begin{aligned} &U(x_1, x_2) = 0 \quad \text{if } x_1 = \infty \text{ or } x_2 = \infty; \\ &U(x_1, x_2) = \nu([x_1, \infty[\times[x_2, \infty[) \quad \text{for } (x_1, x_2) \in [0,\infty[^2 \backslash \{0\}; \\ &U(0,0) = \infty. \end{aligned}$$

On the other hand, modulo a discussion of continuity properties, identical to the one following Definition 5.2, every two-dimensional tail integral defines a positive measure ν on $[0,\infty[^2\backslash\{0\}$. However, to be a Lévy measure, ν must satisfy the integrability condition of Theorem 3.7:

$$\int_{[0,1]^2} |\mathbf{x}|^2 \nu(d\mathbf{x}) = \int_{[0,1]^2} |\mathbf{x}|^2 dU < \infty, \tag{5.12}$$

where the second integral is a Stieltjes integral which exists because U is an increasing function. The following simple lemma shows, when this integrability condition is satisfied.

LEMMA 5.2
Let U be a two-dimensional tail integral with margins U_1 and U_2. U defines a Lévy measure on $[0,\infty[^2\backslash\{0\}$, that is, the integrability condition (5.12) is satisfied if and only if the margins of U correspond to Lévy measures on $[0,\infty[$, that is, for $k = 1,2$,

$$\int_0^1 x^2 dU_k(x) < \infty.$$

PROOF The stated equivalence follows from the estimation:

$$\begin{aligned} \int_{[0,1]^2} |\mathbf{x}|^2 dU(\mathbf{x}) &= \int_{[0,1]^2} \sum_{i=1}^{2} x_i^2 dU(\mathbf{x}) \\ &= \sum_{i=1}^{2} \int_{[0,1]^2} x_i^2 dU(\mathbf{x}) = \sum_{i=1}^{2} \int_0^1 x_i^2 U_i(x_i) - C_1 - C_2, \end{aligned}$$

where the constants

$$C_1 = \int_{[0,1]\times[1,\infty]} x_1^2 dU(\mathbf{x}) \le \int_{[0,1]\times[1,\infty]} dU(\mathbf{x}) \quad \text{and}$$

$$C_2 = \int_{[1,\infty]\times[0,1]} x_2^2 dU(\mathbf{x}) \le \int_{[1,\infty]\times[0,1]} dU(\mathbf{x})$$

are clearly finite. □

Now all the tools are ready to characterize dependence structures of Lévy processes (for the moment, only the ones with positive jumps).

THEOREM 5.4

Let (X_t, Y_t) be a two-dimensional Lévy process with positive jumps having tail integral U and marginal tail integrals U_1 and U_2. There exists a two-dimensional positive Lévy copula F which characterizes the dependence structure of (X_t, Y_t), that is, for all $x_1, x_2 \in [0, \infty]$,

$$U(x_1, x_2) = F(U_1(x_1), U_2(x_2)). \tag{5.13}$$

If U_1 and U_2 are continuous, this Lévy copula is unique. Otherwise it is unique on $\operatorname{Ran} U_1 \times \operatorname{Ran} U_2$.

Conversely, let (X_t) and (Y_t) be two one-dimensional Lévy processes with positive jumps having tail integrals U_1 and U_2 and let F be a two-dimensional positive Lévy copula. Then there exists a two-dimensional Lévy process with Lévy copula F and marginal tail integrals U_1 and U_2. Its tail integral is given by Equation (5.13).

This result is a direct consequence of Theorem 5.3 and Lemma 5.2. The first part of this theorem states that all types of dependence of Lévy processes, including complete dependence and independence, can be represented with Lévy copulas and the second part shows that one can construct multivariate Lévy process models by specifying separately jump dependence structure and one-dimensional laws for the components. The laws of components can have very different structure, in particular, it is possible to couple compound Poisson components with infinite-activity ones.

When the dependence is specified via a Lévy copula and both the copula and the one-dimensional tail integrals are sufficiently smooth, the Lévy density can be computed by differentiation:

$$\nu(x_1, x_2) = \left.\frac{\partial^2 F(y_1, y_2)}{\partial y_1 \partial y_2}\right|_{y_1 = U_1(x_1), y_2 = U_2(x_2)} \nu_1(x_1)\nu_2(x_2)$$

Examples of positive Lévy copulas

Now we will compute the Lévy copulas that correspond to various basic dependence structures.

Example 5.3 Independence
Let (X_t, Y_t) be a Lévy process with independent components. By Proposition 5.3, its Lévy measure is $\nu(A) = \nu_1(A_X) + \nu_2(A_Y)$. The tail integral of this Lévy measure is $U(x_1, x_2) = U_1(x_1)1_{x_2=0} + U_2(x_2)1_{x_1=0}$ and Formula (5.11) allows to compute the Lévy copula of independent processes:

$$F_\perp(x_1, x_2) = x_1 1_{x_2=\infty} + x_2 1_{x_1=\infty}. \tag{5.14}$$

If U_1 and/or U_2 are not continuous, Equation (5.14) gives one of the possible Lévy copulas of (X_t, Y_t). □

To discuss complete jump dependence or comonotonicity of components of a Lévy process with positive jumps, we need the notion of an increasing set.

DEFINITION 5.9 *A subset S of $\bar{\mathbb{R}}^2$ is called* increasing *if for every two vectors $(v_1, v_2) \in S$ and $(u_1, u_2) \in S$ either $v_k < u_k\ \forall k$ or $v_k > u_k\ \forall k$.*

Clearly, an element of an increasing set is completely determined by one coordinate only. This motivates the following definition of jump dependence.

DEFINITION 5.10 *Let $\mathbf{X} \equiv (X_t^1, X_t^2)$ be a Lévy process with positive jumps. Its jumps are said to be* completely dependent *or comonotonic if there exists an increasing subset S of $]0,\infty[^2$ such that every jump $\Delta\mathbf{X}$ of $\mathbf{X}$ is in S.*

Clearly, if the jumps of two pure-jump Lévy processes are completely dependent, the trajectory of one of them can be reconstructed from the trajectory of the other.

PROPOSITION 5.4 Complete dependence
Let $\mathbf{X} \equiv (X_t^1, X_t^2)$ be a Lévy process with positive jumps. If its jumps are completely dependent, then (a possible) Lévy copula of $\mathbf{X}$ is the complete dependence Lévy copula defined by

$$F_{\|}(x_1, x_2) = \min(x_1, x_2).$$

Conversely, if the Lévy copula of $\mathbf{X}$ is given by $F_{\|}$ and the tail integrals of components of $\mathbf{X}$ are continuous, then the jumps of $\mathbf{X}$ are completely dependent.

PROOF The direct statement. The jumps of $\mathbf{X}$ are completely dependent if and only if there exists an increasing subset S of $]0,\infty[^2$ such that the Lévy measure ν of $\mathbf{X}$ is concentrated on S. Therefore, for every $\mathbf{x} = (x_1, x_2)$,

$$U(\mathbf{x}) = \int_{\{\mathbf{y} \geq \mathbf{x}\} \cap S} \nu(d\mathbf{y});$$
$$U_k(x_k) = \int_{\{y_k \geq x_k\} \cap S} \nu(d\mathbf{y}) \quad k = 1, 2.$$

Comparing the above equalities and using the properties of an increasing subset we find:

$$U(\mathbf{x}) = \min(U_1(x_1), U_2(x_2)), \tag{5.15}$$

which proves the first part of the theorem.

The converse statement. In this case the tail integral of $\mathbf{X}$ has the form (5.15) which means that the Lévy measure of $\mathbf{X}$ is concentrated on the set $\{(x_1, x_2) : U_1(x_1) = U_2(x_2)\}$. If the tail integrals U_1 and U_2 are continuous, this set is increasing. □

Example 5.4 Dependence of stable processes

Lévy copulas of stable processes are homogeneous functions of order one, that is,

$$\forall c > 0, \quad F(cx_1, cx_2) = cF(x_1, x_2).$$

Indeed, by formula (3.32) the Lévy measure of an α-stable process satisfies

$$\nu(B) = c^\alpha \nu(cB) \; \forall c > 0 \; \forall B \in \mathcal{B}(\mathbb{R}^2).$$

This means in particular that the tail integral satisfies

$$U(cx_1, cx_2) = c^{-\alpha} U(x_1, x_2).$$

Substituting α-stable margins into this formula, we conclude that the Lévy copula is homogeneous of order 1.

The dependence structure of two stable processes can alternatively be specified via the spherical part of the Lévy measure of the two-dimensional process (see Proposition 3.15). In this case the two-dimensional process will also be stable. Lévy copulas allow a greater variety of possible dependence structures, because using Lévy copulas one can construct a Lévy process that has α-stable margins but is not α-stable itself. To do this, it suffices to take a Lévy copula that is not homogeneous, for example,

$$F(u, v) = \log\left\{\frac{1 - e^{-u-v}}{e^{-u} + e^{-v} - 2e^{-u-v}}\right\}.$$

This example is an Archimedean Lévy copula with generator $\phi(x) = \frac{e^{-x}}{1-e^{-x}}$ (see Proposition 5.6 below). □

Dependence of compound Poisson processes

Dependence of a compound Poisson process can be described using ordinary copulas (compare with Section 5.2). Indeed, a two-dimensional compound Poisson process (X_t, Y_t) can always be split onto dependent and independent parts (this is a consequence of the Lévy-Itô decomposition).

$$X_t = X_t^{\perp} + X_t^{\|}$$
$$Y_t = Y_t^{\perp} + Y_t^{\|}$$

where $(X_t^{\perp})$ and $(Y_t^{\perp})$ are independent from each other and the two other components (independent part of the process) whereas $(X_t^{\|})$ and $(Y_t^{\|})$ are dependent, have the same intensity and jump at the same time (dependent part of the process). Lévy measures of components in this decomposition are related to the Lévy measure ν of (X_t, Y_t):

$$\nu_X^{\perp}(A) = \nu(A \times \{0\}) \quad \forall A \in \mathcal{B}(\mathbb{R}_+)$$
$$\nu_Y^{\perp}(A) = \nu(\{0\} \times A) \quad \forall A \in \mathcal{B}(\mathbb{R}_+)$$
$$\nu^{\|}(A) = \nu(A) - \nu(A_X \times \{0\}) - \nu(\{0\} \times A_Y) \quad \forall A \in \mathcal{B}(\mathbb{R}_+^2)$$

The sets A_X and A_Y were defined in Proposition 5.3.

To fix the model completely, one must specify

- the intensity and jump size distribution of $(X_t^{\perp})$,
- the intensity and jump size distribution of $(Y_t^{\perp})$,
- the intensity of common shocks,
- jump size distributions of $(X_t^{\|})$ and $(Y_t^{\|})$,
- the copula of the last two distributions.

Hence, this approach requires a lot of different quantities, some of which are difficult to observe. It also does not allow to separate the dependence from the margins completely, because the intensities and the distributions of the components are determined both by the dependence structure and by the margins. It is therefore more convenient to use the Lévy copula approach. In this case one must specify two things:

- the margins via the intensity and jump size distribution of (X_t) and (Y_t) (which are easy to observe),
- the dependence structure via the Lévy copula of the process.

All other quantities can be computed from these ones (for the computation we suppose that the copula is continuous and that Lévy measures of components have no atoms, except the one at zero).

Let F be the Lévy copula and let U_X and U_Y be the tail integrals of (X_t) and (Y_t). Then the intensities of (X_t) and (Y_t) are

$$\lambda_X = \lim_{x \downarrow 0} U_X(x) \quad \text{and} \quad \lambda_Y = \lim_{y \downarrow 0} U_Y(y).$$

We cannot simply write $\lambda_X = U_X(0)$ because, according to our definition of tail integral (see Page 147), $U_X(0) = \infty$ for all processes, including compound Poisson. The tail integral of the two-dimensional process (X_t, Y_t) is $U(x, y) = F(U_X(x), U_Y(y))$. Its intensity is equal to the Lévy measure of $\mathbb{R}^2_+ \setminus \{0\}$, which means that

$$\lambda = \lim_{x,y \downarrow 0} (U_X(x) + U_Y(y) - U(x, y)) = \lambda_X + \lambda_Y - F(\lambda_X, \lambda_Y).$$

Since for every positive Lévy copula $0 \leq F(x, y) \leq \min(x, y)$, this intensity is always contained between $\max(\lambda_X, \lambda_Y)$ (strongest possible dependence) and $\lambda_X + \lambda_Y$ (independence). The intensity of the common Poisson shock (intensity of $X_t^{\|}$ and $Y_t^{\|}$) is equal to $F(\lambda_X, \lambda_Y)$. Tail integrals of independent components are

$$U_X^{\perp}(x) = \nu([x, \infty[\times\{0\}) = U(x, 0) - \lim_{y \downarrow 0} U(x, y) = U_X(x) - F(U_X(x), \lambda_Y)$$

$$\text{and} \quad U_Y^{\perp}(y) = U_Y(y) - F(\lambda_X, U_Y(y))$$

for $x, y \in]0, \infty[$. Finally, the tail integral of $(X_t^{\|}, Y_t^{\|})$ is

$$U_{\|}(x, y) = \lim_{x' \downarrow x; y' \downarrow y} U(x', y') = F(\min(U_X(x), \lambda_X), \min(U_Y(y), \lambda_Y))$$

for $(x, y) \in [0, \infty[^2 \setminus \{0\}$, and the survival function[2] of its jump size distribution has the form

$$H(x, y) = \frac{F(\min(U_X(x), \lambda_X), \min(U_Y(y), \lambda_Y))}{F(\lambda_X, \lambda_Y)}.$$

Construction of positive Lévy copulas

Lévy copulas can be computed from multidimensional Lévy processes using formula (5.11). However, this method is not very useful because there are not many multivariate Lévy models available. In this subsection, we give several other methods to construct Lévy copulas.

PROPOSITION 5.5

Let C be an (ordinary) 2-copula and $f(x)$ an increasing convex function from $[0, 1]$ to $[0, \infty]$. Then

$$F(x, y) = f(C(f^{-1}(x), f^{-1}(y))) \tag{5.16}$$

[2] The survival function $H(x)$ of a random variable X is defined by $H(x) = P(X > x)$ and the joint survival function $H(x, y)$ of random variables X and Y is $H(x, y) = P(X > x, Y > y)$.

defines a two-dimensional positive Lévy copula.

PROOF The fact that F is an 2-increasing function follows from properties 1 and 2 of Lemma 5.1. Groundedness and marginal properties can be checked directly. □

There are many functions from $[0,1]$ to $[0,\infty]$ with positive derivatives of all orders that can be used in this proposition, one example is $f(x) = \frac{x}{1-x}$.

By analogy to Archimedean copulas, one can also construct Archimedean Lévy copulas.

PROPOSITION 5.6

Let ϕ be a strictly decreasing convex function from $[0,\infty]$ to $[0,\infty]$ such that $\phi(0) = \infty$ and $\phi(\infty) = 0$. Then

$$F(x,y) = \phi^{-1}(\phi(x) + \phi(y)) \tag{5.17}$$

defines a two-dimensional positive Lévy copula.

PROOF This is again a consequence of properties 1 and 2 of Lemma 5.1, after observing that the function $\phi^{-1}(-u)$ is increasing and convex. □

Example 5.5

For $\phi(u) = u^{-\theta}$ with $\theta > 0$, we obtain the following parametric family of Lévy copulas:

$$F_\theta(u,v) = (u^{-\theta} + v^{-\theta})^{-1/\theta} \tag{5.18}$$

which reminds us of the Clayton family of copulas (5.3). It includes as limiting cases complete dependence (when $\theta \to \infty$) and independence (when $\theta \to 0$). □

Probabilistic interpretation of positive Lévy copulas

Unlike ordinary copulas, Lévy copulas are not distribution functions, but their derivatives have an interesting probabilistic interpretation.

LEMMA 5.3

Let F be a two-dimensional positive Lévy copula. Then for almost all $x \in [0,\infty]$, the function

$$F_x(y) = \frac{\partial}{\partial x} F(x,y)$$

exists and is continuous for all $y \in [0,\infty]$ *outside a countable set. Moreover, it is a distribution function of a positive random variable, that is, it is increasing and satisfies* $F_x(0) = 0$ *and* $F_x(\infty) = 1$.

PROOF For every fixed non-negative y, $F_x(y)$ exists for almost all non-negative x because $F(x,y)$ is increasing in x. It follows that for almost all x, $F_x(y)$ exists for all $y \in \mathbb{Q}_+$, where $\mathbb{Q}_+$ denotes the set of all non-negative rational numbers. Let us fix one such x for the rest of the proof.

$F_x(y)$ is increasing on $\mathbb{Q}_+$ because $F(x,y)$ is 2-increasing. In addition, because F has uniform margins, $F_x(y)$ satisfies $0 \le F_x(y) \le 1$ for all $y \in \mathbb{Q}_+$. Therefore, for every point $y \in \mathbb{R}_+$ and for every two sequences of rational numbers $\{y_n^+\}$ and $\{y_n^-\}$ such that $y_n^+ \downarrow y$ and $y_n^- \uparrow y$, the limits $\lim F_x(y_n^+)$ and $\lim F_x(y_n^-)$ exist. Moreover, for all y outside a countable set C_x these limits are equal. In this case we denote

$$\lim F_x(y_n^+) = \lim F_x(y_n^-) \equiv \tilde{F}_x(y).$$

For every nonnegative $y \in C_x$, every n and every $\Delta \in \mathbb{R}$ we can write:

$$\frac{F(x+\Delta,y) - F(x,y)}{\Delta} - \tilde{F}_x(y) \le \frac{F(x+\Delta,y_n^+) - F(x,y_n^+)}{\Delta} - \tilde{F}_x(y)$$

since $F(x,y)$ is 2-increasing. Now we would like to show that for every $\varepsilon > 0$ there exist $\delta > 0$ such that for all Δ verifying $|\Delta| < \delta$, the left-hand side of the above inequality is smaller than ε. First observe that for every $\varepsilon > 0$ it is possible to choose n such that $|F_x(y_n^+) - \tilde{F}_x(y)| \le \varepsilon/2$ because $\tilde{F}_x(y)$ was definied as the limit of $F_x(y_n^+)$. Moreover, for this fixed n one can choose δ such that for all Δ verifying $|\Delta| \le \delta$,

$$\left|\frac{F(x+\Delta,y_n^+) - F(x,y_n^+)}{\Delta} - F_x(y_n^+)\right| \le \varepsilon/2. \tag{5.19}$$

because y_n^+ is rational and $F_x(y)$ exists for rational y.

Therefore, for every Δ verifying $|\Delta| \le \delta$,

$$\frac{F(x+\Delta,y) - F(x,y)}{\Delta} - F_x(y) \le \varepsilon \tag{5.20}$$

The second sequence, $\{y_n^-\}$, can be used to bound this expression from below, which allows to conclude that for every nonnegative $y \in C_x$, $F(x,y)$ is differentiable and $F_x(y) = \frac{\partial}{\partial x}F(x,y)$ is continuous. The fact that F_x is increasing was also established in the proof and the other claims of the lemma can be verified directly. □

The following theorem shows that F determines the law of the (transformed) jump in the second component of a Lévy process conditionally on the size of jump in the first one. It will be useful in Chapter 6 for simulation of multivariate dependent Lévy processes. Its proof can be found in [372].

THEOREM 5.5
Let (X_t, Y_t) be a two-dimensional Lévy process with positive jumps, having marginal tail integrals U_1, U_2 and Lévy copula F. Let ΔX_t and ΔY_t be the sizes of jumps of the two components at time t. Then, if U_1 has a non-zero density at x, $F_{U_1(x)}$ is the distribution function of $U_2(\Delta Y_t)$ conditionally on $\Delta X_t = x$:

$$F_{U_1(x)}(y) = P\{U_2(\Delta Y_t) \leq y | \Delta X_t = x\} \tag{5.21}$$

Summary of definitions and results for multivariate case

All results of this section can be easily generalized to more than two dimensions (see [372]). Some of the generalizations are given below; others are left to the reader as exercise.

DEFINITION 5.11 Positive Lévy copula *A n-dimensional positive Lévy copula is a n-increasing grounded function $F : [0,\infty]^n \to [0,\infty]$ with margins F_k, $k = 1 \dots n$, which satisfy $F_k(u) = u$ for all u in $[0,\infty]$.*

THEOREM 5.6
Let U be the tail integral of an n-dimensional Lévy process with positive jumps and let $U_1, \cdots, U_n$ be the tail integrals of its components. Then there exists an n-dimensional positive Lévy copula F such that for all vectors $(x_1, \cdots, x_n)$ in $\mathbb{R}^n_+$,

$$U(x_1, \cdots, x_n) = F(U_1(x_1), \cdots, U_n(x_n)).$$

If $U_1, \cdots, U_n$ are continuous then F is unique, otherwise it is unique on $\operatorname{Ran} U_1 \times \cdots \times \operatorname{Ran} U_n$.

Conversely, if F is a n-dimensional positive Lévy copula and $U_1, \cdots, U_n$ are tail integrals of Lévy measures on $[0,\infty[$, then the function U defined above is the tail integral of a n-dimensional Lévy process with positive jumps having marginal tail integrals $U_1, \cdots, U_n$.

The independence Lévy copula in the multivariate case has the form:

$$F_\perp(x_1, \dots, x_n) = x_1 1_{x_2=\infty,\dots,x_n=\infty} + \dots + x_n 1_{x_1=\infty,\dots,x_{n-1}=\infty}$$

and the complete dependence Lévy copula is

$$F_\parallel(x_1, \dots, x_n) = \min(x_1, \dots, x_n)$$

Finally, we give a multidimensional equivalent of Proposition 5.6.

PROPOSITION 5.7
Let ϕ be a strictly decreasing function from $[0,\infty]$ to $[0,\infty]$ such that $\phi(0) = \infty$, $\phi(\infty) = 0$ and ϕ^{-1} has derivatives up to the order n on $]0,\infty[$with alternating signs, that is, $(-1)^k \frac{d^k\phi^{-1}(t)}{dt^k} > 0$. Then

$$F(x_1,\cdots,x_n) = \phi^{-1}(\phi(x_1) + \cdots + \phi(x_n))$$

defines a n-dimensional positive Lévy copula.

5.6 Copulas for general Lévy processes

DEFINITION 5.12 *$F(x,y) : [-\infty,\infty]^2 \to [-\infty,\infty]$ is a Lévy copula*[3] *if it has the following three properties:*

- *F is 2-increasing*
- *$F(0,x) = F(x,0) = 0 \quad \forall x$*
- *$F(x,\infty) - F(x,-\infty) = F(\infty,x) - F(-\infty,x) = x$*

From this definition it is clear that a positive Lévy copula can be extended to a Lévy copula by putting $F(x,y) = 0$ if $x < 0$ or $y < 0$.

Constructing models with Lévy copulas is simple when both the Lévy copula and the marginal tail integrals are sufficiently smooth to allow using Lévy density.

Case of Lévy measures with densities

We start by defining the notion of tail integral for Lévy measures on $\mathbb{R}$. Note that this definition, used in the setting of Lévy measures with densities, is slightly different from the one that we will use in the general case (Definition 5.14).

DEFINITION 5.13 *Let ν be a Lévy measure on $\mathbb{R}$. The* tail integral *of ν is a function $U : \bar{\mathbb{R}} \setminus \{0\} \to \infty$ defined by*

$$\begin{aligned}
&U(x) = \nu([x,\infty[) && \text{for } x \in]0,\infty[,\\
&U(x) = -\nu(]-\infty,-x]) && \text{for } x \in]-\infty,0[,\\
&U(\infty) = U(-\infty) = 0.
\end{aligned}$$

[3]We thank Jan Kallsen for suggesting this method of extending positive Lévy copulas to copulas for general Lévy processes.

The signs are chosen in such way that the tail integral is a decreasing function both on $]0,\infty]$ and on $[-\infty,0[$.

The following result allows to construct two-dimensional Lévy densities from one dimensional ones using Lévy copulas that are sufficiently smooth.

PROPOSITION 5.8
Let F be a two-dimensional Lévy copula, continuous on $[-\infty,\infty]^2$, such that $\frac{\partial F(u,v)}{\partial u \partial v}$ exists on $]-\infty,\infty[^2$ and let U_1 and U_2 be one-dimensional tail integrals with densities ν_1 and ν_2. Then

$$\left.\frac{\partial F(u,v)}{\partial u \partial v}\right|_{u=U_1(x),v=U_2(y)} \nu_1(x)\nu_2(y)$$

is the Lévy density of a Lévy measure with marginal Lévy densities ν_1 and ν_2.

General case

If the Lévy copula and the marginal tail integrals are not sufficiently smooth, multidimensional tail integrals must be used instead of Lévy densities. Because the singularity (zero) is now in the center of the domain of interest, each corner of the Lévy measure must be treated separately, that is, in the one-dimensional case we need two tail integrals: U^+ and U^- and in the two-dimensional case we need four tail integrals: U^{++}, U^{+-}, U^{-+} and U^{--}.

DEFINITION 5.14 *Let ν be a Lévy measure on $\mathbb{R}$. This measure has two tail integrals, $U^+ : [0,\infty] \to [0,\infty]$ for the positive part and $U^- : [-\infty,0] \to [-\infty,0]$ for the negative part, defined as follows:*

$$\begin{aligned}
&U^+(x) = \nu([x,\infty[) && \text{for } x \in]0,\infty[, && U^+(0) = \infty, && U^+(\infty) = 0;\\
&U^-(x) = -\nu(]-\infty,-x]) && \text{for } x \in]-\infty,0[, && U^-(0) = -\infty, && U^-(-\infty) = 0.
\end{aligned}$$

Let ν be a Lévy measure on $\mathbb{R}^2$ with marginal tail integrals U_1^+, U_1^-, U_2^+ and U_2^-. This measure has four tail integrals: U^{++}, U^{+-}, U^{-+} and U^{--}, where each tail integral is defined on its respective quadrant, including the coordinate axes, as follows:

$$\begin{aligned}
&U^{++}(x,y) = \nu([x,\infty[\times[y,\infty[), && \text{if } x \in]0,\infty[\text{ and } y \in]0,\infty[\\
&U^{+-}(x,y) = -\nu([x,\infty) \times (-\infty,y]), && \text{if } x \in]0,\infty[\text{ and } y \in]-\infty,0[\\
&U^{-+}(x,y) = -\nu((-\infty,x] \times [y,\infty)), && \text{if } x \in]-\infty,0[\text{ and } y \in]0,\infty[\\
&U^{--}(x,y) = \nu((-\infty,x] \times (-\infty,y]), && \text{if } x \in]-\infty,0[\text{ and } y \in]-\infty,0[
\end{aligned}$$

If x or y is equal to $+\infty$ or $-\infty$, the corresponding tail integral is zero and if x or y is equal to zero, the tail integrals satisfy the following "marginal" *conditions:*

$$\begin{aligned}
U^{++}(x,0) - U^{+-}(x,0) &= U_1^+(x) \\
U^{-+}(x,0) - U^{--}(x,0) &= U_1^-(x) \\
U^{++}(0,y) - U^{-+}(0,y) &= U_2^+(y) \\
U^{+-}(0,y) - U^{--}(0,y) &= U_2^-(y)
\end{aligned}$$

When a two-dimensional Lévy measure charges the coordinate axes, its tail integrals are not determined uniquely because it is unclear, which of the tail integrals should contain the mass on the axes. Hence, different sets of tail integrals can sometimes correspond to the same Lévy measure.

The relationship between Lévy measures and Lévy copulas is introduced via a representation theorem, analogous to Theorems 5.1 and 5.3.

THEOREM 5.7

Let ν be a Lévy measure on $\mathbb{R}^2$ with marginal tail integrals U_1^+, U_1^-, U_2^+ and U_2^-. There exists a Lévy copula F such that U^{++}, U^{+-}, U^{-+} and U^{--} are tail integrals of ν where

$$\begin{aligned}
U^{++}(x,y) &= F(U_1^+(x), U_2^+(y)) \quad \text{if } x \geq 0 \text{ and } y \geq 0 \\
U^{+-}(x,y) &= F(U_1^+(x), U_2^-(y)) \quad \text{if } x \geq 0 \text{ and } y \leq 0 \\
U^{-+}(x,y) &= F(U_1^-(x), U_2^+(y)) \quad \text{if } x \leq 0 \text{ and } y \geq 0 \\
U^{--}(x,y) &= F(U_1^-(x), U_2^-(y)) \quad \text{if } x \leq 0 \text{ and } y \leq 0
\end{aligned} \tag{5.22}$$

If the marginal tail integrals are absolutely continuous and ν does not charge the coordinate axes, the Lévy copula is unique.

Conversely, if F is a Lévy copula and U_1^+, U_1^-, U_2^+ and U_2^- are tail integrals of one-dimensional Lévy measures then the above formulae define a set of tail integrals of a Lévy measure.

PROOF The Lévy copula can be constructed in each of the four quadrants along the lines of the proof of Theorem 5.3 and Lemma 5.2. With this construction we obtain a function that is 2-increasing on each quadrant but not on the whole $\mathbb{R}^2$. However, the fact that each of the four parts is continuous on its domain including the coordinate axes and equal to zero on the axes entails that F is 2-increasing on $\bar{\mathbb{R}}^2$. □

Example 5.6 Independence of Lévy processes

Since the Lévy measure of a two-dimensional Lévy process with independent components is supported by the coordinate axes, the corresponding Lévy copula is not unique. In this case, Proposition 5.3 entails that for every x and y such that $xy \neq 0$ the corresponding tail integral is zero. Hence, $F(x,y) = 0$

if x and y are finite. Formulae (5.22) show that every Lévy copula that satisfies this property is a Lévy copula of independence. Examples of possible independence Lévy copulas are $x1_{y=\infty} + y1_{x=\infty}$, $-x1_{y=-\infty} - y1_{x=-\infty}$, etc. □

Example 5.7 Complete dependence

In the setting of Lévy processes with jumps of arbitrary sign, one must distinguish two types of complete dependence. In the first case (complete positive dependence) there exists an increasing subset S of $\mathbb{R}^2$ such that every jump $\Delta\mathbf{X}$ of the two-dimensional process is in S. In this situation, using the same method as in Proposition 5.4, we find

$$\begin{aligned}
U^{++}(x,y) &= \min\{U_1^+(x), U_2^+(y)\}\\
U^{--}(x,y) &= \min\{-U_1^-(x), -U_2^-(y)\}\\
U^{+-} &\equiv U^{-+} \equiv 0
\end{aligned}$$

Hence, $F_{\uparrow\uparrow}(x,y) = \min(|x|,|y|)1_{xy\geq 0}$ is a possible copula for this process. In the second case (complete negative dependence) every jump of the two-dimensional process must belong to a *decreasing* subset of $\mathbb{R}^2$, that is, a set S such that for every two vectors $(v_1,v_2) \in S$ and $(u_1,u_2) \in S$ either $v_1 < u_1$ and $v_2 > u_2$ or $v_1 > u_1$ and $v_2 < u_2$. In this case

$$\begin{aligned}
U^{+-}(x,y) &= -\min\{U_1^+(x), -U_2^-(y)\}\\
U^{-+}(x,y) &= -\min\{-U_1^-(x), U_2^+(y)\}\\
U^{++} &\equiv U^{--} \equiv 0
\end{aligned}$$

and $F_{\uparrow\downarrow}(x,y) = -\min(|x|,|y|)1_{xy\leq 0}$. □

Example 5.8

Starting from the positive Lévy copula (5.18) one can define a one-parameter family of Lévy copulas which includes independence and complete positive and negative dependence.

$$F_\theta(u,v) = \begin{cases} (|u|^{-\theta} + |v|^{-\theta})^{-1/\theta}1_{xy\geq 0} & \text{if } \theta > 0\\ -(|u|^{\theta} + |v|^{\theta})^{1/\theta}1_{xy\leq 0} & \text{if } \theta < 0 \end{cases} \tag{5.23}$$

Computing the limits we find that

$$\begin{aligned}
F_\theta &\to F_{\uparrow\downarrow} && \text{when } \theta \to -\infty\\
F_\theta &\to F_{\|} && \text{when } \theta \to 0\\
F_\theta &\to F_{\uparrow\uparrow} && \text{when } \theta \to \infty
\end{aligned}$$

□

Construction of Lévy copulas

Construction of general Lévy copulas is more difficult than that of positive ones because Propositions 5.5 and 5.6 cannot be directly extended to this setting. A possible solution that we will now discuss is to construct general Lévy copulas from positive ones by gluing them together: this amounts to specifying the dependence of jumps of different signs separately. Let F^{++}, F^{--}, F^{-+} and F^{+-} be positive Lévy copulas and consider the following expression:

$$\begin{aligned} F(x,y) = F^{++}(c_1|x|, c_2|y|)1_{x\geq 0, y\geq 0} + F^{--}(c_3|x|, c_4|y|)1_{x\leq 0, y\leq 0} \\ - F^{+-}(c_5|x|, c_6|y|)1_{x\geq 0, y\leq 0} - F^{-+}(c_7|x|, c_8|y|)1_{x\leq 0, y\geq 0} \end{aligned}$$

F is a 2-increasing function if the constants $c_1, \ldots, c_8$ are positive. The marginal conditions impose four additional constraints. For example, for $x > 0$ one has $F(x, \infty) - F(x, -\infty) = c_1 x + c_5 x = x$, therefore $c_1 + c_5 = 1$. The other constraints are $c_7 + c_3 = 1$, $c_2 + c_8 = 1$ and $c_6 + c_4 = 1$. Finally we obtain that

$$\begin{aligned} F(x,y) = F^{++}(c_1|x|, c_2|y|)1_{x\geq 0, y\geq 0} + F^{--}(c_3|x|, c_4|y|)1_{x\leq 0, y\leq 0} \\ -F^{+-}((1-c_1)|x|, (1-c_4)|y|)1_{x\geq 0, y\leq 0} - F^{-+}((1-c_3)|x|, (1-c_2)|y|)1_{x\leq 0, y\geq 0} \end{aligned} \tag{5.24}$$

defines a Lévy copula if $c_1, \ldots, c_4$ are constants between 0 and 1.

To understand the meaning of this construction, let us fix marginal Lévy measures ν_1 and ν_2 and look at the two-dimensional Lévy measure that we obtain with copula (5.24). Its upper right-hand tail integral is

$$U^{++}(x,y) = F^{++}(c_1 U_1^+(x), c_2 U_2^+(y))$$

This means that the upper right-hand quadrant of the Lévy measure corresponds to a Lévy process with positive jumps, Lévy copula F^{++} and marginal Lévy measures $c_1\nu_1(dx)1_{x>0}$ and $c_2\nu_2(dy)1_{y>0}$.

Treating the other quadrants in the same manner, we conclude that a Lévy process with Lévy copula of the form (5.24) is a sum of four independent parts, that correspond to four quadrants of the Lévy measure. The components of the first part that corresponds to the upper right-hand quadrant, jump only upwards, have Lévy measures $c_1\nu_1(dx)1_{x>0}$ and $c_2\nu_2(dx)1_{x>0}$ and are linked with each other via the positive Lévy copula F^{++}. The second independent part of the process corresponds to the lower right-hand quadrant. Its first component jumps only upwards and has Lévy measure $(1 - c_1)\nu_1(dx)1_{x>0}$ whereas its second component jumps downwards and has Lévy measure $(1 - c_4)\nu_2(dx)1_{x<0}$. The two components are linked with each other via positive Lévy copula F^{+-}. The other two independent parts of the Lévy process can be characterized in the same way.

Because the proportion of jumps in the first component that is linked to the positive jumps of the second component does not depend on the size of jumps (it only depends on their sign), we call the Lévy copula defined by (5.24) *constant proportion Lévy copula.*

The class of all Lévy copulas is much larger than the class of constant proportion ones, but the latter one is already sufficiently large for most practical applications. Moreover, Lévy copulas of stable processes can always be represented in the form (5.24). Indeed, following Example 5.4, one can show that Lévy copulas of stable processes are homogeneous functions of order 1. Let $F(x,y)$ be one such Lévy copula. It satisfies $F(x,\infty) = xF(1,\infty)$. This means that

$$\tilde{F}(x,y) := F(\frac{x}{F(1,\infty)}, \frac{y}{F(\infty,1)}), \text{ for } x \geq 0 \text{ and } y \geq 0$$

is a positive Lévy copula. Treating the other three corners of F in the same manner, we obtain the representation (5.24).

Summary of definitions and results for the multivariate case

DEFINITION 5.15 Lévy copula *A n-dimensional Lévy copula is a function $F : [-\infty,\infty]^n \to [-\infty,\infty]$ with the following three properties:*

- *F is n-increasing*
- *F is equal to zero if at least one of its arguments is zero*
- $F(x,\infty\ldots\infty) - F(x,-\infty\ldots-\infty) = x$
  $F(\infty,x,\infty\ldots\infty) - F(-\infty,x,-\infty\ldots-\infty) = x$, *etc.*

To construct multidimensional Lévy measures from Lévy copulas, we introduce a special type of interval:

$$\mathcal{I}(x) = \begin{cases} [x,\infty[, & \text{if } x > 0 \\ ]-\infty,x], & \text{if } x < 0 \end{cases}$$

Using this notation we can now compute tail integrals of the Lévy measure everywhere except on the axes as follows:

$$\nu(\mathcal{I}(x_1)\times\ldots\times\mathcal{I}(x_n)) = (-1)^{\operatorname{sgn} x_1\cdots\operatorname{sgn} x_n} F(U_1^{\operatorname{sgn} x_1}(x_1),\ldots,U_n^{\operatorname{sgn} x_n}(x_n))$$

where $U_1^+, U_1^-, \ldots, U_n^+, U_n^-$ are one-dimensional marginal tail integrals. The nonuniqueness problems on the axes can be solved as it was done in the two-dimensional case in the beginning of this section.

Constructing Lévy copulas in high dimensions using the constant proportion method is not very practical, because in the general case, one needs to specify 2^n positive Lévy copulas plus many additional constants to construct one n-dimensional Lévy copula. A better solution is to use simplified constructions similar to Example 5.8.

5.7 Building multivariate models using Lévy copulas

In previous sections we saw that the multivariate dependence of Lévy processes can be described using the notion of the Lévy copula, which plays for Lévy processes a role analogous to copulas for random variables. In this section we will show using concrete examples that, as in the case of ordinary copulas, the appropriate choice of Lévy copula allows to distinguish the usual concept of linear correlation between returns from the concept of tail dependence and to prescribe different dependence patterns for upward and downward price moves. This distinction is relevant for applications in risk measurement where large comovements in asset prices are of a primary concern. As a result, copulas have been used in applications such as credit risk modelling and quantitative models of operational risk. The framework of the Lévy copulas allows to extend such approaches to truly dynamical settings.

Multivariate variance gamma modelling

Suppose that one wants to model the prices of two stocks, or stock indices, S_t^1 and S_t^2, by exponentials of variance gamma Lévy processes (see Table 4.5). How to describe dependence between them? A simple method, discussed in Section 5.1, consists in representing $\log S_t^1$ and $\log S_t^2$ as two components of a bivariate correlated Brownian motion, time changed by the same gamma subordinator. In this method the dependence structure is described by a single parameter ρ — correlation coefficient of the Brownian motion. The two stocks always have the same jump structure parameter (i.e., κ), are dependent even if the Brownian motions are decorrelated and have the same dependence of upward and downward jumps.

An alternative model can be constructed using Lévy copulas. We have:

$$\log S_t^1 = X_t, \qquad \log S_t^2 = Y_t,$$

where X_t and Y_t are two arbitrary variance gamma processes with dependence specified via a Lévy copula F. It can be taken of the form (5.24) (constant proportion copula). In this case, to specify the dependence structure completely one has to choose four positive Lévy copulas F^{++} (dependence of positive jumps of X and positive jumps of Y), F^{+-} (dependence of positive jumps of X and negative jumps of Y), F^{-+} and F^{--} and four parameters c_1, c_2, c_3 and c_4, which are real numbers between zero and one. If every positive Lévy copula is in some one-parameter family, then in order to specify the dependence in this model completely, one needs eight parameters. In most cases this is too many and some simplifying assumptions can be made.

Example 5.9 Stock price model

Here we assume that positive jumps of X_t are independent from negative jumps of Y_t and vice versa. This assumption is reasonable for stocks but it does not necessarily hold for exchange rates. In addition, dependence structure of positive jumps is described by a Clayton Lévy copula with parameter θ^+ and the dependence structure of negative jumps is given by a Clayton Lévy copula with parameter θ^-. In this case our model takes the form

$$\log S_t^1 = X_t = X_t^+ - X_t^-$$
$$\log S_t^2 = Y_t = Y_t^+ - Y_t^-$$

where X_t^+ and Y_t^+ denote the positive jump parts of the corresponding processes and X_t^- and Y_t^- are the negative jump parts; the couples (X^+, Y^+) and (X^-, Y^-) are independent, X^+ and Y^+ are dependent with copula F_{θ^+} and X^- and Y^- are dependent with copula F_{θ^-}. The dependence structure of (X_t, Y_t) is thus described by two parameters: θ^+ describes positive jumps and θ^- describes negative jumps. This allows to prescribe different dependence structures for positive and negative jumps, which was impossible within the simple model based on correlated Brownian motions. In addition, in this model (X_t) and (Y_t) can follow arbitrary variance gamma models and can be completely independent (this is the limiting case when θ^+ and θ^- approach zero).

If (X_t) is a variance gamma process with parameters σ, θ and κ then (X_t^+) will be a gamma subordinator with parameters $c_+ = 1/\kappa$ and $\lambda_+ = (\sqrt{\theta^2 + 2\sigma^2/\kappa} - \theta)/\sigma^2$ and X_t^- will be a gamma subordinator with parameters $c_- = 1/\kappa$ and $\lambda_- = (\sqrt{\theta^2 + 2\sigma^2/\kappa} + \theta)/\sigma^2$. The Lévy measure of the process (X_t, Y_t) is always supported by the set $\{(x, y) : xy \geq 0\}$, that is by the upper right-hand and lower left-hand quadrants. This means, in particular, that this model does not include negative dependence between jumps (but there may be some negative dependence between returns).

The left graph in Figure 5.2 depicts typical trajectories of two variance gamma processes with dependence structure described by this model with dependence parameters $\theta^- = 5$ (strong dependence of downward jumps) and $\theta^- = 1$ (moderate dependence of upward jumps).

To further compare this model to the model using subordinated Brownian motion, Figure 5.3 presents scatter plots of 50-day returns in the two models. The marginal processes are the same for the two graphs, both are variance gamma with parameters $\sigma = 0.3, \theta = -0.1$ and $\kappa = 0.5$. In the left graph, the dependence is specified via a Lévy copula model with parameters $\theta^+ = 0.5$ and $\theta^- = 10$. This corresponds to a linear correlation of approximately 80%. In the right graph, the two variance gamma processes result from subordinating a correlated Brownian motion with the same gamma subordinator. The parameters were again chosen to have a correlation of 80%. From the left graph it is clear that in models based on Lévy copulas, the dependence structures of positive and negative jumps can be radically different, which is an important issue in risk measurement. In addition to the strong asymmetry of positive and negative jumps, the left graph is characterized by a much stronger tail

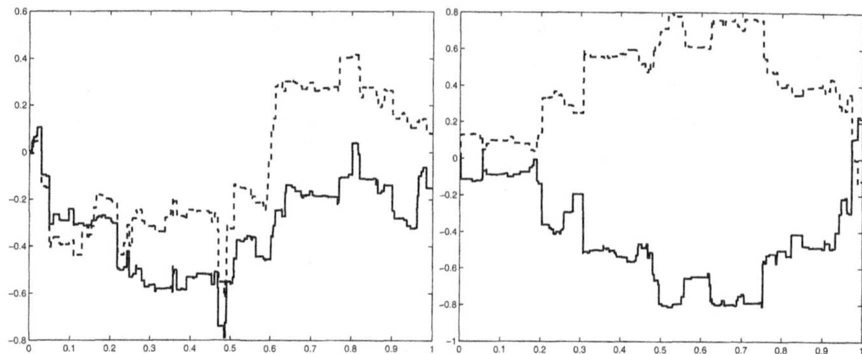

FIGURE 5.2: Left: Trajectories of two variance gamma processes with different dependence of upward and downward jumps. Right: Trajectories of two variance gamma processes with negative dependence.

dependence in the lower tail, than the right one. In particular, for the left graph, the expected jump size in the second component given that the jump in the first component is smaller than -0.4, is equal to -0.55, whereas for the right graph this expectation is only -0.43.

These two graph may remind the reader of the two graphs in Figure 5.1, obtained with ordinary copulas. However, there is a fundamental difference: Lévy copulas allow to construct dynamic models, that is, in a Lévy copula model the dependence is known at all time scales, whereas with ordinary copulas the dependence can only be prescribed for one chosen time scale, and it is hard to say anything about the others.

For comparison, Figure 5.4 provides a scatter plot of weakly returns of two major European stock market indices, DAX and AEX, for the period from January 1, 1990 to November 4, 2002. This graph clearly exhibits tail asymmetry and dependence in the left tail, but both phenomena are not as strong as in the left graph of Figure 5.3. □

Example 5.10 Exchange rate model
In major foreign exchange markets upward and downward movements are more or less symmetric but modelling negative dependence can be an important issue. One can build multivariate jump models with such properties by taking a Lévy copula of the form (5.23). Here one only needs one parameter to specify the dependence structure and the model includes independence and both positive and negative dependence. For positive values of θ the two underlying assets always move in the same direction, and for negative values of θ they always move in opposite directions. The latter situation is represented on the right graph of Figure 5.2. Here both processes are variance gamma

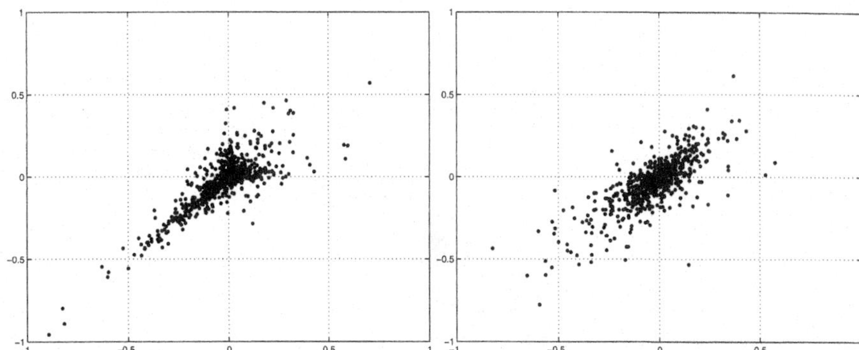

FIGURE 5.3: Scatter plots of 50-day returns of two exp-Lévy models driven by dependent variance gamma processes. Left: the dependence is described with a Lévy copula model. Right: The dependence is modelled by subordinating two correlated Brownian motions with the same gamma subordinator.

with the same parameters and $\theta = -3$. □

5.8 Summary

Many problems in quantitative finance involve the specification of joint models for several assets in a portfolio. In multivariate Gaussian models, dependence among returns of various assets is represented via the corresponding correlation coefficients. Once jump processes are introduced to account for large returns, tools are needed to parameterize the dependence between the jumps: as indicated by empirical studies, jumps often represent systemic risk and cannot be assumed to be independent across assets.

Two solutions proposed in the literature are to represent asset prices by multivariate subordinated Brownian motions or as a factor model driven by Poisson processes. Modelling by Brownian subordination allows to build analytically tractable multivariate Lévy processesbut does not allow a flexible specification of the dependence structure involved. Also, it restricts the choices for the dynamics of individual assets. Modelling dependence by common Poisson shocks is feasible for low-dimensional problems but requires the simultaneity of jumps in all assets involved.

A systematic way to describe the dependence structure of two random variables is to use copula functions, described in Section 5.3. Extending this notion to the case of Lévy processes, we have described in Section 5.6 the notion of Lévy copula which provides a systematic way to build multivariate

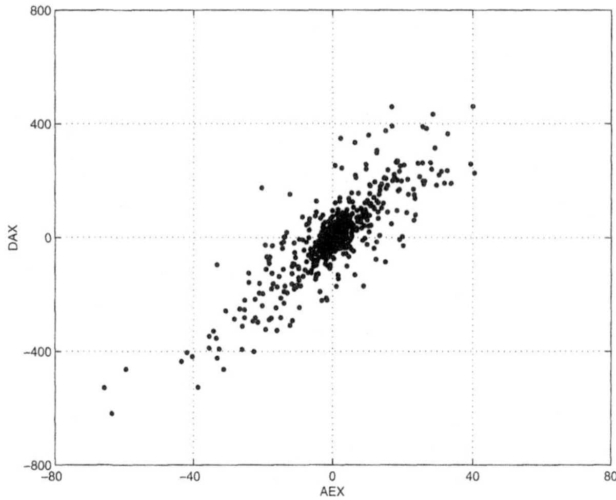

FIGURE 5.4: Scatter plot of weakly returns of DAX and AEX, for the period from January 1, 1990 to November 4, 2002.

Lévy processes from one-dimensional Lévy processes. Model building by Lévy copulas is a general approach which enables a flexible specification of dependence between returns while allowing for an arbitrary marginal distribution of returns for each asset.

Lévy copulas can also be useful outside the realm of financial modelling. Other contexts where modelling dependence in jumps is required are portfolios of insurance claims and models of operational risk.

Consider an insurance company with two subsidiaries, in France and in Germany. The aggregate loss process of the French subsidiary is modelled by the subordinator X_t and the loss process of the German one is Y_t. The nature of processes X_t and Y_t may be different because the subsidiaries may not be working in the same sector and many risks that cause losses are local. However, there will be some dependence between the claims, because there are common risks involved. In this setting it is convenient to model the dependence between X_t and Y_t using a positive Lévy copula. In this case, the two-dimensional Lévy measure of (X_t, Y_t) is known and the overall loss distribution and ruin probability can be computed.

Another example where jump processes naturally appear is given by models of operational risk. The 2001 Basel agreement defines the operational risk as "the risk of direct and indirect loss resulting from inadequate or failed internal processes, people and systems or from external events" and allows banks to use internal loss data to compute regulatory capital requirements. Consequently, reliable measures of risk taking into account the dependence between different business lines are required.[4] Aggregate loss processes from different business lines can be dynamically modelled by subordinators and the dependence between them can be accounted for using a positive Lévy copula.

[4]The internal management approach proposed by the Basel Committee on Banking Supervision does not take into account dependence between different business lines at present, because current industry practice does not permit reliable empirical measurement of such dependence. However, the possibility of accounting for it is likely to appear as banks and supervisors gain more experience.

Part II

Simulation and estimation

Chapter 6

Simulating Lévy processes

Anyone attempting to generate random numbers by deterministic means is, of course, living in a state of sin.

John von Neumann

Lévy processes allow to build more realistic models of price dynamics and offer a more accurate vision of risk than traditional diffusion-based models. However, the price to pay for this is an increased complexity of computations. Although Lévy processes are quite tractable compared to nonlinear diffusion models, analytical pricing methods are only available for European options. For all other applications, such as pricing of exotic options, scenario simulation for risk management, etc., numerical methods are unavoidable. Two possible choices are Monte Carlo methods and the numerical solution of partial integro-differential equations (see Chapter 12). However, as the dimension of the problem grows, PIDE methods become less feasible because computational complexity for fixed precision grows exponentially with dimension. On the contrary, the complexity of Monte Carlo methods for fixed precision grows only linearly with the dimension of the problem. Hence, in higher dimensions there is no alternative to simulation methods and it is very important to develop efficient algorithms for simulating Lévy processes.

For most Lévy processes the law of increments is not known explicitly. This makes it more difficult to simulate a path of a general Lévy process than for example of a Brownian motion. Depending on the type of Lévy process (compound Poisson, stable, etc.) and on the type of problem that must be solved (computing a functional of the entire trajectory, simulating the trajectory at a finite set of dates), the problem of simulating a Lévy process may be split into the following subproblems, which determine the structure of this chapter.

In Section 6.1, we discuss the simulation of compound Poisson processes. Sample paths of compound Poisson processes are piecewise linear and there is a finite number of jumps in every bounded interval. Hence, we can simulate a sample path exactly (without any discretization error) using a finite number of computer operations. To compute every functional of the trajectory it is sufficient to simulate a finite number of jump times and jump sizes.

In Section 6.2, we give examples of classes of Lévy processes that allow exact simulation of increments. Some option pricing problems (European or Bermudan options) and some portfolio management problems (discrete trading) only require the knowledge of the trajectory in a finite number of points. In this case, when one knows how to sample from the law of increments of a Lévy process, he or she can simulate the trajectory at discrete times without approximation. That is, for given times $t_1, \dots, t_n$ one can simulate an n-dimensional random vector $(X(t_1), \dots, X(t_n))$.

When the law of increments of a Lévy process is not known explicitly or when the problem requires an accurate knowledge of the trajectory, the Lévy process must be approximated by a compound Poisson process. A simple approximation consists in truncating the jumps smaller than some level ε. We treat this approximation in Section 6.3.

Such an approximation converges quite slowly when the jumps of the Lévy process are highly concentrated around zero. However, the small jumps of such processes, when properly renormalized, behave like a Brownian motion. In other words, when the precision of the Poisson approximation is low, one can improve it because an explicit expression of the renormalized limiting process is available. This improved approximation is discussed in Section 6.4.

Section 6.5 covers series representations for Lévy processes. Such representations can be seen as a more intelligent way to approximate Lévy processes by compound Poisson ones.

Finally, in Section 6.6, we discuss the simulation of multivariate Lévy processes, using the tools that were introduced in Chapter 5.

Except in a few important cases we do not discuss the methods of simulating random variables with known law. In general, whenever an explicit expression for the density is available, one can construct an appropriate rejection method. Many simulation algorithms for almost all known distributions can be found in the book by Devroye [113].

Table 6.1 lists available simulation methods for various Lévy processes, used in financial modelling. In the rest of this chapter we discuss these methods in detail.

6.1 Simulation of compound Poisson processes

Let $(X_t)_{t\geq 0}$ be a compound Poisson process with Lévy measure ν. For greater generality we also add drift b. Its characteristic function is

$$\phi_t(u) = \exp t[iub + \int_{\mathbb{R}} (e^{iux} - 1)\nu(dx)]$$

TABLE 6.1: Available simulation methods for different Lévy processes

Compound Poisson	See Algorithms 6.1 and 6.2. CP processes can be simulated exactly (Section 6.1); the computational time grows linearly with intensity
Jump diffusion	See Algorithms 6.3 and 6.4. One can simulate a discretized trajectory using the fact that diffusion part is independent from the jump part.
Stable	See Algorithms 6.6 and 6.15. There exist explicit methods for simulating the increments of a stable process (Example 6.3). Series representations are also available (Example 6.15). Their convergence rates may be improved using normal approximation (Example 6.10).
Variance gamma	See Algorithm 6.11. Variance gamma process can be represented either as the difference of two gamma processes or as a Brownian motion subordinated by a gamma process. Since efficient methods for simulating the gamma process are available (see Examples 6.4 and 6.16), the variance gamma process is also easy to simulate.
Normal inverse Gaussian	See Algorithm 6.12. This process can be obtained by subordinating a Brownian motion with an inverse Gaussian subordinator, which is easy to simulate (Example 6.5). Because the Lévy measure of the IG process is explicit, one can construct series representations for both IG and NIG processes. The normal approximation of small jumps is valid.
Generalized hyperbolic	More difficult to simulate than the others because the Lévy measure is not known in explicit form while the probability density is only known for one time scale and even for this time scale the expression involves special functions. One can simulate a discretized trajectory (Example 6.6) using the fact that GH can be obtained by subordinating Brownian motion with a generalized inverse Gaussian subordinator and GIG random variables are easier to simulate (see [113]) because their probability density is analytic.
Tempered stable	See Example 6.9. May be simulated using compound Poisson approximation or series representations (the Lévy measure is explicit). Normal approximation is valid for small jumps.

and the jump intensity is $\lambda = \nu(\mathbb{R})$. A trajectory of this process can be simulated exactly on the interval $[0, T]$ using the following simple algorithm (which uses the fact that waiting times between the jumps are independent exponentially distributed random variables with parameter λ):

ALGORITHM 6.1 Simulation of compound Poisson process

Initialize $k := 0$

REPEAT while $\sum_{i=1}^{k} T_i < T$

Set $k := k + 1$

Simulate $T_k \sim \exp(\lambda)$

Simulate Y_k *from the distribution* $\mu = \nu/\lambda$

The trajectory is given by

$$X(t) = \gamma b + \sum_{i=1}^{N(t)} Y_i \quad \textit{where} \quad N(t) = \sup\{k : \sum_{i=1}^{k} T_i \leq t\}.$$

We will now improve this algorithm using two following observations

- The number of jumps $N(T)$ of a compound Poisson process on the interval $[0, T]$ is a Poisson random variable with parameter λT.
- Conditionally on $N(T)$, the exact moments of jumps on this interval have the same distribution as $N(T)$ independent random numbers, uniformly distributed on this interval, rearranged in increasing order (see Proposition 2.9).

ALGORITHM 6.2 Improved algorithm for compound Poisson process

- *Simulate a random variable* N *from Poisson distribution with parameter* λT. N *gives the total number of jumps on the interval* $[0, T]$.
- *Simulate* N *independent r.v.,* U_i*, uniformly distributed on the interval* $[0, T]$. *These variables correspond to the jump times.*
- *Simulate jump sizes:* N *independent r.v.* Y_i *with law* $\frac{\nu(dx)}{\lambda}$.
  The trajectory is given by

$$X(t) = bt + \sum_{i=1}^{N} 1_{U_i < t} Y_i.$$

Figure 6.1 depicts a typical trajectory of compound Poisson process, simulated using Algorithm 6.2.

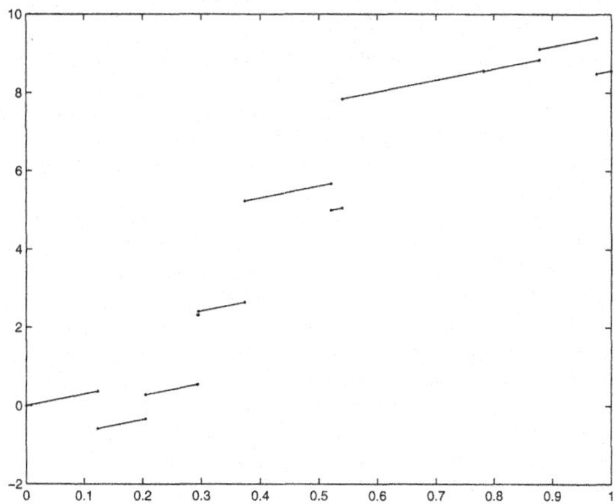

FIGURE 6.1: Typical trajectory of a compound Poisson process. Here jump size distribution is standard normal, the jump intensity is equal to 10 and the drift parameter is equal to 3.

When the Lévy process has a Gaussian component and a jump component of compound Poisson type (in this book, such a process is called a *jump-diffusion*), one can simulate the two independent components separately. The following algorithm gives a discretized trajectory for a process of this type with characteristic triplet (σ^2, ν, b).

ALGORITHM 6.3 Simulating jump-diffusions on a fixed time grid

Simulation of $(X_1, \ldots, X_n)$ *for* n *fixed times* $t_1, \ldots, t_n$.

- *Simulate* n *independent centered Gaussian random variables* G_i *with variances* $\mathrm{Var}(G_i) = (t_i - t_{i-1})\sigma^2$ *where* $t_0 = 0$. *A simple algorithm for simulating Gaussian random variables is described in Example 6.2.*

- *Simulate the compound Poisson part as described in the Algorithm 6.2.*

The discretized trajectory is given by

$$X(t_i) = bt_i + \sum_{k=1}^{i} G_k + \sum_{j=1}^{N} 1_{U_j < t_i} Y_j.$$

A typical trajectory of process simulated by Algorithm 6.3 is shown in Figure 6.2. In Section 6.4, we will see that many infinite activity Lévy processes can be well approximated by a process of such type: the small jumps are truncated and replaced with a properly renormalized Brownian motion.

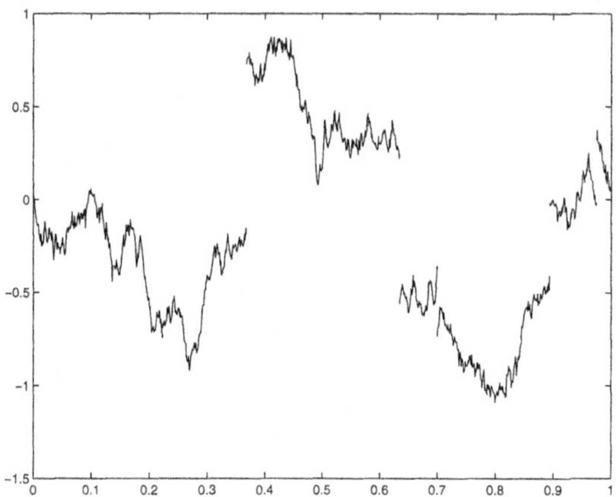

FIGURE 6.2: Typical trajectory of a Lévy process with a Gaussian component and a jump component of compound Poisson type. Here jump size distribution is normal with zero mean and standard deviation 0.5, the jump intensity is 10, the diffusion volatility is 5 and there is no drift.

Note that in the above algorithm the law of increments $X(t_i) - X(t_{i-1})$ is rather complicated because the number of jumps on the interval $[t_{i-1}, t_i]$ is unknown. This makes it difficult to compute functionals like $\sup_{t \leq T} X_t$, which depend on the entire trajectory and not only on its values at discrete times. To overcome this problem, one can exploit the independence of the continuous part and the jump part in the following way.

First, simulate the jump times of the compound Poisson part $\tau_1, \ldots, \tau_N$, the values of the compound Poisson part $N(\tau_i)$ and the values of the continuous part at these times $W(\tau_i)$. Conditionally on this information, the trajectory of the process (X_t) between each two adjacent jump times is continuous. In fact, between two such times τ_i and τ_{i+1}, (X_t) is a Brownian motion, conditioned by its values in τ_i and τ_{i+1}. Now we will be able to compute the required functionals between the jumping times because joint laws of many functionals of the Brownian motion and its final value are well known (see, e.g., [66]). This method is very useful in financial applications like pricing exotic options by Monte Carlo and we illustrate it in the following example.

Example 6.1 Monte Carlo method for jump-diffusion models

In this example, we discuss the pricing of an up-and-out call option in a jump-diffusion model (see Chapter 11 for a description of barrier options and exp-Lévy models). The pricing problem reduces to computing the following expectation (we suppose zero interest rates):

$$C = E\{(e^{X_T} - K)^+ 1_{M_T<b}\}, \tag{6.1}$$

where $(X_t)_{t\geq 0}$ is a jump-diffusion process: $X_t = \gamma t + \sigma W_t + N_t$, such that (e^{X_t}) is a martingale and $M_t = \max_{0\leq s\leq t} X_s$ is the maximum process associated to X. We will show how to compute this expectation using the Monte Carlo method.

The key idea of the method is as follows. First, simulate the jump times τ_i of compound Poisson part, the jump sizes $X_{\tau_i} - X_{\tau_i-}$ and the values of X at the jump times τ_i and at T. If any of these values is beyond the barrier, the payoff for this trajectory is zero. Otherwise, we can analytically compute the probability that this trajectory has gone over the barrier and come back between two consecutive jump times (because between the jump times the trajectory is simply a Brownian bridge). The payoff of this trajectory will then be $(e^{X_T} - K)^+$ *multiplied by this probability.* Now we can repeat the simulation a sufficient number of times to obtain the desired precision.

Put $\mathcal{F}^* = \sigma\{N_t, 0 \leq t \leq T; W(\tau_i), 0 \leq i \leq N\}$ where $\tau_i, 0 \leq i \leq N-1$ are the jump times of the compound Poisson part and $\tau_N = T$. We can rewrite the expectation (6.1) as follows:

$$C = E\{(e^{X_T} - K)^+ E[1_{M_T\leq b}|\mathcal{F}^*]\} \tag{6.2}$$

because X_T is $\mathcal{F}^*$-measurable. In other words, we condition the inner expectation on the trajectory of the compound Poisson part and on the values of the Brownian part at the jump times of the compound Poisson process and at T. The outer expectation in (6.2) will be computed by the Monte Carlo method, and the inner conditional expectation will be computed analytically. To do this we write, using the Markov property of Brownian motion:

$$E[1_{M_T\leq b}|\mathcal{F}^*] = E[\prod_{i=1}^{N} 1_{M_i\leq b}|\mathcal{F}^*] = \prod_{i=1}^{N} P[M_i \leq b|X(\tau_{i-1}), X(\tau_i-)],$$

where $M_i = \max_{\tau_{i-1}\leq t<\tau_i} X_t$. We can further write, using the fact that X_t does not jump on the interval (τ_{i-1}, τ_i):

$$\begin{aligned}
&P[M_i \leq b|X(\tau_{i-1}), X(\tau_i-)] \\
&= P_{X(\tau_{i-1})}[\sup_{0\leq t<\tau_i-\tau_{i-1}} \sigma\hat{W}_t + \gamma t \leq b|\sigma W_{\tau_i-\tau_{i-1}} + \gamma(\tau_i - \tau_{i-1}) = X(\tau_i-)],
\end{aligned}$$

where $\hat{W}$ is a new Brownian motion and P_x denotes the probability under which $\sigma\hat{W}_0 = x$. This expression can be computed analytically [66]:

$$P_x[\sup_{0\leq t\leq l} \sigma W_t + \gamma t \leq b|\sigma W_l + \gamma l = z] = 1 - \exp\left(-\frac{2(z-b)(x-b)}{l\sigma^2}\right).$$

Substituting it expression into (6.2), we find the final result:

$$C =$$

$$E\left[(e^{X_T}-K)^+\prod_{i=1}^{N}1_{X(\tau_i)<b}\left\{1-\exp\left(-\frac{2(X_{\tau_i-}-b)(X_{\tau_{i-1}}-b)}{(\tau_i-\tau_{i-1})\sigma^2}\right)\right\}\right] \quad (6.3)$$

This outer expectation will be evaluated using the Monte Carlo method with Algorithm 6.4. □

ALGORITHM 6.4 Monte Carlo method for jump-diffusions

- *Simulate jump times $\{\tau_i\}$ and values $\{N_{\tau_i}\}$ of the compound Poisson part.*
- *Simulate the values $\{W_{\tau_i}\}$ of the Brownian part at the points $\{\tau_i\}$.*
- *Evaluate the functional under the expectation in (6.3).*
- *Repeat the first three steps a sufficient number of times to compute the average value of the functional with the desired precision.*

Algorithms 6.3 and 6.4 can be generalized to jump-diffusion processes, when the continuous component is not a Brownian motion but an arbitrary Markovian diffusion. In this case increments of the diffusion may be approximated with an Euler scheme and law of the diffusion on some sufficiently small interval conditionally on the values at its ends may still be quite well approximated by a Brownian bridge for the purposes of pricing exotic options.

Algorithm 6.2 allows to simulate the trajectories exactly and to compute all functionals of the path. However, the computational complexity is roughly proportional to the number of jumps on the interval, hence, this algorithm cannot be directly generalized to infinite activity Lévy processes. The alternative is to simulate increments of the process between arbitrary times. This is only possible in some particular cases which are the topic of the next section.

6.2 Exact simulation of increments

To simulate a discretized trajectory of a Lévy process, one must be able to simulate the increments of the process at any time scale. Since the law of increments at a given time scale can be obtained as the convolution power of the law at time 1, simulation is typically possible when the laws of increments lay in the same convolution closed class. There are many distributions with explicit density, that are known to be infinitely divisible but do not lay in a convolution-closed class making the corresponding Lévy processes hard to simulate. This is, e.g., the case for Student distribution and the log-normal distribution. If (X_t) is a Lévy process and X_1 is log-normal, then X_2 is not log-normal. For completeness, we start with the Brownian motion.

Example 6.2 Simulation of Gaussian random variables
There exist many methods to simulate standard normal variables. One of the oldest and the simplest ones (though not the fastest) is the Box-Muller method, which allows to simulate two standard normal variables at the same time. Let U and V be independent random variables, uniformly distributed on $[0,1]$. Then $\sqrt{-2\log(U)}\cos(2\pi V)$ and $\sqrt{-2\log(U)}\sin(2\pi V)$ are independent standard normal. To simulate a normally distributed random variable with arbitrary mean and variance observe that if N is standard normal then $\sigma N + \theta$ has mean θ and variance σ^2. To simulate a d-dimensional Gaussian random vector with covariance matrix Σ and mean vector μ, take a vector $\mathbf{X}$ of d independent standard normal random variables. Then $\mathbf{Y} = \Sigma^{1/2}\mathbf{X} + \mu$, where $\Sigma^{1/2}$ is an arbitrary matrix square root of Σ, has covariance matrix Σ and mean vector μ. □

ALGORITHM 6.5 Discretized trajectory of Brownian motion
Simulation of $(X(t_1),\dots,X(t_n))$ for n fixed times $t_1,\dots,t_n$, where (X_t) is a Brownian motion with volatility σ and drift b.

- *Simulate n independent standard normal variables $N_1 \dots N_n$.*
  Set $\Delta X_i = \sigma N_i \sqrt{t_i - t_{i-1}} + b(t_i - t_{i-1})$ where $t_0 = 0$.

The discretized trajectory is given by $X(t_i) = \sum_{k=1}^{i} \Delta X_i$.

Example 6.3 α-stable Lévy process
Stable processes were defined in Section 3.7. α-stable Lévy processes with α strictly between 0 and 2 have infinite variance, which makes them somewhat inconvenient for financial modelling (though a recent paper by Carr and Wu [84] uses a log-stable process for option pricing; see also [327]), but they are important in other domains including physics, biology, meteorology, etc. Here we use the parameterization (3.34), writing $X \sim S_\alpha(\sigma,\beta,\mu)$ when X is an α-stable random variable with skewness parameter β, scale parameter σ and shift parameter μ.

The class of α-stable random variables is convolution closed: if X_1 and X_2 are independent random variables with $X_i \sim S_\alpha(\sigma_i,\beta_i,\mu_i)$ then $X_1 + X_2 \sim S_\alpha(\sigma,\beta,\mu)$ with

$$\sigma = (\sigma_1^\alpha + \sigma_2^\alpha)^{1/\alpha},\ \beta = \frac{\beta_1\sigma_1^\alpha + \beta_2\sigma_2^\alpha}{\sigma_1^\alpha + \sigma_2^\alpha} \text{ and } \mu = \mu_1 + \mu_2$$

(for the proof see [343, Property 1.2.1]). If X_t is an α-stable Lévy process with $X_1 \sim S_\alpha(\sigma,\beta,\mu)$ then $X_t \sim S_\alpha(\sigma(t)^{1/\alpha},\beta,\mu t)$. The only parameters that are different for different time scales are thus the scale parameter σ and the shift parameter μ. However, a stable variable with arbitrary shift and scale parameters can be obtained from a variable $X \sim S_\alpha(1,\beta,0)$ as follows:

$$\sigma X + \mu \sim S_\alpha(\sigma,\beta,\mu), \quad \text{if } \alpha \neq 1;$$
$$\sigma X + \frac{2}{\pi}\beta\sigma\ln\sigma + \mu \sim S_\alpha(\sigma,\beta,\mu), \quad \text{if } \alpha = 1.$$

Thus, to simulate a discrete skeleton of an α-stable Lévy process with skewness parameter β, one only needs to simulate stable variables of type $S_\alpha(1,\beta,0)$. Chambers, Mallows and Stuck [87] describe a method for generating α-stable random variates with any admissible values of α and β and give a listing of a Fortran program `rstab` which implements this method. This program is now a part of S-PLUS statistical language. We will give an algorithm for simulating symmetric stable processes (with $\beta = 0$ and no drift). Its proof can be found in [343]. This reference also contains a modified version of `rstab` program. □

ALGORITHM 6.6 Discretized trajectory for symmetric α-stable process

Simulation of $(X(t_1),\dots,X(t_n))$ for n fixed times $t_1,\dots,t_n$.

- *Simulate n independent random variables γ_i, uniformly distributed on $(-\pi/2,\pi/2)$ and n independent standard exponential random variables W_i.*

- *Compute ΔX_i for $i = 1\dots n$ using*

$$\Delta X_i = (t_i - t_{i-1})^{1/\alpha}\frac{\sin\alpha\gamma_i}{(\cos\gamma_i)^{1/\alpha}}\left(\frac{\cos((1-\alpha)\gamma_i)}{W_i}\right)^{(1-\alpha)/\alpha} \tag{6.4}$$

with the convention $t_0 = 0$.

The discretized trajectory is given by $X(t_i) = \sum_{k=1}^{i}\Delta X_k$.

Figure 6.3 presents typical trajectories of symmetric stable processes with various stability indices. When α is small (left graph), the process has very fat tails, and the trajectory is dominated by big jumps. Note how this graph resembles the trajectory of a compound Poisson process (Figure 6.1). When α is large (bottom graph), the behavior is determined by small jumps and the trajectory resembles that of a Brownian motion, although occasionally we see some jumps. The right graph corresponds to the Cauchy process ($\alpha = 1$) which is between the two cases. Here both big and small jumps

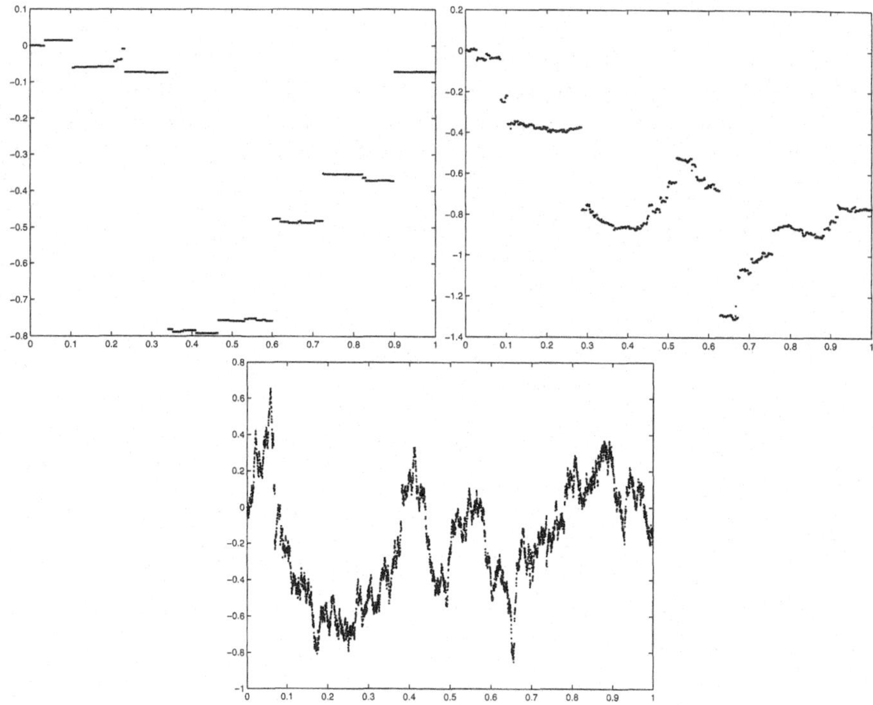

FIGURE 6.3: Simulated trajectories of α-stable processes with $\alpha = 0.5$ (left), $\alpha = 1$ (right) and $\alpha = 1.9$ (bottom).

have a significant effect. Comparing the three graphs with Figure 6.2 makes one think that stable processes may be approximated by a combination of compound Poisson process and Brownian motion. We will see shortly that this is indeed the case for stable processes and many other Lévy processes.

Example 6.4 Gamma process

The gamma subordinator was defined in Table 4.4. At a fixed time t this process has the well-studied gamma distribution with density

$$p_t(x) = \frac{\lambda^{ct}}{\Gamma(ct)} x^{ct-1} e^{-\lambda x}.$$

Gamma process has the following scaling property: if S_t is a gamma process with parameters c and λ then λS_t is a gamma process with parameters c and 1. Therefore it is sufficient to be able to simulate gamma random variables with density of the form

$$p(x) = \frac{x^{a-1}}{\Gamma(a)} e^{-x}.$$

There exist many algorithms for generating such random variables. A survey of available methods can be found in [113]. Below we reproduce two algorithms from this book. The first one should be used if $a \leq 1$ (which is most often the case in applications) and the second one if $a > 1$. Typical trajectories of gamma process with different values of c are shown in Figure 6.4. □

ALGORITHM 6.7 Johnk's generator of gamma variables, $a \leq 1$

REPEAT
  Generate i.i.d. uniform $[0, 1]$ *random variables* U, V
  Set $X = U^{1/a}$, $Y = V^{1/(1-a)}$
UNTIL $X + Y \leq 1$
Generate an exponential random variable E
RETURN $\frac{XE}{X+Y}$

ALGORITHM 6.8 Best's generator of gamma variables, $a \geq 1$

Set $b = a - 1$, $c = 3a - \frac{3}{4}$
REPEAT
  Generate i.i.d. uniform $[0, 1]$ *random variables* U, V
  Set $W = U(1 - U)$, $Y = \sqrt{\frac{c}{W}}(U - \frac{1}{2})$, $X = b + Y$
  If $X < 0$ *go to REPEAT*
  Set $Z = 64W^3V^3$
UNTIL $\log(Z) \leq 2(b\log(\frac{X}{b}) - Y)$
RETURN X

Example 6.5 Inverse Gaussian process
The inverse Gaussian Lévy process gives another example of a subordinator for which both Lévy measure and probability density are known in explicit form (see Table 4.4). The inverse Gaussian density has the form

$$p(x) = \sqrt{\frac{\lambda}{2\pi x^3}} e^{-\frac{\lambda(x-\mu)^2}{2\mu^2 x}} 1_{x>0}. \tag{6.5}$$

Below we reproduce the algorithm of Michael, Schucany and Haas for simulating inverse Gaussian variables (see [113]). □

ALGORITHM 6.9 Generating inverse Gaussian variables

Generate a normal random variable N

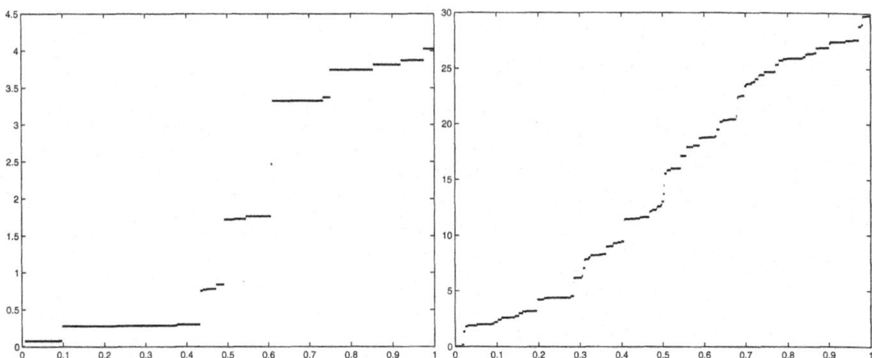

FIGURE 6.4: Two trajectories of gamma process with $c = 3$ (left), $c = 30$ (right) and $\lambda = 1$ for both graphs. These trajectories were simulated using series representation (6.16).

Set $Y = N^2$
Set $X_1 = \mu + \frac{\mu^2 Y}{2\lambda} - \frac{\mu}{2\lambda}\sqrt{4\mu\lambda Y + \mu^2 Y^2}$
Generate a uniform $[0, 1]$ *random variable* U
IF $U \le \frac{\mu}{X_1+\mu}$ *RETURN* X_1 *ELSE RETURN* $\frac{\mu^2}{X_1}$

Example 6.6 Subordinated Brownian motion
A popular class of processes for stock price modelling is obtained by subordinating the standard Brownian motion or a Brownian motion with drift with an independent positive Lévy process. If the subordinator is denoted by V_t then the resulting process will be

$$X_t = W_{V_t} + bV_t, \tag{6.6}$$

where W is standard Brownian motion. When V_t is the gamma process or the inverse Gaussian process, we obtain, respectively, the variance gamma process and the normal inverse Gaussian process. Processes of type (6.6) possess a number of useful properties because they are conditionally Gaussian. In particular, if one knows how to simulate the increments of the subordinator, the increments of X_t can be simulated using the following algorithm. □

ALGORITHM 6.10 Generating the subordinated Brownian motion on a fixed time grid

Simulation of $(X(t_1), \ldots, X(t_n))$ *for* n *fixed times* $t_1, \ldots, t_n$ *where* $X(t) = B(S(t))$ *is Brownian motion with volatility* σ *and drift* b*, time changed with subordinator* (S_t).

- *Simulate increments of the subordinator:* $\Delta S_i = S_{t_i} - S_{t_{i-1}}$ *where* $S_0 = 0$.

- *Simulate n independent standard normal random variables $N_1, \ldots, N_n$. Set $\Delta X_i = \sigma N_i \sqrt{\Delta X_i} + b\Delta S_i$.*

The discretized trajectory is given by $X(t_i) = \sum_{k=1}^{i} \Delta X_k$.

The following two algorithms show how the above method can be used to simulate variance gamma processes and normal inverse Gaussian processes on a fixed time grid:

ALGORITHM 6.11 Simulating a variance gamma process on a fixed time grid
Simulation of $(X(t_1), \ldots, X(t_n))$ for fixed times $t_1, \ldots, t_n$: a discretized trajectory of the variance gamma process with parameters σ, θ, κ.

- *Simulate, using Algorithms 6.7 and 6.8, n independent gamma variables $\Delta S_1, \ldots, \Delta S_n$ with parameters $\frac{t_1}{\kappa}, \frac{t_2 - t_1}{\kappa}, \ldots, \frac{t_n - t_{n-1}}{\kappa}$. Set $\Delta S_i = \kappa \Delta S_i$ for all i.*
- *Simulate n i.i.d. $N(0,1)$ random variables $N_1, \ldots, N_n$. Set $\Delta X_i = \sigma N_i \sqrt{\Delta S_i} + \theta \Delta S_i$ for all i.*

The discretized trajectory is $X(t_i) = \sum_{k=1}^{i} \Delta X_i$.

ALGORITHM 6.12 Simulating normal inverse Gaussian process on a fixed time grid
Simulation of $(X(t_1), \ldots, X(t_n))$ for fixed times $t_1, \ldots, t_n$: a discretized trajectory of the normal inverse Gaussian process with parameters σ, θ, κ.

- *Simulate, using Algorithm 6.9 n independent inverse Gaussian variables $\Delta S_1, \ldots, \Delta S_n$ with parameters $\lambda_i = \frac{(t_i - t_{i-1})^2}{\kappa}$ and $\mu_i = t_i - t_{i-1}$ where we take $t_0 = 0$.*
- *Simulate n i.i.d. $N(0,1)$ random variables $N_1, \ldots, N_n$. Set $\Delta X_i = \sigma N_i \sqrt{\Delta S_i} + \theta \Delta S_i$ for all i.*

The discretized trajectory is $X(t_i) = \sum_{k=1}^{i} \Delta X_i$.

6.3 Approximation of an infinite activity Lévy process by a compound Poisson process

Let $(X_t)_{t \geq 0}$ be an infinite activity Lévy process with characteristic triplet $(0, \nu, \gamma)$. The goal of this and the following section is to find a process $(X_t^\varepsilon)_{t \geq 0}$

which is of compound Poisson type (but possibly with a nontrivial Gaussian component) and which approximates the initial process X in some sense to be specified. We will also give some precision estimates and convergence rates for the approximations that we derive.

From the Lévy-Itô decomposition we know that X can be represented as a sum of a compound Poisson process and an almost sure limit of compensated compound Poisson processes:

$$X_t = \gamma t + \sum_{s\leq t} \Delta X_s 1_{|\Delta X_s|\geq 1} + \lim_{\varepsilon\downarrow 0} N_t^\varepsilon,$$

$$\text{where} \qquad N_t^\varepsilon = \sum_{s\leq t} \Delta X_s 1_{\varepsilon\leq|\Delta X_s|<1} - t\int_{\varepsilon\leq|x|\leq 1} x\nu(dx).$$

Therefore a natural idea is to approximate X_t by

$$X_t^\varepsilon = \gamma t + \sum_{s\leq t} \Delta X_s 1_{|\Delta X_s|\geq 1} + N_t^\varepsilon. \tag{6.7}$$

The residual term (incorporating compensated jumps smaller than ε) is given by

$$R_t^\varepsilon = -N_t^\varepsilon + \lim_{\delta\downarrow 0} N_t^\delta.$$

It is a Lévy process with characteristic triplet $(0, 1_{|x|\leq\varepsilon}\nu(dx), 0)$ satisfying $E[R_t^\varepsilon] = 0$.

In the finite variation case small jumps need not be compensated and one can use zero truncation function in the Lévy-Khinchin representation. The process X can therefore be written as a sum of its jumps plus drift (different from γ because the truncation function is not the same).

$$X_t = bt + \sum_{s\leq t} \Delta X_s$$

$$\text{and} \qquad X_t^\varepsilon = bt + \sum_{s\leq t} \Delta X_s 1_{\varepsilon\leq|\Delta X_s|} + E[\sum_{s\leq t} \Delta X_s 1_{|\Delta X_s|<\varepsilon}].$$

Therefore, in the finite variation case the approximation (6.7) is constructed by replacing small jumps with their expectation.

The process X^ε is of compound Poisson type and may be simulated using Algorithm 6.2. Let us analyze the quality of this approximation. The error process R^ε is an infinite activity Lévy process with bounded jumps and, therefore, finite variance. By Proposition 3.13,

$$\operatorname{Var} R_t^\varepsilon = t\int_{|x|<\varepsilon} x^2\nu(x)dx \equiv t\sigma^2(\varepsilon).$$

Hence, the quality of the approximation depends on the speed at which $\sigma^2(\varepsilon)$ converges to zero as $\varepsilon \to 0$. Suppose that the approximation (6.7) is used to compute functionals of the terminal value of the process using the Monte Carlo method. The following proposition gives a convergence rate for this problem. We only estimate the precision of approximation (6.7) without taking into account the intrinsic error of the Monte Carlo method, that is due to finite number of trajectories.

PROPOSITION 6.1

Let $f(x)$ be a real-valued differentiable function such that $|f'(x)| < C$ for some constant C. Then

$$\left|E[f(X_T^\varepsilon + R_T^\varepsilon)] - E[f(X_T^\varepsilon)]\right| \leq C\sigma(\varepsilon)\sqrt{T}. \tag{6.8}$$

PROOF The difference in question can be estimated as follows:

$$\left|E[f(X_T^\varepsilon + R_T^\varepsilon)] - E[f(X_T^\varepsilon)]\right| = \left|E[R_T^\varepsilon \int_0^1 f'(X_T^\varepsilon + uR_T^\varepsilon)du]\right| \leq CE\left|R_T^\varepsilon\right|.$$

For any random variable Z, Jensen's inequality yields:

$$E[|Z|]^2 \leq E[|Z|^2] = E[Z^2].$$

Applying this to R_T^ε, we conclude that $E\left|R_T^\varepsilon\right| \leq \sigma(\varepsilon)\sqrt{T}$. □

REMARK 6.1 The function f in Proposition 6.1 can correspond to the payoff function of a European put option. Suppose that under the risk neutral probability logarithm of the stock price is a Lévy process. In this case, for a put option $f(x) = (K - S_0 e^x)^+$ where K is the strike of the option and S_0 denotes the initial price of the underlying. For this function $f'(x) = -S_0 e^x 1_{x \leq \log(K/S_0)}$ and $|f'(x)| \leq K$. If the price of this option is approximated by $E[f(X_T^\varepsilon)]$ then the worst case pricing error is given by (6.8). □

REMARK 6.2 Note that the variance of R_t^ε in approximation (6.7) grows linearly with time, and therefore the error of the Monte Carlo method is proportional to the square root of time. □

In Proposition 6.1, we have found that the worst case error of the Monte Carlo method is proportional to the dispersion of the residual process. As ε goes to zero, the pricing error decreases, but the intensity of X_t^ε grows causing a proportional increase in the computer time required to simulate every trajectory. In the examples that follow, we will compute the convergence rates for various processes and see how the pricing error depends on the number of basic computer operations required.

Example 6.7 Convergence rates: symmetric stable process
For symmetric α-stable processes $\sigma(\varepsilon) \sim \varepsilon^{1-\alpha/2}$. This means that for stable processes with a large index of stability that have many small jumps in the neighborhood of zero, convergence rate of the Monte Carlo method may be very low and some kind of improvement is necessary. A simple computation shows that for symmetric stable processes the intensity N of the approximating compound Poisson process, truncated at the level ε, is proportional to $\varepsilon^{-\alpha}$ for fixed α. We denote the intensity by N because it corresponds approximately to the number of operations required to simulate one trajectory. Substituting this expression into the formula for $\sigma(\varepsilon)$, we find that

$$\sigma(\varepsilon) \sim N^{\frac{1}{2}-\frac{1}{\alpha}} \quad \text{for symmetric stable process.}$$

Thus, for stable processes convergence rates range from extremely bad (when there are many jumps in the neighborhood of zero) to very good. □

Example 6.8 Convergence rates: gamma process
For the gamma process, $\sigma(\varepsilon) \sim \varepsilon$, so the quality of the approximation by a compound Poisson process is very good. The intensity of approximating process is $N \sim -\log(\varepsilon)$. Substituting this into the expression for $\sigma(\varepsilon)$, we find that:

$$\sigma(\varepsilon) \sim e^{-N} \quad \text{for gamma process.}$$

This means that for the gamma process the convergence is exponential, that is, by adding ten more jumps we gain a factor of e^{10} in the precision! □

Example 6.9 Simulation of tempered stable processes by compound Poisson approximation
For simplicity we treat the case of tempered stable subordinator, but the method can be easily generalized to processes with jumps of different signs. The method consists of simulating the big jumps and replacing the small ones with their expectation:

$$X_t \approx X_t^\varepsilon = bt + \sum_{s\leq t} \Delta X_s 1_{\Delta X_s \geq \varepsilon} + E\left\{\sum_{s\leq t} \Delta X_s 1_{\Delta X_s < \varepsilon}\right\}.$$

The Lévy density of a tempered stable subordinator is given by $\nu(x) = \frac{ce^{-\lambda x}}{x^{\alpha+1}}$ with $0 < \alpha < 1$. Therefore, the process (X_t^ε) has drift $b^\varepsilon = b + c\int_0^\varepsilon \frac{e^{-\lambda x}dx}{x^\alpha}$ (note that this expression can be easily computed via incomplete gamma function) and finite Lévy measure with density $\nu^\varepsilon(x) = \frac{ce^{-\lambda x}}{x^{\alpha+1}} 1_{x\geq\varepsilon}$. It is a compound Poisson process with intensity $U(\varepsilon) = c\int_\varepsilon^\infty \frac{e^{-\lambda x}dx}{x^{\alpha+1}}$ and jump size distribution $p^\varepsilon(x) = \nu^\varepsilon(x)/U(\varepsilon)$ and it can be simulated using Algorithm 6.2 if one knows how to simulate random variables with distribution $p^\varepsilon(x)$.

Observe that for all x we have

$$p^\varepsilon(x) \leq f^\varepsilon(x)\frac{\varepsilon^{-\alpha}e^{-\lambda\varepsilon}}{\alpha U(\varepsilon)},$$

where $f^\varepsilon(x) = \frac{\alpha\varepsilon^{-\alpha}}{x^{\alpha+1}}1_{x\geq\varepsilon}$ is a probability density. Note that $f(x)$ has survival function $F^\varepsilon(x) = \frac{\varepsilon^\alpha}{x^\alpha}1_{x\geq\varepsilon}$ and inverse survival function $F_\varepsilon^{-1}(u) = \varepsilon u^{-1/\alpha}$. Random variables with distribution $p^\varepsilon(x)$ may be simulated using the rejection method as follows (see [113] for a general description of the rejection method).

REPEAT

Generate W, V: independent and uniformly distributed on $[0,1]$.
Set $X = \varepsilon W^{-1/\alpha}$ (X has distribution f^ε).
Set $T = \frac{f^\varepsilon(X)\varepsilon^{-\alpha}e^{-\lambda\varepsilon}}{p^\varepsilon(X)\alpha U(\varepsilon)}$

UNTIL $VT \leq 1$
RETURN X

The variable X produced by this algorithm has distribution p^ε and the average number of loops needed to generate one random variable is equal to

$$\frac{\varepsilon^{-\alpha}e^{-\lambda\varepsilon}}{\alpha U(\varepsilon)} \sim 1 + \Gamma(1-\alpha)(\lambda\varepsilon)^\alpha \quad \text{when } \varepsilon \to 0,$$

which is good because in applications, ε will typically be close to 0. □

The compound Poisson approximation has very good accuracy if there are not "too many small jumps," i.e., if the growth of the Lévy measure near zero is not too fast, as in the case of the Gamma process. In the next section, we present a method (due to Asmussen and Rosinski [11]) that allows to increase the accuracy of the above approximation by replacing the small jumps by a Brownian motion: this correction will be efficient precisely when there are *many* small jumps, i.e., exactly when compound Poisson approximations are poor, so the two methods are complementary.

6.4 Approximation of small jumps by Brownian motion

In this section, we show that in many cases, the normalized error process $\sigma(\varepsilon)^{-1}R^\varepsilon$ converges in distribution to Brownian motion. First, we will provide an intuitive explanation of this fact and derive a sufficient condition for convergence. Then we will give the exact criterion of convergence, due to Asmussen and Rosinski [11]. Finally, we will prove that this convergence may indeed be used to improve the approximation (6.7), that is, we will show that the approximation

$$\tilde{X}_t^\varepsilon = X_t^\varepsilon + \sigma(\varepsilon)W_t \tag{6.9}$$

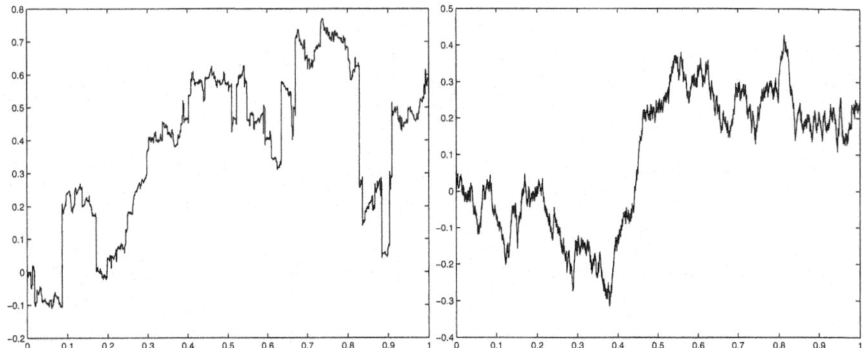

FIGURE 6.5: Convergence of renormalized small jumps of the Cauchy process to Brownian motion. Left graph: Residual process for $\varepsilon = 0.5$. Right graph: Residual process for $\varepsilon = 0.0001$. These trajectories were approximated using a series representation.

has better convergence rates than (6.7) in the same problem of pricing a European option by the Monte Carlo method.

Put $Y_t^\varepsilon = \sigma(\varepsilon)^{-1} R^\varepsilon$. Suppose that the following condition is satisfied

$$\frac{\sigma(\varepsilon)}{\varepsilon} \to \infty \quad \text{as} \quad \varepsilon \to 0. \tag{6.10}$$

Because the jumps of R^ε are bounded by ε, condition (6.10) means that the jumps of Y^ε are bounded by some number that converges to zero. This means that the limiting process has no jumps. Since Y^ε for every ε is a Lévy process with zero mean and the variance of Y_1^ε is equal to one, the limiting process will be a continuous Lévy process with mean zero and variance at time 1 equal to one, hence, a standard Brownian motion.

Figure 6.5 illustrates this convergence: the left graph shows the renormalized residual process for Cauchy process with $\varepsilon = 0.5$ and the right graph shows the residual process for $\varepsilon = 0.0001$.

It turns out that the condition (6.10) is not always necessary; although, as Remark 6.3 shows, it is necessary in most cases of interest. The following theorem gives a necessary and sufficient condition for convergence.

THEOREM 6.1 Asmussen and Rosinski [11]
$\sigma(\varepsilon)^{-1} R^\varepsilon \to W$ in distribution as $\varepsilon \to 0$ if and only if for all $k > 0$

$$\frac{\sigma(k\sigma(\varepsilon) \wedge \varepsilon)}{\sigma(\varepsilon)} \to 1, \quad as \quad \varepsilon \to 0. \tag{6.11}$$

REMARK 6.3 Condition (6.11) is clearly implied by Condition (6.10), which is much easier to check. Moreover, Asmussen and Rosinski[11] prove that the two conditions are equivalent if ν has no atoms in some neighborhood of zero. This is the case for all examples in Table 6.1 and more generally, for all examples considered in this chapter. Hence, for all these examples it is sufficient to check the condition (6.10). □

Example 6.10 Validity of normal approximation for different processes
For symmetric stable processes, $\sigma(\varepsilon) \sim \varepsilon^{1-\alpha/2}$, the condition (6.10) is satisfied and the normal approximation holds. It is easy to check that it also holds for general stable processes and for all Lévy processes with stable-like (power law of type $\frac{1}{|x|^{1+\alpha}}$ with $\alpha > 0$) behavior near the origin, for example, normal inverse Gaussian, truncated stable, etc. The normal approximation *does not* hold for compound Poisson processes ($\sigma(\varepsilon) = o(\varepsilon)$) nor for the gamma process ($\sigma(\varepsilon) \sim \varepsilon$). □

In the following proposition we analyze the effect of using normal approximation of small jumps (6.9) on the worst case error of the Monte Carlo approximation to option price.

PROPOSITION 6.2
Let f be a real-valued differentiable function such that $|f'(x)| < C$ for some constant C. Then

$$\left|E[f(X_T^\varepsilon + R_T^\varepsilon)] - E[f(X_T^\varepsilon + \sigma(\varepsilon)W_T)]\right| \leq A\rho(\varepsilon)C\sigma(\varepsilon), \tag{6.12}$$

where $\rho(\varepsilon) \equiv \frac{\int_{-\varepsilon}^{\varepsilon} |x|^3 \nu(dx)}{\sigma^3(\varepsilon)}$ and A is a constant satisfying $A < 16.5$.

REMARK 6.4 Notice that at least under the stronger condition (6.10) this is a better convergence rate compared to Proposition 6.1 because $\rho(\varepsilon) < \frac{\varepsilon}{\sigma(\varepsilon)}$. □

REMARK 6.5 Contrary to Proposition 6.1, here the error does not grow with T since as T grows, due to a "central limit theorem" effect, the quality of normal approximation also improves because $\frac{R_T^\varepsilon}{\sqrt{\operatorname{Var} R_T^\varepsilon}}$ becomes closer to a standard normal variable. □

PROOF

$$\begin{aligned}
&\left|E[f(X_T^\varepsilon + R_T^\varepsilon)] - E[f(X_T^\varepsilon + \sigma(\varepsilon)W_T)]\right| \\
&\qquad \leq \int p(dx)\left|E[f(x + R_T^\varepsilon)] - E[f(x + \sigma(\varepsilon)W_T)]\right|,
\end{aligned}$$

where $p(dx)$ denotes the law of X_T^ε. We can further write:

$$\left|E[f(x+R_T^\varepsilon)] - E[f(x+\sigma(\varepsilon)W_T)]\right| = \left|\int f(x+y)\{p_R(y) - p_W(y)\}dy\right|$$
$$= \left|\int f'(x+y)\{F_R(y) - F_W(y)\}dy\right| \le C\int \left|F_R(y) - F_W(y)\right|dy,$$

where $p_R(y)$ and $p_W(y)$ are probability density functions of, respectively, R_T^ε and $\sigma(\varepsilon)W_T$ and $F_R(y)$ and $F_W(y)$ are their distribution functions. We now need to find an estimate for $\int |F_R(y) - F_W(y)|dy$.

R_T^ε can be seen as the sum of n i.i.d. random variables with the same law as $R_{T/n}^\varepsilon$. These variables have zero mean and variance $T\sigma^2(\varepsilon)/n$. Now using a nonuniform Berry-Esseen type theorem [315, Theorem V.16] we find that for all n,

$$\int \left|F_R(y) - F_W(y)\right|dy \le A\sigma(\varepsilon)\sqrt{T}\frac{\rho_n}{\sqrt{n}},$$

where A is a constant with $A < 16.5$ and $\rho_n = \frac{n^{3/2}E|R_{T/n}^\varepsilon|^3}{\sigma^3(\varepsilon)T^{3/2}}$. Asmussen and Rosinski show [11, Lemma 3.2] that if the Lévy measure ν of a Lévy process $(Z_t)_{t\ge 0}$ has finite absolute moment of order $p \ge 2$, one has

$$\lim_{n\to\infty} nE|Z_{1/n}|^p = \int_{-\infty}^{\infty} |x|^p \nu(dx).$$

Hence, we find that

$$\int \left|F_R(y) - F_W(y)\right|dy \le A\sigma(\varepsilon)\frac{\int_{-\varepsilon}^{\varepsilon} |x|^3\nu(dx)}{\sigma^3(\varepsilon)}$$

and finally

$$\left|E[f(X_T^\varepsilon + R_T^\varepsilon)] - E[f(X_T^\varepsilon + \sigma(\varepsilon)W_T)]\right| \le AC\sigma(\varepsilon)\frac{\int_{-\varepsilon}^{\varepsilon} |x|^3\nu(dx)}{\sigma^3(\varepsilon)}.$$

□

Example 6.11 Improved convergence rates: symmetric stable process
For symmetric stable processes, $\rho(\varepsilon) \sim \varepsilon^{\alpha/2}$. Since $\sigma(\varepsilon) \sim \varepsilon^{1-\alpha/2}$, the improved convergence rate is independent of the stability index and is always proportional to ε. For processes with small α, the improvement of convergence rate due to normal approximation is small, however, the compound Poisson approximation converges sufficiently fast even without the improvement. When α is large, the direct compound Poisson approximation does not converge fast enough but there are many small jumps which are well

approximated by a Brownian motion. In terms of the number of computer operations required to simulate a given trajectory, the convergence rate is $N^{-1/\alpha}$. Hence, this scheme always converges faster than $N^{-1/2}$, which is the typical convergence rate of the Monte Carlo method □

Example 6.12 No improvement for the gamma process
For gamma process $\rho(\varepsilon) \sim const$ and there is no improvement (this fact is easy to understand since the renormalized residual process does not converge to Brownian motion). However, for this process the compound Poisson approximation converges exponentially fast, therefore the correction is not needed. □

6.5 Series representations of Lévy processes (*)

In this section, we first introduce series representations of Lévy processes through an important example and then give a general result due to Rosinski [337], that allows to construct such series representations and prove their convergence. The example concerns series representations for subordinators (positive Lévy processes).

PROPOSITION 6.3 Series representation of a subordinator
Let $(Z_t)_{t\geq 0}$ be a subordinator whose Lévy measure $\nu(dx)$ has tail integral $U(x) \equiv \int_x^\infty \nu(d\xi)$, let $\{\Gamma_i\}$ be a sequence of jumping times of a standard Poisson process, and $\{V_i\}$ be an independent sequence of independent random variables, uniformly distributed on $[0,1]$. Z is representable in law, on the time interval $[0,1]$, as

$$\{Z_s, 0 \leq s \leq 1\} \stackrel{\mathcal{L}}{=} \{\tilde{Z}_s, 0 \leq s \leq 1\}$$

with

$$\tilde{Z}_s = \sum_{i=1}^{\infty} U^{(-1)}(\Gamma_i) 1_{V_i \leq s}, \tag{6.13}$$

where the generalized inverse $U^{(-1)}$ is defined by

$$U^{(-1)}(y) = \inf\{x > 0 : U(x) < y\}$$

The series in (6.13) converges almost surely and uniformly on $s \in [0,1]$.

For practical simulations, the series (6.13) must be truncated. The right way to truncate it *is not* to keep a fixed number of terms for each simulated trajectory, but to fix some τ and keep a random number $N(\tau)$ of terms where $N(\tau) = \inf\{i : \Gamma_i \geq \tau\}$. In this case, as will become clear from the proof, the truncated series is also a Lévy process (compound Poisson) and this approximation turns out to be equivalent to the one discussed in Section 6.3. One can also use the normal approximation of small jumps to improve the convergence rates of series representations like (6.13), but we do not discuss this issue here since there is no fundamental difference from the case already discussed in Section 6.4.

Proposition 6.3 leads to the following algorithm for approximate simulation of subordinators using series representation. To implement it one needs a closed form expression for the inverse tail integral $U^{(-1)}(x)$ or at least some reasonably fast method to compute it.

ALGORITHM 6.13 Simulation of subordinator on $[0,1]$ by series representation

Fix a number τ depending on the required precision and computational capacity. This number is equal to the average number of terms in the series and it determines the truncation level: jumps smaller than $U^{(-1)}(\tau)$ are truncated.

Initialize $k := 0$

REPEAT WHILE $\sum_{i=1}^{k} T_i < \tau$

Set $k = k + 1$

Simulate T_k*: standard exponential.*

Simulate V_k*: uniform on* $[0,1]$

The trajectory is given by

$$X(t) = \sum_{i=1}^{k} 1_{V_i \leq t} U^{(-1)}(\Gamma_i), \quad \textit{where} \quad \Gamma_i = \sum_{j=1}^{i} T_j.$$

PROOF of Proposition 6.3 Let $N(\tau) = \inf\{i : \Gamma_i \geq \tau\}$. We start by analyzing the truncated series

$$\tilde{Z}_s^\tau = \sum_{i=1}^{N(\tau)} U^{(-1)}(\Gamma_i) 1_{V_i \leq s}. \tag{6.14}$$

First, we will prove that $(\tilde{Z}_s^\tau)$ is a compound Poisson process in s with Lévy measure

$$\nu_\tau(A) = \nu(A \cap [U^{(-1)}(\tau), \infty[).$$

Let $h > s > 0$. Then for all u and v, inserting first the conditional expectation on $\Gamma_i, i \geq 1$ and then on $N(\tau)$, we find:

$$E\Big\{\exp\Big(iu\tilde{Z}_s^\tau + iv(\tilde{Z}_h^\tau - \tilde{Z}_s^\tau)\Big)\Big\}$$
$$= E\left\{\exp\left(\sum_{i=1}^{N(\tau)} U^{(-1)}(\Gamma_i)(iu1_{V_i\leq s} + iv1_{s<V_i\leq h})\right)\right\}$$
$$= E\left\{\prod_{i=1}^{N(\tau)}\left(se^{iuU^{(-1)}(\Gamma_i)} + (h-s)e^{ivU^{(-1)}(\Gamma_i)} + 1 - h\right)\right\}$$
$$= E\left\{\left(1 + \frac{1}{\tau}\int_0^{\tau\wedge U(0)}\left(se^{iuU^{(-1)}(z)} + (h-s)e^{ivU^{(-1)}(z)} - h\right)dz\right)^{N(\tau)}\right\}. \tag{6.15}$$

The next step is to show that the image measure of the Lebesgue measure on $]0, U(0)[$, induced by the mapping $U^{(-1)}$, coincides with ν, in other words, for all $B \in \mathcal{B}(]0,\infty[)$, $\nu(B) = \lambda(\{y : U^{(-1)}(y) \in B\})$, where λ denotes the Lebesgue measure. It is sufficient to show this fact for sets B of the form $[z,\infty[$, that is, we must show that

$$\inf\{x > 0 : U(x) < y\} \geq z \tag{6.16}$$
$$\text{if and only if} \quad y \leq U(z). \tag{6.17}$$

Suppose (6.17). Then for all x such that $U(x) < y$ we have $U(x) < U(z)$ and therefore $x \geq z$ and (6.16) is satisfied. Conversely, suppose that (6.16) is satisfied and $y > U(z)$. Then, because U is left-continuous, there exists an $\varepsilon > 0$ such that $y > U(z-\varepsilon)$. Substituting $x = z - \varepsilon$ into (6.16) we find that $z - \varepsilon \geq z$ which is a contradiction.

The integral in (6.15) can now be transformed using the change of variable formula:

$$\frac{1}{\tau}\int_0^{\tau\wedge U(0)}\left(se^{iuU^{(-1)}(z)} + (h-s)e^{ivU^{(-1)}(z)} - h\right)dz$$
$$= \frac{s}{\tau}\int_{U^{(-1)}(\tau)}^{\infty}(e^{iuz}-1)\nu(dz) + \frac{h-s}{\tau}\int_{U^{(-1)}(\tau)}^{\infty}(e^{ivz}-1)\nu(dz),$$

and finally

$$E\left\{\exp\left(iu\tilde{Z}_s^\tau + iv(\tilde{Z}_h^\tau - \tilde{Z}_s^\tau)\right)\right\}$$
$$= \exp\left\{s\int_{U^{(-1)}(\tau)}^{\infty}(e^{iuz}-1)\nu(dz) + (h-s)\int_{U^{(-1)}(\tau)}^{\infty}(e^{ivz}-1)\nu(dz)\right\}.$$

This computation proves that $(\tilde{Z}^\tau_s)$ has independent increments (because we could have taken any number of increments instead of just two), that it has stationary increments and that the Lévy measure has correct form. This means (see [228, Theorem 13.14]) that $(\tilde{Z}^\tau_t)$ converges in distribution to (Z_t) as $\tau \to \infty$.

Using the same type of computation and the independent increment property of the Poisson process one can prove that $(\tilde{Z}^\tau_t)$ has independent increments with respect to index τ. To prove that series (6.13) converges almost surely, we represent it as follows

$$\sum_{i=1}^{\infty} U^{(-1)}(\Gamma_i) 1_{[0,s]}(V_i) = \sum_{k=1}^{\infty} Z_k \quad \text{with} \quad Z_k = \sum_{N(k-1)}^{N(k)} U^{(-1)}(\Gamma_i) 1_{[0,s]}(V_i).$$

From the independent increments property of $(\tilde{Z}^\tau_t)$ it follows that Z_k are independent. Since they are also positive, by the 0-1 law (see [228, Corollary 2.14]), their sum may either converge with probability one or tend to infinity with probability one. On the other hand, the limit cannot be equal to infinity almost surely, because in this case the distributional convergence would not hold. Finally the uniform convergence follows from the fact that all terms are positive and hence the series for all t converge if $\tilde{Z}_1$ converges. □

Series representations for other types of Lévy processes may be constructed in a similar way. We now give without proof a general result, due to Rosinski [337], which allows to construct many such series representations and prove their convergence.

THEOREM 6.2 Rosinski [337]

Let $\{V_i\}_{i\geq 1}$ be an i.i.d. sequence of random elements in a measurable space S. Assume that $\{V_i\}_{i\geq 1}$ is independent of the sequence $\{\Gamma_i\}_{i\geq 1}$ of jumping times of a standard Poisson process. Let $\{U_i\}_{i\geq 1}$ be a sequence of independent random variables, uniformly distributed on $[0,1]$ and independent from $\{V_i\}_{i\geq 1}$ and $\{\Gamma_i\}_{i\geq 1}$. Let

$$H : (0,\infty) \times S \to \mathbb{R}^d$$

be a measurable function. We define measures on $\mathbb{R}^d$ by

$$\sigma(r,B) = P(H(r,V_i) \in B), \quad r > 0,\ B \in \mathcal{B}(\mathbb{R}^d), \tag{6.18}$$

$$\nu(B) = \int_0^\infty \sigma(r,B)dr.$$

Put

$$A(s) = \int_0^s \int_{|x|\leq 1} x\sigma(r,dx)dr, \quad s \geq 0. \tag{6.19}$$

(i) If ν is a Lévy measure on $\mathbb{R}^d$, that is,

$$\int_{\mathbb{R}^d} (|x|^2 \wedge 1)\nu(dx) < \infty$$

and the limit $\gamma = \lim_{s\to\infty} A(s)$ exists in $\mathbb{R}^d$ then the series $\sum_{i=1}^{\infty} H(\Gamma_i, V_i) 1_{U_i \leq t}$ converges almost surely and uniformly on $t \in [0,1]$ to a Lévy process with characteristic triplet $(0, \gamma, \nu)$, that is, with characteristic function

$$\phi_t(u) = \exp t[iu\gamma + \int_{\mathbb{R}^d} (e^{iux} - 1 - iux1_{|x|\leq 1})\nu(dx)].$$

(ii) If ν is a Lévy measure on $\mathbb{R}^d$ and for each $v \in S$ the function

$$r \mapsto |H(r, v)| \quad \textit{is nonincreasing}$$

then $\sum_{i=1}^{\infty} (H(\Gamma_i, V_i)1_{U_i \leq t} - tc_i)$ converges almost surely and uniformly on $t \in [0,1]$ to a Lévy process with characteristic triplet $(0, 0, \nu)$. Here c_i are deterministic constants given by $c_i = A(i) - A(i-1)$.

REMARK 6.6 It can be proven (see [337]) that the process Y_s^τ defined by

$$Y_s^\tau \equiv \sum_{i:\Gamma_i \leq \tau} (H(\Gamma_i, V_i)1_{U_i \leq t} - tA(\tau)) \tag{6.20}$$

is a compound Poisson process (in s) with characteristic triplet $(0, 0, \nu_s)$ where $\nu_s(A) = \int_0^s \sigma(r, A)dr$. Hence, series representations of Lévy processes obtained using Theorem 6.2 can be seen as compound Poisson approximations. However, the transformation applied to the initial Lévy measure to obtain a finite measure may be more complex than a simple removal of small jumps. □

Example 6.13 Series representation for a subordinator, as a corollary of Theorem 6.2

In this example we obtain the series representation for a subordinator, derived in Proposition 6.3 as a particular case of Theorem 6.2. We use the notation of Proposition 6.3. Consider a family (indexed by r) of probability measures on $\mathbb{R}_+$, defined by $\sigma(r, A) = 1_{U^{(-1)}(r) \in A}$ for all r (each of these measures is a Dirac measure). Then $\int_0^\infty \sigma(r, A)dr = \nu(A)$ and $H(r, V_i) \equiv U^{(-1)}(r)$. Hence, the series $\sum_{i=1}^{\infty} H(\Gamma_i, V_i)1_{U_i \leq t} = \sum_{i=1}^{\infty} U^{(-1)}(\Gamma_i)1_{U_i \leq t}$ converges almost surely and uniformly on $t \in [0,1]$ to a subordinator with characteristic function

$$\phi_t(u) = \exp t[\int_{\mathbb{R}^d} (e^{iux} - 1)\nu(dx)].$$

□

Example 6.14 Series representation for subordinated Brownian motion
Let (S_t) be a subordinator without drift and with continuous Lévy measure ν and let $X_t = W(S_t)$, where W is a Brownian motion, be a subordinated Lévy process. The characteristic triplet of (X_t) is $(0, 0, \nu_X)$ with $\nu_X(A) = \int_0^\infty \mu^s(A)\nu(ds)$ where μ^s is the probability measure of a Gaussian random variable with mean 0 and variance s. See Chapter 4 for details on subordination. Putting $U(x) = \int_x^\infty \nu(dt)$ we have

$$\nu_X(A) = \int_0^\infty \mu^{U^{(-1)}(r)}(B)dr.$$

Hence, we can put $\sigma(r, A) = \mu^{U^{(-1)}(r)}(B)$ and use Theorem 6.2. Due to the symmetry of the normal distribution, $A(s) \equiv 0$. Taking (V_i) to be a series of independent standard normal variables and $H(\Gamma_i, V_i) = \sqrt{U^{(-1)}(\Gamma_i)}V_i$ we conclude that

$$\sum_{i=1}^{\infty} \sqrt{U^{-1}(\Gamma_i)}V_i 1_{U_i \leq t} \stackrel{\mathcal{L}}{=} X_t.$$

□

Example 6.15 Series representation for symmetric stable processes
Now let us obtain a series representation for the symmetric α-stable process. Representations for skewed stable processes may also be computed using Theorem 6.2; see also [343]. Consider a symmetric random variable V such that $E|V|^\alpha < \infty$. Denote the law of V by $\mu(*)$. Then we can write for any measurable set B

$$\int_0^\infty r^{1/\alpha}\mu(r^{1/\alpha}B)dr = C \int_B \frac{dx}{|x|^{1+\alpha}},$$

where $C = \frac{\alpha}{2}E|V|^\alpha$ and the set $r^{1/\alpha}B$ contains all points x such that $r^{-1/\alpha}x \in B$. We can now apply Theorem 6.2 by taking $\sigma(r, B) \equiv r^{1/\alpha}\mu(r^{1/\alpha}B)$ for each r. Because μ is symmetric, $A(s) \equiv 0$ and we find that

$$X_t \equiv \sum_{i=1}^{\infty} \Gamma_i^{-1/\alpha} V_i 1_{U_i \leq t} \tag{6.21}$$

is an α-stable process on the interval $[0, 1]$, where V_i are independent and distributed with the same law as V and U_i are independent uniform on $[0, 1]$. For this series representation one can choose any symmetric variable V with finite moment of order α. For example, a variable taking values 1 and -1 with probability 1/2 could be used. Suppose that this is the case and consider the truncated series

$$X_t^\tau \equiv \sum_{i:\Gamma_i \leq \tau} \Gamma_i^{-1/\alpha} V_i 1_{U_i \leq t}.$$

By Remark 6.6, for each τ, X_t^τ is a compound Poisson process with Lévy measure ν_τ given by

$$\nu_\tau(B) = \int_0^\tau \sigma(r, B)dr = \int_0^\tau r^{1/\alpha} 1_{1 \in r^{1/\alpha}B} dr = \int_B \frac{dx}{|x|^{1+\alpha}} 1_{|x| \geq \tau^{-1/\alpha}}.$$

Hence, the jumps smaller in magnitude than $\tau^{-1/\alpha}$ are truncated.

This series representation can be implemented using the following algorithm. When one only needs to simulate the process at a finite number of points, it is preferable to use Algorithm 6.6. However, when it is necessary to control the size of all jumps, series representation is more convenient.

ALGORITHM 6.14 Simulation of a symmetric stable process by series representation

Fix a number τ depending on the required precision and computational capacity. This number is proportional to the average number of terms in the series and it determines the truncation level: jumps smaller in magnitude than $\tau^{-1/\alpha}$ are truncated.

Initialize $k := 0$

REPEAT WHILE $\sum_{i=1}^{k} T_i < \tau$

Set $k = k + 1$

Simulate T_k: *standard exponential.*

Simulate U_k: *uniform on* $[0, 1]$

Simulate V_k: *takes values* 1 *or* -1 *with probability* 1/2

The trajectory is then given by

$$X(t) = \sum_{i=1}^{k} 1_{U_i \leq t} V_i \Gamma_i^{-1/\alpha} \quad \textit{where} \quad \Gamma_i = \sum_{j=1}^{i} T_j.$$

□

Example 6.16 Series representation for the gamma process

As our last example of this section, we give a convenient series representation for the gamma process, which is both rapidly converging and easy to compute. It is originally due to Bondesson [64] but can also be seen as a simple application of Theorem 6.2. Let (Γ_i) and (U_i) be as above and (V_i) be an independent sequence of standard exponential random variables. Then

$$X_t = \sum_{i=1}^{\infty} \lambda^{-1} e^{-\Gamma_i/c} V_i 1_{U_i \leq t}$$

is a gamma process with parameters c and λ. Two trajectories of the gamma process, simulated using this series representation, are shown in Figure 6.4. □

It can be seen from Remark 6.6 that simulation using series representation is similar to direct compound Poisson approximation treated in Section 6.3. However, series representations are much more general because, as Example 6.15 shows, many different series representations may correspond to the same Lévy process because there are many different ways to truncate small jumps of a Lévy process. Some of them, as in Example 6.16, may be a lot easier to implement and to compute than the direct compound Poisson approximation. Moreover, series representations are more convenient because one is working with the same Lévy measure all the time instead of having to work with different measures of compound Poisson type. Another advantage is that it is easy to extend series representations to multidimensional dependent Lévy processes, when the dependence structure is given by a Lévy copula.

6.6 Simulation of multidimensional Lévy processes

In the multidimensional setting, no closed formulae are available for simulation of increments of Lévy processes, with the exception of multivariate Brownian motion. Thus, one should use approximate methods like compound Poisson approximation and series representations. These methods can be extended to multidimensional framework without significant changes, since Theorem 6.2 is already stated in multivariate setting. As an application of this theorem, we now give a method due to LePage [249] for constructing series representations, which is especially useful for multivariate stable processes.

Example 6.17 LePage's series representation for multivariate Lévy processes
Consider the following radial decomposition of a Lévy measure ν:

$$\nu(A) = \int_{S^{d-1}} \int_0^{\infty} 1_A(xv)\mu(dx, v)\lambda(dv),$$

where λ is a probability measure on the unit sphere S^{d-1} of $\mathbb{R}^d$ and $\mu(*, v)$ is a Lévy measure on $(0, \infty)$ for each $v \in S^{d-1}$. Put $U(x, v) = \int_x^{\infty} \mu(d\xi, v)$. Let Γ_i be a sequence of arrival times of a standard Poisson process, V_i be an independent sequence of independent random variables, distributed on S^{d-1} with distribution λ and U_i be an independent sequence of independent random variables, uniformly distributed on the interval $[0,1]$. Then

$$\sum_{i=1}^{\infty}(U^{(-1)}(\Gamma_i, V_i)V_i 1_{U_i \le t} - tc_i)$$

is a series representation for Lévy process with Lévy measure ν, where

$$U^{(-1)}(z,v) = \inf\{x > 0 : U([x,\infty[, v) < z\}.$$

To prove this representation one can use Theorem 6.2 with measure $\sigma(r,*)$ in (6.18) given by

$$\sigma(r,B) = \int_{S^{d-1}} \lambda(dv) 1_B(U^{(-1)}(r,v)v).$$

The deterministic constants $c_i = A(i) - A(i-1)$ may be computed using (6.19).

For stable processes the corresponding radial decomposition is

$$\nu(A) = \int_{S^{d-1}} \int_0^\infty 1_A(xv) \frac{dx}{x^{1+\alpha}} \lambda(dv).$$

□

A class of multivariate Lévy models, quite popular in financial applications, is constructed by subordinating a multivariate Brownian motion with a univariate subordinator (see Section 5.1). Processes of this type are easy to simulate because simulating multivariate Brownian motion is straightforward.

When the dependence of components of a multidimensional Lévy process is specified via a Lévy copula (see Chapter 5), series representations can also be constructed using Theorem 6.2 and the probabilistic interpretation of Lévy copulas (Theorem 5.5). To make the method easier to understand we give a series representation for two-dimensional processes with positive jumps of finite variation. Generalization to other Lévy processes of finite variation is straightforward by treating different corners of Lévy measure separately.

THEOREM 6.3

Let (Z_s) be an two-dimensional Lévy process with positive jumps with marginal tail integrals U_1 and U_2 and Lévy copula $F(x,y)$. If F is continuous on $[0,\infty]^2$ then the process Z is representable in law, on the time interval $[0,1]$, as

$$\{Z_s, 0 \le s \le 1\} \stackrel{\mathcal{L}}{=} \{\tilde{Z}_s, 0 \le s \le 1\}$$

with

$$\tilde{Z}_s^1 = \sum_{i=1}^{\infty} U_1^{(-1)}(\Gamma_i^{(1)}) 1_{[0,s]}(V_i) \quad \textit{and}$$
$$\tilde{Z}_s^2 = \sum_{i=1}^{\infty} U_2^{(-1)}(\Gamma_i^{(2)}) 1_{[0,s]}(V_i), \tag{6.22}$$

where (V_i) *are independent and uniformly distributed on* $[0,1]$, $(\Gamma_i^{(1)})$ *is an independent sequence of jump times of a standard Poisson process and for every* i, $\Gamma_i^{(2)}$ *conditionally on* $\Gamma_i^{(1)}$ *is independent from all other variables and has distribution function* $\frac{\partial}{\partial x} F(x,y)\big|_{x=\Gamma_i^{(1)}}$ *(viewed as a function of* y*).*

All the series in (6.22) converge almost surely and uniformly on $s \in [0,1]$.

REMARK 6.7 Here we enumerate the jumps of the two-dimensional process in such a way that the jump sizes of the first component are in descending order, and we simulate the jumps in the other component conditionally on the size of jump in the first one. This explains the special role of the first component in the theorem. □

REMARK 6.8 F satisfies the Lipschitz condition and is therefore continuous on $[0,\infty[^2$ because it has uniform margins. Therefore the nontrivial continuity condition imposed in this theorem is "continuity at infinity" which means for example, that $F(x,\infty)$ must be equal to $\lim_{y\to\infty} F(x,y)$. This means that the Lévy copula of independence $F_\perp(x,y) = x1_{y=\infty} + y1_{x=\infty}$ cannot be used. However, this is not a very important restriction since Lévy processes with independent components may be simulated directly. □

PROOF Proof of Theorem 6.3 Let $\{W_i\}_{i\geq 1}$ be a sequence of independent random variables, uniformly distributed on $[0,1]$. Then for each $r \in [0,\infty[$ there exists a function $h_r(v) : [0,1] \to [0,\infty[$ such that for each i, $h_r(W_i)$ has distribution function $\frac{\partial}{\partial x} F(x,y)|_{x=r}$ (see, e.g., [228, Lemma 2.22]). We can now define the function $H(r,v) : [0,\infty[\times[0,1] \to [0,\infty[^2$ component by component as follows:

$$H^{(1)}(r,v) = U_1^{(-1)}(r);$$
$$H^{(2)}(r,v) = U_2^{(-1)}(h_r(v)).$$

It remains to check that $H(r,v)$, when integrated as in (6.18), yields the Lévy measure of the process that we want to simulate. It is sufficient to consider the sets A of the form $A = [x,\infty[\times[y,\infty[$. Recall that in the proof of Proposition 6.3, we have shown that if U is a tail integral of a Lévy measure on $]0,\infty[$

then $U^{(-1)} \geq z$ if and only if $y \geq U(z)$.

$$P[H(r, W_i) \in A] = 1_{U_1^{(-1)}(r) \in [x,\infty[} P\{U_2^{(-1)}(h_r(W_i)) \in [y, \infty[\}$$
$$= 1_{r \in]0, U_1(x)]} P\{h_r(W_i) \in]0, U_2(y)]\}.$$

The second factor in this expression is nothing but the probability distribution function of $h_r(W_i)$, which is known by construction, hence,

$$P[H(r, W_i) \in A] = 1_{r \in]0, U_1(x)]} \frac{\partial}{\partial r} F(r, U_2(y)).$$

We now conclude that because F is absolutely continuous with respect to the first variable in all finite points and continuous on $[0, \infty]^2$,

$$\int_0^\infty P(H(r, W_i) \in A) dr = \int_0^{U_1(x)} \frac{\partial}{\partial r} F(r, U_2(y)) dr = F(U_1(x), U_2(y))$$

yields the tail integral of the Lévy measure of subordinator that we want to simulate and we can apply Theorem 6.2. □

This theorem justifies the following algorithm.

ALGORITHM 6.15 Simulation of two-dimensional subordinator with dependent components by series representation

Fix a number τ depending on the required precision and computational capacity. This number is equal to the average number of terms in the series and it determines the truncation level: jumps in the first component smaller than $U^{(-1)}(\tau)$ are truncated.

Initialize $k = 0$, $\Gamma_0^{(1)} = 0$
REPEAT WHILE $\Gamma_k^{(1)} < \tau$
- *Set* $k = k + 1$
- *Simulate* T_k*: standard exponential.*
- *Set* $\Gamma_k^{(1)} = \Gamma_{k-1}^{(1)} + T_k$ *(transformed jump in the first component)*
- *Simulate* $\Gamma_k^{(2)}$ *from distribution function* $F_1(y) = \frac{\partial F(x,y)}{\partial x}\big|_{x=\Gamma_k^{(1)}}$ *(transformed jump in the second component)*
- *Simulate* V_k*: uniform on* $[0, 1]$ *(jump time)*

The trajectory is then given by

$$X(t) = \sum_{i=1}^{k} 1_{V_i \leq t} U_1^{(-1)}(\Gamma_i^{(1)});$$
$$Y(t) = \sum_{i=1}^{k} 1_{V_i \leq t} U_2^{(-1)}(\Gamma_i^{(2)}).$$

Example 6.18 Simulation of a two-dimensional subordinator with dependence given by Clayton Lévy copula
When the dependence Lévy copula is in Clayton family (5.18), the conditional distribution function takes a particularly simple form

$$F(v|u) = \frac{\partial F_\theta(u,v)}{\partial u} = \left\{1 + \left(\frac{u}{v}\right)^\theta\right\}^{-1-1/\theta}$$

and can be easily inverted:

$$F^{-1}(y|u) = u\left(y^{-\frac{\theta}{1+\theta}} - 1\right)^{-1/\theta}, \tag{6.23}$$

which makes simulation of jumps very simple. The two-dimensional subordinator has the following representation

$$X_s = \sum_{i=1}^{\infty} U^{(-1)}(\Gamma_i)1_{[0,s]}(V_i);$$
$$Y_s = \sum_{i=1}^{\infty} U^{(-1)}(F^{-1}(W_i|\Gamma_i))1_{[0,s]}(V_i),$$

where (W_i) and (V_i) are independent sequences of independent random variables, uniformly distributed on $[0,1]$ and (Γ_i) is an independent sequence of jump times of a standard Poisson process.

Figure 6.6 depicts the simulated trajectories of a subordinator with $\frac{1}{2}$-stable margins and dependence given by Clayton Lévy copula with parameter θ for different modes of dependence (different values of θ). On the left graph the dependence is very weak; the shocks in the two components are almost completely independent. On the right graph the dependence is stronger; here big shocks in the two components tend to arrive at the same time; however their exact magnitude may be very different. Finally, on the bottom graph we have almost complete dependence: the shocks arrive together and have approximately the same magnitude. □

Further reading

Algorithms for simulating random variables with explicit density can be found in Devroye [113]. Simulation methods for general α-stable random variables were first published by Chambers, Mallows, and Stuck [87]. The computer program which implements their method, can also be found in Samorodnitsky and Taqqu [343]. For details on approximating small jumps

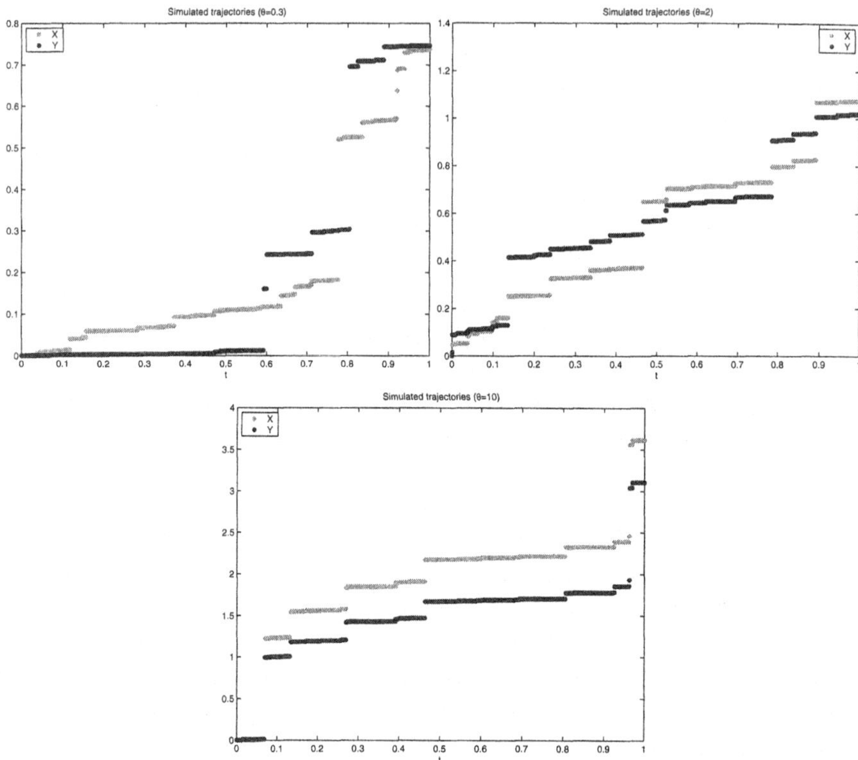

FIGURE 6.6: Simulated trajectories of a two-dimensional subordinator under weak (left), medium (right) and strong (bottom) dependence.

of Lévy processes by a Brownian motion, the interested reader should consult the original article of Asmussen and Rosinski [11] as well as Wiktorsson [380]. Rosinski [337] gives a comprehensive survey of series representations for Lévy processes. The case of stable processes is also treated in Samorodnitsky and Taqqu [343]. Details on multivariate Lévy processes can be found in Chapter 5 of this book.

In this chapter we did not discuss Euler schemes for Lévy processes because we do not use them in this book and the scope of their application in finance is somewhat limited. Nevertheless it is a rapidly evolving field of research. Interested readers are referred to Jacod [211, 212] for a survey of recent results on Euler schemes for Lévy processes. Rubenthaler [339] also discusses Euler schemes for Lévy processes, treating the more realistic case when increments of the driving Lévy process cannot be simulated directly.

Louis BACHELIER

Louis Bachelier has been called the founder of mathematical finance and the "father of modern option pricing theory". Born in Le Havre in 1870, Louis Bachelier moved to Paris around 1892 and worked for some time in the Paris Bourse, where he became familiar with the workings of financial markets. Bachelier then undertook a doctoral thesis in mathematical physics at the Sorbonne under the supervision of Poincaré. His thesis, published in 1900 under the title *Théorie de la Spéculation* [17], dealed with the probabilistic modelling of financial markets and marked the beginning of two scientific theories: the theory of Brownian motion and the mathematical modelling of financial markets.

Five years before Einstein's famous 1905 paper on Brownian Motion, Bachelier worked out in his doctoral thesis the distribution function for what is now known as the Wiener process (the stochastic process that underlies Brownian Motion), the integral equation verified by the distribution (later called the Chapman Kolmogorov equation) and linked it mathematically to Fourier's heat equation. Bachelier's work was initially underestimated by the academic community. It appears that Einstein in 1905 ignored the work of Bachelier, but Kolmogorov knew of it and brought it to the knowledge of Paul Lévy many years later. In 1931, Lévy wrote a letter of apology to Bachelier recognizing his work. Bachelier's treatment and understanding of the theory of Brownian Motion is more mathematical than in Einstein's 1905 paper, which was more focused on Brownian motion of physical particles. In his thesis, Bachelier also derived the distribution of the maximum of Brownian motion on an interval and uses it to study barrier options. Later, Paul Lévy and William Feller called the Brownian motion process the Bachelier-Wiener Process.

Bachelier, whose name is frequently quoted today in works of probability and mathematical finance, was rejected by the mathematical community of his time. Bachelier's works are formulated in the language of a physicist and his mathematics was not rigorous (since many of the mathematical techniques necessary to formulate it had not been developed at the time) but the results he obtained were original and basically correct. William Feller writes ([141], p 323):"Credit for discovering the connection between random walks and diffusion is due principally to Louis Bachelier. His work is frequently of a heuristic nature but he derived many new results. Kolmogorov's theory of stochastic processes of the Markov type is based largely on Bachelier's ideas."

Bachelier's work on stochastic modelling of financial markets was unearthed in the 1950s by L.J. Savage and P. Samuelson in the United States and an English translation of his thesis subsequently appeared in [100]. Inspired by his work, Samuelson formulated the log-normal model for stock prices which formed the basis for the Black-Scholes option pricing model. However, Bachelier died in 1946 and did not live to see the development of modern mathematical finance.

Bachelier's intuitions were ahead of his time and his work was not appreciated in his lifetime. His contribution has by now been amply recognized: the centenary of his thesis was celebrated in 2000 by a international congress held in Paris. Interesting material on Bachelier's life and scientific work may be found in [374, 101].

Chapter 7

Modelling financial time series with Lévy processes

> Si [. . .] j'ai comparé les résultats de l'observation à ceux de la théorie, ce n'était pas pour vérifier des formulaes établies par des méthodes mathématiques mais pour montrer que le marché, à son insu, obéit à une loi qui le domine: la loi de la probabilité.
>
> Louis Bachelier *Théorie de la spéculation* (1900)

> The purpose of models is not to fit the data but to sharpen the questions
>
> Samuel Karlin
> *11th R. A. Fisher Memorial Lecture*, Royal Society, April 1983

As mentioned in the introduction, one of the principal motivations for departing from Gaussian models in finance has been to take into account some of the observed empirical properties of asset returns which disagree with these models. Lévy processes entered financial econometrics in 1963 when Mandelbrot [278] proposed α-stable Lévy processes as models for cotton prices. Since then a variety of models based on Lévy processes have been proposed as models for asset prices and tested on empirical data. In this chapter we discuss some of these models and examine how they fare in reproducing stylized properties of asset prices.

Market prices are observed in the form of time series of prices, trading volumes and other quantities observed at a discrete set of dates. What is then the motivation for using continuous time model to represent such data? As long as one considers prices sampled at a single — say daily — frequency, the usual approach is to represent the time series as a discrete-time stochastic process where the time step corresponds to the interval between observations: there is a long tradition of time series modelling using this approach in financial econometrics [76, 361, 63]. However applications involve different time horizons, ranging from intraday (minutes) to weekly or longer and, as we shall see below, the statistical properties of asset returns corresponding to different time intervals can be quite different! Therefore, a model may perform well based on returns on a given time horizon, say Δ but fail to do so for another horizon Δ'. Also, most time series models in popular use are not stable under time aggregation: for example, if returns computed at interval Δ follow a GARCH(1,1) process, returns computed at interval 2Δ do not in general follow a GARCH process [117]. Defining a different time series model for every sampling frequency is definitely a cumbersome approach. By contrast, when price dynamics are specified by a continuous time process, the distribution and properties of log-returns at all sampling intervals are embedded in the definition of the model. Moreover, all discrete-time quantities will have a well-defined limit as the sampling interval becomes small.

In this chapter we will review some of the statistical properties of asset returns and compare them with class of *exponential-Lévy models* in which the asset price S_t is represented as

$$S_t = S_0 \exp(X_t),$$

where X is a Lévy process, i.e., a process with stationary independent increments (see Chapters 3 and 4). These models are simple to study and we will examine to what extent they can accommodate observed features of financial data.

Section 7.1 presents some commonly observed statistical properties of asset returns, known as *stylized empirical facts* in the financial econometrics literature. In order to examine the adequacy of a statistical model, first one estimates model parameters from time series of asset returns and then compares the properties of the estimated model with the statistical properties observed in the returns. While it is possible to define a variety of models which can potentially reproduce these empirical properties, *estimating* these models from empirical data is not always an obvious task. We will present an overview of estimation methods and some of their pitfalls in Section 7.2. Distributional properties of asset returns at a given time horizon are discussed in Section 7.3. We shall observe that Lévy processes enable flexible modelling of the distribution of returns at a given time horizon, especially when it comes to modelling the tails of the distribution. Less obvious is to model the time-aggregation properties of returns: the dependence of their statistical properties with respect to the time horizon. This issue is discussed in Section 7.4: we define in particular the notion of self-similarity and discuss its relevance to the study of asset returns.

A lot of recent studies have been devoted to the study of realized volatility of financial assets and its empirical properties. In Section 7.5 we define this notion and examine what exponential Lévy models have to say about realized volatility.

Empirical observations consist of a single trajectory of the price process and the profit/loss of an investor is actually determined by the behavior of this single sample path, rather than by properties averaged over paths. Pathwise properties such as measure of smoothness of sample paths are discussed in Section 7.6.

In the following, S_t will denote the price of a financial asset — a stock, an exchange rate or a market index — and $X_t = \ln S_t$ its logarithm. Given a *time scale* Δ, the log return at scale Δ is defined as:

$$r_t(\Delta) = X_{t+\Delta} - X_t. \tag{7.1}$$

Δ may vary between a minute (or even seconds) for tick data to several days. We will conserve the variable Δ to stress the fact that the statistical properties of the returns depend on Δ in a nontrivial way. Observations are sampled at discrete times $t_n = n\Delta$. Time lags will be denoted by the Greek letter τ; typically, τ will be a multiple of Δ in estimations. For example, if Δ =1 day, $\text{corr}[r_{t+\tau}(\Delta), r_t(\Delta)]$ denotes the correlation between the daily return at period s and the daily return τ periods later. When Δ is small — for example of the order of minutes — one speaks of "fine" scales whereas if Δ is large we will speak of "coarse-grained" returns. Given a sample of N observations $(r_{t_n}(\Delta), n = 1 \ldots N)$, the sample average of a function $f(.)$ of returns is defined as:

$$< f(r(\Delta)) > = \frac{1}{N} \sum_{n=1}^{N} f(r_{t_n}(\Delta)). \tag{7.2}$$

Sample averages are, of course, not to be confused with expectations, which are denoted in the usual way by $E[f(r(\Delta))]$.

7.1 Empirical properties of asset returns

The viewpoint of most market analysts in "explaining" market fluctuations, conveyed by most financial newspapers and journals, has been and remains an *event-based* approach in which one attempts to rationalize a given market movement by relating it to an economic or political event or announcement. From this point of view, one could easily imagine that, since different assets are not necessarily influenced by the same events or information sets, price series obtained from different assets and — *a fortiori* — from different markets will exhibit different properties. After all, why should properties of corn futures be similar to those of IBM shares or the Dollar/Yen exchange rate? Nevertheless, the result of more than half a century of empirical studies on financial time series indicates that this is the case if one examines their properties from a

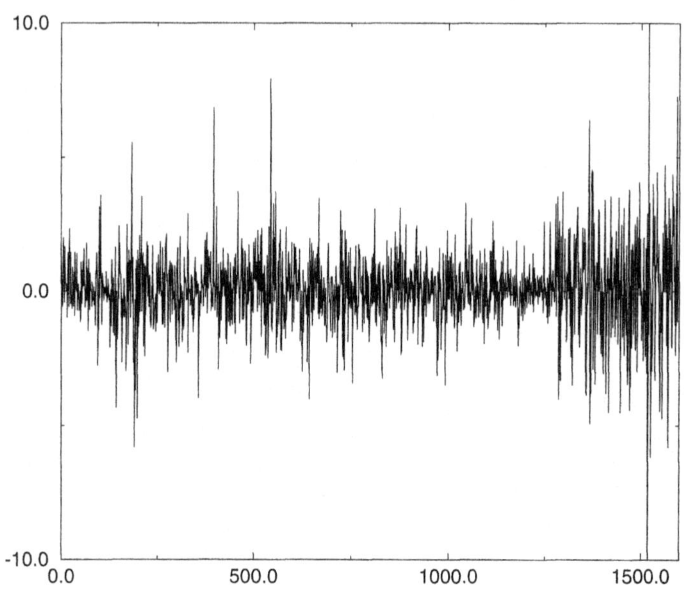

FIGURE 7.1: BMW daily log-returns.

statistical point of view. The seemingly random variations of asset prices *do* share some quite nontrivial statistical properties. Such properties, common across a wide range of instruments, markets and time periods are called *stylized empirical facts* [92].

Stylized facts are thus obtained by taking a common denominator among the properties observed in studies of different markets and instruments. Obviously by doing so one gains in generality but tends to lose in precision of the statements one can make about asset returns. Indeed, stylized facts are usually formulated in terms of *qualitative properties* of asset returns and may not be precise enough to distinguish among different parametric models. Nevertheless, we will see that, albeit qualitative, these stylized facts are so constraining that it is not even easy to exhibit an (ad hoc) stochastic process which possesses the same set of properties and stochastic models have gone to great lengths to reproduce these stylized facts.

We enumerate here some of these empirical facts; for further details readers are referred to [92, 76, 313]:

1. **Heavy tails:** the (unconditional) distribution of returns seems to display a heavy tail with positive excess kurtosis with a tail index which is

finite, higher than two and less than five for most data sets studied. In particular this excludes stable laws with infinite variance and the normal distribution. Power-law or Pareto tails reproduce such behavior, but the precise form of the tails is difficult to determine and some authors have suggested models with exponential tails ("semiheavy tails") as an alternative.

2. **Absence of autocorrelations:** (linear) autocorrelations of asset returns are often insignificant, except for very small intraday time scales ($\simeq$ 20 minutes) for which microstructure effects come into play. Figure 7.3 illustrates this fact for a stock and an exchange rate.

3. **Gain/loss asymmetry:** one observes large drawdowns in stock prices and stock index values but not equally large upward movements.[1]

4. **Aggregational normality:** as one increases the time scale Δ over which returns are calculated, their distribution looks more and more like a normal distribution. In particular, the shape of the distribution is not the same at different time scales: the heavy-tailed feature becomes less pronounced as the time horizon is increased.

5. **Volatility clustering:** "large changes tend to be followed by large changes, of either sign, and small changes tend to be followed by small changes" [278]. A quantitative manifestation of this fact is that, while returns themselves are uncorrelated, absolute returns $|r_t(\Delta)|$ or their squares display a positive, significant and slowly decaying autocorrelation function.

6. **Conditional heavy tails:** even after correcting returns for volatility clustering (e.g., via GARCH-type models), the residual time series still exhibit heavy tails. However, the tails are less heavy than in the unconditional distribution of returns.

7. **Slow decay of autocorrelation in absolute returns:** the autocorrelation function of absolute (or squared) returns decays slowly as a function of the time lag, roughly as a power law with an exponent $\beta \in [0.2, 0.4]$. An example is shown in Figure 7.3. This is sometimes interpreted as a sign of long-range dependence in volatility.

8. **"Leverage" effect:** most measures of volatility of an asset are negatively correlated with the returns of that asset.

9. **Volume/volatility correlation:** trading volume is positively correlated with all measures of volatility. The same holds for other measures of market activity such as the number of trades.

[1]This property is not true for exchange rates where there is a higher symmetry in up/down moves.

10. **Asymmetry in time scales:** coarse-grained measures of volatility predict fine-scale volatility better than the other way round [181].

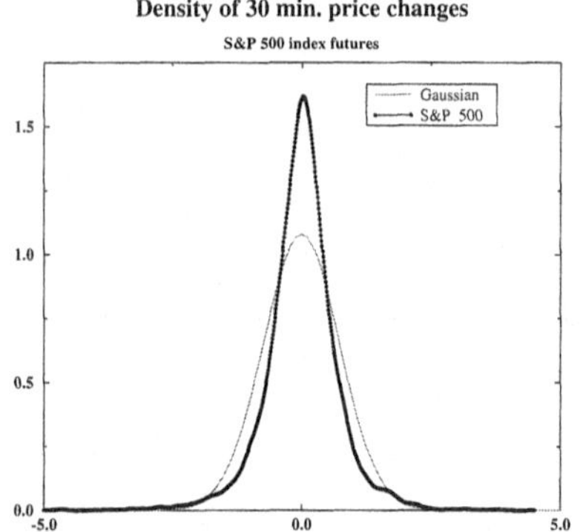

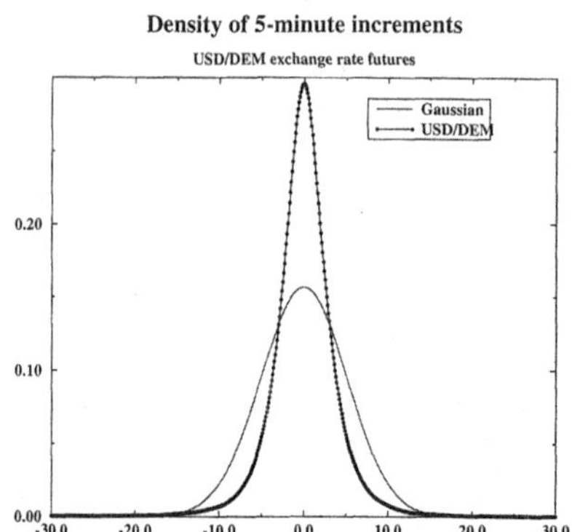

FIGURE 7.2: Probability density of log-returns compared with the Gaussian distribution. Left: 30 minute returns for S&P500 index. Right: 5 minute log-returns for Deutschemark-US Dollar exchange rates.

7.2 Statistical estimation methods and their pitfalls

All estimation approaches are based on choosing model parameters in order to optimize a certain criterion computed using the observed returns $(r_t, t = 1 \ldots N)$. In the maximum likelihood method, this criterion is the likelihood function (see 7.2.1). In the (generalized) method of moments, the criterion is based on the difference of some moments of the distribution and the corresponding empirical moments. In most cases of interest, the resulting optimization problem cannot be solved analytically and numerical methods must be used.

7.2.1 Maximum likelihood estimation

The most common method for estimating a parametric model for the distribution of returns is the *maximum likelihood* method: given a functional form $f(x; \theta)$ for the density of log-returns and observations $(r_t, t = 1 \ldots N)$, we choose the model parameter θ to maximize the *likelihood* that the observed

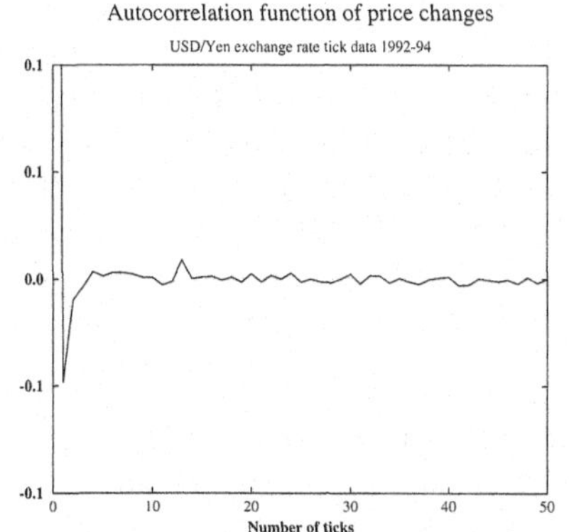

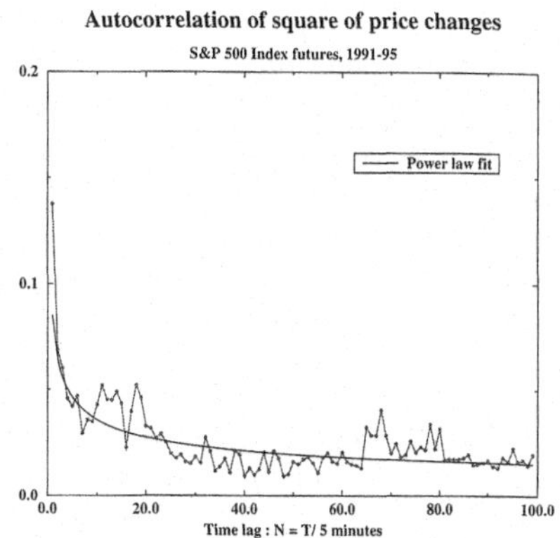

FIGURE 7.3: Left: Autocorrelation function of log-returns, USD/Yen exchange rate, $\Delta = 5$ minutes. Right: Autocorrelation function of squared log-returns: S&P500 futures, $\Delta = 30$ minutes.

data have been generated from the model:

$$\max_{\theta} \prod_{t=1}^{N} f(r_t; \theta). \tag{7.3}$$

This is of course equivalent to maximizing the log-likelihood function

$$l(\theta) = \sum_{t=1}^{N} \ln f(r_t; \theta). \tag{7.4}$$

If the functional form of $f(.|\theta)$ is known, one can derive an equation for the maximizers of $l(\theta)$ by differentiating l and try to solve it numerically. Unfortunately, for many of the Lévy processes seen in Chapter 4, the likelihood function (which is here simply the density of the process at time Δ) is not known in closed form but involves special functions which have to be computed numerically. Therefore even the *computation* of the likelihood function must be done numerically. If the log-likelihood function is a concave function of the parameters, then (7.4) has a unique maximizer which can be computed by a gradient descent algorithm such as Newton-type methods or the BFGS method. But if the log-likelihood is not concave, then (7.3) may or may not have a unique maximum: typically, it may have several *local* maxima. Let us stress that numerical optimization of non-convex functions in several dimensions is far from trivial and is a topic of ongoing research. This point is often disregarded in many empirical studies, where some black-box optimization software (often gradient based) is used to compute maximum likelihood estimators. Even if the likelihood does have a unique global maximum, such algorithms may not converge to it: indeed, depending on how it is initialized, a gradient-based optimization algorithm will typically converge to a local maximum.

In many of the models described in Chapter 4, while the density or likelihood function of the process is not known in closed form, the Lévy measure of the process has a simple parametric form. Using the relation between the probability density and the Lévy density given in Chapter 3, one can approximate the likelihood function of increments on a small time interval Δ. Denote by $\rho_t(.)$ the density of X_t. The following result is shown in [341, Theorem 1]:

PROPOSITION 7.1 Small time expansion of densities [341]
Assume that the Lévy process X_t has a C^∞ density $\rho_t(x)$, the density $\nu(x)$ of the Lévy measure is C^∞ and verifies:

$$\forall \varepsilon > 0, \quad \int_{|x|\geq\epsilon} \frac{|\nu'(x)|^2}{\nu(x)} dx < \infty \quad \textit{and}$$
$$\exists\, h \in C^\infty \quad \textit{with} \quad h(x) \leq c|x|^2 \quad \textit{and} \quad h(x) > 0 \quad \textit{if} \quad \nu(x) > 0$$
$$\textit{such that} \quad \int_{|x|\leq 1} \left| \frac{d}{dx} h(x)\nu(x) \right|^2 \frac{dx}{\nu(x)} < \infty. \tag{7.5}$$

Denote by $\nu_\epsilon(x) = \nu(x)1_{|x|\geq\epsilon}$ the truncated Lévy density. Then for any $N \geq 1$ and any $x_0 > 0$ there exist $\epsilon_0 > 0$ and $t_0 > 0$ such that for $0 < \epsilon \leq \epsilon_0$ and any $t < t_0$:

$$|x| > x_0 > 0 : \qquad \rho_t(x) = e^{-t\int \nu_\epsilon(y)dy} \sum_{i=1}^{N-1} \frac{t^i}{i!} \nu_\epsilon^{*i}(x) + O(t^N). \tag{7.6}$$

A condition which ensures that a Lévy process has a C^∞ density was given in Proposition 3.12: this is the case for example if $\sigma > 0$ or if $\sigma = 0$ and X is a tempered stable process with $\alpha_\pm > 0$. ν_ϵ is explicitly known in most cases, but its higher convolution powers ν_ϵ^{*i} are not easy to compute so in practice one uses a first order expansion to approximate the density ρ_t. In the case of finite intensity, one does not need the truncation and $\epsilon = 0$.

When the log-price is a compound Poisson process [320] with intensity λ and jump size density ν_0 the first order expansion above leads to the following approximation for the likelihood/density of increments:

$$\rho_\Delta(x) \simeq \lambda\Delta\nu_0(x) + (1 - \lambda\Delta)\delta_0. \tag{7.7}$$

In this case the "likelihood function" is actually singular (it is a distribution) so the above equation must be interpreted in the following way: $X_{t+\Delta} - X_t = 0$ with probability $\simeq 1 - \lambda\Delta$ and, conditionally on $X_{t+\Delta} - X_t \neq 0$, the distribution of (nonzero) returns is the jump size distribution ν_0 whose parameters can be estimated by a maximum likelihood procedure. The maximum likelihood estimate of λ using the approximation (7.7) is then simply the proportion of zeroes in the log-returns: if N_0 is the number of zeroes observed in the series $(r_t(\Delta))$ then

$$\hat{\lambda} = \frac{1 - \frac{N_0}{N}}{\Delta}.$$

However this approximation is valid only if $\lambda\Delta << 1$ which means that a large fraction of the returns are zero! This is not the case in empirical data, even at very high frequencies, pointing to an inadequacy of the compound Poisson model. In fact, since the proportion of zero returns remains finite even at the tick level, $\hat{\lambda}$ is typically of order $1/\Delta$ and tends to increase without bound as $\Delta \to 0$ (see for examples the empirical study in [199]), indicating that either the continuous time limit has an infinite number of jumps per unit time (infinite activity, in the sense defined in Chapter 3) or that a diffusion component has been left out.[2]

The jump intensity λ can also be estimated from the time between trades: in the compound Poisson model the interval between price changes should follow an exponential distribution. Figure 7.4 compares the distribution of time intervals between trades for KLM tick data (NYSE:KLM) from 1993 to 1998 with an exponential distribution estimated by a maximum likelihood method. In the case of the exponential distribution the maximum likelihood estimate $\hat{\lambda}$ of the parameter corresponds to the sample mean: $\hat{\lambda}$ is simply the average time between two trades, in this case 859 seconds. Therefore the estimated exponential model and the empirical distribution have the same mean. However the sample has a kurtosis of 24.6, much higher than the kurtosis of an exponential distribution which is equal to 9.

The log-log plot clearly illustrates that the distribution of time intervals between trades does not resemble an exponential; this is further confirmed by the quantile-quantile plot in Figure 7.4 which shows that their tails behave very differently. All these remarks point to the fact that a compound Poisson process with drift is not a good representation of price dynamics: either a diffusion component must be added or the process should be allowed to have infinite activity.

In jump-diffusion models, the diffusion components are supposed to represent "normal" market returns and the jump component serves to capture rare events, i.e., abnormally large returns of either sign. In the case of a jump-diffusion model $X_t = \sigma W_t + \sum_{j=1}^{N_t} Y_j$, $Y_j \sim \nu_0$ with finite jump intensity λ the small time approximation becomes:

$$\rho_\Delta(x; \lambda, \theta) \simeq \lambda\Delta f_1(x;\theta) + (1 - \lambda\Delta) f_0(x;\theta), \qquad (7.8)$$

where f_0 is the normal $N(\mu\Delta, \sigma^2\Delta)$ density and $f_1 = f_0 * \nu_0$ is the transition density given that one jump has occurred during $[t, t+\Delta]$, that is, the density of $W_\Delta + Y_1$, given by the convolution of f_0 with the jump size distribution ν_0.

[2]Note that this problem can go unnoticed if one estimates the model at a single sampling interval.

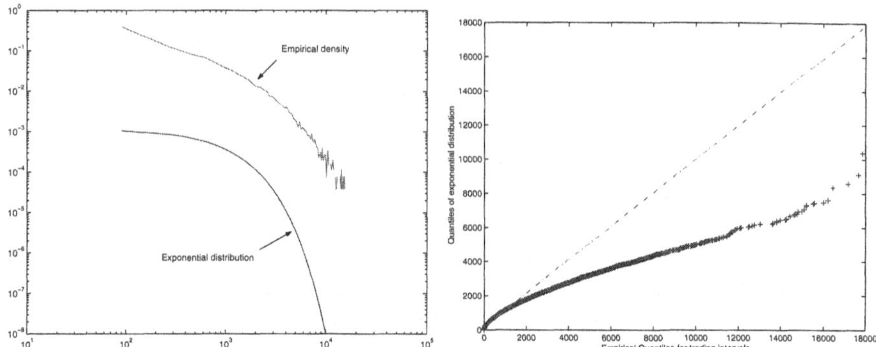

FIGURE 7.4: Left: The distribution of time intervals between trades, compared with an exponential distribution estimated by maximum likelihood. KLM tick data, NYSE, 1993–1998. Right: Quantiles of the exponential distribution plotted against the quantiles of the empirical distribution of trading intervals (durations). KLM tick data, NYSE, 1993–1998.

In order for this approximation to be useful one needs to know in closed form the density $f_0 * \nu_0$. This is the case of course if the jump size is Gaussian (the Merton model) or constant (Brownian motion with Poisson jumps) but there are not many other cases where f_1 is known. Here θ denotes the vector of parameters in f_0, f_1; these do not depend on the intensity λ. The likelihood function therefore takes the form

$$l(\lambda, \theta) = \prod_{i=1}^{N} [\rho_\Delta(x; \lambda, \theta)] \simeq \prod_{i=1}^{N} [f_0(x_i; \theta) + \lambda\Delta(f_1(x_i; \theta) - f_0(x_i; \theta))]. \quad (7.9)$$

We notice that the dependence in λ is a polynomial of degree N whose coefficients depend on the sample data. Since the above approximation is valid for $\lambda\Delta << 1$, $\hat{\lambda}$ is obtained by minimizing (7.9) over $\lambda \in]0, 1/\Delta[$. In a finite sample this frequently leads to a "overestimated" jump rate $\hat{\lambda}$ which increase as Δ becomes smaller [224, 199]. The interpretation of jumps as rare events is then not clear: to maintain a coherent vision as Δ varies one is naturally led to formulate a model allowing infinite jump rates.

In the case of infinite activity Lévy processes we must rely either on the knowledge of the analytical expression of the density as a function of the parameters or use expansions such as (7.6). If the derivatives of the likelihood function with respect to parameters are known, a gradient based algorithm can be used to numerically compute the maximum likelihood estimator. This is the case for the five parameter class of generalized hyperbolic models introduced in Section 4.6, for which the log-likelihood function is given by:

$$l(\lambda,\alpha,\beta,\delta) = \ln C(\lambda,\alpha,\beta,\delta) + (\frac{\lambda}{2} - \frac{1}{4}) \sum_{t=1}^{N} \ln(\delta^2 + (r_t - \mu)^2)$$

$$+ \sum_{t=1}^{N} [\ln K_{\lambda-\frac{1}{2}}(\alpha\sqrt{\delta^2 + (r_t-\mu)^2}) + \beta(r_t - \mu)], \qquad (7.10)$$

where $C(\lambda,\alpha,\beta,\delta)$ is given in (4.38). Computing (7.10) involves N numerical evaluations of the modified Bessel function $K_{\lambda-1/2}$. For this model Prause [319] gives analytical expressions (again involving special functions) of the derivatives of l with respect to the parameters ([319], p. 10–11). Unfortunately the corresponding first order conditions cannot be solved analytically and a numerical optimization procedure must be used. Let us emphasize again that log-likelihood functions are not necessarily concave functions of parameters in these models, leading to multiple maxima which are not easy to locate by gradient-based numerical methods, a point overlooked in most empirical studies. Barndorff-Nielsen and Blaesild [34] point out the flatness of the likelihood function for the case of the hyperbolic distribution. Blaesild and Sørensen [61] provide a gradient-based algorithm for maximum likelihood estimation of hyperbolic distributions but due to this lack of concavity their algorithm often fails to converge or gives a δ very close to zero [319, 329].

Given the possibility of nesting various models in larger and larger parametric families, it is tempting to nest everything into a huge family (such as generalized hyperbolic) and estimate parameters for this hyper-family. The point is that by increasing the dimension of parameter space the maximization problem becomes less and less easy. For example, [319] points out that the likelihood landscape of the generalized hyperbolic distribution is even flatter than the subfamily of hyperbolic distributions. Consequently, using the full generalized hyperbolic family leads to imprecise parameter estimates, unless some ad hoc parametric restrictions are made [319, 329]) which amounts to choosing a subfamily of models.

7.2.2 Generalized method of moments

While there are few Lévy processes for which the likelihood functions are available in closed form, expression for moments and cumulants are almost always available in closed form as a function of model parameters: they are easily obtained by differentiating the characteristic function (see Equation 2.21). This enables to construct *method of moments estimators* by matching empirical moments with theoretical moments and solving for the parameters.

The generalized method of moments (GMM) [187] consists in choosing the parameter θ to match — in a least squares sense — a given set of sample averages:

$$\theta^* = \arg\min_\theta {}^tV(\theta)WV(\theta), \tag{7.11}$$

$$V(\theta) = (< f_j(r_t) > -E[f_j(r_t)|\theta]\ )_{j=1\ldots m}, \tag{7.12}$$

where the $f_j(.)$ are a set of quantities chosen to generate the moment conditions and W is a symmetric positive definite weighting matrix.

The "generalized moments" $< f_j(r_t) >$ can represent the sample average of any quantity. This includes the usual moments and cumulants — sample moments of any order for returns, sample autocovariances of returns, absolute returns — but also empirical probabilities ($f_j = 1_{[a,b]}$) and quantiles, represented by the empirical distribution (histogram) of returns. Thus, "curve fitting" methods where model parameters are estimated by performing least squares fit on a histogram or empirical autocorrelation function are particular cases of the GMM method and therefore the asymptotic theory developed for GMM applies to them.

Since sample moments are not independent from each other, we should expect that the errors on the various moments are correlated so it might not be desirable to used a diagonal weighting matrix W: ideally, W should be a data dependent matrix. An "optimal" choice of the weighting matrix is discussed in [187], see also [175].

The choice of moment conditions for a GMM estimator can have very pronounced effects upon the efficiency of the estimator. A poor choice of moment conditions may lead to very inefficient estimators and can even cause identification problems. The uniqueness of the minimizer in (7.11) is also an issue: choosing moment conditions $f_j(.)$ that ensure identification may be a nontrivial problem.

In the case of a model specified by a Lévy process for the log-prices, the simplest moment conditions could be the first four moments of the returns $r_t(\Delta)$ (in order for mean, variance, skewness and kurtosis to be correctly reproduced), but these may not be enough to identify the model. In particular, if the moment conditions are specified only in terms of returns at a given frequency Δ — say, daily — then this may result in overfitting of the distribution of daily returns while the model gives a poor representation of distributions at other time scales. It is thus better to use moment conditions involving returns from several time horizons, in order to ensure the coherence of the model across different time resolutions.

An issue important for applications is capturing the behavior of the tails. Including tail-sensitive quantities such as probabilities to exceed a certain value ($f_j = 1_{[b,\infty[}$) prevents from overfitting the center of the distribution while poorly fitting the tails. Most four parameter families of Lévy processes such as the ones presented in Chapter 4 allow flexible tail behavior. Of course the number of parameters (four or more) is not the right criterion here: the parameterization should allow different types of tail decay as well as asymmetry in the left and right tail. The generalized hyperbolic family, presented in Section 4.6 allows such freedom. Another flexible parametric family is that of tempered stable processes, presented in Section 4.5.

Since Lévy processes have independent increments, there is no point in including autocovariances of returns, absolute returns or their squares as moment conditions: if they are not zero, the corresponding moment conditions cannot be satisfied. This is of course a limitation of Lévy processes and not of GMM.

7.2.3 Discussion

In some cases the density of returns is completely characterized by (all of) its moments $(m_n, n \geq 1)$. One may then interpret this by saying that the maximum likelihood estimate is a GMM estimator which uses as moment condition *all* moments of the distribution, weighted in a particular way. The question is then: why do we use moment conditions instead of maximizing the likelihood? The answer is that there are several advantages to using GMM:

- Robustness: GMM is based upon a limited set of moment conditions. For consistency, only these moment conditions need to be correctly specified, whereas MLE in effect requires correct specification of *every conceivable* moment condition. In this sense, GMM is *robust with respect to distributional misspecification.* The price for robustness is loss of efficiency with respect to the MLE estimator. Keep in mind that the true distribution is *not known* so if we erroneously specify a distribution and estimate by MLE, the estimator will be inconsistent in general (not always).

- Feasibility: in many models based on Lévy processes the MLE estimator is not available, because we are not able to compute the likelihood function while GMM estimation is still feasible since characteristic functions (therefore moments) are often known in closed form.

A large part of the econometrics literature is concerned with convergence properties of the estimators such as the ones described above when the sample size N is large. Two kinds of properties of estimators are of interest here: consistency — whether the estimators θ_N converge to their true value θ_0 when the sample becomes large — and the distribution of the error $\theta_N - \theta_0$. Consistency is usually obtained by applying a law of large numbers and in some cases, the estimation error $\theta_N - \theta_0$ can be shown to be asymptotically normal, i.e., verify a central limit theorem when $N \to \infty$ which then enables to derive confidence intervals by approximating its distribution by a normal distribution.

Convergence of maximum likelihood estimators is a classical topic and Lévy processes do not introduce any new issue here. Large sample properties of GMM estimators have been studied in [187]. Contrarily to maximum likelihood methods that require the data to be generated from i.i.d. noise, GMM is only based on the convergence of the (generalized) sample moments which is valid under weaker conditions. While for Lévy processes increments are indeed i.i.d., as mentioned in Section 7.1 there is ample evidence for non-linear dependence in asset returns so robustness to departure from the i.i.d. case is clearly an advantage with GMM. However the consistency and asymptotic normality of the GMM estimator clearly require the chosen moments to be *finite*. In the presence of heavy tails, this is far from being obvious: as we will see in the next section, empirical studies indicate that even the fourth moment of returns is not finite for most data sets which have been studied so even the second sample moment may not give an asymptotically normal estimator of the second moment of the distribution of returns [267, 268]. In this case higher order sample moments $n \geq 4$ are not even consistent estimators: they diverge when $N \to \infty$.

7.3 The distribution of returns: a tale of heavy tails

Empirical research in financial econometrics in the 1970s mainly concentrated on modelling the unconditional distribution of returns, defined as:

$$F_\Delta(u) = P(r_t(\Delta) \leq u). \tag{7.13}$$

The probability density function (PDF) is then defined as its derivative $f_\Delta = F'_\Delta$. As early as the 1960s, Mandelbrot [278] pointed out the insufficiency of the normal distribution for modelling the marginal distribution of asset returns and their heavy tails. Since then, the non-Gaussian character of the distribution of price changes has been repeatedly observed in various market data. One way to quantify the deviation from the normal distribution is by using the kurtosis of the distribution F_Δ defined as

$$\hat{\kappa}(\Delta) = \frac{\langle (r_t(\Delta) - \langle r_t(\Delta)\rangle)^4\rangle}{\hat{\sigma(\Delta)}^4} - 3, \tag{7.14}$$

where $\hat{\sigma}(\Delta)^2$ is the sample variance of the log-returns $r_t(\Delta) = X(t+\Delta) - X(t)$. The kurtosis is defined such that $\kappa = 0$ for a Gaussian distribution, a positive value of κ indicating a "fat tail," that is, a slow asymptotic decay of the PDF. The kurtosis of the increments of asset prices is far from its Gaussian value: typical values for $T = 5$ minutes are [76, 92, 94, 313]: $\kappa \simeq 74$ (USD/DM exchange rate futures), $\kappa \simeq 60$ (US$/Swiss Franc exchange rate futures), $\kappa \simeq 16$ (S&P500 index futures) . One can summarize the empirical results by saying that the distribution f_Δ tends to be non-Gaussian, sharp peaked and heavy tailed, these properties being more pronounced for intraday time scales ($\Delta < 1$ day). This feature is consistent with a description of the log-price as a

Lévy process: all Lévy processes generate distributions with positive kurtosis and the Lévy measure can be chosen to generate heavy tails such as to fit the tails of return for any given horizon Δ. Once the kurtosis is fixed for a given Δ, the kurtosis for other maturities decays as $1/\Delta$ (see remarks after Proposition 3.13). However in empirical data the kurtosis $\kappa(\Delta)$ decays much more slowly [94], indicating an inconsistency of the Lévy model across time scales.

These features are not sufficient for identifying the distribution of returns and leave a considerable margin for the choice of the distribution. Fitting various functional forms to the distribution of stock returns and stock price changes has become a popular pastime. Dozens of parametric models have been proposed in the literature: α-stable distributions [278], the Student t distribution [62], hyperbolic distributions [125], normal inverse Gaussian distributions [30], exponentially tempered stable distributions [67, 94, 80] are some of them. All of these distributions are infinitely divisible and are therefore compatible with a Lévy process for the log-price X. In coherence with the empirical facts mentioned above, these studies show that in order for a parametric model to successfully reproduce all the above properties of the marginal distributions it must have at least four degrees of freedom governed by different model parameters: a location parameter, a scale (volatility) parameter, a parameter describing the decay of the tails and eventually an asymmetry parameter allowing the left and right tails to have different behavior. Normal inverse Gaussian distributions [30], generalized hyperbolic distributions [125] and tempered stable distributions [67, 94, 80] meet these requirements. The choice among these classes is then a matter of analytical and numerical tractability. However a quantitative assessment of the adequacy of these models requires a closer examination of the tails of returns.

7.3.1 How heavy tailed is the distribution of returns?

One of the important characteristics of financial time series is their high variability, as revealed by the heavy tailed distributions of their increments and the non-negligible probability of occurrence of violent market movements. These large market movements, far from being discardable as simple outliers, focus the attention of market participants since their magnitude may be such that they account for an important fraction of the total return over a long period. Not only are such studies relevant for risk measurement but they are rendered necessary for the calculation of the quantiles of the profit-and-loss distribution, baptized *Value-at-Risk*, which is required to determine regulatory capital. Value-at-Risk (VaR) is defined as a high quantile of the loss distribution of a portfolio over a certain time horizon Δ:

$$P(W_0(r_t(\Delta) - 1) \leq \mathrm{VaR}(p, t, \Delta)) = p, \tag{7.15}$$

where W_0 is the present market value of the portfolio, $r_t(\Delta)$ its (random) return between t and $t + \Delta$. Δ is typically taken to be one day or ten days and $p = 1\%$ or 5%. VaR can be computed by estimating a stochastic model for the returns and then computing or simulating the VaR of a portfolio within the model. Calculating VaR implies a knowledge of the tail behavior of the distribution of returns.

The non-Gaussian character of the distribution makes it necessary to use other measures of dispersion than the standard deviation in order to capture the variability of returns. One can consider, for example, higher order moments or cumulants as measures of dispersion and variability. However, given the heavy tailed nature of the distribution, one has to know beforehand whether such moments are well defined otherwise the corresponding sample moments may be meaningless. The tail index k of a distribution may be defined as the order of the highest absolute moment which is finite. The higher the tail index, the thinner is the tail; for a Gaussian or exponential tail, $k = +\infty$ (all moments are finite) while for a power-law distribution with density

$$f(x) \underset{|x|\to\infty}{\sim} \frac{C}{|x|^{1+\alpha}} \tag{7.16}$$

the tail index is equal to α. However a distribution may have a finite tail index α without having a power-law tail as in (7.16): in fact any function f verifying

$$f(x) \underset{|x|\to\infty}{\sim} \frac{L(|x|)}{|x|^{1+\alpha}} \text{ where } \forall u > 0,\ \frac{L(ux)}{L(x)} \overset{x\to\infty}{\to} 1 \tag{7.17}$$

has tail index α. Such functions are said to be regularly varying with index α. A function L such as the one in (7.17) is said to be slowly varying: examples are $L(x) = \ln(x)$ or any function with a finite limit at ∞. Knowing the tail index of a distribution gives an idea of how heavy the tail is but specifies tail behavior only up to a "slowly varying" function L. This may be good enough to get an idea of the qualitative behavior but $L(x)$ definitely does influence finite sample behavior and thus makes estimation of tail behavior very tricky.

A simple method is to represent the sample moments (or cumulants) as a function of the sample size n. If the theoretical moment is finite then the sample moment will eventually settle down to a region defined around its theoretical limit and fluctuate around that value. In the case where the true value is infinite, the sample moment will either diverge as a function of sample size or exhibit erratic behavior and large fluctuations. Applying this method to time series of cotton prices, Mandelbrot [278] conjectured that the theoretical variance of returns may be infinite since the sample variance did not converge to a particular value as the sample size increased and continued to fluctuate incessantly.

Figure 7.5 indicates an example of the behavior of the sample variance as a function of sample size. The behavior of sample variance suggests that the variance of the distribution is indeed finite: the sample variance settles down to a limit value after a transitory phase of wild oscillations.

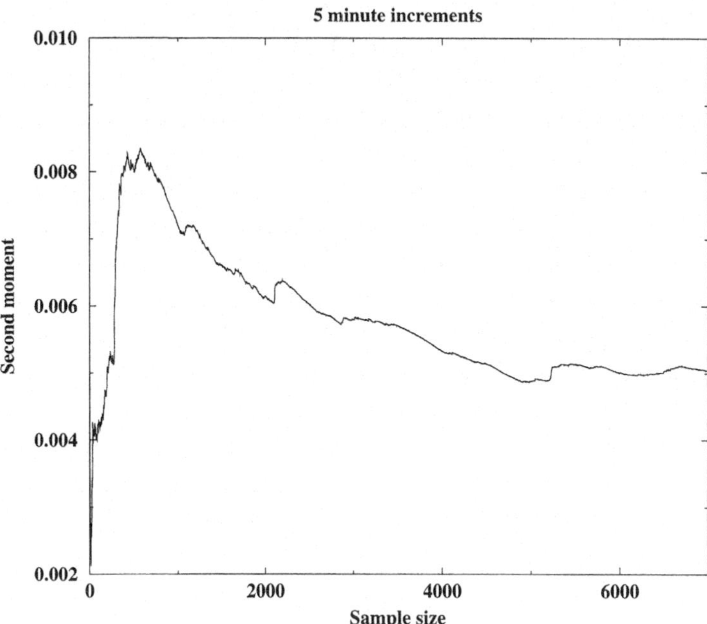

FIGURE 7.5: Empirical second moment of log-returns as a function of sample size: S&P500.

One way to go beyond the graphical analysis described above, is to use the tools of *extreme value theory*, a branch of probability theory dealing precisely with the probabilities of extreme events. Given a time series of n nonoverlapping returns $r_t(\Delta), t = 0, \Delta, 2\Delta, \ldots, n\Delta$, the extremal (minimal and maximal) returns are defined as:

$$m_n(\Delta) = \min\{r_{t+k\Delta}(\Delta), k \in [1 \ldots n]\}, \tag{7.18}$$

$$M_n(\Delta) = \max\{r_{t+k\Delta}(\Delta), k \in [1 \ldots n]\}. \tag{7.19}$$

In economic terms, $m_n(\Delta)$ represents the worst relative loss over a time horizon Δ of an investor holding the portfolio $P(t)$. A relevant question is to know the properties of these extremal returns, for example the distribution of $m_n(\Delta)$ and $M_n(\Delta)$. In this approach, one looks for a distributional limit of $m_n(\Delta)$ and $M_n(\Delta)$ as the sample size n increases. If such a limit exists, then it is described by the Fisher–Tippett theorem in the case where the returns are i.i.d:

THEOREM 7.1 Extreme value theorem for i.i.d. sequences [142]
Assume the log-returns $(r_t(\Delta))_{t=0,\Delta t,2\Delta t,..}$ form an i.i.d. sequence with distri-

bution F_Δ. If there exist normalizing constants (λ_n, σ_n) and a non-degenerate limit distribution H for the normalized maximum return:

$$\mathbb{P}\left(\frac{M_n - \lambda_n}{\sigma_n} \leq x\right) \underset{x\to\infty}{\to} H(x) \tag{7.20}$$

then the limit distribution H is either a Gumbel distribution

$$G(x) = \exp[\exp(-\frac{x-\lambda}{\gamma})], \qquad x \in \mathbb{R}, \tag{7.21}$$

a Weibull distribution

$$W_\alpha(x) = \exp[-(\frac{x}{\gamma})^\alpha], \qquad x < 0,\ \alpha > 0, \tag{7.22}$$

or a Fréchet distribution:

$$H(x) = \exp[-(\frac{x}{\gamma})^{-\alpha}], \qquad x > 0,\ \alpha > 0. \tag{7.23}$$

The three distributional forms can be parameterized in the following unified form, called the Cramer–von Mises parameterization:

$$H_\xi(x) = \exp[-(1+\xi x)^{-1/\xi}], \tag{7.24}$$

where the sign of the shape parameter ξ determines the extremal type: $\xi > 0$ for Fréchet, $\xi < 0$ for Weibull and $\xi = 0$ for Gumbel. Obviously if one knew the stochastic process generating the returns, one could also evaluate the distribution of the extremes, but the extreme value theorem implies that one need not know the exact parametric form of the marginal distribution of returns F_Δ to evaluate the distribution of extremal returns. The value of ξ *only* depends on the tail behavior of the distribution F_Δ of the returns: a distribution F_Δ with finite support gives $\xi < 0$ (Weibull) while a distribution F_Δ with a power-law tail with exponent α falls in the Fréchet class with $\xi = 1/\alpha > 0$. The Fréchet class therefore contains all regularly varying distributions, which correspond to heavy tails. All other distributions fall in the Gumbel class $\xi = 0$ which plays a role for extreme values analogous to that of the normal distribution for sums of random variables: it is the typical limit for the distribution of i.i.d. extremes. For example, the normal, log-normal and exponential distribution fall in the Gumbel class, as well as most distributions with an infinite tail index (see Table 7.1).

This theorem also provides a theoretical justification for using a simple parametric family of distributions for estimating the extremal behavior of asset returns. The estimation may be done as follows: one interprets the asymptotic result above as

$$P(M_n \leq u) = H_\xi\left(\frac{u-\lambda_n}{\sigma_n}\right) = H_{\xi,\lambda_n,\sigma_n}(x). \tag{7.25}$$

TABLE 7.1: Tail properties of some Lévy processes

Lévy process	**Tail index**	**Tail decay**
α-stable	$\alpha \in]0,2[$	$\sim \lvert x\rvert^{-(1+\alpha)}$
Variance gamma	∞	Exponential
NIG	∞	$\sim \lvert x\rvert^{-3/2}\exp(-\alpha\lvert x\rvert + \beta x)$
Hyperbolic	∞	$\sim \exp(-a\lvert x\rvert)$
Student t (see Equation (7.44))	$\beta > 0$	$\sim A\lvert x\rvert^{-1-\beta}$

The estimation of the distribution of maximal returns then is reduced to a parameter estimation problem for the three-parameter family $H_{\xi,\lambda,\sigma}$. One can estimate these parameters by the so-called *block* method [331, 265]: one divides the data into N subperiods of length n and takes the extremal returns in each subperiod, obtaining a series of N extremal returns $(x_i)_{i=1,\dots,N}$, which is then assumed to be an i.i.d. sequence with distribution $H_{\xi,\lambda,\sigma}$. A maximum likelihood estimator of (ξ, λ, σ) can be obtained by maximizing the log-likelihood function:

$$L(\lambda, \sigma, \xi) = \sum_{i=1}^{N} l(\lambda, \sigma, \xi, x_i), \tag{7.26}$$

where l is the log density obtained by differentiating Equation (7.24) and taking logarithms:

$$l(\lambda, \sigma, \xi, x_i) = -\ln\sigma - \left(1 + \frac{1}{\xi}\right)\ln\left[1 + \xi\left(\frac{x_i - \lambda}{\sigma}\right)\right] - \left[1 + \xi\left(\frac{x_i - \lambda}{\sigma}\right)\right]^{1/\xi}. \tag{7.27}$$

If $\xi > -1$ (which covers the Gumbel and Fréchet cases), the maximum likelihood estimator is consistent and asymptotically normal [131].

To our knowledge, the first application of extreme value theory to financial time series was given by Jansen and de Vries [221], followed by Longin [265], Dacorogna *et al* [301], Lux [268] and others. Applying the techniques above to daily returns of stocks, market indices and exchange rates these empirical studies yield a positive value of ξ between 0.2 and 0.4, which means a tail index $2 < \alpha(T) \leq 5$. In all cases, ξ is bounded away from zero, indicating heavy tails belonging to the Fréchet domain of attraction but the tail index is often found to be larger than two [221, 265, 92] — which means that the variance is finite and the tails lighter than those of stable Lévy distributions [139], but compatible with a power-law (Pareto) tail with (the same) exponent $\alpha(T) = 1/\xi$. These studies seem to validate the power-law nature of the distribution of returns, with an exponent around three. Similar results are obtained using a direct log-log regression on the histogram of returns [174] or using semi-parametric methods [232]. Note however that these studies do not allow us to pinpoint the tail index with more than a single significant digit. Also, a positive value of ξ does not *imply* power-law tails but is compatible with any regularly varying tail with exponent $\alpha = 1/\xi$:

$$F_\Delta(x) \underset{x\to\infty}{\sim} \frac{L(x)}{x^\alpha}, \tag{7.28}$$

where $L(.)$ is a slowly-varying function as in Equation(7.17). Any choice of L will give a different distribution F_Δ of returns but the *same* extremal type $\xi = 1/\alpha$, meaning that, in the Fréchet class, the extremal behavior only identifies the tail behavior up to a (unknown!) slowly-varying function which may in turn considerably influence the results of log-log fits on the histogram! A more detailed study on high-frequency data using different methods [301, 105] indicates that the tail index slightly increases when the time resolution moves from intraday (30 minutes) to a daily scale. However, the i.i.d. hypothesis underlying these estimation procedures has to be treated with caution given the dependence present in asset returns [92]: dependence in returns can cause large biases in estimates of the tail index [275].

7.4 Time aggregation and scaling

Although most empirical studies on the distribution of asset returns have been done on daily returns, applications in trading and risk management involve various time scales from a few minutes for intraday traders to several months for portfolio managers. It is therefore of interest to see how the statistical properties of returns $r_t(\Delta)$ vary as Δ varies. Moving from small time scales (say, intraday) to larger time scales (say, daily or weekly) — an operation known as time aggregation — corresponds to adding up returns at high frequency to obtain those at a lower frequency: the series $r_t(k\Delta)$ is obtained by taking partial sums of blocks of k consecutive elements in the series $r_t(\Delta)$. We describe in this section how time aggregation affects the statistical properties of returns.

7.4.1 Self-similarity

The ideas of self-similarity and scaling correspond to the quest for statistical quantities which remain unchanged under time aggregation. Since the pio-

TABLE 7.2: A comparison of stylized empirical properties of asset returns with statistical properties of Lévy processes

Log-prices	**Lévy processes**
Absence of autocorrelation in increments	True for all Lévy processes.
Heavy/semiheavy tails	Possible by choosing Lévy measure with heavy/semiheavy tails.
Finite variance	True for any Lévy process with finite second moment.
Aggregational normality	True for any Lévy process with finite second moment.
Jumps in price trajectories	Always true.
Asymmetric distribution of increments	Possible by choosing asymmetric Lévy measure/distribution.
Volatility clustering: clustering of large increments	Not true: large events occur at independent random intervals.
Positive autocorrelation in absolute returns	Not true: increments are independent.
"Leverage" effect: $\text{Cov}(r_t^2 r_{t+\Delta t}) < 0$	Not true: increments are independent.

neering work of Mandelbrot on cotton prices [278], a large body of literature has emerged on self-similarity and fractal properties of market prices. The 1990s witnessed a regain of interest in this topic with the availability of high-frequency data and a large number of empirical studies on asset prices have investigated self-similarity properties of asset returns and various generalizations of them, under the names of scale invariance, fractality and multiscaling. We attempt here to define some of the concepts involved and summarize the evidence found in empirical studies.

Given a random process $(X_t)_{t\geq 0}$, the fundamental idea behind the notion of scaling is that an observer looking at the process at various time resolutions will observe the same statistical properties, up to a rescaling of units. This is formalized in the concept of self-similarity:

DEFINITION 7.1 Self-similarity *A stochastic process is said to be* self-similar *if there exists $H > 0$ such that for any scaling factor $c > 0$, the* processes $(X_{ct})_{t\geq 0}$ *and* $(c^H X_t)_{t\geq 0}$ *have the same law:*

$$(X_{ct})_{t\geq 0} \stackrel{d}{=} (c^H X_t)_{t\geq 0}. \tag{7.29}$$

H is called the self-similarity exponent *of the process X.*

X is said to be self-affine *if there exists $H > 0$ such that for any $c > 0$ the* processes $(X_{ct})_{t\geq 0}$ *and* $(c^H X_t)_{t\geq 0}$ *have the same law up to centering:*

$$\exists b_c : [0, \infty[\rightarrow \mathbb{R}, \qquad (X_{ct})_{t\geq 0} \stackrel{d}{=} (b_c(t) + c^H X_t)_{t\geq 0}. \tag{7.30}$$

In particular X_t must be defined for $t \in [0, \infty[$ in order for the definition to make sense. It is easy to see that a self-similar process cannot be stationary. Note that we require that the two *processes* $(X_{ct})_{t\geq 0}$ and $(c^H X_t)_{t\geq 0}$ are identical in distribution: their sample paths are not equal but their *statistical properties* (not just their marginal distributions!) are the same. Brownian motion (without drift!) is an example of a self-similar process with self-similarity exponent $H = 1/2$. Brownian motion with drift is self-affine but not self-similar. In the sequel we will focus on self-similar processes since self-affine processes are obtained from self-similar ones by centering.

A consequence of Definition 7.1 is that for any $c, t > 0$, X_{ct} and $c^H X_t$ have the same distribution. Choosing $c = 1/t$ yields

$$\forall t > 0, \quad X_t \stackrel{d}{=} t^H X_1, \tag{7.31}$$

so the distribution of X_t, for any t, is completely determined by the distribution of X_1:

$$F_t(x) = \mathbb{P}(t^H X_1 \leq x) = F_1\left(\frac{x}{t^H}\right). \tag{7.32}$$

In particular if the tail of F_1 decays as a power of x, then the tail of F_t decays in the same way:

$$\mathbb{P}(X_1 \geq x) \underset{x\to\infty}{\sim} \frac{C}{x^\alpha} \Rightarrow [\forall t > 0,\ \mathbb{P}(X_1 \geq x) \underset{x\to\infty}{\sim} C\frac{t^{\alpha H}}{x^\alpha} = \frac{C(t)}{x^\alpha}]. \quad (7.33)$$

If F_t has a density ρ_t we obtain, by differentiating (7.32), the following relation for the densities:

$$\rho_t(x) = \frac{1}{t^H}\rho_1\left(\frac{x}{t^H}\right). \quad (7.34)$$

Substituting $x = 0$ in (7.34) yields the following scaling relation:

$$\forall t > 0,\ \rho_t(0) = \frac{\rho_1(0)}{t^H}. \quad (7.35)$$

Let us now consider the moments of X_t. From (7.31) it is obvious that $E[|X_t|^k] < \infty$ if and only if $E[|X_1|^k] < \infty$ in which case

$$E[X_t] = t^H E[X_1], \quad \mathrm{Var}(X_t) = t^{2H}\,\mathrm{Var}(X_1), \quad (7.36)$$

$$E[|X_t|^k] = t^{kH} E[|X_1|^k]. \quad (7.37)$$

Can a Lévy process be self-similar? First note that for a Lévy process with finite second moment, independence and stationarity of increments implies $\mathrm{Var}(X_t) = t\,\mathrm{Var}(X_1)$, which contradicts (7.36) unless $H = 1/2$, which is the case of Brownian motion. So a self-similar Lévy process is either a Brownian motion ($H = 1/2$) or must have infinite variance. As observed in Chapter 3, if X is a Lévy process the characteristic function of X_t is expressed as: $\Phi_t(z) = \exp[-t\psi(z)]$ where $\psi(.)$ is the characteristic exponent of the Lévy process. The self-similarity of X then implies following scaling relation for ϕ:

$$\forall t > 0, \quad X_t \overset{d}{=} t^H X_1 \iff \forall t > 0, \forall z \in \mathbb{R}, \psi(t^H z) = t\psi(z). \quad (7.38)$$

The only solutions of (7.38) are of the form $\psi(z) = C|z|^{1/H}$ where C is a constant; in turn, this defines a characteristic function iff $H \geq 1/2$. For $H = 1/2$ we recover Brownian motion and for $H > 1/2$ we recover a symmetric α-stable Lévy process with $\alpha = 1/H$:

$$\Phi_t(z) = \exp[-\sigma^\alpha |z|^\alpha], \qquad \alpha = \frac{1}{H} \in]0, 2[.$$

Therefore the only self-similar Lévy processes are the symmetric α-stable Lévy processes (also called Lévy flights): an α-stable Lévy process has self-similarity exponent $H = 1/\alpha \in [0.5, +\infty[$. Note that we have assumed stationary of increments here: there are other self-similar processes with independent but not stationary increments, see [346]. However stationarity of returns is a crucial working hypothesis in statistical estimation.

Let us emphasize that self-similarity has nothing to do with independent increments. An important class of self-similar processes are fractional Brownian motions [283, 344]: a fractional Brownian motion with self-similarity exponent $H \in]0,1[$ is a real centered Gaussian process with stationary increments $(B_t^H)_{t\geq 0}$ with covariance function:

$$\mathrm{cov}(B_t^H, B_s^H) = \frac{1}{2}(|t|^{2H} + |s|^{2H} - |t+s|^{2H}). \tag{7.39}$$

For $H = 1/2$ we recover Brownian motion. For $H \neq 1/2$, the covariance of the increments decays very slowly, as a power of the lag; for $H > 1/2$ this leads to long-range dependence in the increments [283, 344].

Comparing fractional Brownian motions and α-stable Lévy processes shows that self-similarity can have very different origins: it can arise from high variability, in situations where increments are independent and heavy-tailed (stable Lévy processes) or it can arise from *strong dependence* between increments even in absence of high variability, as illustrated by the example of fractional Brownian motion. These two mechanisms for self-similarity have been called the "Noah effect" and the "Joseph effect" by Mandelbrot [279]. By mixing these effects, one can construct self-similar processes where both long range dependence and heavy tails are present: fractional stable processes [344] offer such examples. The relation between self-similar processes, Lévy processes and Gaussian processes is summarized in Figure 7.4.1: the only self-similar Lévy processes are the (symmetric) α-stable Lévy processes (also called Lévy flights, see Chapter 3) for which $H = 1/\alpha \in [0.5, \infty[$. In particular Brownian motion is self-similar with exponent $H = 1/2$.

7.4.2 Are financial returns self-similar?

Let us now briefly explain how the properties above can be tested empirically in the case of asset prices. One should distinguish general tests for self-similarity from tests of particular parametric models (such as α-stable Lévy processes).

Assume that the log-price $X_t = \ln S_t$ is a process with *stationary increments*. Since $X_{t+\Delta} - X_t$ has the same law as X_Δ, the density and moments of X_Δ can be estimated from a sample of increments.

The relation (7.35) has been used by several authors to test for self-similarity and estimate H from the behavior of the density of returns at zero: first one estimates $\rho_t(0)$ using the empirical histogram or a kernel estimator and then obtains an estimate of H as the regression coefficient of $\ln \rho_t(0)$ on $\ln t$:

$$\ln \hat{\rho}_t(0) = H \ln \frac{t}{\Delta} + \ln \hat{\rho}_\Delta(0) + \epsilon. \tag{7.40}$$

Applying this method to S&P 500 returns, Mantegna and Stanley [285] obtained $H \simeq 0.55$ and concluded towards evidence for an α-stable model with

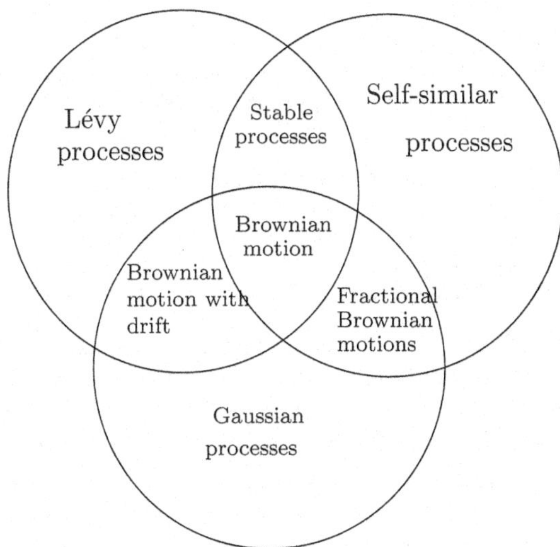

FIGURE 7.6: Self-similar processes and their relation to Lévy processes and Gaussian processes.

$\alpha = 1/H \simeq 1.75$. However, the scaling relation (7.35) holds for any self-similar process with exponent H and does not imply in any way that the process is a (stable) Lévy process. For example, (7.35) also holds for a fractional Brownian motion with exponent H — a Gaussian process with correlated increments having long range dependence! Scaling behavior of $\rho_t(0)$ is simply a necessary but not a sufficient condition for self-similarity: even if (7.35) is verified, one cannot conclude that the data generating process is self-similar and even less that it is an α-stable process.

Another method which has often been used in the empirical literature to test self-similarity is the "curve collapsing" method: one compares the aggregation properties of empirical densities with (7.34). Using asset prices sampled at interval Δ, one computes returns at various time horizons $n\Delta, n = 1 \ldots M$ and estimates the marginal density of these returns (via a histogram or a smooth kernel estimator). The scaling relation (7.34) then implies that the densities $\hat{\rho}_{n\Delta}(x)$ and $\frac{1}{n^H}\hat{\rho}_{\Delta}(\frac{x}{n^H})$ should coincide, a hypothesis which can be tested graphically and also more formally using a Kolmogorov–Smirnov test.

Although self-similarity is not limited to α-stable Lévy processes, rejecting self-similarity also leads to reject the α-stable Lévy process as a model for log-returns. If the log-price follows an α-stable Lévy process, daily, weekly and monthly returns should also be α-stable (with the same α). Empirical estimates [3, 62] show a value of α which increases with the time horizon. Finally, as noted in Section 7.3.1, various estimates of tail indices for most stocks and exchange rates [3, 105, 221, 267, 198, 265, 268] are often found to be larger than 2, which rules out infinite variance and stable distributions.

McCulloch ([289], page 417) argues against this point, claiming that such estimators are highly biased in samples of realistic size generated from α-stable distributions with α close to 2, see also [387].

7.5 Realized variance and "stochastic volatility"

One of the objectives of econometric models is to quantify the notion of "market volatility." It should be understood from the outset that "volatility" is not a model-free notion: in a parametric model, various measures of risk and variability of prices can be computed from model parameters. However given the vast choice of models and the availability of large databases of returns, many authors have recently turned to using the model-free notion of quadratic variation — known in the financial literature as "realized volatility" — as a measure of market volatility [63, 38, 9, 8].

Given a sample of N returns $(r_t(\Delta), t = 0, \Delta, \ldots, N\Delta)$ observed at intervals Δ over the period $[0, T = N\Delta]$, the realized variance on $[0, T]$ is defined as

$$v_\Delta(T) = \sum_{t=1}^{N} |r_t(\Delta)|^2 = \sum_{t=1}^{N} |X_{t+\Delta} - X_t|^2. \tag{7.41}$$

Note that the realized volatility is different from the sample variance of the returns, defined by:

$$\hat{\sigma}^2(\Delta) = \frac{1}{N}\sum_{t=1}^{N} |r_t(\Delta)|^2 - [\frac{1}{N}\sum_{t=1}^{N} |r_t(\Delta)|]^2. \tag{7.42}$$

Since returns are serially uncorrelated the sample variance scales linearly with time: $\hat{\sigma}^2(\Delta) = \hat{\sigma}^2(1)\Delta$ and does not display a particularly interesting behavior. This is of course also the case for the (theoretical) variance of log-returns in exponential-Lévy models: $\sigma^2(\Delta) = \mathrm{var}[r_t(\Delta)] = \sigma^2(1)\Delta$ because of the independence of log-returns. On the other hand, the realized variance $v_\Delta(T)$ is not an average but a cumulative measure of volatility and does contain interesting information when computed at fine scales ($\Delta \to 0$), even when returns are independent: we will see in Chapter 8 that the process $(v_\Delta(t))_{t\in[0,T]}$ converges in probability to a (nontrivial) stochastic process $([X,X]_t)_{t\in[0,T]}$ called the quadratic variation[3] of X. In a model where X is a Lévy process with characteristic triplet (σ^2, ν, γ) the quadratic variation process is given by

$$[X]_t = \sigma^2 t + \sum_{0\leq s\leq t} |\Delta X_s|^2, \tag{7.43}$$

[3]This notion is studied in more detail in Section 8.2.

where the sum runs over the jumps of X between 0 and t. Note that $[X]_t$ is a random variable: it is deterministic only if there are no jumps, i.e., in the case where the log-price is a Brownian motion with drift. Thus, in a model where log-prices follow a (non-Gaussian) Lévy process, realized volatility is always stochastic: one does *not* need to insert an additional source of randomness in the form of a volatility variable in order to obtain such effects. In fact,

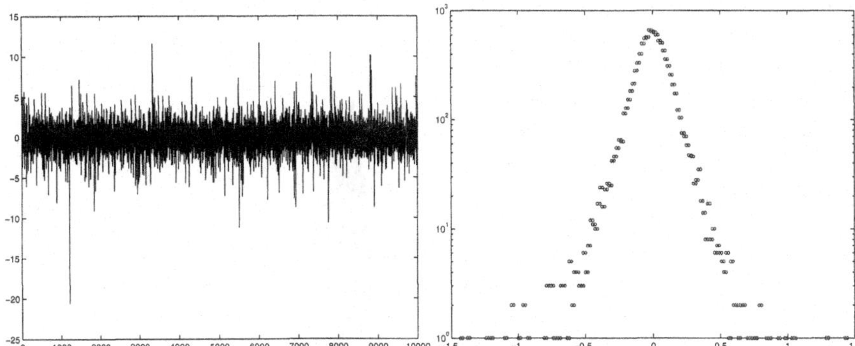

FIGURE 7.7: Left: Daily returns simulated from the Student t random walk model described in Section 7.5. Right: Distribution of daily returns: parameters have been chosen such that this distribution has heavy tails with finite variance and a tail index $\alpha = 3.8$.

even when the (conditional / unconditional) variance of returns is constant, the realized variance can display very high variability when returns have reasonably heavy tails, as illustrated by the example below. Consider a random walk model where the daily returns are i.i.d. random variables with a Student t distribution with parameter β, whose density is given by:

$$r_t \overset{\text{i.i.d.}}{\sim} f_\beta(x) = \frac{\Gamma((\beta+1)/2)}{\sqrt{\beta\pi}\,\Gamma(\beta/2)}(1+\frac{x^2}{\beta})^{-(\beta+1)/2}. \tag{7.44}$$

The Student t distribution has heavy tails with tail index equal to β and has been advocated as a model for daily returns [62]. In the discussion below we will take $\beta = 3.8$, which is coherent with the estimation results presented in Section 7.3, but our arguments will apply for any $2 < \beta < 4$. Thus, the daily increments have finite variance but infinite fourth moments. In fact as noted in Section 4.6, the Student t distribution is infinitely divisible [177] so there exists an exponential-Lévy model (belonging to the generalized hyperbolic class) whose daily increments follow the Student t distribution: the Student t random walk can also be seen as an exponential-Lévy model for the log-price. Since the Student t law has a known density, this model is easy to simulate. A simulation of daily returns from this model is shown in Figure 7.7. From these returns we compute "realized variance" using a moving average over $N = 1000$ days and annualizing (normalizing) to obtain numbers comparable to annual volatilities:

$$\hat{\sigma}_t^2 = \frac{N_{\text{year}}}{N} \sum_{j=1}^{N} |r_{t-j\Delta}(\Delta)|^2.$$

The evolution of the estimated "realized volatility" $\hat{\sigma}_t$ is shown in Figure 7.8

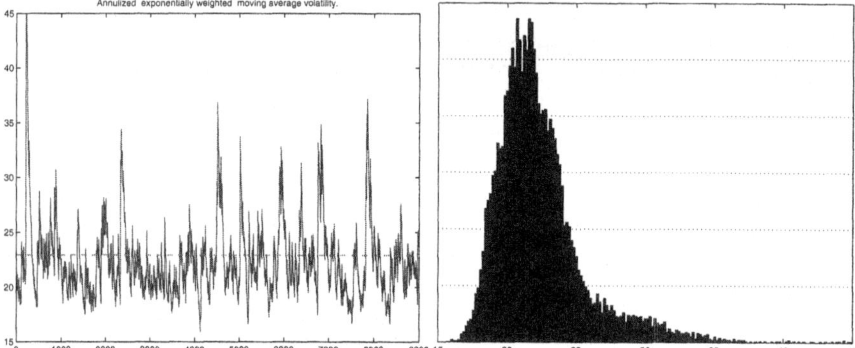

FIGURE 7.8: Left: Realized (1000-day moving average) volatility as a function of time in the Student random walk model. The true standard deviation of returns is indicated by a horizontal line. Notice the high degree of variability. Right: Histogram of realized 1000-day volatility. Notice the high dispersion around the theoretical value of 32%.

(left). Since $\beta > 2$, the squared returns are i.i.d. variables with $E[|r_t|^2] < \infty$ and the law of large numbers applies to $\hat{\sigma}^2$:

$$\forall t, \quad \mathbb{P}(\hat{\sigma}_t^2 \overset{\Delta \to 0}{\to} \sigma^2) = 1.$$

Thus the realized volatility $\hat{\sigma}$ is a consistent estimator of the "true volatility" σ. However, due to the heavy tails of the returns, the estimation error $\hat{\sigma}^2 - \sigma^2$ has infinite variance: the quadratic estimation error involves the fourth power of the returns. Therefore, instead of verifying a central limit theorem leading to asymptotic normality, the estimator $\hat{\sigma}^2 - \sigma^2$ verifies the infinite variance version of the central limit theorem:

PROPOSITION 7.2 Stable central limit theorem

Let $(Y_n)_{n \geq 1}$ be an i.i.d. sequence of centered random variables with a regularly varying distribution of index $\alpha \in]0, 2[$. In particular $E[Y_n^2] = \infty$. Then the *partial sum, normalized by $N^{1/\alpha}$, converged in distribution to an α-stable random variable with infinite variance:*

$$\frac{1}{N^{1/\alpha-1}} \sum_{n=1}^{N} Y_n \to S_\alpha, \qquad \alpha = \beta - 2. \tag{7.45}$$

In this case, the squared returns have a distribution which is regularly varying (with power law tails) with exponent $\alpha = \beta - 2$, which implies that for large sample size N, $\hat{\sigma}^2 - \sigma^2$ behaves like $N^{1/\alpha} S_\alpha$ where S_α is a standard α-stable random variable. In particular, even though the returns have finite variance and well-defined "volatility," the volatility estimation error has heavy tails and infinite variance. Figure 7.9 illustrates a quantile-quantile plot of the realized volatility $\hat{\sigma}$ computed over 1000 days (4 years); the quantiles of this estimator are compared to those of a normal distribution with same mean and variance using a Monte Carlo simulation. While the sample median is 32.5, the 95% confidence interval obtained with 10 000 Monte Carlo trials is [27.9%, 41.0%]; the 99% confidence interval is [26.3%,51.4%]. In simple terms, it means that even though the returns have a true standard deviation of $\sigma =$ 32%, if we try to estimate this value by using realized variance we have a 5% chance of getting something either lower than 28% (not bad) or higher than 41% (serious overestimation).

This heavy tailed asymptotic distribution has drastic consequences for the interpretation of "realized variance": large or highly fluctuating values of realized volatility do not imply that the variance of returns in the underlying process is large or time varying, nor does it constitute evidence for "stochastic volatility." It simply indicates that when faced with heavy tails in returns, one should not be using "realized volatility" as a measure of volatility!

Figure 7.8 (right) shows the sample distribution of the "realized volatility" $\hat{\sigma}$ which is shown to display a considerable dispersion ("volatility of volatility"), as expected from the results above. In fact the histogram of realized volatility is quite similar to the one observed in exchange rate data [9]! Observing these figures, it is tempting to model these observations with a "stochastic volatility model," introducing the "volatility of volatility" to reproduce the observations in Figure 7.8. In fact, the realized volatility in Figure 7.8 (left) even displays a "mean reverting" behavior: the estimator $\hat{\sigma}$ oscillates around the true value σ!

Figure 7.9 compares the quantiles of "realized volatility" $\hat{\sigma}$ to those of a normal distribution: as expected, it is far from being normal and exhibits heavy tails. The heavy tails in the distribution of $\hat{\sigma}$ indicate that the "realized volatility" $\hat{\sigma}$ in a given sample can be very different from σ and will be itself quite variable, but the strong variation in realized volatility does not reflect any information on the actual volatility of the returns process but arises for purely statistical reasons, because of the heavy tails in the returns. In particular Figure 7.9 shows that "confidence intervals" derived using asymp-

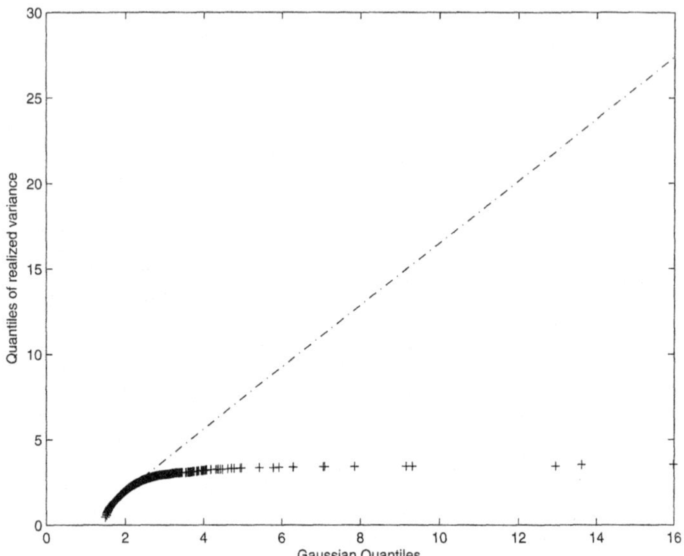

FIGURE 7.9: Quantile of 1000-day realized volatility compared to quantiles of the normal distribution.

totic normality of $\hat{\sigma^2}$ and often used in testing for the presence of stochastic volatility can be meaningless.

This simple example shows that in the presence of heavy tails in the returns (and we know that they are present!), common estimators for volatility can lead to erroneous conclusions on the nature of market volatility. These conclusions hold when the tail index is less than 4, which has been suggested by many empirical studies [174, 221, 268, 265]. In particular, one can easily attribute to heteroscedasticity, nonstationarity or "stochastic volatility" what is in fact due to heavy tails in returns. These issues will be discussed in the framework of stochastic volatility models in Chapter 15.

7.6 Pathwise properties of price trajectories (*)

Although Lévy processes can easily accommodate the heavy tails, skewness and in fact any other distributional feature of asset returns, distributional properties alone are not sufficient to distinguish them from (nonlinear) diffusion models. In fact, (Brownian) diffusion processes with nonlinear dependence of the local volatility can also generate heavy tails and skewness [54]. In fact any infinitely divisible distribution satisfying a weak regularity condition can be obtained as the marginal distribution of a stationary diffusion process with linear drift and the diffusion coefficients corresponding to many common probability distributions are found explicitly in [367].

The most important feature which distinguishes jump processes from diffusion processes and more generally, processes with continuous sample paths is, of course, the presence of discontinuities in price behavior. No matter what ingredient one puts into a diffusion model — time and state dependent local volatility, stochastic volatility — it will generate, with probability 1, prices which are continuous functions of time. This argument alone should be sufficient to rule out diffusion models as realistic models for price dynamics. Indeed, even after correcting for intraday effects, formal tests on intraday data reject the diffusion hypothesis.

But the importance of the issue of (dis)continuity also suggests that statistical averages such as the variance of returns or even quantiles of returns are not the only relevant quantities for representing risk: after all, an investor will only be exposed to a single sample path of the price and even if her portfolio minimizes some the variance averaged across sample paths it may have a high variability in practice. The risky character of a financial asset is therefore directly related to the lack of smoothness of its sample path $t \mapsto S_t$ and this is one crucial aspect of empirical data that one would like a mathematical model to reproduce. But how can we quantify the smoothness of sample paths?

7.6.1 Hölder regularity and singularity spectra

The usual definitions of smoothness for functions are based on the number of times a function can be differentiated: if $f : [0, T] \to \mathbb{R}$ admits k continuous derivatives at a point t, it is said to be C^k. A more refined notion is that of Hölder regularity: the local regularity of a function may be characterized by its *local Hölder exponents.* A function f is h-Hölder continuous at point t_0 iff there exists a polynomial of degree $< h$ such that

$$|f(t) - P(t - t_0)| \leq K_{t_0} |t - t_0|^h \tag{7.46}$$

in a neighborhood of t_0, where K_{t_0} is a constant. Let $C^h(t_0)$ be the space of (real-valued) functions which verify the above property at t_0. A function f is said to have local Hölder exponent α if for $h < \alpha, f \in C^h(t_0)$ and for $h > \alpha, f \notin C^h(t_0)$. Let $h_f(t)$ denote the local Hölder exponent of f at point t. If $h_f(t_0) \geq 1$ then f is differentiable at point t_0, whereas a discontinuity of f at t_0 implies $h_f(t_0) = 0$. More generally, the higher the value of $h_f(t_0)$, the greater is the local regularity of f at t_0.

In the case of a sample path $X_t(\omega)$ of a stochastic process X_t, $h_{X(\omega)}(t) = h_\omega(t)$ depends on the particular sample path considered, i.e., on ω. There are however some famous exceptions: for example for fractional Brownian motion with self-similarity parameter H, $h_B(t) = 1/H$ almost everywhere with probability one, i.e., for almost all sample paths. Note however that no such results hold for sample paths of Lévy processes or even stable Lévy motion.

Given that the local Hölder exponent may vary from sample path to sample path in the case of a stochastic process, it is not a robust statistical tool for characterizing the roughness of a function: the notion of *singularity spectrum* of a function was introduced to give a less detailed but more stable characterization of the local smoothness of a function in a "statistical" sense.

DEFINITION 7.2 Singularity spectrum *Let $f : R \to R$ be a real-valued function and for each $\alpha > 0$ define the set of points at which f has local Hölder exponent h:*

$$\Omega(\alpha) = \{t, h_f(t) = \alpha\}. \tag{7.47}$$

The singularity spectrum of f is the function $D : R^+ \to R$ which associates to each $\alpha > 0$ the Hausdorff-Besicovich dimension[4] *of $\Omega(\alpha)$:*

$$D_f(\alpha) = \dim_{\mathrm{HB}} \Omega(\alpha). \tag{7.48}$$

Using the above definition, one may associate to each sample path $X_t(\omega)$ of a stochastic process X_t its singularity spectrum $d_\omega(\alpha)$. If d_ω depends on ω then the empirical estimation of the singularity spectrum is not likely to give much information about the properties of the process X_t. Fortunately, this turns out not to be the case: it has been shown that, for large classes of stochastic processes, the singularity spectrum is the same for almost all sample paths. A result due to Jaffard [218] shows that a large class of Lévy processes verify this property: their singularity spectrum is the same for almost all sample paths and depends only on the behavior of the Lévy measure near the origin, as measured by the Blumenthal-Getoor index:

$$\beta = \inf\{\gamma > 0, \int_{|x|\leq 1} x^\gamma \nu(dx) < \infty\}. \tag{7.49}$$

PROPOSITION 7.3 Singularity spectrum of Lévy processes[218] *Let X be a Lévy process with Lévy triplet (σ^2, ν, b) and Blumenthal-Getoor index β.*

- *If $2 > \beta > 0$ and $\sigma = 0$ then for almost every sample path*

$$\dim \Omega(\alpha) = \beta\alpha \qquad \text{for } \alpha \leq \frac{1}{\beta} \tag{7.50}$$

and $\Omega(\alpha) = \emptyset$ for $\alpha > 1/\beta$.

[4] The Hausdorff–Besicovich dimension is one of the numerous mathematical notions corresponding to the general concept of "fractal" dimension, see [138, 277].

- *If* $2 > \beta > 0$ *and* $\sigma \neq 0$ *then for almost every sample path*

$$\dim \Omega(\alpha) = \beta\alpha \qquad \text{for } \alpha < \frac{1}{2}$$

$$\dim \Omega\left(\frac{1}{2}\right) = 1 \tag{7.51}$$

and $\Omega(\alpha) = \emptyset$ *for* $\alpha > 1/2$.

- *If* $\beta = 0$ *then for each* $\alpha > 0$ *with probability 1,* $\dim \Omega(\alpha) = 0$.

Notice that since ν a Lévy measure $2 \geq \beta \geq 0$ so Jaffard's result covers all cases of interest, the other cases being already known. This result shows in particular that the smoothness/roughness of paths is measured by the behavior of the Lévy measure near 0, i.e., the small jumps: is is not influenced by the tails. In particular, α-stable Lévy processes and tempered stable processes have the same singularity spectra.

7.6.2 Estimating singularity spectra

As defined above, the singularity spectrum of a function does not appear to be of any practical use since its definition involves first the continuous time ($\Delta \to 0$) limit for determining the local Hölder exponents and second the determination of the Hausdorff dimension of the sets $\Omega(\alpha)$ which, as remarked already by Halsey et al. [185], may be intertwined fractal sets with complex structures and impossible to separate on a point by point basis. The interest of physicists and empirical researchers in singularity spectra was ignited by the work of Parisi and Frisch [152] who, in the context or fluid turbulence, proposed a formalism for empirically computing the singularity spectrum from sample paths of the process. This formalism, called the multi-fractal formalism [152, 185, 216, 217, 279], enables the singularity spectrum to be computed from sample moments (called "structure functions" in the turbulence literature) of the increments: if the sample moments of the returns verify a scaling property

$$< |r_t(\Delta)|^q >= K_q \Delta^{\zeta(q)} \tag{7.52}$$

then the singularity spectrum $D(\alpha)$ is given by the Legendre transform of the scaling exponent $\zeta(q)$:

$$\zeta(q) = 1 + \inf(q\alpha - D(\alpha)). \tag{7.53}$$

$\zeta(q)$ may be obtained by regressing $\log < |r_t(T)|^q >$ against $\log T$. When the scaling in Equation (7.52) holds exactly, the Legendre transform (7.53) may be inverted to obtain $D(\alpha)$ from $\zeta(q)$. This technique was subsequently refined using the wavelet transform leading to an algorithm (WTMM method) for determining the singularity spectrum from the modulus of its wavelet transform [19, 10].

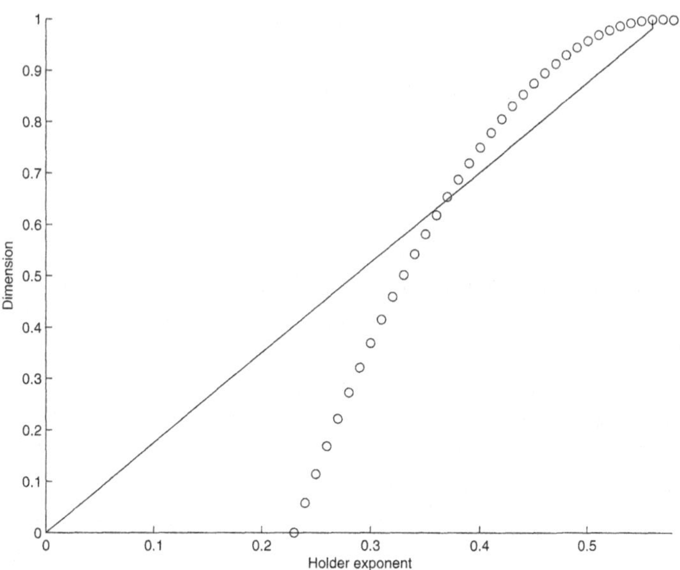

FIGURE 7.10: Singularity spectrum estimated for S&P500 tick data using the wavelet transform (WTMM) method (circles) compared to the singularity spectrum of a Lévy process with same (almost-everywhere) Hölder regularity $\alpha \simeq 0.57$.

These methods provide a framework to investigate pathwise regularity of price trajectories. Figure 7.10 shows the singularity spectrum estimated using the wavelet (WTMM) method [19] from S&P500 futures tick data. Remarkable points are the intercept — the lowest Hölder exponent — and the maximum at $\alpha \simeq 0.57$, which represents the almost everywhere Hölder exponent. Global estimators (as opposed to wavelet estimators which are local in time) are only sensitive to the almost-everywhere Hölder exponent which explains the often reported values of the "Hurst exponent" in the range 0.55 - 0.6. Various empirical studies on time series of market indices and exchange rates [282, 302] report a similar inverted parabola shape for other data sets. It should be noted that this non-trivial spectrum is very different from what one would expect from diffusion processes, Lévy processes or jump-diffusion processes used in continuous-time finance, for which the singularity spectrum is theoretically known (see, e.g., Proposition 7.3). In particular, the empirical spectrum in Figure 7.10 indicates a less smooth behavior than diffusion processes but no discontinuous ("jump") component in the signal since the Hölder exponent does not extend down to zero. The rare examples of stochastic processes for which the singularity spectrum resembles the one observed in financial data are stochastic cascades [279, 282] or their causal versions, the multi-fractal random walks [302, 18]. One drawback of these estimation methods is that their finite sample properties are not well known. The only currently feasible approach is, as in [282], to supplement such studies by Monte Carlo simulations of various stochastic models used in finance in order to check whether the peculiar shape of the spectra obtained are not artifacts due either to small sample size or discretization.

7.7 Summary: advantages and shortcomings of Lévy processes

Let us now summarize the advantages and shortcomings of representing log-prices as Lévy processes: how far can they go in capturing empirical features of asset returns?

The first category of properties we examined were related to the distribution of returns at a given horizon (Section 7.3). Given that in building a Lévy process we have the freedom of choosing the distribution of the process at a given time resolution, the distributional properties of returns for a given time horizon can always be perfectly matched by a Lévy process. Closer examination of the shape of the distribution and its tail properties enables to choose a suitable class of Lévy processes for this purpose using the methods described in Section 7.2.

The second category of properties concerned the time-aggregation properties of returns, namely how the distribution of returns varies with the horizon on which returns are computed. As observed in Section 7.4, the marginal distribution of a Lévy process has a shape which does vary with time in a qualitatively similar way to the log-price, reproducing the main features of the flow of marginal distributions in empirical data. However the stationarity of increments leads to rigid scaling properties for cumulants of Lévy processes which are not observed in log-prices. These discrepancies can be overcome by considering processes with independent but time-inhomogeneous increments, discussed in Chapter 14.

We also observed that, unlike the Brownian model where "realized volatility" has a deterministic continuous-time limit, models based on Lévy processes lead to a realized volatility which remains stochastic when computed on fine time grids. In particular, Lévy processes with heavy tailed increments — with a finite tail index — can lead to high variability of realized volatility: we get "stochastic volatility" effects for free, even in absence of any additional random factors.

However when it comes to *dependence* properties of returns across time — especially "volatility clustering" effects — exponential-Lévy models have nothing to say: having independent increments, they are not capable of mimicking volatility clustering and similar phenomena linked to nonlinear dependence in (absolute) returns. In particular, since the time aggregation properties of asset returns are related to their (nonlinear dependence) structure, models with independent increments, even if they do reproduce some time series properties of returns in a certain range of time scales, cannot give a "structural" explanation for the time aggregation properties of returns. A summary of these points is given in Table 7.2.

The shortcomings pointed out here have prompted the development of more sophisticated models, incorporating time inhomogeneity and/or extra sources of randomness to represent the volatility clustering effect. Some of these models will be discussed in Chapters 14 and 15.

Further reading

A vast literature dealing with the statistical properties of financial time series has developed during the last fifty years. Surveys of stylized facts of financial time series are given in [180] for foreign exchange rates, [140, 313, 76, 92] for stocks and indices. Lévy processes were first introduced in financial econometrics by Mandelbrot [278, 139, 100] and regained popularity in the 1990s with the advent of high frequency data and computer tools for data processing [94, 285, 36, 123, 123, 125]. The α-stable Lévy process was introduced as a model for cotton prices by Mandelbrot in [278]. Empirical tests on α-stable models include [3, 62], a review is given in [289]; see also [39, 236]. While empirical evidence has consistently rejected this model for most data sets of indices, exchange rates and major stocks, some applications may be found in commodity markets and volatile emerging markets [387]. A parallel literature on α-stable models in finance continues to flourish, see [328, 327]. Jorion [224] performs a maximum likelihood estimation of jump-diffusion models on the US/DEM exchange rate and a US equity index, see also [222]. Ait Sahalia [2] develops a nonasymptotic test for the presence of jumps based on the knowledge of transition densities of a Markov process at discrete time intervals; applying this test to option-implied densities he concludes towards the presence of jumps. The empirical performance of hyperbolic and generalized hyperbolic models are discussed in [123, 125, 319]; normal inverse Gaussian processes are discussed in [30, 342]. Empirical performance of tempered stable models is studied in [94, 67, 80].

Scaling and time aggregation properties for high frequency data were studied in [180, 301, 181] for foreign exchange rates. Numerous empirical studies on scaling properties of returns can also be found in the physics literature, see [67] for a review on this literature. An alternative view is given in [35] where scaling is claimed to be a spurious effect due to heavy tails.

A good reference on self-similarity is [374] (see also Chapter 7 in [344]). The hypothesis of self-similarity of market prices was first proposed by Mandelbrot

[278] (see also [279]). The α-stable model for the distribution of returns was subsequently investigated for many data sets [139, 62, 184, 285, 268, 94, 328]. Empirical studies on self-similarity of prices include [278, 280, 281, 279, 285, 94, 388].

The statistical estimation methods discussed in Section 7.2 are reviewed in [175]. The generalized method of moments is discussed in detail in [187]; see also Chapter 14 in [186]. An interesting discussion on the way to specify moment conditions is given in [155]. A more advanced discussion of estimation methods for financial time series can be found in [361], [274] and [186]. These references deal with parametric approaches: with the availability of computers and large data sets, the recent tendency in financial econometrics has been to shift to nonparametric methods [188, 189].

Part III

Option pricing in models with jumps

Chapter 8

Stochastic calculus for jump processes

> If one disqualifies the Pythagorean Theorem from contention, it is hard to think of a mathematical result which is better known and more widely applied in the world today than "Ito's Lemma." This result holds the same position in stochastic analysis that Newton's fundamental theorem holds in classical analysis. That is, it is the sine qua non of the subject.

Description of Kiyosi Ito's scientific work by the US National Academy of Sciences.

Given a financial asset whose price is represented by a stochastic process $S = (S_t)_{t\in[0,T]}$, there are two types of objects related to S that naturally arise in a financial modelling context: trading strategies involving S and derivatives written on the underlying asset S.

To describe trading strategies, one needs to consider dynamic portfolios resulting from buying and selling the assets. If an investor trades at times $T_0 = 0 < T_1 < \cdots < T_n < T_{n+1} = T$, detaining a quantity ϕ_i of the asset during the period $]T_i, T_{i+1}]$ then the capital gain resulting from fluctuations in the market price is given by

$$\sum_{i=0}^{n} \phi_i (S_{T_{i+1}} - S_{T_i}).$$

This quantity, which represents the capital gain of the investor following the strategy ϕ, is called the stochastic integral of ϕ with respect to S and denoted by $\int_0^T \phi_t dS_t$. Stochastic integrals with respect to processes with jumps are discussed in Section 8.1. We have followed the approach proposed in [324]; other references are given at the end of the chapter.

If S_t is built from a Lévy process, the positions and the amplitudes of its jumps are described by a Poisson random measure and various quantities involving the jump times and jump sizes can be expressed as integrals with respect to this measure. Stochastic integrals with respect to Poisson random measures are discussed in Section 8.1.4.

The second fundamental problem is to describe the time evolution of a derivative instrument whose value $V_t = f(t, S_t)$ depends on S_t. The key tool here is the change of variable formula — the Itô formula — which relates the local behavior of V_t to the behavior of S_t. This formula forms the basis of the stochastic calculus for processes with jumps and we will give several forms of it which are useful in various applications.

In the preceding chapters we have worked with Lévy processes, which provide analytically tractable examples of jump processes. However, as we will see below, if X is a Lévy process or a diffusion process, quantities such as $\int_0^t \phi dX$ or $f(t, X_t)$ are, in general, not Lévy processes or diffusions any more: the class of Lévy processes is not stable under stochastic integration or nonlinear transformations. Similarly, starting from a Markov process X these transformations will not give us Markov processes anymore. Therefore, even if we are only interested in Markovian models or models based on Lévy processes, we are naturally led to consider a larger class of stochastic processes, which contains our objects of interest and is stable under the operations considered above. The class of *semimartingales* provides such a framework: not only this class is stable under stochastic integration and (smooth) nonlinear transformations, it is also stable under other operations such as change of measure, change of filtration and "time change," which we will encounter in the following chapters. Although we will not present the results in their most general form, most of them hold in a general semimartingale setting.

The main goal of this chapter is to present some useful results on stochastic integration and stochastic calculus, using an elementary approach accessible to the nonspecialist. Stochastic calculus for processes with jumps is usually presented in the general framework of semimartingales, which can be quite difficult for the beginner. Instead of following this path we propose a pedestrian approach, starting from the simple case of the Poisson process in Section 8.3.1 and progressively increasing the complexity of the processes involved. In fact, we will try to convince the reader that stochastic calculus for (pure) jump processes is more intuitive and easy to understand than for diffusion processes!

This chapter could have been entitled "A beginner's guide to stochastic calculus for processes with jumps." We refer to [215, 324, 323] for complements and to [194] or [110] for more detailed material concerning stochastic integrals and their properties. If Dellacherie and Meyer [110] is your favorite bedtime reading, you might as well skip this chapter and go quickly to the next one!

8.1 Trading strategies and stochastic integrals

Consider a market with d assets whose prices are modelled by a (vector) stochastic process $S_t = (S_t^1, \ldots, S_t^d)$, which is supposed to be cadlag. A portfolio is a vector describing the amount of each asset held by the investor: $\phi = (\phi^1, \dots, \phi^d)$. The value of a such a portfolio at time t is then given by

$$V_t(\phi) = \sum_{k=1}^{d} \phi^k S_t^k \equiv \phi.S_t. \tag{8.1}$$

A trading strategy consists of maintaining a dynamic portfolio ϕ_t by buying and selling assets at different dates. Let us denote the transaction dates by $T_0 = 0 < T_1 < T_2 < \dots < T_n < T_{n+1} = T$. Between two transaction dates T_i and T_{i+1} the portfolio remains unchanged and we will denote its composition by ϕ_i. The portfolio ϕ_t held at date t may then be expressed as:

$$\phi_t = \phi_0 1_{t=0} + \sum_{i=0}^{n} \phi_i 1_{]T_i, T_{i+1}]}(t). \tag{8.2}$$

The transaction dates (T_i) can be fixed but, more realistically, they are not known in advance and an investor will decide to buy or sell at T_i depending on the information revealed before T_i. For example, in the case of a limit order T_i is the first time when S_t crosses a certain value (the limit price). Therefore, the transaction times (T_i) should be defined as nonanticipating random times (also called stopping times, see Section 2.4.3): they are said to form a random partition of $[0, T]$. Since the new portfolio ϕ_i is chosen based on information available at T_i, ϕ_i is $\mathcal{F}_{T_i}$-measurable. When the broker decides to transact at $t = T_i$, the portfolio is still described by ϕ_{i-1}; it takes its new value ϕ_i *right after* the transaction, i.e., for $t > T_i$. Therefore, the indicator function in (8.2) is of the form $1_{]T_i, T_{i+1}]}$ (left-continuous) as opposed to $1_{[T_i, T_{i+1}[}$ (right-continuous): $\phi_{t-} = \phi_t$. Being left-continuous (caglad, in fact), $(\phi_t)_{t \in [0,T]}$ is therefore a predictable process.[1] Stochastic processes of the form (8.2) are called simple predictable processes:

DEFINITION 8.1 Simple predictable process *A stochastic process* $(\phi_t)_{t \in [0,T]}$ *is called a* simple predictable process *if it can be represented as*

$$\phi_t = \phi_0 1_{t=0} + \sum_{i=0}^{n} \phi_i 1_{]T_i, T_{i+1}]}(t),$$

where $T_0 = 0 < T_1 < T_2 < \dots < T_n < T_{n+1} = T$ *are nonanticipating random times and each* ϕ_i *is bounded random variable whose value is revealed at* T_i *(it is* $\mathcal{F}_{T_i}$*-measurable).*

The set of simple predictable processes on $[0, T]$ will be denoted by $\mathbb{S}([0, T])$. Any realistic and implementable strategy should be given by a simple predictable process or, at least, one should be able to approximate it by strategies

[1]See Section 2.4.5 for a definition.

of the form (8.2). One of the reasons of this is that in a market with jumps strategies that are not predictable can generate arbitrage opportunities, as emphasized by the following example.

Example 8.1 Trading strategies have to be predictable.
Let $S_t = \lambda t - N_t$, where N_t denotes a Poisson process with intensity λ. $-S$ is a compensated Poisson process thus S is a martingale (see Section 2.5.4). Denote by T_1 the time of its first jump: $S_{T_1} = S_{T_1-} - 1$. T_1 is an exponential random variable with parameter λ. Consider now the strategy which consists in buying one unit of the asset S at $t = 0$ (at zero price!) and selling it right before the price falls down ("selling right before the crash"): $\phi_t = 1_{[0,T_1[}$. Contrarily to the strategies defined in (8.2), $t \mapsto \phi_t$ is not left-continuous but right-continuous (cadlag). The capital gain associated to this strategy is then given by

$$G_t(\phi) = \int_0^t \phi_u dS_u = \lambda t \quad \text{for } t < T_1,$$
$$= \lambda T_1 \text{ for } t \geq T_1.$$

Therefore this strategy requires zero initial investment and has an almost surely nonnegative gain, which is strictly positive with nonzero probability. Hence, it is an arbitrage opportunity. Such strategies should therefore should be ruled out in an arbitrage-free model. □

Obviously, the "strategy" ϕ proposed in this example is impossible to implement unless one knows beforehand that the price is going to fall: ϕ_t is not a predictable process. This example motivates us to restrict integrands/strategies to predictable processes. Note also that in this example we have used in an essential way the fact that the process S is discontinuous: if S is continuous then changing the strategy at one point does not have any effect on the resulting gain of the investor, so one could allow in that case for right-continuous strategies.

Between T_i and T_{i+1}, the quantity of asset in the portfolio is ϕ_i and the asset moves by $(S_{T_{i+1}} - S_{T_i})$ so the capital gain of the portfolio is given by $\phi_i.(S_{T_{i+1}} - S_{T_i})$. Hence an investor starting with a portfolio ϕ_0 and following the strategy ϕ will have accumulated at time $t > 0$ a capital equal to:

$$G_t(\phi) = \phi_0.S_0 + \sum_{i=0}^{j-1} \phi_i.(S_{T_{i+1}} - S_{T_i}) + \phi_j.(S_t - S_{T_j}) \text{ for } T_j < t \leq T_{j+1}.$$

The stochastic process $(G_t(\phi))_{t\in[0,T]}$ thus defined is called the gain process of the strategy ϕ. Using the stopping time notation defined in (2.50), the gain process can be rewritten as:

$$G_t(\phi) = \phi_0.S_0 + \sum_{i=0}^{n} \phi_i.(S_{T_{i+1}\wedge t} - S_{T_i\wedge t}). \tag{8.3}$$

For the stochastic integral to be interpreted as the gain process of the strategy ϕ, the portfolio ϕ_i should be constituted at the beginning of the period, T_i (therefore ϕ_i is revealed at T_i while the variation of the assets $S_{T_{i+1}} - S_{T_i}$ is only revealed at the end of the period, T_{i+1}). The stochastic process defined by (8.3) is called the *stochastic integral*[2] of the predictable process ϕ with respect to S and denoted by:

$$\int_0^t \phi_u dS_u := \phi_0.S_0 + \sum_{i=0}^{n} \phi_i.(S_{T_{i+1}\wedge t} - S_{T_i \wedge t}). \tag{8.4}$$

Thus, the stochastic integral $\int_0^t \phi dS$ represents the capital accumulated between 0 and t by following the strategy ϕ. On the other hand, the portfolio is worth $V_t(\phi) = \phi_t.S_t$ at time t. The difference between these two quantities represents the cost of the strategy up to time t:

$$C_t(\phi) = V_t(\phi) - G_t(\phi) = \phi_t.S_t - \int_0^t \phi_u dS_u. \tag{8.5}$$

$C_t(\phi)$ is called the cost process associated to the strategy ϕ. A strategy $(\phi_t)_{t\in[0,T]}$ is said to be self-financing if the cost is (almost surely) equal to zero: the value $V_t(\phi)$ of the portfolio is then equal to the initial value plus the capital gain between 0 and t:

$$V_t(\phi) = \int_0^t \phi_u dS_u = \phi_0 S_0 + \int_{0+}^t \phi_u dS_u. \tag{8.6}$$

Equation (8.6) simply means that the only source of variation of the portfolio's value is the variation of the asset values: all trading operations are financed by capital gains, all capital gains are reinvested into the portfolio and no cash is added to or withdrawn from the account.

The gain process/stochastic integral associated to a strategy has the following fundamental property: if S_t is a martingale then the gain process associated to any strategy is also a martingale:

PROPOSITION 8.1 Martingale-preserving property

If $(S_t)_{t\in[0,T]}$ is a martingale then for any simple predictable process ϕ the stochastic integral $G_t = \int_0^t \phi dS$ is also a martingale.

PROOF Consider a simple predictable process ϕ as in 8.2. By construction its stochastic integral G_t is a cadlag nonanticipating process. Since the ϕ_i are

[2]Since the stochastic integral of a vector process is defined as the sum of integrals of its components, in the sequel we will mostly discuss the one-dimensional case. Multidimensional extensions are, in most cases, straightforward.

bounded and $E|S_{T_i}| < \infty$, it is easy to show that $E|G_t| < \infty$. We will now prove that $E[G_T|\mathcal{F}_s] = G_t$. It is sufficient to show that for each i, $E[\phi_i(S_{T_{i+1}} - S_{T_i})|\mathcal{F}_t] = \phi_i(S_{T_{i+1}\wedge t} - S_{T_i\wedge t})$.

$$\begin{aligned} E[\phi_i(S_{T_{i+1}} - S_{T_i})|\mathcal{F}_t] &= E[1_{t>T_{i+1}}\phi_i(S_{T_{i+1}} - S_{T_i})|\mathcal{F}_t] \\ &+ E[1_{]T_i,T_{i+1}]}(t)\phi_i(S_{T_{i+1}} - S_{T_i})|\mathcal{F}_t] \\ &+ E[1_{t\leq T_i}\phi_i(S_{T_{i+1}} - S_{T_i})|\mathcal{F}_t]. \end{aligned}$$

Since T_i, T_{i+1} are stopping times, $1_{t>T_{i+1}}$, $1_{t\leq T_i}$ and $1_{]T_i,T_{i+1}]}(t)$ are $\mathcal{F}_t$-measurable and can be taken out of the conditional expectation. The first two terms are simple to compute:

$$\begin{aligned} E[1_{t>T_{i+1}}\phi_i(S_{T_{i+1}} - S_{T_i})|\mathcal{F}_t] &= 1_{t>T_{i+1}}\phi_i(S_{T_{i+1}} - S_{T_i}) \\ E[1_{]T_i,T_{i+1}]}(t)\phi_i(S_{T_{i+1}} - S_{T_i})|\mathcal{F}_t] &= 1_{]T_i,T_{i+1}]}(t)\phi_i E[S_{T_{i+1}} - S_{T_i}|\mathcal{F}_t] \\ &= 1_{]T_i,T_{i+1}]}(t)\phi_i(S_t - S_{T_i}). \end{aligned}$$

For the last term we use the law of iterated expectations and the fact that ϕ_i is $\mathcal{F}_{T_i}$-measurable:

$$\begin{aligned} E[1_{t\leq T_i}\phi_i(S_{T_{i+1}} - S_{T_i})|\mathcal{F}_t] &= 1_{t\leq T_i}E[E[\phi_i(S_{T_{i+1}} - S_{T_i})|\mathcal{F}_{T_i}]|\mathcal{F}_t] \\ &= 1_{t\leq T_i}E[\phi_i E[S_{T_{i+1}} - S_{T_i}|\mathcal{F}_{T_i}]|\mathcal{F}_t] = 0, \end{aligned}$$

the last equality resulting from the sampling theorem (Theorem 2.7) applied to S. So:

$$\begin{aligned} E[\phi_i(S_{T_{i+1}} - S_{T_i})|\mathcal{F}_t] &= 1_{t>T_{i+1}}\phi_i(S_{T_{i+1}} - S_{T_i}) + 1_{]T_i,T_{i+1}]}(t)\phi_i(S_t - S_{T_i}) \\ &= \phi_i(S_{T_{i+1}\wedge t} - S_{T_i\wedge t}). \end{aligned}$$

□

Since for a self-financing strategy $V_t(\phi) = G_t(\phi)$, we conclude that if the underlying asset follows a martingale $(S_t)_{t\in[0,T]}$ then the value of any self-financing strategy is a martingale.

Apart from computing the gain associated to a strategy, stochastic integrals can also be used as a means of building new stochastic processes (in particular, new martingales) from old ones: given a nonanticipating cadlag process $(X_t)_{t\in[0,T]}$ one can build new processes $\int_0^t \sigma_u dX_u$ by choosing various (simple) predictable processes $(\sigma_t)_{t\in[0,T]}$. Here X_t is interpreted as a "source of randomness" and σ_t as a "volatility coefficient." Starting with a simple stochastic process X such as a Lévy process, this procedure can be used to build stochastic models with desired properties. The following result shows that if the asset price is modelled as a stochastic integral $S_t = \int_0^t \sigma dX$ with respect to a "source of randomness" then the gain process of any strategy involving S can also be expressed as a stochastic integral with respect to X.

PROPOSITION 8.2 Associativity
Let $(X_t)_{t\in[0,T]}$ be a real-valued nonanticipating cadlag process and $(\sigma_t)_{t\geq 0}$ and $(\phi_t)_{t\geq 0}$ be real-valued simple predictable processes. Then $S_t = \int_0^t \sigma dX$ is a nonanticipating cadlag process and

$$\int_0^t \phi_u dS_u = \int_0^t \phi_u \sigma_u dX_u.$$

The relation $S_t = \int_0^t \sigma_t dX_t$ is often abbreviated to a "differential" notation $dS_t = \sigma_t dX_t$, which should be understood as a shorthand for the integral notation. The associativity property then leads to a composition rule for "stochastic differentials": if $dS_t = \sigma_t dX_t$ and $dG_t = \phi_t dS_t$ then $dG_t = \phi_t \sigma_t dX_t$.

8.1.1 Semimartingales

For the moment we have not required any specific property for the process S in order to define gain processes/stochastic integrals for simple predictable processes: in fact, (8.3) makes sense for any cadlag process S. A reasonable requirement is a stability property: a small change in the portfolio should lead to a small change in the gain process. A mathematical formulation of this idea is to require that, if $\phi^n \to \phi$ in $\mathbb{S}([0,T])$ (for example, in the sense of uniform convergence) then $\int_0^t \phi^n dS \to \int_0^t \phi dS$ in some appropriate sense (for example, convergence in probability). Unfortunately, this (reasonable) stability property does not hold for any stochastic process (S_t) and those who verify it deserve a special name.

DEFINITION 8.2 Semimartingale *A nonanticipating cadlag process S is called a* semimartingale[3] *if the stochastic integral of simple predictable processes with respect to S:*

$$\phi = \phi_0 1_{t=0} + \sum_{i=0}^{n} \phi_i 1_{]T_i,T_{i+1}]} \mapsto \int_0^T \phi dS = \phi_0 S_0 + \sum_{i=0}^{n} \phi_i (S_{T_{i+1}} - S_{T_i}),$$

verifies the following continuity property: for every $\phi^n, \phi \in \mathbb{S}([0,T])$ if

$$\sup_{(t,\omega)\in[0,T]\times\Omega} |\phi_t^n(\omega) - \phi_t(\omega)| \underset{n\to\infty}{\to} 0 \text{ then } \int_0^T \phi^n dS \underset{n\to\infty}{\overset{\mathbb{P}}{\to}} \int_0^T \phi dS. \quad (8.7)$$

If the continuity property above does not hold, it means that, if an asset is modelled by (S_t), a very small error in the composition of a strategy can

[3]While mathematicians have no lack of imagination when it comes to proving theorems, finding names for new concepts seems to be a problem: martingales, supermartingales, submartingales, semimartingales, quasimartingales, local martingales, sigma-martingales, etc., are neither exciting nor very explanatory!

bring about a large change in the portfolio value! It is therefore preferable, in models of continuous time trading, to use stochastic processes which are semimartingales, otherwise our model may produce results which are difficult to use or interpret.

The continuity property (8.7) involves the convergence of the random variables $\int_0^T \phi^n dS$ (for T fixed). The following technical result [324, Theorem 11] shows that a stability property also holds for the process defined by the stochastic integral:

PROPOSITION 8.3

If $(S_t)_{t\in[0,T]}$ is a semimartingale then for every $\phi^n, \phi \in \mathbb{S}([0,T])$

$$\text{if} \qquad \sup_{(t,\omega)\in[0,T]\times\Omega} |\phi_t^n(\omega) - \phi_t(\omega)| \underset{n\to\infty}{\to} 0 \tag{8.8}$$

$$\text{then} \qquad \sup_{t\in[0,T]} \left| \int_0^t \phi^n dS - \int_0^t \phi\, dS \right| \underset{n\to\infty}{\overset{\mathbb{P}}{\to}} 0. \tag{8.9}$$

The convergence in (8.8) is uniform convergence on $[0,T]\times\Omega$; the convergence in (8.9) is uniform convergence in probability on $[0,T]$ (sometimes called ucp convergence, meaning "uniformly on compacts in probability").

Definition 8.2 may appear very difficult to apply in practice — indeed, how do we know that a given process satisfies the stability property mentioned above? However, we will give two simple examples which will allow us to see that almost all processes that we encounter in this book are semimartingales.

Example 8.2 Every finite variation process is a semimartingale

For a finite variation process S, denoting by $TV(S)$ its total variation on $[0,T]$, we always have:

$$\sup_{t\in[0,T]} \int_0^t \phi dS \leq TV(S) \sup_{(t,\omega)\in[0,T]\times\Omega} |\phi_t(\omega)|, \tag{8.10}$$

from which the property (8.8)–(8.9) can be deduced. □

Example 8.3 Every square integrable martingale is a semimartingale
For a square integrable martingale M we can write:

$$E\left[\left(\int_0^t \phi dM\right)^2\right] = E\left[\left(\phi_0 M_0 + \sum_{i=0}^{n} \phi_i (M_{T_{i+1}\wedge t} - M_{T_i \wedge t})\right)^2\right]$$
$$= E\left[\phi_0^2 M_0^2 + \sum_{i=0}^{n} \phi_i^2 (M_{T_{i+1}\wedge t} - M_{T_i \wedge t})^2\right]$$
$$\leq \sup_{s,\omega} |\phi_s(\omega)| E\left[M_0^2 + \sum_{i=0}^{n} (M_{T_{i+1}\wedge t} - M_{T_i \wedge t})^2\right]$$
$$\leq \sup_{s,\omega} |\phi_s(\omega)| E\left[M_0^2 + \sum_{i=0}^{n} (M^2_{T_{i+1}\wedge t} - M^2_{T_i \wedge t})\right]$$
$$\leq \sup_{s,\omega} |\phi_s(\omega)| \sup_s E[M_s^2],$$

where we have used Doob's sampling theorem (Theorem 2.7) several times. The above inequality implies that when the strategies converge uniformly, the stochastic integrals converge in L^2, uniformly in t. But the L^2 convergence implies convergence in probability hence the result. □

In addition, it is clear from Definition 8.2 that the semimartingales form a vector space: any linear combination of a finite number of semimartingales is a semimartingale. This remark and the two simple but important examples above allow to conclude that the following familiar processes are semimartingales:

- The Wiener process (because it is a square integrable martingale).
- The Poisson process (because it is a finite variation process).
- All Lévy processes are semimartingales because a Lévy process can be split into a sum of a square integrable martingale and a finite variation process : this the Lévy-Itô decomposition (Proposition 3.7).

It is also possible (but by no means straightforward) to show that every (local) martingale is a semimartingale in the sense of Definition 8.2 (see [324]). On the other hand, a deterministic process is a semimartingale if and only if it is of finite variation (see [215]) so all infinite variation deterministic processes are examples of processes that are not semimartingales. More interesting examples of processes that are not semimartingales are provided by fractional Brownian motions ("$1/f$ noise") discussed in Section 7.4.1.

More generally any semimartingale can be represented as a (local) martingale plus a finite variation process but this is a difficult result, see [55, 110] or [324, Chapter 4].

The associativity property (Proposition 8.2) allows to show that a stochastic integral (with respect to a semimartingale) is again a semimartingale. Therefore all new processes constructed from semimartingales using stochastic integration will again be semimartingales. The class of semimartingales is quite large and convenient to work with, because it is stable with respect to many operations (stochastic integration is not the only example).

8.1.2 Stochastic integrals for caglad processes

Realistic trading strategies are always of the form (8.2) so it might seem that defining stochastic integrals for simple predictable processes is sufficient for our purpose. But as we will see later, hedging strategies for options cannot in general be expressed in the form (8.2) but can only be approximated in some sense by simple predictable processes. Also, if X is interpreted not as an asset price but as a source of randomness, one can be interested in defining asset price processes $S_t = \int_0^t \sigma_u dX_u$ where the instantaneous volatility σ_t does not necessarily move in a stepwise manner as in (8.2). These remarks motivate the extension of stochastic integrals beyond the setting of simple predictable processes.

Example 8.1 showed that when S has jumps we cannot hope to define stochastic integrals with respect to S for right-continuous integrands while still conserving the martingale property (as in Proposition 8.1). We therefore have to content ourselves with left-continuous (caglad) integrands.

It turns out that any caglad process $\phi \in \mathbb{L}([0,T])$ can be uniformly approximated by a sequence $(\phi^n) \in \mathbb{S}([0,T])$ of simple predictable process ([324], Theorem 10) in the sense of (8.9). Using the continuity property (8.8)–(8.9) of the stochastic integral then allows to define the stochastic integral $\int \phi dS$ as the limit — in the sense of (8.9) — of $\int \phi^n dS$. In particular, one may choose the usual "Riemann sums" as discrete approximations, as shown by the following result [324, Theorem 21].

PROPOSITION 8.4 Stochastic integrals via Riemann sums

Let S be a semimartingale, ϕ be a caglad process and $\pi^n = (T_0^n = 0 < T_1^n < \cdots < T_{n+1}^n = T)$ a sequence of random partitions of $[0,T]$ such that $|\pi^n| = \sup_k |T_k^n - T_{k-1}^n| \to 0$ a.s. when $n \to \infty$. Then

$$\phi_0 S_0 + \sum_{k=0}^{n} \phi_{T_k}(S_{T_{k+1}\wedge t} - S_{T_k \wedge t}) \underset{n\to\infty}{\overset{\mathbb{P}}{\to}} \int_0^t \phi_{u-} dS_u \tag{8.11}$$

uniformly in t on $[0,T]$.

While this looks very much like the definition of a Riemann integral, one important difference should be noted: in the sum (8.11) the variation of S is multiplied by the value of ϕ_t at the *left* endpoint of the interval, otherwise the sum (and the limit) is not a nonanticipating process anymore and, even if it does converge, will define a process with different properties. This phenomenon has no equivalent in the case of ordinary (Riemann) integrals: one can compute ϕ at any point in the interval $[T_i, T_{i+1}]$ including both endpoints or the midpoint, without changing the limit. The key idea, due to K. Itô, is thus to use *nonanticipating* Riemann sums.

Proposition 8.4 shows that using trading strategies given by a caglad process makes sense since the gain process of such a strategy may be approximated by a simple predictable (therefore realistic) strategy. No such result holds for more general (not caglad) processes so their interpretation as "trading strategies" is not necessarily meaningful.

By approximating a caglad process by a sequence of simple predictable processes, one can show that the properties described above for simple predictable integrands continue to hold for caglad integrands. If S is a semimartingale then for any nonanticipating caglad process $(\sigma_t)_{t\in[0,T]}$ on $(\Omega, \mathcal{F}, (\mathcal{F}_t), \mathbb{P})$ the following properties hold:

- Semimartingale property: $S_t := \int_0^t \sigma dX$ is also a semimartingale.
- Associativity: if ϕ_t is another nonanticipating caglad process then

$$\int_0^t \phi dS = \int_0^t (\phi\sigma) dX.$$

- Martingale preservation property: if $(X_t)_{t\in[0,T]}$ is a square integrable martingale and ϕ is bounded then the stochastic integral $M_t = \int_0^t \phi dX$ is a square integrable martingale.

These properties were shown in Section 8.1 for simple predictable integrands; their proof for caglad integrands is based on limit arguments.[4]

8.1.3 Stochastic integrals with respect to Brownian motion

Since the Wiener process is a semimartingale, stochastic integrals with respect to the Wiener process may be defined as above: given a simple predictable process ϕ:

$$\phi_t = \phi_0 1_{t=0} + \sum_{i=0}^{n} \phi_i 1_{]T_i, T_{i+1}]}(t),$$

the Brownian stochastic integral $\int \phi dW$ is defined as

$$\int_0^T \phi_t \, dW_t = \sum_{i=0}^{n} \phi_i \; (W_{T_{i+1}} - W_{T_i}).$$

[4]See [324, Chapter 2].

Since W_t is a martingale, $\int_0^t \phi dW$ is also a martingale and $E[\int_0^T \phi_t dW_t] = 0$. Due to the independent increments property of W, we can also compute the second moment:

$$\begin{aligned}
E\left[\left|\int_0^T \phi_t \, dW_t\right|^2\right] &= \text{Var}\left[\sum_{i=0}^{n} \phi_i \, (W_{T_i} - W_{T_{i+1}})\right] \\
&= \sum_{i=0}^{n} E[\phi_i^2 \, (W_{T_i} - W_{T_{i+1}})^2] \\
&\quad + 2\sum_{i>j} \text{Cov}(\phi_i(W_{T_{i+1}} - W_{T_i}), \phi_j(W_{T_{j+1}} - W_{T_j})) \\
&= \sum_{i=0}^{n} E[E[\phi_i^2 \, (W_{T_{i+1}} - W_{T_i})^2|\mathcal{F}_{T_i}] \,] \\
&\quad + \sum_{i>j} E[E[\phi_i\phi_j(W_{T_{i+1}} - W_{T_i})(W_{T_{j+1}} - W_{T_j})|\mathcal{F}_{T_j}] \,] \\
&= \sum_{i=0}^{n} E[\phi_i^2 E[(W_{T_{i+1}} - W_{T_i})^2|\mathcal{F}_{T_i}] \,] + 0 \\
&= \sum_{i=0}^{n} E[\phi_i^2](T_{i+1} - T_i) = E\int_0^t \phi_t^2 dt.
\end{aligned}$$

This relation is very useful in computations and is called the *isometry formula.*

PROPOSITION 8.5 Isometry formula: simple integrands
Let $(\phi_t)_{0\leq t\leq T}$ be a simple predictable process and $(W_t)_{0\leq t\leq T}$ be a Wiener process. Then

$$E\left[\int_0^T \phi_t \, dW_t\right] = 0, \tag{8.12}$$

$$E\left[\left|\int_0^T \phi_t \, dW_t\right|^2\right] = E\left[\int_0^T |\phi_t|^2 dt\right]. \tag{8.13}$$

Using this isometry, one can construct stochastic integrals with respect to the Wiener process for predictable processes $(\phi_t)_{t\in[0,T]}$ verifying

$$E\left[\int_0^T |\phi_t|^2 dt\right] < \infty \tag{8.14}$$

by approximating φ with a sequence of simple predictable processes (ϕ^n) in the L^2 sense:

$$E\left[\int_0^T |\phi_t^n - \phi_t|^2 dt\right] \stackrel{n\to\infty}{\to} 0.$$

Using the isometry relation (8.13) one can then show that $\int_0^T \phi_t^n dW_t$ converges in $L^2(\mathbb{P})$ towards a limit denoted as $\int_0^T \phi dW_t$. The martingale and isometry properties are conserved in this procedure [205].

PROPOSITION 8.6 Isometry formula for Brownian integrals
Let ϕ be a predictable process verifying

$$E\left[\int_0^T |\phi_t|^2 dt\right] < \infty. \tag{8.15}$$

Then $\int_0^t \phi\, dW$ is a square integrable martingale and

$$E\left[\int_0^T \phi_t\, dW_t\right] = 0, \tag{8.16}$$

$$E\left[\left|\int_0^T \phi_t\, dW_t\right|^2\right] = E\left[\int_0^T |\phi_t|^2 dt\right]. \tag{8.17}$$

Note that, although this construction gives a well-defined random variable $\int_0^T \phi dW$, ϕ cannot be interpreted as a "trading strategy": it is not necessarily caglad and its integral cannot necessarily be represented as a limit of Riemann sums.

8.1.4 Stochastic integrals with respect to Poisson random measures

Poisson random measures were defined in Section 2.6.1 and we have already encountered integrals of deterministic functions with respect to a Poisson random measure: this notion was used to study path properties of Lévy processes in Chapter 3. We now define the notion of stochastic integral of a random predictable function with respect to a Poisson random measure, following [242, 205].

Let M be a Poisson random measure on $[0,T]\times\mathbb{R}^d$ with intensity $\mu(dt\, dx)$. The compensated random measure $\tilde{M}$ is defined as the centered version of M: $\tilde{M}(A) = M(A) - \mu(A) = M(A) - E[M(A)]$. Recall from Section 2.6.1 that for each measurable set $A \subset \mathbb{R}^d$ with $\mu([0,T]\times A) < \infty$, $M_t(A) = M([0,t]\times A)$ defines a counting process, $\tilde{M}_t(A) = M([0,t]\times A) - \mu([0,t]\times A)$ is a martingale and if $A \cap B = \emptyset$ then $M_t(A)$ and $M_t(B)$ are independent.

By analogy with simple predictable processes defined in (8.2), we consider simple predictable functions $\phi : \Omega \times [0,T] \times \mathbb{R}^d \to \mathbb{R}$:

$$\phi(t,y) = \sum_{i=1}^{n}\sum_{j=1}^{m} \phi_{ij} 1_{]T_i,T_{i+1}]}(t) 1_{A_j}(y), \tag{8.18}$$

where $T_1 \leq T_2 \leq \cdots \leq T_n$ are nonanticipating random times, $(\phi_{ij})_{j=1\dots m}$ are bounded $\mathcal{F}_{T_i}$-measurable random variables and $(A_j)_{i=1\dots m}$ are disjoint subsets of $\mathbb{R}^d$ with $\mu([0,T]\times A_j) < \infty$. The stochastic integral $\int_{[0,T]\times\mathbb{R}^d} \phi(t,y)M(dt\ dy)$ is then defined as the random variable

$$\int_0^T \int_{\mathbb{R}^d} \phi(t,y)M(dt\ dy) = \sum_{i,j=1}^{n,m} \phi_{ij} M(\ ]T_i, T_{i+1}] \times A_j)$$
$$= \sum_{i=1}^{n} \sum_{j=1}^{m} \phi_{ij}[M_{T_{i+1}}(A_j) - M_{T_i}(A_j)]. \quad (8.19)$$

Similarly, one can define the process $t \mapsto \int_0^t \int_{\mathbb{R}^d} \phi(t,y)M(dt\ dy)$ by

$$\int_0^t \int_{\mathbb{R}^d} \phi(s,y)M(ds\ dy) = \sum_{i,j=1}^{n,m} \phi_{ij}[M_{T_{i+1}\wedge t}(A_j) - M_{T_i\wedge t}(A_j)]. \quad (8.20)$$

The stochastic integral $t \mapsto \int_0^t \int_{\mathbb{R}^d} \phi(s,y)M(ds\ dy)$ is a cadlag, nonanticipating process. Similarly, the compensated integral $\int_{[0,T]\times\mathbb{R}^d} \phi(t,y)\tilde{M}(dt\ dy)$ is defined as the random variable:

$$\int_0^T \int_{\mathbb{R}^d} \phi(s,y)\tilde{M}(ds\ dy) = \sum_{i,j=1}^{n,m} \phi_{ij}\tilde{M}(\ ]T_i, T_{i+1}] \times A_j) \quad (8.21)$$
$$= \sum_{i,j=1}^{n,m} \phi_{ij}\ [M(]T_i, T_{i+1}] \times A_j) - \ \mu(]T_i, T_{i+1}] \times A_j)].$$

By restricting to terms with $T_i \leq t$ (i.e., stopping at t), we obtain a stochastic process:

$$\int_0^t \int_{\mathbb{R}^d} \phi(s,y)\tilde{M}(ds\ dy) = \sum_{i=1}^{n} \phi_i[\tilde{M}_{T_{i+1}\wedge t}(A_j) - \tilde{M}_{T_i\wedge t}(A_j)]. \quad (8.22)$$

The notion of compensated integral is justified by the following result:

PROPOSITION 8.7 Martingale preserving property
For any simple predictable function $\phi : \Omega \times [0,T] \times \mathbb{R}^d \to \mathbb{R}$ the process $(X_t)_{t\in[0,T]}$ defined by the compensated integral

$$X_t = \int_0^t \int_{\mathbb{R}^d} \phi(s,y)\tilde{M}(ds\ dy)$$

is a square integrable martingale and verifies the isometry formula*:*

$$E[\ |X_t|^2] = E\left[\int_0^t \int_{\mathbb{R}^d} |\phi(s,y)|^2 \mu(ds\ dy)\right]. \quad (8.23)$$

PROOF For $j = 1 \dots m$ define $Y_t^j = \tilde{M}(]0,t] \times A_j) = \tilde{M}_t(A_j)$. From Proposition 2.16, $(Y_t^j)_{t\in[0,T]}$ is a martingale with independent increments. Since the A_j are disjoint, the processes Y^j are mutually independent.

Writing $\tilde{M}(]T_i \wedge t, T_{i+1} \wedge t] \times A_j) = Y^j_{T_{i+1}\wedge t} - Y^j_{T_i \wedge t}$, the compensated integral X_t can be expressed as a sum of stochastic integrals:

$$X_t = \sum_{j=1}^{m} \sum_{i=1}^{n} \phi_{ij} (Y^j_{T_{i+1}\wedge t} - Y^j_{T_i\wedge t})$$
$$= \sum_{j=1}^{m} \int_0^t \phi^j dY^j \quad \text{where} \quad \phi^j = \sum_{i=1}^{n} \phi_{ij} 1_{]T_i, T_{i+1}]}.$$

Since ϕ^j are simple predictable processes, by Proposition 8.1, $\int_0^t \phi^j dY^j$ are martingales, which entails that X is a martingale. By conditioning each term on $\mathcal{F}_{T_i}$ and applying the sampling theorem to Y^j, we obtain that all terms in the sum have expectation zero, hence $EX_t = 0$. Finally, since Y^j are independent, their variance adds up:

$$E|X_T|^2 = \text{Var}\left[\int_0^T \int_{\mathbb{R}^d} \phi(s,y) \tilde{M}(ds\ dy)\right] = \sum_{i,j} E\left[|\phi_{ij}|^2 (Y^j_{T_{i+1}\wedge t} - Y^j_{T_i\wedge t})^2\right]$$
$$= \sum_{i,j} E\left[E\left\{|\phi_{ij}|^2 (Y^j_{T_{i+1}} - Y^j_{T_i})^2 | \mathcal{F}_{T_i}\right\}\right] = \sum_{i,j} E\left[|\phi_{ij}|^2 E[(Y^j_{T_{i+1}} - Y^j_{T_i})^2 | \mathcal{F}_{T_i}]\right]$$
$$= \sum_{i,j} E\left[|\phi_{ij}|^2 \mu(]T_i, T_{i+1}] \times A_j)\right],$$

which yields (8.23). Since $E|X_t|^2 \leq E|X_T|^2 < \infty$, X is a square integrable martingale. □

The isometry formula (8.23) can be used to extend the compensated integral to square integrable predictable functions [48, Section 5.4]. Given a predictable random function verifying

$$E \int_0^T \int_{\mathbb{R}^d} |\phi(t,y)|^2 \mu(dt\ dy) < \infty,$$

there exists a sequence (ϕ^n) of simple predictable functions of the form (8.18) such that

$$E\left[\int_0^T \int_{\mathbb{R}^d} |\phi^n(t,y) - \phi(t,y)|^2 \mu(dt\ dy)\right] \stackrel{n\to\infty}{\to} 0.$$

Using the isometry relation (8.23), one can then show that $\int_0^T \int_{\mathbb{R}^d} \phi^n d\tilde{M}$ converges in $L^2(\mathbb{P})$ towards a limit denoted as $\int_0^T \int_{\mathbb{R}^d} \phi d\tilde{M}$. The martingale and isometry properties are conserved in this procedure [205].

PROPOSITION 8.8 Compensated Poisson integrals
For any predictable random function $\phi : \Omega \times [0,T] \times \mathbb{R}^d \to \mathbb{R}$ *verifying*

$$E\left[\int_0^T \int_{\mathbb{R}^d} |\phi(t,y)|^2 \mu(dt\ dy)\right] < \infty,$$

the following properties hold:

$$t \mapsto \int_0^t \int_{\mathbb{R}^d} \phi(s,y) \tilde{M}(ds\ dy) \quad \textit{is a square integrable martingale,}$$

$$E\left[\left|\int_0^t \int_{\mathbb{R}^d} \phi(s,y)\tilde{M}(ds\ dy)\right|^2\right] = E\left[\int_0^t \int_{\mathbb{R}^d} |\phi(s,y)|^2 \mu(ds\ dy)\right]. \tag{8.24}$$

An important case for our purpose is the case where the Poisson random measure M describes the jump times and jump sizes of a stochastic process $(S_t)_{t\geq 0}$, that is, M is the jump measure defined by Equation (2.97):

$$M = J_S(\omega, .) = \sum_{t\in[0,T]}^{\Delta S_t \neq 0} \delta_{(t,\Delta S_t)}.$$

An example is given by Lévy processes: if S is a Lévy process with Lévy measure ν then the jump measure J_S is a Poisson random measure with intensity $\mu(dt\ dx) = dt\nu(dx)$. Then for a predictable random function ϕ the integral in (8.19) is a sum of terms involving jumps times and jump sizes of S_t:

$$\int_0^T \int_{\mathbb{R}^d} \phi(s,y) M(ds\ dy) = \sum_{t\in[0,T]}^{\Delta S_t \neq 0} \phi(t, \Delta S_t). \tag{8.25}$$

REMARK 8.1 Integration with respect to the jump measure of a jump process In the special case where S is a pure jump process with jump times $T_1 < T_2 < \ldots$ and $\phi(\omega, s, y) = \psi_s(\omega) y$ where $\psi = \sum \psi_i 1_{]T_i, T_{i+1}]}$ is constant between two jumps, the integral with respect to J_S is just a stochastic integral with respect to S:

$$\int_0^T \int_{\mathbb{R}^d} \phi(s,y) M(ds\ dy) = \int_0^T \int_{\mathbb{R}^d} \psi_s y M(ds\ dy) = \sum_{t\in[0,T]}^{\Delta S_t \neq 0} \psi_t \Delta S_t = \int_0^T \psi_t dS_t$$

and can then be interpreted as the capital gain resulting from a strategy ψ with transactions taking place at the jump times (T_i) of S, the investor holding a portfolio ψ_i between the jump times T_i and T_{i+1}. But this is a special case: for a general integrand ϕ, the Poisson integral (8.19) with respect to the jump measure J_S *cannot* be expressed as a stochastic integral[5] with respect to S. This example shows that, when X is a jump process whose jump measure J_X is a Poisson random measure, integration with respect to J_X and integration with respect to X are two *different* concepts. This remark will play a role in Chapter 10 during the discussion of hedging strategies in markets with jumps. □

8.2 Quadratic variation

8.2.1 Realized volatility and quadratic variation

The concept of realized volatility was defined in Section 7.5 in the context of volatility measurement. Given a process observed on a time grid $\pi = \{t_0 = 0 < t_1 < \cdots < t_{n+1} = T\}$, the "realized variance" was defined as:

$$V_X(\pi) = \sum_{t_i \in \pi} (X_{t_{i+1}} - X_{t_i})^2. \tag{8.26}$$

Rewriting each term in the sum as

$$(X_{t_{i+1}} - X_{t_i})^2 = X_{t_{i+1}}^2 - X_{t_i}^2 - 2X_{t_i}(X_{t_{i+1}} - X_{t_i}),$$

the realized volatility can be rewritten as a Riemann sum:

$$X_T^2 - X_0^2 - 2\sum_{t_i \in \pi} X_{t_i}(X_{t_{i+1}} - X_{t_i}). \tag{8.27}$$

Consider now the case where X is a semimartingale with $X_0 = 0$. By definition it is a nonanticipating right-continuous process with left limits (cadlag) and one may define the process $X_- = (X_{t-})_{t\in[0,T]}$ which is caglad. Proposition (8.4) then shows that the Riemann sums in (8.27) uniformly converge in probability to the random variable:

$$[X, X]_t := |X_T|^2 - 2\int_0^T X_{u-} dX_u, \tag{8.28}$$

called the *quadratic variation* of X on $[0, T]$. Note that the quadratic variation is a random variable, not a number. Repeating the same procedure over $[0, t]$, one can define the quadratic variation process:

---

[5] It can, however, be expressed as a sum of integrals with respect to the martingales $\tilde{M}_t^j$ defined above.

DEFINITION 8.3 Quadratic variation *The* quadratic variation *process of a semimartingale X is the nonanticipating cadlag process defined by:*

$$[X, X]_t = |X_t|^2 - 2\int_0^t X_{u-} dX_u. \tag{8.29}$$

If $\pi^n = (t_0^n = 0 < t_1^n < \cdots < t_{n+1}^n = T)$ is a sequence of partitions of $[0, T]$ such that $|\pi^n| = \sup_k |t_k^n - t_{k-1}^n| \to 0$ as $n \to \infty$ then

$$\sum_{\substack{0 \le t_i < t \\ t_i \in \pi^n}} (X_{t_{i+1}} - X_{t_i})^2 \underset{n\to\infty}{\overset{\mathbb{P}}{\to}} [X, X]_t,$$

where the convergence is uniform in t. Since $[X, X]$ is defined as a limit of positive sums, $[X, X]_t \ge 0$, and for $t > s$, since $[X, X]_t - [X, X]_s$ is again a limit of positive sums, $[X, X]_t \ge [X, X]_s$, and we conclude that $[X, X]$ is an increasing process. This allows to define integrals $\int_0^t \phi\, d[X, X]$.

If X is continuous and has paths of finite variation then, denoting by $TV(X)$ the total variation of X on $[0, T]$, we obtain:

$$\begin{aligned}\sum_{t_i \in \pi^n} (X_{t_{i+1}} - X_{t_i})^2 &\le \sup_i |X_{t_{i+1}} - X_{t_i}| \sum_{t_i \in \pi} |X_{t_{i+1}} - X_{t_i}| \\ &\le \sup_i |X_{t_{i+1}} - X_{t_i}|\, TV(X) \underset{|\pi|\to 0}{\to} 0,\end{aligned}$$

therefore, $[X, X] = 0$. In particular, for a smooth (C^1) function, $[f, f] = 0$. This result is no longer true for processes with discontinuous sample paths since in this case $|X_{t_{i+1}} - X_{t_i}|$ will not go to zero when $|t_{i+1} - t_i| \to 0$. The following proposition summarizes important properties of quadratic variation.

PROPOSITION 8.9 Properties of quadratic variation

- $([X, X]_t)_{t\in[0,T]}$ *is an increasing process.*
- *The jumps of $[X, X]$ are related to the jumps of X by: $\Delta[X, X]_t = |\Delta X_t|^2$. In particular, $[X, X]$ has continuous sample paths if and only if X does.*
- *If X is continuous and has paths of finite variation then $[X, X] = 0$.*
- *If X is a martingale and $[X, X] = 0$ then $X = X_0$ almost surely.*

For a proof of the last point see, e.g., Propositions 4.13 and 4.50 in [215] or Chapter II in [324].

REMARK 8.2 Martingales vs. drifts The last point of the preceding proposition has an important consequence: if X is a continuous martingale with paths of finite variation then $[X,X] = 0$, so $X = X_0$ almost surely. Thus *a continuous square-integrable martingale with paths of finite variation is constant with probability 1.* This remark draws a distinction between two classes of processes: martingales (such as Brownian motion or the compensated Poisson process), which are typical examples of "noise processes" and continuous processes of finite variation, which may be interpreted as "drifts." There is no nontrivial process which belongs to both classes:

$$\text{Martingales} \bigcap \text{Continuous processes of finite variation} = \text{constants}.$$

This point will be very useful in the sequel: it allows to assert that if a process is decomposed into as the sum of a square-integrable martingale term and a continuous process with finite variation

$$X_t = M_t + \int_0^t a(t)dt,$$

then this decomposition is unique (up to a constant). In fact, this remark can be generalized to the case where M is a *local* martingale [215, Proposition 14.50]. □

Note that the quadratic variation, contrarily to the variance, is not defined by taking expectations: it is a "sample path" property. The quadratic variation is a well-defined quantity for all semimartingales, including those with infinite variance. We now turn to some fundamental examples of processes for which the quadratic variation can be explicitly computed.

PROPOSITION 8.10 Quadratic variation of Brownian motion
If $B_t = \sigma W_t$ where W is a standard Wiener process then $[B,B]_t = \sigma^2 t$.

PROOF Let $\pi^n = (t_0^n = 0 < t_1^n < \cdots < t_{n+1}^n = T)$ be a sequence of partitions of $[0,T]$ such that $|\pi^n| = \sup_k |t_k^n - t_{k-1}^n| \to 0$. First observe that $V_B(\pi^n) - \sigma^2 T = \sum_{\pi^n} (B_{t_{i+1}} - B_{t_i})^2 - \sigma^2(t_{i+1} - t_i)$ is a sum of independent terms with mean zero. Therefore

$$\begin{aligned}
E\left|V_B(\pi^n) - \sigma^2 T\right|^2 &= \sum_{\pi^n} E\left[(B_{t_{i+1}} - B_{t_i})^2 - \sigma^2(t_{i+1} - t_i)\right]^2 \\
&= \sum_{\pi^n} \sigma^4 |t_{i+1} - t_i|^2 E\left[\left(\frac{(B_{t_{i+1}} - B_{t_i})^2}{\sigma^2(t_{i+1} - t_i)} - 1\right)^2\right] \\
&= \sigma^4 \sum_{\pi^n} |t_{i+1} - t_i|^2 E\left[(Z^2 - 1)^2\right] \text{ where } Z \sim N(0,1) \\
&\leq E\left[(Z^2 - 1)^2\right] \sigma^4 T |\pi^n| \to 0.
\end{aligned}$$

Thus $E|V_B(\pi^n) - \sigma^2 T|^2 \to 0$ which implies convergence in probability of $V_B(\pi^n)$ to $\sigma^2 T$. □

Therefore, when X is a Brownian motion, $[X,X]_t$ is indeed equal to the variance of X_t, which explains the origin of the term "realized volatility." However, the Brownian case is very special. In general $[X,X]_t$ is a random process, as is shown by the following examples:

Example 8.4 Quadratic variation of a Poisson process
If N is a Poisson process then it is easy to see by definition that $[N,N]_t = N_t$. More generally, if N_t is a counting process with jump times T_i and Z_i are random variables revealed at T_i, denoting the jump sizes:

$$X_t = \sum_{i=1}^{N_t} Z_i \Rightarrow [X,X]_t = \sum_{i=1}^{N_t} |Z_i|^2 = \sum_{0\leq s\leq t} |\Delta X_s|^2.$$

Even more generally, it is easy to show that the same formula holds for every finite variation process X:

$$[X,X]_t = \sum_{0\leq s\leq t} |\Delta X_s|^2. \tag{8.30}$$

□

Example 8.5 Quadratic variation of a Lévy process
If X is a Lévy process with characteristic triplet (σ^2,ν,γ), its quadratic variation process is given by

$$[X,X]_t = \sigma^2 t + \sum_{\substack{s\in[0,t]\\ \Delta X_s\neq 0}} |\Delta X_s|^2 = \sigma^2 t + \int_{[0,t]}\int_{\mathbb{R}} y^2 J_X(ds\, dy). \tag{8.31}$$

In particular, if X is a symmetric α-stable Lévy process, which has infinite variance, the quadratic variation is a well-defined process, even though the variance is not defined. The quadratic variation of a Lévy process again a Lévy process: it is a subordinator (see Proposition 3.11). □

Example 8.6 Quadratic variation of Brownian stochastic integrals
Consider the process defined by $X_t = \int_0^t \sigma_t dW_t$, where $(\sigma_t)_{t\in[0,T]}$ is a caglad process. Then

$$[X,X]_t = \int_0^t \sigma_s^2 ds. \tag{8.32}$$

In particular if a Brownian stochastic integral has zero quadratic variation, it is almost surely equal to zero: $[X,X] = 0$ a.s.$\Rightarrow$ $X = 0$ a.s. This implication is a special case of the last point in Proposition 8.9. □

Example 8.7 Quadratic variation of a Poisson integral
Consider a Poisson random measure M on $[0,T]\times\mathbb{R}^d$ with intensity $\mu(ds\,dy)$ and a simple predictable random function $\psi : [0,T]\times\mathbb{R}^d \mapsto \mathbb{R}$. Define the process X as the integral of ψ with respect to M as in (8.20):

$$X_t = \int_0^t \int_{\mathbb{R}^d} \psi(s,y) M(ds\,dy).$$

The quadratic variation of X is then given by:

$$[X,X]_t = \int_0^t \int_{\mathbb{R}^d} |\psi(s,y)|^2 M(ds\,dy). \tag{8.33}$$

□

8.2.2 Quadratic covariation

The concept of "realized volatility" has a multidimensional counterpart which one can define as follows: given two processes X, Y and a time grid $\pi = \{t_0 = 0 < t_1 < \cdots < t_{n+1} = T\}$ we can define the "realized covariance" by:

$$\sum_{t_i\in\pi} (X_{t_{i+1}} - X_{t_i})(Y_{t_{i+1}} - Y_{t_i}). \tag{8.34}$$

Rewriting each term above as

$$X_{t_{i+1}}Y_{t_{i+1}} - X_{t_i}Y_{t_i} - Y_{t_i}(X_{t_{i+1}} - X_{t_i}) - X_{t_i}(Y_{t_{i+1}} - Y_{t_i}),$$

the sum in (8.34) can be rewritten as a Riemann sum:

$$X_TY_T - X_0Y_0 - \sum_{t_i\in\pi} \{Y_{t_i}(X_{t_{i+1}} - X_{t_i}) + X_{t_i}(Y_{t_{i+1}} - Y_{t_i})\}.$$

When X, Y are semimartingales, by Proposition 8.4, the expression above converges in probability to the random variable:

$$X_TY_T - X_0Y_0 - \int_0^T X_{t-}dY_t - \int_0^T Y_{t-}dX_t,$$

called the *quadratic covariation* of X and Y on $[0,T]$.

DEFINITION 8.4 Quadratic covariation *Given two semimartingales X, Y, the quadratic covariation process $[X,Y]$ is the semimartingale defined by*

$$[X,Y]_t = X_tY_t - X_0Y_0 - \int_0^t X_{s-}dY_s - \int_0^t Y_{s-}dX_s. \tag{8.35}$$

The quadratic covariation has the following important properties [324]:

- $[X, Y]$ is a nonanticipating cadlag process with paths of finite variation.
- Polarization identity:

$$[X, Y] = \frac{1}{4}([X + Y, X + Y] - [X - Y, X - Y]). \tag{8.36}$$

- The discrete approximations (8.34) converge in probability to $[X, Y]$ uniformly on $[0, T]$:

$$\sum_{t_i \in \pi}^{t_i < t} (X_{t_{i+1}} - X_{t_i})(Y_{t_{i+1}} - Y_{t_i}) \underset{|\pi| \to 0}{\overset{\mathbb{P}}{\to}} [X, Y]_t. \tag{8.37}$$

- The covariation $[X, Y]$ is not modified if we add to X or Y continuous processes with finite variation ("random drift terms"): it is only sensitive to the martingale parts ("noise terms") or jumps in X and Y. It is thus more relevant in a financial modelling context than measures of correlation between increments.
- If X, Y are semimartingales and ϕ, ψ integrable predictable processes then

$$\left[\int \phi dX, \int \psi dY\right]_t = \int_0^t \phi\psi \, d[X, Y]. \tag{8.38}$$

The following result is simply a restatement of the definition of $[X, Y]$.

PROPOSITION 8.11 Product differentiation rule
If X, Y are semimartingales then

$$X_t Y_t = X_0 Y_0 + \int_0^t X_{s-} dY_s + \int_0^t Y_{s-} dX_s + [X, Y]_t. \tag{8.39}$$

In the following example the quadratic covariation may be computed using the polarization identity (8.36).

Example 8.8 Quadratic covariation of correlated Brownian motions
If $B_t^1 = \sigma^1 W_t^1$ and $B_t^2 = \sigma^2 W_t^2$ where W^1, W^2 are standard Wiener processes with correlation ρ then $[B^1, B^2]_t = \rho\sigma_1\sigma_2 t$. □

The next example uses Property (8.38) of quadratic covariation.

Example 8.9 Brownian stochastic integrals
Let $X_t = \int_0^t \sigma_s^1 dW_s^1$ and $Y_t = \int_0^t \sigma_s^2 dW_s^2$ where $(\sigma_t^i)_{t \in [0,T]}$ are predictable processes and W^1, W^2 are correlated Wiener processes with $\mathrm{Cov}(W_t^1, W_t^2) =$

ρt. Then

$$[X, Y]_t = \int_0^t \sigma_s^1 \sigma_s^2 \rho ds. \tag{8.40}$$

□

The above examples show that, for Brownian stochastic integrals, quadratic covariation is a kind of "integrated instantaneous covariance": however, it is a more general concept. Here is another case where quadratic covariation can be computed:

Example 8.10
Let M be a Poisson random measure on $[0, T] \times \mathbb{R}$ and $(W_t)_{t \in [0,T]}$ a Wiener process, independent from M. If

$$X_t^i = X_0^i + \int_0^t \phi_s^i dW_s + \int_0^t \int_{\mathbb{R}^d} \psi^i(s, y) \tilde{M}(ds\ dy) \quad i = 1, 2,$$

then the quadratic covariation $[X^1, X^2]$ is equal to

$$[X^1, X^2]_t = \int_0^t \phi_s^1 \phi_s^2 ds + \int_0^t \int_{\mathbb{R}^d} \psi^1(s, y) \psi^2(s, y) M(ds\ dy). \tag{8.41}$$

□

8.3 The Itô formula

If $f : \mathbb{R} \to \mathbb{R}$, $g : [0, T] \to \mathbb{R}$ are smooth (say, C^1) functions then from the change of variables formula for smooth functions we know that

$$f(g(t)) - f(g(0)) = \int_0^t f'(g(s)) g'(s) ds = \int_0^t f'(g(s)) dg(s). \tag{8.42}$$

Applying this to $f(x) = x^2$ we get:

$$g(t)^2 - g(0)^2 = 2 \int_0^t g(s) dg(s).$$

However, when X is a semimartingale we observed in Section 8.2 that

$$X_t^2 - X_0^2 = 2 \int_0^t X_{s-} dX_s + [X, X]_t,$$

where, in general, $[X, X] \neq 0$. Therefore, stochastic integrals with respect to semimartingales do not seem to obey the usual change of variable formulae for smooth functions. The goal of this section is to give formulae analogous to (8.42) for $f(t, X_t)$ when f is a smooth function and X a semimartingale with jumps.

The reader is probably familiar with the Itô formula for Brownian integrals, which states that if f is a C^2 function and $X_t = \int_0^t \sigma_s dW_s$ then

$$f(X_t) = f(0) + \int_0^t f'(X_s)\sigma_s dW_s + \int_0^t \frac{1}{2}\sigma_s^2 f''(X_s)ds. \tag{8.43}$$

We will discuss such representations when the Wiener process is replaced by a process with jumps.

8.3.1 Pathwise calculus for finite activity jump processes

Let us begin with some simple remarks which have nothing to do with probability or stochastic processes. Consider a function $x : [0, T] \to \mathbb{R}$ which has (a finite number of) discontinuities at $T_1 \leq T_2 \leq \cdots \leq T_n \leq T_{n+1} = T$, but is smooth on each interval $]T_i, T_{i+1}[$. We can choose x to be cadlag at the discontinuity points by defining $x(T_i) := x(T_i+)$. Such a function may be represented as:

$$x(t) = \int_0^t b(s)ds + \sum_{\{i, T_i \leq t\}} \Delta x_i \quad \text{where } \Delta x_i = x(T_i) - x(T_i-), \tag{8.44}$$

where the sum takes into account the discontinuities occurring between 0 and t. For instance, if b is continuous then x is piecewise C^1. Consider now a C^1 function $f : \mathbb{R} \to \mathbb{R}$. Since on each interval $]T_i, T_{i+1}[$, x is smooth, $f(x(t))$ is also smooth. Therefore we can apply the change of variable formula for smooth functions and write for $i = 0 \ldots n$ with the convention $T_0 = 0$:

$$f(x(T_{i+1}-)) - f(x(T_i)) = \int_{T_i}^{T_{i+1}-} f'(x(t))x'(t)dt = \int_{T_i}^{T_{i+1}-} f'(x(t))b(t)dt.$$

At each discontinuity point, $f(x(t))$ has a jump equal to

$$f(x(T_i)) - f(x(T_i-)) = f(x(T_i-) + \Delta x_i) - f(x(T_i-)).$$

Adding these two contributions together, the overall variation of f between 0 and t can be written as:

$$\begin{aligned} f(x(T)) - f(x(0)) &= \sum_{i=0}^{n}\{f(x(T_{i+1})) - f(x(T_i))\} \\ &= \sum_{i=0}^{n}\{f(x(T_{i+1})) - f(x(T_{i+1}-)) + f(x(T_{i+1}-)) - f(x(T_i))\} \\ &= \sum_{i=1}^{n+1}\{f(x(T_i-) + \Delta X_i) - f(x(T_i-))\} + \sum_{i=0}^{n}\int_{T_i}^{T_{i+1}-} b(t)f'(x(t))dt. \end{aligned}$$

The integrals in the last sum can be grouped together provided that we replace $f(x(t))$ by $f(x(t-))$, so we finally obtain:

PROPOSITION 8.12 Change of variable formula for piecewise smooth functions
If x is a piecewise C^1 function given by:

$$x(t) = \int_0^t b(s)ds + \sum_{\{i=1\dots n+1, T_i \le t\}} \Delta x_i \quad \text{where } \Delta x_i = x(T_i) - x(T_i-),$$

then for every C^1 function $f : \mathbb{R} \to \mathbb{R}$:

$$\begin{aligned} f(x(T)) - f(x(0)) &= \int_0^T b(t)f'(x(t-))dt \\ &+ \sum_{i=1}^{n+1} f(x(T_i-) + \Delta x_i) - f(x(T_i-)). \end{aligned}$$

Note also that, if $b = 0$ (i.e., x is piecewise constant) then the integral term is equal to zero and the formula becomes valid for a continuous (or even a measurable) function with no further smoothness requirement.

Of course, this formula has nothing to do with stochastic processes. But consider a stochastic process $(X_t)_{t\in[0,T]}$ whose sample paths $t \mapsto X_t(\omega)$ are (almost surely) of the form (8.44):

$$X_t(\omega) = X_0 + \int_0^t b_s(\omega)ds + \sum_{i=1}^{N_T(\omega)} \Delta X_i(\omega), \tag{8.45}$$

where $\Delta X_i = X(T_i) - X(T_i-)$ are the jump sizes and $N_t(\omega)$ is the (random) number of jumps that can be represented as the value at t of a counting process. We have included ω to emphasize that we are looking at a given sample path. Then by Proposition 8.12, the following change of variable formula holds almost surely:

$$f(X_T) - f(X_0) = \int_0^T b(t) f'(X_t) dt + \sum_{\{i, T_i \leq t\}} f(X_{T_i -} + \Delta x_i) - f(X_{T_i -})$$
$$= \int_0^T b(t-) f'(X_{t-}) dt + \sum_{\substack{\Delta X_t \neq 0 \\ 0 \leq t \leq T}} f(X_{t-} + \Delta X_t) - f(X_{t-}), \quad (8.46)$$

where the sum now runs over the (random) jump times (T_i) of X. This change of variable formula is valid independently of the probabilistic structure of the process X: the jump times and jump sizes may have dependence between them and with respect to the past, possess or not possess finite moments, etc. It is a *pathwise* formula. However, if we are interested in computing expectations, then we should introduce some structure in the process. In the special case where $T_{i+1} - T_i$ are i.i.d. exponential random variables (which means that the counting process N_t is a Poisson process) and ΔX_i are i.i.d. random variables with distribution F, the jumps are described by a compound Poisson process. In this case, of course, the above formula is still valid but now we can decompose $f(X_t)$ into a "martingale" part and a "drift" part as follows. First, we introduce the random measure on $[0, T] \times \mathbb{R}$ which describes the locations and sizes of jumps of X:

$$J_X = \sum_{n \geq 1} \delta_{(T_n, \Delta X_{T_n})}.$$

J_X is a Poisson random measure with intensity $\mu(dt\, dy) = \lambda dt F(dy)$. The jump term in (8.46) may now be rewritten as:

$$\int_0^t \int_{\mathbb{R}} [f(X_{s-} + y) - f(X_{s-})] \; J_X(ds\; dy).$$

Using the compensated jump measure $\tilde{J}_X(dt\; dy) = J_X(dt\, dy) - \lambda dt F(dy)$, the jump term above can be rewritten as:

$$\int_0^t \int_{\mathbb{R}} [f(X_{s-} + y) - f(X_{s-})] \; \tilde{J}_X(ds\; dy)$$
$$+ \int_0^t \lambda ds \int_{\mathbb{R}} F(dy) [f(X_{s-} + y) - f(X_{s-})].$$

The first term above can be interpreted as the martingale or "noise" component and the second one, which is an ordinary Lebesgue integral, represents the "signal" or drift part. These results are summarized in the following proposition:

PROPOSITION 8.13 Itô formula for finite activity jump processes

Let X be a jump process with values in $\mathbb{R}$ defined by:

$$X_t = \int_0^t b_s ds + \sum_{i=1}^{N_t} Z_i,$$

where b_s is a nonanticipating cadlag process, (N_t) is a counting process representing the number of jumps between 0 and t and Z_i is the size of the i-th jump. Denote by $(T_n)_{n\geq 1}$ the jump times of X_t and J_X the random measure on $[0,T]\times\mathbb{R}$ associated to the jumps of X:

$$J_X = \sum_{\{n\geq 1, T_n\leq T\}} \delta_{(T_n,Z_n)}.$$

Then for any measurable function $f:[0,T]\times\mathbb{R}\to\mathbb{R}$:

$$\begin{aligned} f(t,X_t)-f(0,X_0) &= \int_0^t [\frac{\partial f}{\partial s}(s,X_{s-}) + b_s\frac{\partial f}{\partial x}(s,X_{s-})]ds \\ &+ \sum_{\{n\geq 1, T_n\leq t\}} [f(s,X_{s-}+\Delta X_s) - f(s,X_{s-})] \\ &= \int_0^t [\frac{\partial f}{\partial s}(s,X_{s-}) + b\frac{\partial f}{\partial x}(s,X_{s-})]ds \\ &+ \int_0^t\int_{-\infty}^{\infty} [f(s,X_{s-}+y) - f(s,X_{s-})] J_X(ds\ dy). \end{aligned}$$

Furthermore, if N_t is a Poisson process with $EN_t=\lambda t$, $Z_i\sim F$ are i.i.d. and f is bounded then $Y_t = f(t,Y_t) = V_t + M_t$ where M is the martingale part:

$$M_t = \int_0^t\int_{-\infty}^{\infty} [f(s,X_{s-}+y) - f(s,X_{s-})]\tilde{J_X}(ds\ dy), \tag{8.47}$$

where $\tilde{J_X}$ denotes the compensated Poisson random measure:

$$\tilde{J_X}(dt\ dy) = J_X(dt\ dy) - \lambda F(dy)dt,$$

and V is a continuous finite variation drift:

$$\begin{aligned} V_t &= \int_0^t [\frac{\partial f}{\partial s}(s,X_{s-}) + b_s\frac{\partial f}{\partial x}(s,X_{s-})]ds \\ &+ \int_0^t ds \int_{\mathbb{R}^d} F(dy)[f(s,X_{s-}+y) - f(s,X_{s-})]. \end{aligned} \tag{8.48}$$

8.3.2 Itô formula for diffusions with jumps

Consider now a jump-diffusion process

$$X_t = \sigma W_t + \mu t + J_t = X^c(t) + J_t, \tag{8.49}$$

where J is a compound Poisson process and X^c is the continuous part of X:

$$J_t = \sum_{i=1}^{N_t} \Delta X_i, \qquad X_t^c = \mu t + \sigma W_t.$$

Define $Y_t = f(X_t)$ where $f \in C^2(\mathbb{R})$ and denote by $T_i, i = 1 \ldots N_T$ the jump times of X. On $]T_i, T_{i+1}[$, X evolves according to

$$dX_t = dX_t^c = \sigma dW_t + \mu dt,$$

so by applying the Itô formula in the Brownian case we obtain

$$\begin{aligned} Y_{T_{i+1}-} - Y_{T_i} &= \int_{T_i}^{T_{i+1}-} \frac{\sigma^2}{2} f''(X_t)dt + \int_{T_i}^{T_{i+1}-} f'(X_t)dX_t \\ &= \int_{T_i}^{T_{i+1}-} \left\{ \frac{\sigma^2}{2} f''(X_t)dt + f'(X_t)dX_t^c \right\} \end{aligned}$$

since $dX_t = dX^c(t)$ on this interval. If a jump of size ΔX_t occurs then the resulting change in Y_t is given by $f(X_{t-} + \Delta X_t) - f(X_{t-})$. The total change in Y_t can therefore be written as the sum of these two contributions:

$$\begin{aligned} f(X_t) - f(X_0) &= \int_0^t f'(X_s)dX_s^c + \int_0^t \frac{\sigma^2}{2} f''(X_s)ds \\ &+ \sum_{0 \le s \le t,\ \Delta X_s \neq 0} [f(X_{s-} + \Delta X_s) - f(X_{s-})]. \end{aligned} \tag{8.50}$$

REMARK 8.3 Replacing dX_s^c by $dX_s - \Delta X_s$ one obtains an equivalent expression:

$$\begin{aligned} f(X_t) - f(X_0) &= \int_0^t f'(X_{s-})dX_s + \int_0^t \frac{\sigma^2}{2} f''(X_s)ds \\ &+ \sum_{0 \le s \le t,\ \Delta X_s \neq 0} [f(X_{s-} + \Delta X_s) - f(X_{s-}) - \Delta X_s f'(X_{s-})]. \end{aligned} \tag{8.51}$$

When the number of jumps is finite, this form is equivalent to (8.50). However, as we will see below, the form (8.51) is more general: both the stochastic integral and the sum over the jumps in (8.51) are well-defined for any semimartingale, even if the number of jumps is infinite, while the sum in Equation (8.50) may not converge if jumps have infinite variation. □

Here we have only used the Itô formula for diffusions, which is of course still valid if σ is replaced by a nonanticipating square-integrable process $(\sigma_t)_{t\in[0,T]}$.

PROPOSITION 8.14 Itô formula for jump-diffusion processes
Let X be a diffusion process with jumps, defined as the sum of a drift term, a Brownian stochastic integral and a compound Poisson process:

$$X_t = X_0 + \int_0^t b_s ds + \int_0^t \sigma_s dW_s + \sum_{i=1}^{N_t} \Delta X_i,$$

where b_t and σ_t are continuous nonanticipating processes with

$$E[\int_0^T \sigma_t^2 dt] < \infty.$$

Then, for any $C^{1,2}$ function. $f : [0,T] \times \mathbb{R} \to \mathbb{R}$, the process $Y_t = f(t, X_t)$ can be represented as:

$$\begin{aligned} f(t, X_t) - f(0, X_0) = & \int_0^t \left[\frac{\partial f}{\partial s}(s, X_s) + \frac{\partial f}{\partial x}(s, X_s) b_s\right] ds \\ & + \frac{1}{2}\int_0^t \sigma_s^2 \frac{\partial^2 f}{\partial x^2}(s, X_s) ds + \int_0^t \frac{\partial f}{\partial x}(s, X_s)\sigma_s dW_s \\ & + \sum_{\{i\geq 1, T_i \leq t\}} [f(X_{T_i-} + \Delta X_i) - f(X_{T_i-})]. \end{aligned}$$

In differential notation:

$$\begin{aligned} dY_t = & \frac{\partial f}{\partial t}(t, X_t)dt + b_t \frac{\partial f}{\partial x}(t, X_t)dt + \frac{\sigma_t^2}{2}\frac{\partial^2 f}{\partial x^2}(t, X_t)dt \\ & + \frac{\partial f}{\partial x}(t, X_t)\sigma_t dW_t + [f(X_{t-} + \Delta X_t) - f(X_{t-})]. \end{aligned}$$

8.3.3 Itô formula for Lévy processes

Let us now turn to the case of a general Lévy process X. The difficulty which prevents us from applying the above results directly is that, in the infinite activity case, an infinite number of jumps may occur in each interval. In this case, the number of jump terms in the sum

$$\sum_{0\leq s\leq t,\ \Delta X_s \neq 0} [f(X_{s-} + \Delta X_s) - f(X_{s-}) - \Delta X_s f'(X_{s-})]$$

becomes infinite and one should investigate the conditions, under which the series converges. Furthermore, the jump times form a dense subset of $[0, T]$

and one cannot separate the evolution due to jumps from the one that is due to the Brownian component so the reasoning used in Section 8.3.2 does not apply anymore. However, albeit with more sophisticated methods, the same result as before can be shown for general Lévy processes and even, as we will see later, for general semimartingales.

PROPOSITION 8.15 Itô formula for scalar Lévy processes
Let $(X_t)_{t\geq 0}$ be a Lévy process with Lévy triplet (σ^2, ν, γ) and $f : \mathbb{R} \to \mathbb{R}$ a C^2 function. Then

$$f(X_t) = f(0) < + \int_0^t \frac{\sigma^2}{2} f''(X_s)ds + \int_0^t f'(X_{s-})dX_s + \sum_{\substack{0\leq s\leq t\\ \Delta X_s\neq 0}} [f(X_{s-} + \Delta X_s) - f(X_{s-}) - \Delta X_s f'(X_{s-})]. \quad (8.52)$$

PROOF The proof of the Itô formula for a general semimartingale X (see Proposition 8.19) consists in rewriting the difference $f(X_t) - f(X_0)$ as a sum of small increments of f:

$$f(X_t) - f(X_0) = \sum_{i=1}^{n} (f(X_{t_{i+1}}) - f(X_{t_i})),$$

then expanding each small increment using a Taylor formula and proving the convergence of resulting Riemann sums to (8.52). For Lévy processes however, a simpler approach can be used. Suppose that f and its two derivatives are bounded by a constant C. Then

$$\left| f(X_{s-} + \Delta X_s) - f(X_{s-}) - \Delta X_s f'(X_{s-}) \right| \leq C \Delta X_s^2.$$

This means that the sum in (8.52) is finite (see Proposition 3.11). Now recall that every Lévy process may be represented as $X_t = X_t^\varepsilon + R_t^\varepsilon$ such that X_t^ε is a jump-diffusion process and R_t^ε is a mean-zero square integrable martingale and $\operatorname{Var} R_t^\varepsilon \to 0$ as $\varepsilon \to 0$ (see the proof of Proposition 3.7). Since the first derivative of f is bounded,

$$\left| f(X_t) - f(X_t^\varepsilon) \right|^2 \leq C^2 (R_t^\varepsilon)^2.$$

This estimate implies that $f(X_t) = \lim_{\varepsilon\to 0} f(X_t^\varepsilon)$ in $L^2(\mathbb{P})$. But X_t^ε is a jump-diffusion process, so we can use the formula (8.51) and taking the limits in this formula we obtain exactly the expression (8.52). □

When computing expectations, it is useful to decompose the semimartingale $Y_t = f(X_t)$ into a martingale part and a drift term. This can be done by plugging the Lévy-Itô decomposition for X into the stochastic integral part of (8.52) and rearranging the terms.

PROPOSITION 8.16 Martingale-drift decomposition of functions of a Lévy process

Let $(X_t)_{t\geq 0}$ be a Lévy process with Lévy triplet (σ^2, ν, γ) and $f : \mathbb{R} \to \mathbb{R}$ a C^2 function such that f and its two derivatives are bounded by a constant C. Then $Y_t = f(X_t) = M_t + V_t$ where M is the martingale part given by

$$M_t = f(X_0) + \int_0^t f'(X_s)\sigma dW_s + \int_{[0,t]\times\mathbb{R}^d} \tilde{J}_X(ds\ dy)[f(X_{s-} + y) - f(X_{s-})],$$

and V_t a continuous finite variation process:

$$V_t = \int_0^t \frac{\sigma^2}{2} f''(X_s)ds + \int_0^t \gamma\ f'(X_s)ds$$
$$+ \int_{[0,t]\times\mathbb{R}} ds\ \nu(dy)[f(X_{s-} + y) - f(X_{s-}) - yf'(X_{s-})1_{|y|\leq 1}]. \quad (8.53)$$

In differential form, Formula (8.52) can be expressed as follows:

$$df(X_t) = \frac{\sigma^2}{2} f''(X_t)dt + f'(X_{t-})dX_t + f(X_t) - f(X_{t-}) - \Delta X_t f'(X_{t-}).$$

In applications it is often useful to generalize formula (8.52) to the case where f explicitly depends on time. Suppose once again that $(X_t)_{t\in[0,T]}$ is a Lévy process with Lévy triplet (σ^2, ν, γ) and that $f \in C^{1,2}([0,T] \times \mathbb{R}, \mathbb{R})$. Then

$$f(t,X_t) - f(0, X_0) = \int_0^t \frac{\partial f}{\partial x}(s, X_{s-})dX_s$$
$$+ \int_0^t [\frac{\partial f}{\partial s}(s, X_s) + \frac{\sigma^2}{2}\frac{\partial^2 f}{\partial x^2}(s, X_s)]ds$$
$$+ \sum_{\substack{0\leq s\leq t \\ \Delta X_s \neq 0}} [f(s, X_{s-} + \Delta X_s) - f(s, X_{s-}) - \Delta X_s \frac{\partial f}{\partial x}(s, X_{s-})]. \quad (8.54)$$

The reader can easily derive the generalization of martingale-drift representation in this case.

If the Lévy process is of finite variation, there is no need to subtract $\Delta X_s f'(X_{s-})$ from each term of the sum in (8.52). In this case the Itô formula can be simplified and is valid under weaker regularity assumptions on f.

PROPOSITION 8.17 Itô formula for Lévy processes: finite variation jumps

Let X be a finite variation Lévy process with characteristic exponent

$$\psi_X(u) = ibu + \int_{\infty}^{\infty} (e^{iuy} - 1)\nu(dy), \tag{8.55}$$

where the Lévy measure ν verifies $\int |y|\nu(dy) < \infty$. Then for any $C^{1,1}$ function $f : [0,T] \times \mathbb{R} \to \mathbb{R}$,

$$f(t, X_t) - f(0, X_0) = \int_0^t [\frac{\partial f}{\partial s}(s, X_{s-}) + b\frac{\partial f}{\partial x}(s, X_{s-})]ds + \sum_{\substack{0 \leq s \leq t \\ \Delta X_s \neq 0}} [f(X_{s-} + \Delta X_s) - f(X_{s-})].$$

If f and its first derivative in x are bounded, then $Y_t = f(t, Y_t)$ is the sum of a martingale part given by

$$\int_0^t \int_{-\infty}^{\infty} [f(s, X_{s-} + y) - f(s, X_{s-})] \tilde{J}_X(ds\ dy) \tag{8.56}$$

and a "drift" part given by:

$$\int_0^t \left[\frac{\partial f}{\partial s}(s, X_{s-}) + b\frac{\partial f}{\partial x}(s, X_{s-})\right] ds + \int_0^t ds \int_{\mathbb{R}} \nu(dy)[f(s, X_{s-} + y) - f(s, X_{s-})]. \tag{8.57}$$

The next step in generalizing the Itô formula is to allow f to depend on a multidimensional Lévy process.

PROPOSITION 8.18 Itô formula for multidimensional Lévy processes
Let $X_t = (X_t^1, \ldots, X_t^d)$ be a multidimensional Lévy process with characteristic triplet (A, ν, γ). Then for any $C^{1,2}$ function $f : [0,T] \times \mathbb{R}^d \to \mathbb{R}$,

$$f(t, X_t) - f(0,0) = \int_0^t \sum_{i=1}^d \frac{\partial f}{\partial x_i}(s, X_{s-}) dX_s^i + \int_0^t \frac{\partial f}{\partial s}(s, X_s) ds$$
$$+ \frac{1}{2} \int_0^t \sum_{i,j=1}^d A_{ij} \frac{\partial^2 f}{\partial x_i \partial x_j}(s, X_s) ds$$
$$+ \sum_{0 \leq s \leq t}^{\Delta X_s \neq 0} \left[f(s, X_{s-} + \Delta X_s) - f(s, X_{s-}) - \sum_{i=1}^d \Delta X_s^i \frac{\partial f}{\partial x_i}(s, X_{s-}) \right].$$

8.3.4 Itô formula for semimartingales

If X is a Lévy process then $Y_t = f(t, X_t)$ is not a Lévy process anymore; however, it can be expressed in terms of stochastic integrals so it is still a semimartingale. Therefore, if (Y_t) is a random process driven by the Lévy process (X_t), in order to consider quantities like $f(t, Y_t)$, we need to have a change of variable formula for discontinuous semimartingales such as (Y_t).

On the other hand, as already mentioned, the formulae presented above are *pathwise* in nature: they do not exploit the probabilistic properties of the processes involved but only the local structure of their sample paths. In particular, the formulae for functions of Lévy processes have nothing to do with the independence and stationarity of their increments! This implies that these results remain valid for functions of more general semimartingales, with time inhomogeneities and complex dependence structures in their increments.

Let X be a semimartingale with quadratic variation process $[X, X]$. Since the quadratic variation is an increasing process, it can be decomposed into jump part and continuous part. The continuous part will be denoted by $[X, X]^c$.

PROPOSITION 8.19 Itô formula for semimartingales
Let $(X_t)_{t\geq 0}$ be a semimartingale. For any $C^{1,2}$ function $f : [0,T]\times\mathbb{R}\to\mathbb{R}$,

$$\begin{aligned} f(t,X_t)-f(0,X_0) &= \int_0^t \frac{\partial f}{\partial s}(s,X_s)ds + \int_0^t \frac{\partial f}{\partial x}(s,X_{s-})dX_s \\ &+ \frac{1}{2}\int_0^t \frac{\partial^2 f}{\partial x^2}(s,X_{s-})d[X,X]^c_s \\ &+ \sum_{\substack{0\leq s\leq t\\ \Delta X_s\neq 0}} \left[f(s,X_s)-f(s,X_{s-})-\Delta X_s\frac{\partial f}{\partial x}(s,X_{s-})\right]. \end{aligned}$$

PROOF For simplicity, we will give a proof in the case where f has no time dependence. Our proof follows the one given in [323]. Consider a random partition $T_0 = 0 < T_1 < \cdots < T_{n+1} = t$. The idea is to write $f(X_t)$ as the sum of its increments, then use a second order Taylor expansion:

$$f(y) = f(x) + f'(x)(y-x) + f''(x)(y-x)^2/2 + r(x,y)$$

and let $\sup|T_i - T_{i-1}|$ go to zero almost surely.

$$\begin{aligned} f(X_t)-f(X_0) &= \sum_{i=0}^{n}\{f(X_{T_{i+1}})-f(X_{T_i})\} = \sum_{i=0}^{n} f'(X_{T_i})(X_{T_{i+1}}-X_{T_i}) \\ &+ \frac{1}{2}\sum_{i=0}^{n} f''(X_{T_i})(X_{T_{i+1}}-X_{T_i})^2 + \sum_{i=0}^{n} r(X_{T_{i+1}},X_{T_i}). \end{aligned}$$

The difficulty is to control the influence of jumps in each term. The key observation is that X has a well defined quadratic variation so $\sum\Delta X_s^2$ converges almost surely. For $\epsilon>0$, let $A\subset[0,T]\times\Omega$ such that $\sum_{0\leq s\leq T}\Delta X_s^2 < \epsilon$ on A and $B = \{(s,\omega)\notin A, \Delta X_s \neq 0\}$. The sum above can be rewritten as:

$$\begin{aligned} f(X_t)-f(X_0) &= \sum_{i=0}^{n} f'(X_{T_i})(X_{T_{i+1}}-X_{T_i}) + \frac{1}{2}\sum_{i=0}^{n} f''(X_{T_i})(X_{T_{i+1}}-X_{T_i})^2 \\ &+ \sum_{B\cap]T_i,T_{i+1}]\neq\emptyset} \Big\{ f(X_{T_{i+1}})-f(X_{T_i})-f'(X_{T_i})(X_{T_{i+1}}-X_{T_i}) \\ &- \frac{1}{2}f''(X_{T_i})(X_{T_{i+1}}-X_{T_i})^2\Big\} + \sum_{B\cap]T_i,T_{i+1}]=\emptyset} r(X_{T_i},X_{T_{i+1}}). \end{aligned}$$

When $\sup|T_{i+1}-T_i|\to 0$ a.s. the first two terms are Riemann sums which, by Proposition 8.4, converge to the following expressions:

$$\sum_{i=0}^{n} f'(X_{T_i})(X_{T_{i+1}} - X_{T_i}) \to \int_0^t f'(X_s)dX_s,$$
$$\frac{1}{2}\sum_{i=0}^{n} f''(X_{T_i})(X_{T_{i+1}} - X_{T_i})^2 \to \frac{1}{2}\int_0^t f''(X_s)d[X,X]_s.$$

The third term converges to:

$$\sum_{B} \{f(X_s) - f(X_{s-}) - \Delta X_s f'(X_{s-}) - f''(X_{s-})|\Delta X_s|^2\}. \qquad (8.58)$$

The remainder term verifies: $r(x,y) \leq (y-x)^2\alpha(|x-y|)$ with $\alpha(u) \to 0$ as $u \to 0$. Since the fourth sum only contains terms with $B\cap]T_i, T_{i+1}] = \emptyset$, $|X_{T_{i+1}} - X_{T_i}| \leq \epsilon$ for $|T_{i+1} - T_i|$ small enough, so when $\sup|T_{i+1} - T_i| \to 0$, each term is smaller than $\alpha(2\epsilon)$:

$$\sum_{B\cap]T_i,T_{i+1}]=\emptyset} |r(X_{T_i}, X_{T_{i+1}})| \leq \alpha(2\epsilon)\sum_i (X_{T_{i+1}} - X_{T_i})^2$$
$$\leq \alpha(2\epsilon)[X,X]_t \to 0 \text{ as } \sup|T_{i+1} - T_i| \to 0.$$

Let us now show that (8.58) converges absolutely when $\epsilon \to 0$. Assume first that X is a bounded process: $|X| \leq K$. Since f'' is continuous on $[-K, K]$, there exists a constant c such that $|f''(x)| \leq c$ for $|x| \leq K$. Therefore,

$$\sum_{\substack{0\leq s\leq t\\ \Delta X_s\neq 0}} |f(X_s) - f(X_{s-}) - \Delta X_s f'(X_{s-})| \leq c \sum_{\substack{0\leq s\leq t\\ \Delta X_s\neq 0}} |\Delta X_s|^2 \leq c[X,X]_t < \infty,$$
$$\sum_{\substack{0\leq s\leq t\\ \Delta X_s\neq 0}} |f''(X_{s-})\Delta X_s^2| \leq c[X,X]_t < \infty.$$

If X is not bounded, we can replace X by $X1_{[0,T_K[}$ where $T_K = \inf\{t > 0, |X_t| \geq K\}$ and repeat the same analysis for any $K > 0$. Therefore, (8.58) converges to

$$\sum_{0\leq s\leq t} \{f(X_s) - f(X_{s-}) - \Delta X_s f'(X_{s-}) - f''(X_{s-})|\Delta X_s|^2\}.$$

Summing up all the terms we get:

$$f(X_t) - f(X_0) = \int_0^t f'(X_{s-})dX_s + \frac{1}{2}\int_0^t f''(X_s)d[X,X]_s$$
$$+ \sum_{0\leq s\leq t} \{f(X_s) - f(X_{s-}) - \Delta X_s f'(X_{s-}) - f''(X_{s-})|\Delta X_s|^2\}.$$

Since

$$\int_0^t f''(X_{s-})d[X,X]_s = \int_0^t f''(X_{s-})d[X,X]_s^c + \sum_{0\leq s\leq t} f''(X_{s-})|\Delta X_s|^2,$$

we obtain (8.58). □

Martingale — drift decompositions similar to Proposition 8.16 can also be obtained for general semimartingales, but these results involve some new notions (predictable characteristics of semimartingales) that we do not discuss here; curious readers are referred to [215]. In Chapter 14, we discuss such decompositions for additive processes, which are generalizations of Lévy processes allowing for time dependent (but deterministic) characteristics.

REMARK 8.4 The Itô formula can be extended to convex functions which are continuous but not differentiable, a typical example being the call option payoff function $f(x) = (x - K)^+$. Such a convex function always has a left derivative which we denote by f' but its second derivative f'' is given by a Radon measure ρ, not necessarily a function (in the example above it is a Dirac measure δ_K). The extended change of variable formula, called the Tanaka-Meyer formula, then reads:

$$\begin{aligned} f(X_t) - f(X_0) = &\int_0^t f'(X_s)dX_s + \frac{1}{2}\int_0^t L_t^x \rho(dx) \\ &+ \sum_{0\leq s\leq t} \{f(X_s) - f(X_{s-}) - \Delta X_s f'(X_{s-})\}, \end{aligned} \tag{8.59}$$

where L_t^x is a continuous, nonanticipating, increasing process called the local time of X at x [194, Theorem 9.46]. □

8.4 Stochastic exponentials vs. ordinary exponentials

In the Black-Scholes model, the evolution of an asset price was described by the exponential of a Brownian motion with drift:

$$S_t = S_0 \exp(B_t^0), \tag{8.60}$$

where $B_t^0 = \mu t + \sigma W_t$ is a Brownian motion with drift. Applying the Itô formula we obtain:

$$\frac{dS_t}{S_t} = (\mu + \frac{\sigma^2}{2})dt + \sigma dW_t = dB_t^1 \quad \text{or} \quad S_t = S_0 + \int_0^t S_u dB_u^1, \tag{8.61}$$

where $B^1_t = (\mu+\sigma^2/2)t+\sigma W_t$ is another Brownian motion with drift. We will now study these two ways of constructing "exponential processes" in the case where the Brownian motions B^0, B^1 are replaced by a Lévy process. Replacing B^0_t by a Lévy process in (8.60) we obtain the class of *exponential-Lévy models*: $S_t = S_0 \exp X_t$ where X is Lévy process. Replacing B^1_t by a Lévy process in (8.61) we obtain the *stochastic exponential*, introduced by Doléans-Dade [115]. This process satisfies a stochastic differential equation of the same form as the classical ordinary differential equation for the exponential function:

$$S_t = S_0 + \int_0^t S_{u-} dX_u.$$

8.4.1 Exponential of a Lévy process

Let $(X_t)_{t\geq 0}$ be a Lévy process with jump measure J_X. Applying the Itô formula to $Y_t = \exp X_t$ yields:

$$\begin{aligned} Y_t &= 1 + \int_0^t Y_{s-} dX_s + \frac{\sigma^2}{2}\int_0^t Y_{s-} ds \\ &\qquad + \sum_{0\leq s\leq t; \Delta X_s \neq 0} (e^{X_{s-}+\Delta X_s} - e^{X_{s-}} - \Delta X_s e^{X_{s-}}) \\ &= 1 + \int_0^t Y_{s-} dX_s + \frac{\sigma^2}{2}\int_0^t Y_{s-} ds + \int_{[0,t]\times\mathbb{R}} Y_{s-}(e^z - 1 - z) J_X(ds\, dz) \end{aligned}$$

or, in differential notation:

$$\frac{dY_t}{Y_{t-}} = dX_t + \frac{\sigma^2}{2} dt + (e^{\Delta X_t} - 1 - \Delta X_t). \tag{8.62}$$

Making an additional assumption that $E[|Y_t|] = E[\exp(X_t)] < \infty$, which is by Proposition 3.14 equivalent to saying that $\int_{|y|\geq 1} e^y \nu(dy) < \infty$, we can decompose Y_t into a martingale part and a drift part, where the martingale part is the sum of an integral with respect to the Brownian component of X and a compensated sum of jump terms:

$$M_t = 1 + \int_0^t Y_{s-}\sigma dW_s + \int_{[0,t]\times\mathbb{R}} Y_{s-}(e^z - 1)\tilde{J}_X(ds\, dz), \tag{8.63}$$

while the drift term is given by:

$$\int_0^t Y_{s-}\left[\gamma + \frac{\sigma^2}{2} + \int_{-\infty}^{\infty} (e^z - 1 - z1_{|z|\leq 1})\nu(dz)\right] ds. \tag{8.64}$$

Therefore, $Y_t = \exp(X_t)$ is a martingale if and only if the drift term vanishes, that is,

$$\gamma + \frac{\sigma^2}{2} + \int_{-\infty}^{\infty} (e^z - 1 - z1_{|z|\leq 1})\nu(dz) = 0. \tag{8.65}$$

This result can of course be retrieved in a simpler way: by Proposition 3.17, Y is a martingale if and only if $E[\exp X_t] = E[Y_t] = 1$. But $E[\exp X_t] = \exp[t\psi_X(-i)]$, where ψ_X is the characteristic exponent of X. Therefore, we obtain once again:

$$\psi_X(-i) = \gamma + \frac{\sigma^2}{2} + \int_{-\infty}^{\infty} (e^z - 1 - z1_{|z|\leq 1})\nu(dz) = 0.$$

These properties are summarized in the following proposition:

PROPOSITION 8.20 Exponential of a Lévy process
Let $(X)_{t\geq 0}$ be a Lévy process with Lévy triplet (σ^2, ν, γ) verifying

$$\int_{|y|\geq 1} e^y \nu(dy) < \infty.$$

Then $Y_t = \exp X_t$ is a semimartingale with decomposition $Y_t = M_t + A_t$ where the martingale part is given by

$$M_t = 1 + \int_0^t Y_{s-}\sigma dW_s + \int_{[0,t]\times\mathbb{R}} Y_{s-}(e^z - 1)\tilde{J}_X(ds\ dz), \qquad (8.66)$$

and the continuous finite variation drift part is given by

$$A_t = \int_0^t Y_{s-}\left[\gamma + \frac{\sigma^2}{2} + \int_{-\infty}^{\infty} (e^z - 1 - z1_{|z|\leq 1})\nu(dz)\right] ds. \qquad (8.67)$$

(Y_t) is a martingale if and only if

$$\gamma + \frac{\sigma^2}{2} + \int_{-\infty}^{\infty} (e^z - 1 - z1_{|z|\leq 1})\nu(dz) = 0. \qquad (8.68)$$

8.4.2 Stochastic (Doléans-Dade) exponential

PROPOSITION 8.21 Doléans-Dade exponential
Let $(X)_{t\geq 0}$ be a Lévy process with Lévy triplet (σ^2, ν, γ). There exists a unique cadlag process $(Z)_{t\geq 0}$ such that

$$dZ_t = Z_{t-}dX_t \qquad Z_0 = 1. \qquad (8.69)$$

Z is given by:

$$Z_t = e^{X_t - \frac{\sigma^2 t}{2}} \prod_{0 \leq s \leq t} (1 + \Delta X_s) e^{-\Delta X_s}. \tag{8.70}$$

If $\int_{-1}^{1} |x|\nu(dx) < \infty$ then the jumps of X have finite variation and the stochastic exponential of X can be expressed as

$$Z_t = e^{X_t^c - \frac{\sigma^2 t}{2}} \prod_{0 \leq s \leq t} (1 + \Delta X_s). \tag{8.71}$$

PROOF Our proof partly follows [215]. Let

$$V_t = \prod_{0 \leq s \leq t; \Delta X_s \neq 0} (1 + \Delta X_s) e^{-\Delta X_s}.$$

The first step is to show that this process exists and is of finite variation. We decompose V into a product of two terms: $V_t = V_t' V_t''$, where

$$V_t' = \prod_{\substack{0 \leq s \leq t \\ |\Delta X_s| \leq 1/2}} (1 + \Delta X_s) e^{-\Delta X_s} \quad \text{and} \quad V_t'' = \prod_{\substack{0 \leq s \leq t \\ |\Delta X_s| > 1/2}} (1 + \Delta X_s) e^{-\Delta X_s}.$$

V'' for every t is a product of finite number of factors, so it is clearly of finite variation and there are no existence problems. V' is positive and we can consider its logarithm.

$$\ln V_t' = \sum_{0 \leq s \leq t; |\Delta X_s| \leq 1/2} (\ln(1 + \Delta X_s) - \Delta X_s).$$

Note that each term of this sum satisfies

$$0 > \ln(1 + \Delta X_s) - \Delta X_s > -\Delta X_s^2.$$

Therefore, the series is decreasing and bounded from below by $-\sum_{0 \leq s \leq t} \Delta X_s^2$, which is finite for every Lévy process (see Proposition 3.11). Hence, $(\ln V_t')$ exists and is a decreasing process. This entails that (V_t) exists and has trajectories of finite variation.

The second step is to apply the Itô formula for semimartingales to the function $Z_t \equiv f(t, X_t, V_t) \equiv e^{X_t - \sigma^2 t/2} V_t$. This yields (in differential form)

$$\begin{aligned} dZ_t = & -\frac{\sigma^2}{2} e^{X_{t-} - \frac{\sigma^2 t}{2}} V_{t-} dt + e^{X_{t-} - \frac{\sigma^2 t}{2}} V_{t-} dX_t + e^{X_{t-} - \frac{\sigma^2 t}{2}} dV_t \\ & + \frac{\sigma^2}{2} e^{X_{t-} - \frac{\sigma^2 t}{2}} V_{t-} dt + e^{X_{t} - \frac{\sigma^2 t}{2}} V_t - e^{X_{t-} - \frac{\sigma^2 t}{2}} V_{t-} \\ & - e^{X_{t-} - \frac{\sigma^2 t}{2}} V_{t-} \Delta X_t - e^{X_{t-} - \frac{\sigma^2 t}{2}} \Delta V_t. \end{aligned}$$

Now observe that since V_t is a pure jump process,

$$dV_t \equiv \Delta V_t = V_{t-}(e^{\Delta X_t}(1+\Delta X_t) - 1).$$

Substituting this into the above equality and making all the cancellations yields the Equation (8.69).

To understand why the solution is unique, observe that if $(Z_t^{(1)})$ and $(Z_t^{(2)})$ satisfy the Equation (8.69), then their difference $\tilde{Z}_t = Z_t^{(1)} - Z_t^{(2)}$ satisfies the same equation with initial condition $\tilde{Z}_0 = 0$. From the form of this equation, it is clear that if the solution is equal to zero at some point, it will remain zero. □

Z is called the stochastic exponential or the Doléans-Dade exponential of X and is denoted by $Z = \mathcal{E}(X)$. Notice that we could have defined it for an arbitrary semimartingale and not only for a Lévy process: the proof does not use the independence or stationarity of increments.

8.4.3 Relation between ordinary and stochastic exponential

It is clear from the above results that the ordinary exponential and the stochastic exponential of a Lévy process are two different notions: they do not correspond to the same stochastic process. In fact, contrarily to the ordinary exponential $\exp(X_t)$, which is obviously a positive process, the stochastic exponential $Z = \mathcal{E}(X)$ is not necessarily positive. It is easy to see from the explicit solution (8.70) that the stochastic exponential is always nonnegative if all jumps of X are greater than -1, or, equivalently, $\nu((-\infty, -1]) = 0$.

It is therefore natural to ask, which of the two processes is more suitable for building models for price dynamics. However the following result, due to Goll and Kallsen [170], shows that the two approaches are equivalent: if $Z > 0$ is the stochastic exponential of a Lévy process then it is also the ordinary exponential of another Lévy process and vice versa. Therefore, the two operations, although they produce different objects when applied to the same Lévy process, end up by giving us the *same class of positive processes.* In the continuous case this result is very simple to understand: if $X_t = \sigma W_t$ is a Wiener process then the ordinary and stochastic exponential are respectively given by

$$Y_t = e^{\sigma W_t} \text{ and } Z_t = e^{\sigma W_t - \frac{\sigma^2 t}{2}}, \tag{8.72}$$

therefore the stochastic exponential of a Brownian motion is also the ordinary exponential of *another* Lévy process L, that is in this case a Brownian motion with drift: $L_t = \sigma W_t - \frac{\sigma^2 t}{2}$.

The following result generalizes this remark to Lévy processes with jumps.

PROPOSITION 8.22 Relation between ordinary and stochastic exponentials [170]

1. *Let $(X)_{t\geq 0}$ be a real valued Lévy process with Lévy triplet (σ^2, ν, γ) and $Z = \mathcal{E}(X)$ its stochastic exponential. If $Z > 0$ a.s. then there exists another Lévy process $(L_t)_{t\geq 0}$ such that $Z_t = e^{L_t}$ where*

$$L_t = \ln Z_t = X_t - \frac{\sigma^2 t}{2} + \sum_{0\leq s\leq t} \{\ln(1+\Delta X_s) - \Delta X_s\}. \tag{8.73}$$

Its Lévy triplet $(\sigma_L^2, \nu_L, \gamma_L)$ is given by:

$$\sigma_L = \sigma,$$
$$\nu_L(A) = \nu(\{x : \ln(1+x) \in A\}) = \int 1_A(\ln(1+x))\nu(dx), \tag{8.74}$$
$$\gamma_L = \gamma - \frac{\sigma^2}{2} + \int \nu(dx)\left\{\ln(1+x)1_{[-1,1]}(\ln(1+x)) - x1_{[-1,1]}(x)\right\}.$$

2. *Let $(L)_{t\geq 0}$ be a real valued Lévy process with Lévy triplet $(\sigma_L^2, \nu_L, \gamma_L)$ and $S_t = \exp L_t$ its exponential. Then there exists a Lévy process $(X)_{t\geq 0}$ such that S_t is the stochastic exponential of X: $S = \mathcal{E}(X)$ where*

$$X_t = L_t + \frac{\sigma^2 t}{2} + \sum_{0\leq s\leq t} \left\{1 + \Delta L_s - e^{\Delta L_s}\right\}. \tag{8.75}$$

The Lévy triplet (σ^2, ν, γ) of X is given by:

$$\sigma = \sigma_L,$$
$$\nu(A) = \nu_L(\{x : e^x - 1 \in A\}) = \int 1_A(e^x - 1)\nu_L(dx), \tag{8.76}$$
$$\gamma = \gamma_L + \frac{\sigma_L^2}{2} + \int \nu_L(dx)\left\{(e^x-1)1_{[-1,1]}(e^x-1) - x1_{[-1,1]}(x)\right\}.$$

PROOF 1. The condition $Z > 0$ a.s. is equivalent to $\Delta X_s > -1$ for all s a.s., so taking the logarithm is justified here. In the proof of Proposition 8.21 we have seen that the sum $\sum_{0\leq s\leq t}\{\ln(1+\Delta X_s) - \Delta X_s\}$ converges and is a finite variation process. Then it is clear that L is a Lévy process and that $\sigma_L = \sigma$. Moreover, $\Delta L_s = \ln(1+\Delta X_s)$ for all s. This entails that

$$J_L([0,t]\times A) = \int_{[0,t]\times\mathbb{R}} 1_A(\ln(1+x))J_X(ds\,dx)$$

and also $\nu_L(A) = \int 1_A(\ln(1+x))\nu(dx)$. It remains to compute γ_L. Substituting the Lévy-Itô decomposition for (L_t) and (X_t) into (8.73), we obtain

$$\gamma_L t - \gamma t + \frac{\sigma^2 t}{2} + \int_{s\in[0,t],|x|\leq 1} x\tilde{J}_L(ds\,dx) + \int_{s\in[0,t],|x|>1} xJ_L(ds\,dx)$$
$$- \int_{s\in[0,t],|x|\leq 1} x\tilde{J}_X(ds\,dx) - \int_{s\in[0,t],|x|>1} xJ_X(ds\,dx)$$
$$- \sum_{0\leq s\leq t}\{\ln(1+\Delta X_s) - \Delta X_s\} = 0.$$

Observing that

$$\int_{s\in[0,t],|x|\leq 1} x(J_L(ds\,dx) - J_X(ds\,dx))$$
$$= \sum_{0\leq s\leq t}(\Delta X_s 1_{[-1,1]}(\Delta X_s) - \ln(1+\Delta X_s)1_{[-1,1]}(\ln(1+\Delta X_s)))$$

converges, we can split the above expression into jump part and drift part, both of which must be equal to zero. For the drift part we obtain:

$$\gamma_L - \gamma + \frac{\sigma^2}{2} - \int_{-1}^{1}\{x\nu_L(dx) - x\nu(dx)\} = 0,$$

which yields the correct formula for γ_L after a change of variable.

2. The second part can be proven in a similar fashion. □

As a corollary of this result, we will now derive a very important property of stochastic exponentials: without any supplementary conditions, the stochastic exponential of any martingale Lévy process is again a martingale.

PROPOSITION 8.23 Martingale preserving property

If $(X)_{t\geq 0}$ is a Lévy process and a martingale, then its stochastic exponential $Z = \mathcal{E}(X)$ is also a martingale.

PROOF Let $(X_t)_{t\geq 0}$ be a Lévy process with characteristic triplet (σ^2, ν, γ) such that $\gamma + \int_{|x|\geq 1} x\nu(dx) = 0$ (this is the martingale condition). First, suppose that $|\Delta X_s| \leq \varepsilon < 1$ a.s. Then by Proposition 8.22 there exists a Lévy process L_t such that $e^{L_t} = Z_t$. Moreover, this process has bounded jumps and therefore admits all exponential moments. Again, by Proposition

8.22, we can write:

$$\gamma_L + \frac{\sigma_L^2}{2} + \int_{-\infty}^{\infty} (e^z - 1 - z1_{|z|\le 1})\nu_L(dz) = \gamma + \int_{-1}^{1} \{z\nu_L(dz) - z\nu(dz)\}$$
$$+ \int_{-\infty}^{\infty} (e^z - 1 - z1_{|z|\le 1})\nu_L(dz) = \int_{-\infty}^{\infty} \{(e^z - 1)\nu_L(dz) - z\nu(dz)\} = 0$$

because $\Delta X_s = e^{\Delta L_s} - 1$ for all s. Therefore, by Proposition 8.20, $Z_t = e^{L_t}$ is a martingale.

The second step is to prove the proposition when X is a compensated compound Poisson process. In this case, the stochastic exponential has a very simple form:

$$Z_t = e^{bt} \prod_{0\le s\le t} (1 + \Delta X_s),$$

where $b = -\int_{-\infty}^{\infty} x\nu(dx)$. Denoting the intensity of X by λ, we obtain

$$E[Z_t] = e^{-\lambda t + bt} \sum_{n=0}^{\infty} \frac{(\lambda t)^n}{n!}(1 + E[\Delta X])^n = 1.$$

Together with the independent increments property of X this proves that Z is a martingale.

Now let X be an arbitrary martingale Lévy process. It can be decomposed into a sum of a compensated compound Poisson process X' and an independent martingale Lévy process with jumps smaller than ε, denoted by X''. Since these two processes never jump together, $\mathcal{E}(X' + X'') = \mathcal{E}(X')\mathcal{E}(X'')$. Moreover, each of the factors is a martingale and they are independent, therefore we conclude that $\mathcal{E}(X' + X'')$ is a martingale. □

$\mathcal{E}(X)$ is an example of an "exponential martingale" associated to X. One can use this correspondence to define the notion of stochastic logarithm of a Lévy process [229], i.e., a process $\mathcal{L}(X)$ such that $\mathcal{E}(\mathcal{L}(X)) = X$.

Further reading

The relation between the theory of stochastic integrals and continuous time trading strategies was pointed out by Harrison and Pliska in [191] and [192], which remain one of the most readable accounts on this topic. These papers also contain examples of market models driven by Poisson processes.

Most introductory (and less introductory) texts on stochastic calculus almost exclusively deal with Brownian motion and continuous semimartingales, see, e.g., [335]. Stochastic integration for discontinuous processes is presented in [55, 110, 210, 215, 324, 194, 205]. In the "classical" approach to stochastic integration, semimartingales are defined as processes which can be expressed as the sum of a (local) martingale and a process with finite variation and the integration theory is subsequently developed. While Dellacherie and Meyer[110] remains the classical reference on this approach, a detailed account can be found in [194] and the main results are summarized in [215]. This approach is less intuitive for our purpose since there is no a priori reason to model an asset price as the sum of a martingale and a finite variation process.

The approach we have adopted here is due to Protter [324, 323]: semimartingales are first defined as processes with good integration properties (Definition 8.2) and then one proceeds to show that usual stochastic processes are semimartingales. A very readable account of this approach is given in [323].

Stroock [370] discusses in detail the construction of stochastic integrals in the framework of Markov processes using Itô's original approach. The difference between componentwise and vector stochastic integration is carefully discussed in [89] and [90]. Stochastic integration with respect to Poisson random measures is discussed in [205], see also [242, 306]. Stochastic integration with respect to more general random measures is discussed in [215].

Stochastic integrals can be extended to integrands which are more general than caglad, but as pointed out before, in order to interpret the stochastic integral of a trading strategy as the gain process, we need to be able to approximate the stochastic integral pathwise by Riemann sums, hence we have limited integrands to caglad processes. This point is further discussed in [382].

There exist alternative approaches to stochastic integration, some of them having pathwise interpretations, the most well known being the Stratonovich integral, so one may wonder whether alternative models for continuous trading could be built from these constructions. But these approaches do not necessarily yield nonanticipating gain processes when applied to a trading strategy, which makes it difficult to give them a financial interpretation. The relation between Itô and Stratonovich integrals is discussed in [207].

Stochastic exponentials were introduced in [115] and are linked to "Wick" exponentials used in theoretical physics. They are important for constructing various exponential martingales and studying transformations of probability measures on path space, as will be pointed out in the next chapter. A discussion of stochastic exponentials can be found in [215, Chapter II].

Chapter 9

Measure transformations for Lévy processes

> It might be possible to prove certain theorems [about probability], but they might not be of any interest since, in practice, it would be impossible to verify whether the assumptions are fulfilled.
>
> Emile Borel

As noted in Chapter 3, a Lévy process $(X_t)_{t\in[0,T]}$ can be considered as a random variable on the space $\Omega = D([0,T])$ of discontinuous (cadlag) paths, equipped with its σ-algebra $\mathcal{F}$ telling us which events are measurable or, in other words, which statements can be made about these paths. The probability distribution of X then defines a probability measure $\mathbb{P}^X$ on this space of paths. Now consider another Lévy process $(Y_t)_{t\in[0,T]}$ and $\mathbb{P}^Y$ its distribution on the path space Ω. We will discuss in this chapter conditions under which $\mathbb{P}^X$ and $\mathbb{P}^Y$ are equivalent probability measures, in the sense that they define the same set of possible scenarios:

$$\mathbb{P}^X(A) = 1 \iff \mathbb{P}^Y(A) = 1.$$

If $\mathbb{P}^X$ and $\mathbb{P}^Y$ are equivalent, then the stochastic models X and Y define the same set of possible evolutions. The construction of a new process on the same set of paths by assigning new probabilities to events (i.e., reweighting the probabilities) is called a *change of measure.* To emphasize the fact that we are working under a specified measure, in this chapter we will often write $(X, \mathbb{P})$ to denote the process X and the corresponding distribution it defines over scenarios.

Given a probability measure $\mathbb{P}$ on the path space $\Omega = D([0,T])$, equivalent measures may be generated in many ways: given any (path-dependent) random variable $Z > 0$ on Ω with $E^{\mathbb{P}}[Z] = 1$, the new probability measure defined by adjusting the probability each path $\omega \in \Omega$ by $Z(\omega)$:

$$\frac{d\mathbb{Q}}{d\mathbb{P}} = Z, \quad \text{i.e.,} \quad \forall A \in \mathcal{F}, \quad \mathbb{Q}(A) = E^{\mathbb{P}}[Z 1_A],$$

is equivalent to $\mathbb{P}$. If we restrict our attention to events occurring between 0 and t then each path between 0 and t is reweighted by $Z_t(\omega) = E[Z|\mathcal{F}_t]$:

$$\frac{d\mathbb{Q}_{\mathcal{F}_t}}{d\mathbb{P}_{\mathcal{F}_t}} = Z_t \qquad \forall A \in \mathcal{F}_t, \quad \mathbb{Q}(A) = E^{\mathbb{P}}[Z_t 1_A]. \tag{9.1}$$

By construction, (Z_t) is a strictly positive martingale verifying $E[Z_t] = 1$. Conversely, any strictly positive martingale $(Z_t)_{t\in[0,T]}$ with $E[Z_t] = 1$ defines a new measure on path space given by (9.1). Though the processes defined by $\mathbb{P}$ and $\mathbb{Q}$ share the same paths, they can have quite different analytical and statistical properties. For example, if $\mathbb{P}$ defines a Lévy process X, the process Y defined by $\mathbb{Q}$ is not necessarily a Lévy process: it may have increments which are neither independent nor stationary. Given a class of processes, such as Lévy processes, it is investigate to investigate *structure-preserving* changes of measure which leave a given process within the class after a change of measure.

When X and Y are both Lévy processes, the equivalence of their measures gives relations between their parameters. As an example, take a Poisson process X with jump size equal to 1 and intensity λ. Then, with probability 1, the paths of X are piecewise constant with jumps equal to 1. Let Y be another Poisson process on the same path space, with intensity λ and jump size equal to 2. The measures corresponding to X and Y are clearly not equivalent since all the trajectories of Y that have jumps have zero probability of being trajectories of X and vice versa. However, if Y has the same jump size as X but a different intensity $\tilde{\lambda}$ then every trajectory of X on $[0,T]$ can also be a possible trajectory of Y and vice versa, so the two measures have a chance of being equivalent (we will see later that this is indeed the case).

This example also shows that two stochastic processes can define equivalent measures on scenarios while having different statistical properties. We will distinguish properties that are invariant under equivalent change of measure from those that are not. This will allow us to differentiate between "pathwise" or "almost sure" properties — properties of the process that can be deduced from the observation of a typical sample path — and the "statistical" properties, which are not preserved by an equivalent change of measure. This distinction is particularly important in financial modelling since when dealing with empirical data we observe a single trajectory of the price; considering the price history to be a typical sample path of a stochastic model we can infer its pathwise properties but its statistical properties may be harder to detect unless stronger hypotheses are made. In the above example, a Poisson process X is transformed into another Poisson process with a different intensity. This is an example of a structure-preserving transformation. Structure preserving changes of measure are interesting since they enable us to stay within an analytically tractable family of models. We will see more examples of such structure-preserving changes of measure for various types of Lévy processes in this chapter.

Equivalent changes of measure play an important role in arbitrage pricing theory. Two important concepts in the mathematical theory of contingent claim pricing are the *absence of arbitrage*, which imposes constraints on the way instruments are priced in a market and the notion of *risk-neutral pricing*, which represents the price of any instrument in an arbitrage-free market as its discounted expected payoff under an appropriate probability measure called the "risk-neutral" measure. Both of these notions are expressed in mathematical terms using the notion of equivalent change of measure: we will see that in a market model defined by a probability measure $\mathbb{P}$ on market scenarios there is one-to-one correspondence between arbitrage-free valuation rules and risk-neutral probability measures $\mathbb{Q}$ equivalent to $\mathbb{P}$ verifying a martingale property.

In this chapter, we will discuss equivalent changes of measure, their relation to arbitrage pricing and market completeness and give examples in the case of Lévy processes. Section 9.1 discusses the notions of risk-neutral pricing, absence of arbitrage and equivalent martingale measures in a general setting. In the following sections, these notions are examined in the context of models where randomness is generated by a Lévy process: Section 9.3 considers the simple cases of Poisson processes, compound Poisson processes and Brownian motion with drift and Section 9.4 gives the results for general Lévy processes. Section 9.5 gives an example of structure-preserving change of measure for Lévy processes, the Esscher transform.

Equivalent probability measures are defined on the same set of scenarios and are comparable. The notion of relative entropy, discussed in Section 9.6, defines a "distance" between two equivalent probability measures. It turns out that relative entropy can in fact be explicitly computed in the case of Lévy processes; this fact will be used in Chapter 13 when we will discuss model calibration.

9.1 Pricing rules and martingale measures

In this section we will attempt to explain the fundamental concepts behind risk-neutral pricing, absence of arbitrage and equivalent martingale measures using a minimum of jargon and technicality. Readers thirsty for more technical expositions are referred to [190, 191, 226, 109].

Consider a market whose possible evolutions between 0 and T are described by a scenario space $(\Omega, \mathcal{F})$: $\mathcal{F}$ contains all statements which can be made about behavior of prices between 0 and T. Underlying assets may then be described by a nonanticipating (cadlag) process:

$$\begin{aligned} S : [0,T] \times \Omega &\mapsto \mathbb{R}^{d+1} \\ (t,\omega) &\mapsto (S_t^0(\omega), S_t^1(\omega), \ldots, S_t^d(\omega)), \end{aligned} \tag{9.2}$$

where $S_t^i(\omega)$ represents the value of asset i at time t in the market scenario

ω and S_t^0 is a numeraire. A typical example of numeraire is a cash account with interest rate r: $S_t^0 = \exp(rt)$. Discounting is done using the numeraire S_t^0: for any portfolio with value V_t, the discounted value is defined by

$$\hat{V}_t = \frac{V_t}{S_t^0}$$

and $B(t,T) = S_t^0/S_T^0$ is called the discount factor. In the case where the numeraire is $S_t^0 = \exp(rt)$, the discount factor is simply given by $B(t,T) = \exp[-r(T-t)]$. We denote by $(\mathcal{F}_t)_{t\in[0,T]}$ the information generated by the history of assets up to t. $\mathcal{F}_0$ contains no information and $\mathcal{F}_T = \mathcal{F}$ is the history of all assets up to T. One could include other sources of information than just price histories but this is the framework assumed in the majority of pricing models. A contingent claim with maturity T may be represented by specifying its terminal payoff $H(\omega)$ in each scenario: since H is revealed at T it is a $\mathcal{F}_T$-measurable map $H : \Omega \to \mathbb{R}$. In practice, one does not consider all such contingent claims but a subclass of payoffs with some properties and various choices are possible. We will denote the set of contingent claims of interest by $\mathcal{H}$. Of course, the underlying assets themselves can be viewed as particular contingent claims whose payoff is given by the terminal value S_T^i: it is therefore natural to assume $S_T^i \in \mathcal{H}$. Other examples are European calls and puts: $H = (K - S_T^i)^+$, $H = (K - S_T^i)^+$ and path dependent options, where H can depend on the whole path of the underlying $H(\omega) = h(S_t(\omega), t \in [0,T])$.

A central problem in this context is the valuation problem: how can we attribute a notion of "value" to each contingent claim $H \in \mathcal{H}$? A *pricing rule* (also called a valuation operator) is a procedure which attributes to each contingent claim $H \in \mathcal{H}$ a *value* $\Pi_t(H)$ at each point in time. There are some minimal requirements that $\Pi_t(H)$ should verify to qualify as a pricing rule. First, if the pricing rule is to be of any use, one should be able to compute the value $\Pi_t(H)$ using the information given at t: $\Pi_t(H)$ should be a nonanticipating process. A second requirement is positiveness: a claim with a positive payoff should naturally have a positive value:

$$\forall \omega \in \Omega, H(\omega) \geq 0 \Rightarrow \forall t \in [0,T], \Pi_t(H) \geq 0. \tag{9.3}$$

Another requirement is linearity: the value of a portfolio is given by the sum of the values of its components:

$$\Pi_t(\sum_{j=1}^{J} H_j) = \sum_{j=1}^{J} \Pi_t(H_j). \tag{9.4}$$

Linearity may actually fail to hold for large portfolios: large block trades may be given discount prices on the market. We will encounter below some examples of pricing rules which are in fact nonlinear. But let us focus for the moment on linear pricing rules.

For any event $A \in \mathcal{F}$, the random variable 1_A represents the payoff of a contingent claim which pays 1 at T if A occurs and zero otherwise: it is a *bet* on A (also called a lottery). We will assume that $1_A \in \mathcal{H}$: such contingent claims are priced on the market. In particular $1_\Omega = 1$ is just a zero-coupon bond paying 1 at T. Its value $\Pi_t(1)$ represents the present value of 1 unit of currency paid at T, i.e., the discount factor:

$$\Pi_t(1) = e^{-r(T-t)}.$$

Define now $\mathbb{Q} : \mathcal{F} \mapsto \mathbb{R}$ by

$$\mathbb{Q}(A) = \frac{\Pi_0(1_A)}{\Pi_0(1)} = e^{rT}\Pi_0(1_A). \tag{9.5}$$

$\mathbb{Q}(A)$ is thus the value of a bet of nominal $\exp(rT)$ on the event A. Then, the linearity and positiveness of Π entail the following properties for $\mathbb{Q}$:

- $1 \geq \mathbb{Q}(A) \geq 0$, since $1 \geq 1_A \leq 0$.
- If A, B are disjoint events $1_{A\cup B} = 1_A + 1_B$ so by linearity of the valuation operator: $\mathbb{Q}(A \cup B) = \mathbb{Q}(A) + \mathbb{Q}(B)$.

If one extends the linearity (9.4) to infinite sums, then $\mathbb{Q}$ defined by (9.5) is nothing else but a probability measure over the scenario space $(\Omega, \mathcal{F})$! So, starting from a valuation rule Π, we have constructed a probability measure $\mathbb{Q}$ over scenarios. Conversely, Π can be retrieved from $\mathbb{Q}$ in the following way: for random payoffs of the form $H = \sum c_i 1_{A_i}$ which means, in financial terms, portfolios of cash-or-nothing options, by linearity of Π we have $\Pi_0(H) = E^{\mathbb{Q}}[H]$. Now if Π verifies an additional continuity property (i.e., if a dominated convergence theorem holds on $\mathcal{H}$) then we can conclude that for any random payoff $H \in \mathcal{H}$,

$$\Pi_0(H) = e^{-rT}E^{\mathbb{Q}}[H]. \tag{9.6}$$

Therefore there is a one-to-one correspondence between *linear valuation rules* Π verifying the properties above and *probability measures* $\mathbb{Q}$ on scenarios: they are related by

$$\Pi_0(H) = e^{-r(T-t)}E^{\mathbb{Q}}[H] \text{ and } \mathbb{Q}(A) = e^{r(T-t)}\Pi_0(1_A). \tag{9.7}$$

The relation (9.6) is sometimes called a *risk-neutral pricing* formula: the value of a random payoff is given by its discounted expectation under $\mathbb{Q}$. We have shown above that *any linear valuation rule* Π verifying the properties above is given by a "risk-neutral" pricing rule: there are no others! It is important to understand that $\mathbb{Q}$ has nothing to do with the actual/objective probability of occurrence of scenarios: in fact, we have not defined any objective probability measure on the scenarios yet! $\mathbb{Q}$ is called a *risk-neutral* measure or a *pricing* measure. Although it is, mathematically speaking, a probability measure on the set of scenarios, $\mathbb{Q}(A)$ should not be interpreted as the probability that A happens in the real world but as the value of a bet on A. A risk-neutral measure is just a convenient representation of the pricing rule Π: it is not obtained by an econometric analysis of time series or anything of the sort, but by looking at prices of contingent claims at $t = 0$.

Similarly for each t, $A \mapsto A = e^{rt}\Pi_t(1_A)$ defines a probability measure over scenarios between 0 and t, i.e., a probability measure $\mathbb{Q}_t$ on $(\Omega, \mathcal{F}_t)$. If we require that the pricing rule Π is time consistent, i.e., the value at 0 of the payoff H at T is the same as the value at 0 of the payoff $\Pi_t(H)$ at t then $\mathbb{Q}_t$ should be given by the restriction of $\mathbb{Q}$, defined above, to $\mathcal{F}_t$ and $\Pi_t(H)$ is given by the discounted conditional expectation with respect to $\mathbb{Q}$:

$$\Pi_t(H) = e^{-r(T-t)} E^{\mathbb{Q}}[H|\mathcal{F}_t]. \tag{9.8}$$

Therefore we have argued that *any time consistent linear pricing rule* Π verifying some continuity property is given by a discounted conditional expectation with respect to some probability measure $\mathbb{Q}$. We will now consider such a pricing rule given by a probability measure $\mathbb{Q}$ and examine what restrictions are imposed on $\mathbb{Q}$ by the requirement of absence of arbitrage.

9.1.1 Arbitrage-free pricing rules and martingale measures

Assume now that, in addition to the market scenarios $(\Omega, \mathcal{F})$ and the information flow $\mathcal{F}_t$, we know something about the probability of occurrence of these scenarios, represented by a probability measure $\mathbb{P}$. $\mathbb{P}$ represents here either the "objective" probability of future scenarios or the subjective view of an investor. What additional constraints should a pricing rule given by (9.8) verify in order to be compatible with this statistical view of the future evolution of the market?

A fundamental requirement for a pricing rule is that it does not generate arbitrage opportunities. An arbitrage opportunity is a self-financing strategy ϕ which can lead to a positive terminal gain, without any probability of intermediate loss:

$$\mathbb{P}(\forall t \in [0,T], V_t(\phi) \geq 0) = 1, \qquad \mathbb{P}(V_T(\phi) > V_0(\phi)) \neq 0.$$

Of course such strategies have to be realistic, i.e., of the form (8.2) to be of any use. Note that the definition of an arbitrage opportunity involves $\mathbb{P}$ but $\mathbb{P}$ is only used to specify whether the profit is possible or impossible, not to compute its probability of occurring: only events with probability 0 or 1 are involved in this definition. Thus the reasoning in the sequel will not require a precise knowledge of probabilities of scenarios. The self-financing property is important: it is trivial to exhibit strategies which are not self-financing verifying the property above, by injecting cash into the portfolio right before maturity. A consequence of absence of arbitrage is the *law of one price*: two self-financing strategies with the same terminal payoff must have the same value at all times, otherwise the difference would generate an arbitrage.

Consider now a market where prices are given by a pricing rule as in (9.8) represented by some probability measure $\mathbb{Q}$. Consider an event A such that $\mathbb{P}(A) = 0$ and an option which pays the holder 1 (unit of currency) if the event A occurs. Since the event A is considered to be impossible, this option is worthless to the investor. But the pricing rule defined by $\mathbb{Q}$ attributes to this option a value at $t = 0$ equal to

$$\Pi_0(1_A) = e^{-rT} E^{\mathbb{Q}}[1_A] = e^{-rT}\mathbb{Q}(A).$$

So the pricing rule $\mathbb{Q}$ is coherent with the views of the investor only if $\mathbb{Q}(A) = 0$. Conversely if $\mathbb{Q}(A) = 0$ then the option with payoff $1_A \geq 0$ is deemed worthless; if $\mathbb{P}(A) \neq 0$ then purchasing this option (for free) would lead to an arbitrage. So the compatibility of the pricing rule $\mathbb{Q}$ with the stochastic model $\mathbb{P}$ means that $\mathbb{Q}$ and $\mathbb{P}$ are equivalent probability measures: they define the same set of (im)possible events

$$\mathbb{P} \sim \mathbb{Q} : \forall A \in \mathcal{F} \qquad \mathbb{Q}(A) = 0 \iff \mathbb{P}(A) = 0. \tag{9.9}$$

Consider now an asset S^i traded at price S^i_t. This asset can be held until T, generating a terminal payoff S^i_T, or be sold for S_t: the resulting sum invested at the interest rate r will then generate a terminal wealth of $e^{r(T-t)}S^i_t$. These two buy-and-hold strategies are self-financing and have the same terminal payoff so they should have the same value at t:

$$E^{\mathbb{Q}}[S^i_T|\mathcal{F}_t] = E^{\mathbb{Q}}[e^{r(T-t)}S^i_t|\mathcal{F}_t] = e^{r(T-t)}S^i_t. \tag{9.10}$$

Dividing by $S^0_T = e^{rT}$ we obtain:

$$E^{\mathbb{Q}}[\frac{S^i_T}{S^0_T}|\mathcal{F}_t] = \frac{S^i_t}{S^0_t}, \quad \text{i.e.,} \quad E^{\mathbb{Q}}[\hat{S}^i_T|\mathcal{F}_t] = \hat{S}^i_t. \tag{9.11}$$

Therefore absence of arbitrage implies that discounted values $\hat{S}^i_t = e^{-rt}S^i_t$ of all traded assets are martingales with respect to the probability measure $\mathbb{Q}$. A probability measure verifying (9.9) and (9.11) is called an *equivalent martingale measure*. We have thus shown that any arbitrage-free pricing rule is given by an equivalent martingale measure.

Conversely, it is easy to see that any equivalent martingale measure $\mathbb{Q}$ defines an arbitrage-free pricing rule via (9.8). Consider a self-financing strategy $(\phi_t)_{t\in[0,T]}$. Of course a realistic strategy must be represented by a simple (piecewise constant) predictable process as in (8.2). Since $\mathbb{Q}$ is a martingale measure $\hat{S}_t$ is a martingale under $\mathbb{Q}$ so, as observed in Chapter 8, the value of the portfolio $V_t(\phi) = V_0 + \int_0^t \phi dS$ is a martingale so in particular $E^{\mathbb{Q}}[\int_0^t \phi dS] = 0$. The random variable $\int_0^t \phi dS$ must therefore take both positive and negative values: $\mathbb{Q}(V_T(\phi) - V_0 = \int_0^T \phi_t dS_t \geq 0) \neq 1$. Since $\mathbb{P} \sim \mathbb{Q}$ this entails $\mathbb{P}(\int_0^T \phi_t dS_t \geq 0) \neq 1$: ϕ cannot be an arbitrage strategy[1].

There is hence a one-to-one correspondence between arbitrage-free pricing rules and equivalent martingale measures:

> Specifying an arbitrage-free pricing rule on $(\Omega, \mathcal{F}, (\mathcal{F}_t), \mathbb{P})$
>
> $\Updownarrow$
>
> Specifying a probability measure $\mathbb{Q} \sim \mathbb{P}$ on market scenarios such that the prices of traded assets are martingales.

Let us now summarize these results:

PROPOSITION 9.1 Risk-neutral pricing
In a market described by a probability measure $\mathbb{P}$ on scenarios, any arbitrage-free linear pricing rule Π can be represented as

$$\Pi_t(H) = e^{-r(T-t)}\ E^{\mathbb{Q}}[H|\mathcal{F}_t], \tag{9.12}$$

where $\mathbb{Q}$ is an equivalent martingale measure*: a probability measure on the market scenarios such that*

$$\begin{aligned} \mathbb{P} \sim \mathbb{Q}: \quad & \forall A \in \mathcal{F}\ \ \mathbb{Q}(A) = 0 \iff \mathbb{P}(A) = 0 \\ \text{and} \quad & \forall i = 1 \dots d, \qquad E^{\mathbb{Q}}[\hat{S}_T^i|\mathcal{F}_t] = \hat{S}_t^i. \end{aligned}$$

Up to now we have assumed that such an arbitrage-free pricing rule/equivalent martingale measures does indeed exist, which is not obvious in a given model. The above arguments show that *if* an equivalent martingale measure exists, then the market is arbitrage-free. The converse result [190, 191, 109], more difficult to show, is sometimes called the Fundamental theorem of asset pricing:

PROPOSITION 9.2 Fundamental theorem of asset pricing
The market model defined by $(\Omega, \mathcal{F}, (\mathcal{F}_t), \mathbb{P})$and asset prices $(S_t)_{t\in[0,T]}$ is arbitrage-free if and only if there exists a probability measure $\mathbb{Q} \sim \mathbb{P}$ such that the discounted assets $(\hat{S}_t)_{t\in[0,T]}$ are martingales with respect to $\mathbb{Q}$.

[1]This argument can be extended to strategies which are not piecewise constant but more complex, as long as the corresponding value process $\int_0^t \phi dS$ is a martingale, e.g., by requiring boundedness, square-integrability, etc.

A proper mathematical statement requires a careful specification of the set of admissible strategies and in fact it is quite hard to give a precise version of this theorem: in fact for a general, unbounded semimartingale (such as an exponential-Lévy models with unbounded jumps) "martingale measure" should be replaced by the notion of "σ-martingale measure," the definition of arbitrage opportunity should be slightly modified to "no free lunch with vanishing risk," etc. We refer the reader to the academic literature on this topic [109, 90, 191, 190, 349]. This theorem establishes an equivalence between the financial concept of absence of arbitrage and the mathematical notion of equivalent martingale measure.

9.2 Market completeness

Besides the idea of absence of arbitrage, another important concept originating in the Black-Scholes model is the concept of *perfect hedge*: a self-financing strategy (ϕ_t^0, ϕ_t) is said to be a perfect hedge (or a replication strategy) for a contingent claim H if

$$H = V_0 + \int_0^T \phi_t dS_t + \int_0^T \phi_t^0 dS_t^0 \qquad \mathbb{P} - a.s. \tag{9.13}$$

By absence of arbitrage, if a replicating strategy exists, then V_0 is unique since two replicating strategies with different initial capital could lead to an arbitrage.

A market is said to be *complete* if any contingent claim admits a replicating portfolio: for any $H \in \mathcal{H}$ there exists a self-financing strategy (ϕ_t^0, ϕ_t) such that (9.13) holds with probability 1 under $\mathbb{P}$. As usual, ϕ_t should either be a simple predictable strategy as in (8.2) or be approximated by such strategies.

If (9.13) holds with probability 1, it also holds with probability 1 under any equivalent martingale measure $\mathbb{Q} \sim \mathbb{P}$. The discounted values then verify

$$\hat{H} = V_0 + \int_0^T \phi_t d\hat{S}_t \qquad \mathbb{Q} - a.s. \tag{9.14}$$

Taking expectations with respect to $\mathbb{Q}$ and assuming that the strategy (ϕ_t) verifies conditions such that $\int_0^t \phi dS$ is a martingale (for example, it is bounded) we obtain $E^{\mathbb{Q}}[\hat{H}] = V_0$: the value attributed by the pricing rule $\mathbb{Q}$ is given by the initial capital of the hedging strategy. Since this is true for any equivalent martingale measure $\mathbb{Q}$ we conclude (up to the boundedness assumptions we have made on the replicating strategies) that in a complete market there is only one way to define the value of a contingent claim: the value of any contingent claim is given by the initial capital needed to set up a perfect hedge for H. In particular, all equivalent martingale measures give the same pricing rules: they are the same! Therefore market completeness seems to imply the uniqueness of pricing rules/equivalent martingale measures. In fact, the converse result also holds but it much more difficult to show:

PROPOSITION 9.3 Second Fundamental Theorem of Asset Pricing

A market defined by the assets $(S_t^0, S_t^1, \ldots, S^d)_{t\in[0,T]}$, *described as stochastic processes on* $(\Omega, \mathcal{F}, (\mathcal{F}_t), \mathbb{P})$ *is complete if and only if there is a unique martingale measure* $\mathbb{Q}$ *equivalent to* $\mathbb{P}$.

This theorem establishes the equivalence between the financial notion of market completeness — the possibility to perfectly hedge any contingent claim — and the uniqueness of equivalent martingale measure, which is a mathematical property of the underlying stochastic model. The theorem holds as stated above in discrete time models. In continuous time models one has to carefully define the set of admissible strategies, contingent claims and the notion of "martingale measure." Unfortunately in the case where S has unbounded jumps, which is the case of most exponential-Lévy models, a rigorous formulation is quite difficult and requires to use the notion of "σ-martingale" [109]. Cherny and Shiryaev [90, Example 5.3.] present an example (albeit not a very realistic one) of a complete model for which an equivalent martingale measure (or local martingale measure) does not exist. We will not make use of these subtleties in the sequel; the curious reader is referred to [226] or [90] for a review. This suggests the following "equivalence," to be understood in a loose sense:

Market completeness:
any contingent claim $H \in \mathcal{H}$ can be represented as the final value of a self-financing strategy: $H = E[H] + \int_0^T \phi_t dS_t$

$$\Updownarrow$$

Uniqueness of equivalent martingale measure:
There exists a unique probability measure $\mathbb{Q} \sim \mathbb{P}$ such that discounted assets are $\mathbb{Q}$-martingales.

While most stochastic models used in option pricing are arbitrage-free, only a few of these models are complete: stochastic volatility models and as we will see shortly, exponential-Lévy models, jump-diffusion models fall into the category of incomplete models. By contrast, one dimensional diffusion models (in particular, the Black-Scholes model) define complete markets.

REMARK 9.1 Martingale representation property In mathematical terms, (9.14) means that for any random variable $H \in \mathcal{H}$ depending on the history of S_t between 0 and T, $\hat{H}$ can be represented as the sum of a constant and a stochastic integral of a predictable process with respect to $\hat{S}$. If this property holds for *all* terminal payoffs with finite variance, i.e., any $H \in L^2(\mathcal{F}_T, \mathbb{Q})$ can be represented as

$$\hat{H} = E[H] + \int_0^T \phi_s \, d\hat{S}_t \tag{9.15}$$

for some predictable process ϕ, the martingale $(\hat{S}_t)_{t\in[0,T]}$ is said to have the *predictable representation property.* Thus market completeness is often identified with the predictable representation property, which has been studied for many classical martingales. The predictable representation property can be shown to hold when $\hat{S}$ is (geometric) Brownian motion or a Brownian stochastic integral [335], but it fails to hold for most discontinuous models used in finance. For example, it is known to fail for all non-Gaussian Lévy processes except the (compensated) Poisson process: Chou and Meyer [91] show that if the jump size of a process with independent increments can take more than a single value then the predictable representation property fails. Yor and de Lazaro [385] generalize this result by showing that a martingale $\hat{S}$ with stationary increments has the predictable representation property (with respect to its own history) if and only if X is either a Brownian motion or a compensated Poisson process. We will show in Chapter 10 that this property also fails in exponential-Lévy models by a direct computation. Nevertheless one *can* construct price models with jumps where the predictable representation property holds: an example is the Azéma martingale, see [326].

Even if the predictable representation property holds it does not automatically lead to "market completeness": as argued in Chapter 8, any predictable process ϕ_t cannot be interpreted as a trading strategy. For this interpretation to hold we must be able, in some way, to approximate its value process using an implementable (piecewise constant in time) portfolio of the form (8.2) so predictable processes which can be reasonably interpreted as "trading strategies" are simple predictable processes as in (8.2) or caglad processes. Ma et al. [269] give conditions on the random variable H under which the representation involves a process with regular (caglad) paths.

Finally let us note that we are looking for a representation of $\hat{H}$ in terms of a stochastic integral *with respect to* $\hat{S}$. In fact the following theorem [205, Chapter 2] shows that when the source of randomness is a Brownian motion and a Poisson random measure, a random variable with finite variance can be always represented as a stochastic integral:

PROPOSITION 9.4
Let $(W_t)_{t\in[0,T]}$ be a d-dimensional Wiener process and M a Poisson random measure on $[0,T]\times\mathbb{R}^d$, independent from W. Then any random variable $\hat{H}$ with finite variance depending on the history $(\mathcal{F}_t)_{t\in[0,T]}$ of W and M between 0 and T can be represented as the sum of a constant, a stochastic integral

with respect to W *and a* compensated *Poisson integral with respect to* M*: there exists a predictable process* $\phi : \Omega \times [0,T] \mapsto \mathbb{R}$ *and a predictable random function* $\psi : \Omega \times [0,T] \times \mathbb{R}^d \mapsto \mathbb{R}$ *such that*

$$\hat{H} = E[\hat{H}] + \int_0^t \phi_s dW_s + \int_0^t \int_{\mathbb{R}^d} \psi(s,y)\tilde{M}(ds\ dy). \tag{9.16}$$

This property is also called a predictable representation property[2] by many authors but has nothing to do with market completeness. As noted in Remark 8.1, even when S is driven by the same sources of randomness W and M and $M = J_S$ represents the jump measure of the process S, the expression (9.16) cannot be represented as a stochastic integral with respect to S. Such representations are nevertheless useful for discussing hedging strategies, as we will see in Chapter 10. □

Is market completeness a desirable property in an option pricing model? Unfortunately, perfect hedges do not exist in practice. So if a model asserts that all options can be perfectly hedged, it is likely to give a distorted view of the risky nature of option trading, underestimating the risk inherent in writing an option. Of course, one could always argue that "dynamic hedging" is a continuous time concept and in a market with frictions, transaction costs, etc., perfect hedging is not achievable: these questions have stimulated a literature on corrections induced by such "market imperfections" in diffusion models: residual risk, increased hedging costs,... However, the real risks induced by taking positions in options are market risks — gamma risk, vega risk — whose order of magnitude is *much larger* than "corrections" induced by transaction costs and other imperfections. In complete market models, these risks are simply absent: they are reducible to "delta risk", which can be hedged away to zero! This is certainly a much rougher approximation than assuming zero transaction costs. Thus in our view it makes more sense to use incomplete market models where the risk of hedging can be quantified rather than sticking to complete market models where the risk of hedging is by definition zero and then elaborate on second-order effects such as market frictions. As most discontinuous price models generically lead to incomplete markets, this is not really a constraint on the model, while completeness is.

There is thus a one-to-one correspondence between arbitrage-free valuation methods and equivalent martingale measures: the concept of equivalent martingale of measure thus plays an important role. The next section will describe the construction of such equivalent martingale measures and the issue of their uniqueness in models where the randomness is described by a Lévy process.

[2]More precisely one should say "predictable representation with respect to W, M" in this case.

9.3 Equivalence of measures for Lévy processes: simple cases

Having seen the role played by equivalent changes of measure in defining arbitrage-free pricing models, we will now study such changes of measure in the case where the source of randomness is given by a Lévy process.

Recall that if the $\mathbb{P}$ and $\mathbb{Q}$ are equivalent probability measures then, as we have seen in Section 2.1.3, there exists a positive random variable, called the density of $\mathbb{Q}$ with respect to $\mathbb{P}$ and denoted $\frac{d\mathbb{Q}}{d\mathbb{P}}$ such that for any random variable Z

$$E^{\mathbb{Q}}[Z] = E^{\mathbb{P}}\left\{Z\frac{d\mathbb{Q}}{d\mathbb{P}}\right\}.$$

Let us start by considering a Poisson process defined by its jump size a and its jump intensity λ. The following result shows that if one wants two Poisson processes to define equivalent measures on the set of sample paths, one can freely change the intensity but the jump size must remain the same:

PROPOSITION 9.5 Equivalence of measure for Poisson processes
Let $(N, \mathbb{P}_{\lambda_1})$ and $(N, \mathbb{P}_{\lambda_2})$ be Poisson processes on $(\Omega, \mathcal{F}_T)$ with intensities λ_1 and λ_2 and jump sizes a_1 and a_2.

1. *If $a_1 = a_2$ then $\mathbb{P}_{\lambda_1}$ is equivalent to $\mathbb{P}_{\lambda_2}$ with Radon-Nikodym density*

$$\frac{d\mathbb{P}_{\lambda_1}}{d\mathbb{P}_{\lambda_2}} = \exp\left[(\lambda_2 - \lambda_1)T - N_T \ln\frac{\lambda_2}{\lambda_1}\right]. \tag{9.17}$$

2. *If $a_1 \neq a_2$ then the measures $\mathbb{P}_{\lambda_1}$ and $\mathbb{P}_{\lambda_2}$ are not equivalent.*

The result above means that changing the intensity of jumps amounts to "reweighting" the probabilities on paths: no new paths are generated by simply shifting the intensity. However changing the jump size generates different kind of paths: while the paths of N_1 are step functions with step size a_1, the paths of N_2 are step functions with step size a_2. Therefore the intensity of a Poisson process can be modified without changing the "support" of the process, but changing the size of jumps generates a new measure which assigns nonzero probability to some events which were impossible under the old one.

PROOF

1. Let $B \in \mathcal{F}_T$. We must show that

$$\mathbb{P}_{\lambda_1}\{B\} = E^{\mathbb{P}_{\lambda_2}}\{1_B\frac{d\mathbb{P}_{\lambda_1}}{d\mathbb{P}_{\lambda_2}}\}$$

holds with the Radon-Nikodym derivative given by (9.17). For the left-hand side of this equation we have, using the properties of Poisson process:

$$\mathbb{P}_{\lambda_1}\{B\} = \sum_{k=0}^{\infty} \frac{e^{-\lambda_1 T}(\lambda_1 T)^k}{k!} E^{\mathbb{P}_{\lambda_1}}\{1_B \big| N_T = k\},$$

and the right-hand side simplifies to

$$\begin{aligned} &E^{\mathbb{P}_{\lambda_2}}\{1_B \frac{d\mathbb{P}_{\lambda_1}}{d\mathbb{P}_{\lambda_2}}\} \\ &= \sum_{k=0}^{\infty} \frac{e^{-\lambda_2 T}(\lambda_2 T)^k}{k!} \left(\frac{\lambda_1}{\lambda_2}\right)^k e^{(\lambda_2-\lambda_1)T} E^{\mathbb{P}_{\lambda_2}}\{1_B \big| N_T = k\} \\ &= \sum_{k=0}^{\infty} \frac{e^{-\lambda_1 T}(\lambda_1 T)^k}{k!} E^{\mathbb{P}_{\lambda_2}}\{1_B \big| N_T = k\}. \end{aligned}$$

Since conditionally on the number of jumps in the interval, the jump times of a Poisson process are uniformly distributed on this interval, $E^{\mathbb{P}_\lambda}\{1_B \big| N_T = k\}$ does not depend on λ, which proves the assertion of the proposition.

2. This is straightforward since all nonconstant paths possible under $\mathbb{P}_1$ are impossible under $\mathbb{P}_2$. Note that these measures are not singular because they both assign a nonzero probability to the constant function.

□

REMARK 9.2 Two Poisson processes with different intensities define equivalent measures only on a finite time interval. If $T = \infty$, formula (9.17) shows that when intensities of two Poisson processes are different, the Radon-Nikodym derivative is either zero or infinity, which means that the measures are mutually singular. This corresponds to the fact that the intensity cannot be "read" from a trajectory of finite length but it can be estimated in an almost sure way from an infinite trajectory. □

It is often interesting to restrict the change of measure to a shorter time interval than the original one. Let us consider the information flow $\mathcal{F}_t$ generated by the history of the process, defined in Section 2.4.2. The restriction of probability measure $\mathbb{P}$ to $\mathcal{F}_t$, denoted by $\mathbb{P}\big|_{\mathcal{F}_t}$, is a probability measure on $\mathcal{F}_t$ which assigns to all events in $\mathcal{F}_t$ the same probability as $\mathbb{P}$. The Radon-Nikodym derivative $D_t = \frac{d\mathbb{Q}|_{\mathcal{F}_t}}{d\mathbb{P}|_{\mathcal{F}_t}}$ is thus an $(\mathcal{F}_t)$-adapted process. Moreover, it is a $\mathbb{P}$-martingale because for every $B \in \mathcal{F}_s$ we have

$$E^{\mathbb{P}}\{D_t 1_B\} = E^{\mathbb{Q}}\{1_B\} = E^{\mathbb{P}}\{D_s 1_B\}.$$

We now turn to the case of compound Poisson processes.

PROPOSITION 9.6 Equivalence of measures for compound Poisson processes
Let $(X, \mathbb{P})$ and $(X, \mathbb{Q})$ be compound Poisson processes on $(\Omega, \mathcal{F}_T)$ with Lévy measures $\nu^{\mathbb{P}}$ and $\nu^{\mathbb{Q}}$. $\mathbb{P}$ and $\mathbb{Q}$ are equivalent if and only if $\nu^{\mathbb{P}}$ and $\nu^{\mathbb{Q}}$ are equivalent. In this case the Radon-Nikodym derivative is

$$D_T = \frac{d\mathbb{Q}}{d\mathbb{P}} = \exp\left(T(\lambda^{\mathbb{P}} - \lambda^{\mathbb{Q}}) + \sum_{s \leq T} \phi(\Delta X_s)\right), \tag{9.18}$$

where $\lambda^{\mathbb{P}} \equiv \nu^{\mathbb{P}}(\mathbb{R})$ and $\lambda^{\mathbb{Q}} \equiv \nu^{\mathbb{Q}}(\mathbb{R})$ are the jump intensities of the two processes and $\phi \equiv \ln\left(\frac{d\nu^{\mathbb{Q}}}{d\nu^{\mathbb{P}}}\right)$.

PROOF The if part. Suppose that $\nu^{\mathbb{P}}$ and $\nu^{\mathbb{Q}}$ are equivalent. Conditioning the trajectory of X on the number of jumps in $[0, T]$ we have:

$$\begin{aligned}
E^{\mathbb{P}}\{D_T\} &= E^{\mathbb{P}}\{e^{T(\lambda^{\mathbb{P}} - \lambda^{\mathbb{Q}}) + \sum_{s \leq T} \phi(\Delta X_s)}\} \\
&= e^{-\lambda^{\mathbb{Q}} T} \sum_{k=0}^{\infty} \frac{(\lambda^{\mathbb{P}} T)^k}{k!} E^{\mathbb{P}}\{e^{\phi(\Delta X)}\}^k = 1.
\end{aligned}$$

Therefore $D_T \mathbb{P}$ is a probability. To prove the if part of the proposition it is now sufficient to show that if X is a compound Poisson process under $\mathbb{P}$ with Lévy measure $\nu^{\mathbb{P}}$ then it is a compound Poisson process under $\mathbb{Q}$ with Lévy measure $\nu^{\mathbb{Q}}$. To show this we will first check that X has $\mathbb{Q}$-independent increments and second that the law of X_T under $\mathbb{Q}$ is that of a compound Poisson random variable with Lévy measure $T\nu^{\mathbb{Q}}$. Let f and g be two bounded measurable functions and let $s < t \leq T$. Using the fact that X and $\ln D$ are $\mathbb{P}$-Lévy processes, and that D is a $\mathbb{P}$-martingale we have:

$$\begin{aligned}
&E^{\mathbb{Q}}\{f(X_s)g(X_t - X_s)\} = E^{\mathbb{P}}\{f(X_s)g(X_t - X_s)D_t\} \\
&= E^{\mathbb{P}}\{f(X_s)D_s\}E^{\mathbb{P}}\{g(X_t - X_s)\frac{D_t}{D_s}\} = E^{\mathbb{P}}\{f(X_s)D_s\}E^{\mathbb{P}}\{g(X_t - X_s)D_t\} \\
&\qquad = E^{\mathbb{Q}}\{f(X_s)\}E^{\mathbb{Q}}\{g(X_t - X_s)\},
\end{aligned}$$

which proves $\mathbb{Q}$-independence of increments. Furthermore, by conditioning the trajectory of X as above, we find

$$\begin{aligned}
&E^{\mathbb{P}}\{e^{iuX_T} e^{T(\lambda^{\mathbb{P}} - \lambda^{\mathbb{Q}}) + \sum_{s \leq T} \phi(\Delta X_s)}\} \\
&= e^{-\lambda^{\mathbb{Q}} T} \sum_{k=0}^{\infty} \frac{(\lambda^{\mathbb{P}} T)^k}{k!} E^{\mathbb{P}}\{e^{iu\Delta X + \phi(\Delta X)}\}^k = \exp\left(T \int (e^{iux} - 1)\nu^{\mathbb{Q}}(dx)\right).
\end{aligned}$$

The only if part. Suppose that $\nu^{\mathbb{P}}$ and $\nu^{\mathbb{Q}}$ are not equivalent. Then we can find either a set B such that $\nu^{\mathbb{P}}(B) > 0$ and $\nu^{\mathbb{Q}}(B) = 0$ or a set B' such that $\nu^{\mathbb{P}}(B) = 0$ and $\nu^{\mathbb{Q}}(B) > 0$. Suppose that we are in the first case. Then the set of trajectories having at least one jump the size of which is in B has positive $\mathbb{P}$-probability and zero $\mathbb{Q}$-probability, which shows that these two measures are not equivalent. □

Before discussing measure change for general Lévy processes, we treat the case of Brownian motion with drift.

PROPOSITION 9.7 Equivalence of measures for Brownian motion with drift

Let $(X, \mathbb{P})$ and $(X, \mathbb{Q})$ be two Brownian motions on $(\Omega, \mathcal{F}_T)$ with volatilities $\sigma^{\mathbb{P}} > 0$ and $\sigma^{\mathbb{Q}} > 0$ and drifts $\mu^{\mathbb{P}}$ and $\mu^{\mathbb{Q}}$. $\mathbb{P}$ and $\mathbb{Q}$ are equivalent if $\sigma^{\mathbb{P}} = \sigma^{\mathbb{Q}}$ and singular otherwise. When they are equivalent the Radon-Nikodym derivative is

$$\frac{d\mathbb{Q}}{d\mathbb{P}} = \exp\left\{\frac{\mu^{\mathbb{Q}} - \mu^{\mathbb{P}}}{\sigma^2} X_T - \frac{1}{2}\frac{(\mu^{\mathbb{Q}})^2 - (\mu^{\mathbb{P}})^2}{\sigma^2} T\right\}.$$

The proof of this proposition is similar to the previous ones and is left to the reader. Note that the Radon-Nikodym derivative can be rewritten as an exponential martingale:

$$\frac{d\mathbb{Q}}{d\mathbb{P}} = \exp\left\{\frac{\mu^{\mathbb{Q}} - \mu^{\mathbb{P}}}{\sigma} W_T - \frac{1}{2}\frac{(\mu^{\mathbb{Q}} - \mu^{\mathbb{P}})^2}{\sigma^2} T\right\},$$

where $W_t = \frac{X_t - \mu^{\mathbb{P}} t}{\sigma}$ is a standard Brownian motion under $\mathbb{P}$.

The above result — known as the Cameron-Martin theorem — shows that the drift and the volatility play very different roles in specifying a diffusion model. While modifying the drift amounts to reweighting the scenarios (paths of X), changing the volatility will generate a completely different process, leading to new scenarios which were initially impossible. Note also that we have restricted the Brownian paths to $[0, T]$ as usual: as in the case of Poisson processes, this result does not hold if $T = \infty$. A more general version of this result, valid for diffusion processes with random drift and volatility, is given by the Girsanov theorem (see [215, 335]).

REMARK 9.3 Up to now we have been exploring conditions under which two exponential-Lévy models define equivalent probabilities on the space of paths. However, it is important to note that for a given exponential-Lévy model the class of equivalent models is very large and includes many models which are *not* exponential Lévy models. Consider a compound Poisson process $(X, \mathbb{P})$ on $(\Omega, \mathcal{F}_T)$ with Lévy measure $\nu^{\mathbb{P}}$. Then from Equation (9.18) we know that the Radon-Nikodym derivative of the form

$$D_T = \frac{d\mathbb{Q}}{d\mathbb{P}} = \exp\left(T(\lambda^{\mathbb{P}} - \lambda^{\mathbb{Q}}) + \sum_{s \leq T} \phi(\Delta X_s)\right)$$

defines a measure change such that the new process $(X, \mathbb{Q})$ is again a compound Poisson process. Let us now rewrite this formula in a different form: denoting $\Psi(x) = \frac{d\nu^Q}{d\nu^P}(x) - 1$, we find that

$$D_t = \mathcal{E}\left(t(\lambda^P - \lambda^Q) + \sum_{s \leq t} \Psi(\Delta X_s)\right) = \mathcal{E}\left(\int_{[0,t]\times\mathbb{R}} \Psi(x)\tilde{J}_X(ds\ dx)\right),$$

where $\mathcal{E}$ stands for the stochastic exponential (defined in Section 8.4) and $\tilde{J}_X$ is the compensated jump measure of X. It is this last form of the Radon-Nikodym derivative D_t that is most interesting for us now. In order to define a measure change on the path space, D must be a positive martingale. However, to obtain a positive martingale Ψ need not be a deterministic and time independent function. If we only take such functions, the class of processes that can be obtained via measure change is restricted to compound Poisson processes. it More generally choices for Ψ lead to other classes of processes which live on the same set of paths. For example, if Ψ is deterministic but time-dependent then

$$\frac{d\mathbb{Q}}{d\mathbb{P}}\Big|_t = \mathcal{E}\left(\int_{[0,t]\times\mathbb{R}} \Psi(s,x)\tilde{J}_X(ds\ dx)\right) \tag{9.19}$$

defines a measure $\mathbb{Q}$ under which the increments of X will still be independent but will no longer be stationary. If we go even further and allow Ψ to be a random predictable function of x and s than we can obtain a process with dependent increments. All these processes may be very different from each other in their statistical properties but they will always have piecewise constant trajectories with a finite number of jumps in every interval, because this is a pathwise (almost sure) property preserved under equivalent measure changes. In particular, choosing a risk-neutral pricing rule given by an exponential-Lévy model can be compatible with dependence and heteroskedasticity of actual log-returns remarked in Chapter 7. □

9.4 Equivalence of measures for Lévy processes: general results

In this section we present without proof (but with a discussion) a general result of equivalence of measures for Lévy processes which includes the three propositions of the preceding section as particular cases. Proof of the following statement can be found in [345]. Similar results and generalizations are given in [215]. An important consequence of this result is that in presence of jumps, even if we restrict our attention to structure preserving measures, the class of probabilities equivalent to a given one is surprisingly large.

PROPOSITION 9.8 (see Sato [345], Theorems 33.1 and 33.2)
Let (X_t,P) and (X_t,P') be two Lévy processes on $\mathbb{R}$ with characteristic triplets (σ^2,ν,γ) and (σ'^2,ν',γ'). Then $P|_{\mathcal{F}_t}$ and $P'|_{\mathcal{F}_t}$ are equivalent for all t (or equivalently for one $t>0$) if and only if three following conditions are satisfied:

1. $\sigma=\sigma'$.
2. *The Lévy measures are equivalent with*

$$\int_{-\infty}^{\infty}(e^{\phi(x)/2}-1)^2\nu(dx)<\infty, \tag{9.20}$$

where $\phi(x)=\ln\left(\frac{d\nu'}{d\nu}\right)$.

3. *If $\sigma=0$ then we must in addition have*

$$\gamma'-\gamma=\int_{-1}^{1}x(\nu'-\nu)(dx). \tag{9.21}$$

When $\mathbb{P}$ and $\mathbb{Q}$ are equivalent, the Radon-Nikodym derivative is

$$\frac{dP'|_{\mathcal{F}_t}}{dP|_{\mathcal{F}_t}}=e^{U_t} \tag{9.22}$$

with

$$U_t=\eta X_t^c-\frac{\eta^2\sigma^2 t}{2}-\eta\gamma t + \lim_{\varepsilon\downarrow 0}\left(\sum_{s\le t,\ |\Delta X_s|>\varepsilon}\phi(\Delta X_s)-t\int_{|x|>\varepsilon}(e^{\phi(x)}-1)\nu(dx)\right).$$

Here (X_t^c) is the continuous part of (X_t) and η is such that

$$\gamma'-\gamma-\int_{-1}^{1}x(\nu'-\nu)(dx)=\sigma^2\eta$$

if $\sigma>0$ and zero if $\sigma=0$.

U_t is a Lévy process with characteristic triplet (a_Y, ν_U, γ_U) given by:

$$a_U = \sigma^2\eta^2, \tag{9.23}$$
$$\nu_U = \nu\phi^{-1}, \tag{9.24}$$
$$\gamma_U = -\frac{1}{2}a\eta^2 - \int_{-\infty}^{\infty}(e^y - 1 - y1_{|y|\leq 1})(\nu\phi^{-1})(dy). \tag{9.25}$$

The above result shows an interesting feature of models with jumps compared to diffusion models: we have considerable freedom in changing the Lévy measure, while retaining the equivalence of measures, but, unless a diffusion component is present, we cannot freely change the drift. The following example shows, to what extent the Lévy measure can be modified.

Example 9.1 Change of measure for tempered stable process
The tempered stable process defined in Chapter 4 has a Lévy measure of the following form:

$$\nu(x) = \frac{c^+ e^{-\lambda^+ x}}{x^{1+\alpha^+}}1_{x\geq 0} + \frac{c^- e^{-\lambda^-|x|}}{|x|^{1+\alpha^-}}1_{x<0}, \tag{9.26}$$

where $\lambda^\pm$ and $c^\pm$ are positive, and $0 < \alpha^\pm < 2$. The measures on the path space corresponding to two tempered stable processes are mutually absolutely continuous if and only if their coefficients $\alpha^\pm$ and $c^\pm$, which describe the behavior of Lévy measure near zero, coincide. In fact, the left-hand side of condition (9.20) for Lévy measure on the positive half-axis writes:

$$c_1\int_0^\infty\left(\frac{e^{-\frac{1}{2}(\beta_2^+ - \beta_1^+)x}\sqrt{c_2^+/c_1^+}}{x^{\frac{\alpha_2^+ - \alpha_1^+}{2}}} - 1\right)^2 \frac{e^{-\beta_1^+ x}}{x^{1+\alpha_1^+}}dx.$$

When $\alpha_2^+ < \alpha_1^+$ or when $\alpha_1^+ = \alpha_2^+$ but $c_1^+ \neq c_2^+$, the integrand is equivalent to $\frac{1}{x^{1+\alpha_1^+}}$ near zero and, hence, is not integrable; the case $\alpha_2^+ > \alpha_1^+$ is symmetric. However, when $\alpha_2^+ = \alpha_1^+$ and $c_1^+ = c_2^+$, the integrand is equivalent to $\frac{1}{x^{\alpha_1^+ - 1}}$ and is always integrable. □

This simple example shows that one can change freely the distribution of large jumps (as long as the new Lévy measure is absolutely continuous with respect to the old one) but one should be very careful with the distribution of small jumps (which is determined by the behavior of the Lévy measure near zero). This is a good property since large jumps are the ones which are important from the point of view of option pricing: they affect the tail of the return distribution and option prices in an important way.

9.5 The Esscher transform

In the Black-Scholes model, an equivalent martingale measure could be obtained by changing the drift. In models with jumps, if the Gaussian component is absent, we cannot change the drift but we can obtain a much greater variety of equivalent measures by altering the distribution of jumps. A convenient transformation, which is somewhat analogous to the drift change for geometric Brownian motion is the Esscher transform, that we have already encountered in Chapter 4.

Let X be a Lévy process with characteristic triplet (σ^2, ν, γ), θ a real number and assume that the Lévy measure ν is such that $\int_{|x|\geq 1} e^{\theta x}\nu(dx) < \infty$. Applying a measure transformation with the function $\phi(x)$ in Proposition 9.8 given by $\phi(x) = \theta x$ we obtain an equivalent probability under which X is a Lévy process with zero Gaussian component, Lévy measure $\tilde{\nu}(dx) = e^{\theta x}\nu(dx)$ and drift $\tilde{\gamma} = \gamma + \int_{-1}^{1} x(e^{\theta x} - 1)\nu(dx)$. This transformation is known in the literature as the Esscher transform. Using Proposition 9.8, the Radon-Nikodym derivative corresponding to this measure change is found to be

$$\frac{d\mathbb{Q}|_{\mathcal{F}_t}}{d\mathbb{P}|_{\mathcal{F}_t}} = \frac{e^{\theta X_t}}{E[e^{\theta X_t}]} = \exp(\theta X_t + \gamma(\theta) t), \tag{9.27}$$

where $\gamma(\theta) = -\ln E[\exp(\theta X_1)]$ is the log of the moment generating function of X_1 which, up to the change of variable $\theta \leftrightarrow -i\theta$, is given by the characteristic exponent of the Lévy process X.

The Esscher transform can be used to construct equivalent martingale measures in exponential-Lévy models.

PROPOSITION 9.9 Absence of arbitrage in exp-Lévy models

Let $(X, \mathbb{P})$ be a Lévy process. If the trajectories of X are neither almost surely increasing nor almost surely decreasing, then the exp-Lévy model given by $S_t = e^{rt+X_t}$ is arbitrage-free: there exists a probability measure $\mathbb{Q}$ equivalent to $\mathbb{P}$ such that $(e^{-rt}S_t)_{t\in[0,T]}$ is a $\mathbb{Q}$-martingale, where r is the interest rate.

In other words, the exponential-Lévy model is arbitrage-free in the following (not mutually exclusive) cases:

- X has a nonzero Gaussian component: $\sigma > 0$.
- X has infinite variation $\int_{-1}^{1} |x|\nu(dx) = \infty$.
- X has both positive and negative jumps.
- X has positive jumps and negative drift or negative jumps and positive drift.

PROOF Let X have characteristic triplet (σ^2, ν, γ). If $\sigma > 0$, an equivalent martingale measure can be obtained by changing the drift, as in the Black-Scholes model, without changing the Lévy measure. Hence, we only need to consider the case $\sigma = 0$. Furthermore, by applying a measure transformation with the function $\phi(x)$ in Proposition 9.8 given by $\phi(x) = -x^2$ we obtain an equivalent probability under which X is a Lévy process with zero Gaussian component, the same drift coefficient and Lévy measure $\tilde{\nu}(dx) = e^{-x^2}\nu(dx)$, which has exponential moments of all orders. Therefore we can suppose that ν has exponential moments of all orders to begin with.

We will now show that an equivalent martingale measure can be obtained from $\mathbb{P}$ by Esscher transform under the conditions of this theorem. After such transform with parameter θ the characteristic triplet of X becomes $(0, \tilde{\nu}, \tilde{\gamma})$ with $\tilde{\nu}(dx) = e^{\theta x}\nu(dx)$ and $\tilde{\gamma} = \gamma + \int_{-1}^{1} x(e^{\theta x} - 1)\nu(dx)$. For $\exp(X)$ to be a martingale under the new probability, the new triplet must satisfy

$$\tilde{\gamma} + \int_{-\infty}^{\infty} (e^x - 1 - x_{|x|\leq 1})\tilde{\nu}(dx) = 0.$$

To prove the theorem we must now show that the equation $-\gamma = f(\theta)$ has a solution under the conditions of the theorem, where

$$f(\theta) = \int_{-\infty}^{\infty} (e^x - 1 - x_{|x|\leq 1})e^{\theta x}\nu(dx) + \int_{-1}^{1} x(e^{\theta x} - 1)\nu(dx).$$

By dominated convergence we have that f is continuous and that $f'(\theta) = \int_{-\infty}^{\infty} x(e^x - 1)e^{\theta x}\nu(dx) \geq 0$, therefore, $f(\theta)$ is an increasing function. Moreover, if $\nu((0, \infty)) > 0$ and $\nu((-\infty, 0)) > 0$ then f' is everywhere bounded from below by a positive number. Therefore in this case $f(+\infty) = +\infty$, $f(-\infty) = -\infty$ and we have a solution.

It remains to treat the case when ν is concentrated on one of the half-lines and by symmetry it is enough to treat the case $\nu((-\infty, 0)) = 0$. In this case we still have $f(+\infty) = +\infty$ but the limit of f when $\theta \to -\infty$ need not be equal to $-\infty$. Observe that the first term in the definition of f always converges to zero. As for the second term, if $\int_0^{\infty} x\nu(dx) = \infty$, then it goes to $-\infty$ as $\theta \to \infty$, which means that in this case we have a solution. Otherwise it converges to $-\int_0^{\infty} x\nu(dx)$ which is exactly the difference between γ and the drift of X (with respect to the zero truncation function) in the finite variation case. Therefore, in the finite variation case a solution exists if X has negative drift. To sum everything up, we have proven that a solution exists unless $\nu((-\infty, 0)) = 0$, $\int_0^{\infty} x\nu(dx) < \infty$ and the drift is positive. This is exactly the case when X has almost surely increasing trajectories (Proposition 3.10). By symmetry we can treat the case of decreasing trajectories and complete the proof. □

9.6 Relative entropy for Lévy processes (*)

The notion of *relative entropy* or *Kullback-Leibler distance* is often used as measure of proximity of two equivalent probability measures. In this section we recall its definition and properties and compute the relative entropy of the measures generated by two risk-neutral exp-Lévy models. We will use this notion in Chapter 13 as a model selection criterion for calibrating an exponential-Lévy model to a set of option prices.

Define $(\Omega, \mathcal{F})$ as the space of real-valued discontinuous cadlag functions defined on $[0, T]$, $\mathcal{F}_t$ the history of paths up to t and $\mathbb{P}$ and $\mathbb{Q}$ be two equivalent probability measures $(\Omega, \mathcal{F})$. The relative entropy of $\mathbb{Q}$ with respect to $\mathbb{P}$ is defined as

$$\mathcal{E}(\mathbb{Q}, \mathbb{P}) = E^{\mathbb{Q}}\left[\ln \frac{d\mathbb{Q}}{d\mathbb{P}}\right] = E^{\mathbb{P}}\left[\frac{d\mathbb{Q}}{d\mathbb{P}} \ln \frac{d\mathbb{Q}}{d\mathbb{P}}\right].$$

If we introduce the strictly convex function $f(x) = x \ln x$, we can write the relative entropy

$$\mathcal{E}(\mathbb{Q}, \mathbb{P}) = E^{\mathbb{P}}\left[f\left(\frac{d\mathbb{Q}}{d\mathbb{P}}\right)\right].$$

It is readily observed that the relative entropy is a convex functional of $\mathbb{Q}$. Jensen's inequality shows that it is always nonnegative $\mathcal{E}(\mathbb{Q}, \mathbb{P}) \geq 0$, with $\mathcal{E}(\mathbb{Q}, \mathbb{P}) = 0$ *if and only if* $\frac{d\mathbb{Q}}{d\mathbb{P}} = 1$ almost surely. The following result shows that if the measures are generated by exponential-Lévy models, relative entropy can be expressed in terms of the Lévy measures:

PROPOSITION 9.10 Relative entropy of Lévy processes
Let $\mathbb{P}$ and $\mathbb{Q}$ be equivalent measures on $(\Omega, \mathcal{F})$ generated by exponential-Lévy models with Lévy triplets $(\sigma^2, \nu^P, \gamma^P)$ and $(\sigma^2, \nu^Q, \gamma^Q)$. Assume $\sigma > 0$. The relative entropy $\mathcal{E}(\mathbb{Q}, \mathbb{P})$ is then given by:

$$\mathcal{E}(\mathbb{Q}|\mathbb{P}) = \frac{T}{2\sigma^2}\left\{\gamma^Q - \gamma^P - \int_{-1}^{1} x(\nu^Q - \nu^P)(dx)\right\}^2 + \\ T\int_{-\infty}^{\infty}\left(\frac{d\nu^Q}{d\nu^P} \ln \frac{d\nu^Q}{d\nu^P} + 1 - \frac{d\nu^Q}{d\nu^P}\right)\nu^P(dx). \quad (9.28)$$

If $\mathbb{P}$ and $\mathbb{Q}$ correspond to risk-neutral exponential-Lévy models, the relative entropy reduces to:

$$\mathcal{E}(\mathbb{Q}|\mathbb{P}) = \frac{T}{2\sigma^2}\left\{\int_{-\infty}^{\infty} (e^x - 1)(\nu^Q - \nu^P)(dx)\right\}^2 \\ + T\int_{-\infty}^{\infty}\left(\frac{d\nu^Q}{d\nu^P} \ln \frac{d\nu^Q}{d\nu^P} + 1 - \frac{d\nu^Q}{d\nu^P}\right)\nu^P(dx). \quad (9.29)$$

PROOF Let (X_t) be a Lévy process and $S_t = \exp(X_t)$. From the bijectivity of the exponential it is clear that the histories generated by (X_t) and (S_t) coincide. We can therefore equivalently compute the relative entropy of the log-price processes (which are Lévy processes). To compute the relative entropy of two Lévy processes we will use expression (9.22) for Radon-Nikodym derivative:

$$\mathcal{E} = \int \frac{d\mathbb{Q}}{d\mathbb{P}} \ln \frac{d\mathbb{Q}}{d\mathbb{P}} d\mathbb{P} = E^P[U_T e^{U_T}].$$

where (U_t) is a Lévy process with characteristic triplet given by formulae (9.23)–(9.25). Let $\phi_t(z)$ denote its characteristic function and $\psi(z)$ its characteristic exponent, that is,

$$\phi_t(z) = E^P[e^{izU_t}] = e^{t\psi(z)}.$$

Then we can write:

$$\begin{aligned} E^P[U_T e^{U_T}] &= -i\frac{d}{dz}\phi_T(-i) = -iTe^{T\psi(-i)}\psi'(-i) \\ &= -iT\psi'(-i)E^P[e^{U_T}] = -iT\psi'(-i). \end{aligned}$$

From the Lévy-Khinchin formula we know that

$$\psi'(z) = -a^U z + i\gamma^U + \int_{-\infty}^{\infty} (ixe^{izx} - ix1_{|x|\leq 1})\nu^U(dx).$$

We can now compute the relative entropy as follows:

$$\begin{aligned} \mathcal{E} &= a^U T + \gamma^U T + T\int_{-\infty}^{\infty} (xe^x - x1_{|x|\leq 1})\nu^U(dx) \\ &= \frac{\sigma^2 T}{2}\eta^2 + T\int (ye^y - e^y + 1)(\nu^P\phi^{-1})(dy) \\ &= \frac{\sigma^2 T}{2}\eta^2 + T\int \left(\frac{d\nu^Q}{d\nu^P}\ln\frac{d\nu^Q}{d\nu^P} + 1 - \frac{d\nu^Q}{d\nu^P}\right)\nu^P(dx), \end{aligned}$$

where η is chosen such that

$$\gamma^Q - \gamma^P - \int_{-1}^{1} x(\nu^Q - \nu^P)(dx) = \sigma^2\eta.$$

Since we have assumed $\sigma > 0$, we can write

$$\frac{1}{2}\sigma^2\eta^2 = \frac{1}{2\sigma^2}\left\{\gamma^Q - \gamma^P - \int_{-1}^{1} x(\nu^Q - \nu^P)(dx)\right\}^2.$$

which leads to (9.28). If P and Q are martingale measures, we can express the drift γ using σ and ν:

$$\frac{\sigma^2}{2}\eta^2 = \frac{1}{2\sigma^2}\left\{\int_{-\infty}^{\infty} (e^x - 1)(\nu^Q - \nu^P)(dx)\right\}^2.$$

Substituting the above in (9.28) yields (9.29). □

Observe that, due to time homogeneity of the processes, the relative entropy (9.28) or (9.29) is a linear function of T: the relative entropy per unit time is finite and constant. The first term in the relative entropy (9.28) of the two Lévy processes penalizes the difference of drifts and the second one penalizes the difference of Lévy measures.

In the risk-neutral case the relative entropy only depends on the two Lévy measures ν^P and ν^Q. For a given reference measure ν^P the expression (9.29) viewed as a function of ν^Q defines a positive (possibly infinite) functional on the set of Lévy measures $\mathcal{L}(\mathbb{R})$:

$$\begin{aligned} H : \mathcal{L}(\mathbb{R}) &\to [0, \infty] \\ \nu^Q &\to H(\nu^Q) = \mathcal{E}(\mathbb{Q}(\nu^Q, \sigma)), \mathbb{P}(\nu^P, \sigma)). \end{aligned} \tag{9.30}$$

We shall call H the relative entropy functional. Its expression is given by (9.29). It is a positive convex functional of ν^Q, equal to zero only when $\nu^Q \equiv \nu^P$.

Example 9.2 Relative entropy for tempered stable processes
Consider two tempered stable processes that are mutually absolutely continuous and have Lévy densities given by:

$$\begin{aligned} \nu^Q(x) &= \frac{ce^{(-\beta_1-1)x}}{x^{1+\alpha}} 1_{x \geq 0} + \frac{ce^{(-\beta_1+1)|x|}}{|x|^{1+\alpha}} 1_{x<0}, \\ \nu^P(x) &= \frac{ce^{(-\beta_2-1)x}}{x^{1+\alpha}} 1_{x \geq 0} + \frac{ce^{(-\beta_2+1)|x|}}{|x|^{1+\alpha}} 1_{x<0} \end{aligned}$$

with $\beta_1 > 1$ and $\beta_2 > 1$ imposed by the no-arbitrage property. The relative entropy of $\mathbb{Q}$ with respect to $\mathbb{P}$ will always be finite because we can write for the first term in (9.29) (we consider for definiteness the positive half-axis):

$$\int_0^\infty (e^x - 1)(\nu^Q - \nu^P)(dx) = c \int_0^\infty dx \frac{(1 - e^{-x})(e^{-\beta_1 x} - e^{-\beta_2 x})}{x^\alpha},$$

which is finite because for small x the numerator is equivalent to x^2 and for large x it decays exponentially. For the second term in (9.29) on the positive half-axis we have:

$$\begin{aligned} &\int_0^\infty \left(\frac{d\nu^Q}{d\nu^P} \ln \frac{d\nu^Q}{d\nu^P} + 1 - \frac{d\nu^Q}{d\nu^P} \right) \nu^P(dx) \\ &\qquad = c \int_0^\infty \frac{e^{-(\beta_2+1)x} - e^{-(\beta_1+1)x} - x(\beta_1 - \beta_2)e^{-(\beta_1+1)x}}{x^\alpha}, \end{aligned}$$

which is again finite because for small x the numerator is equivalent to x^2 and for large x we have exponential decay. □

TABLE 9.1: Preservation of properties under equivalent changes of measure.

Property	**Preserved or not?**
Continuity/discontinuity of sample paths	Yes
Cadlag property for sample paths	Yes
Quadratic variation of sample paths	Yes
Intensity of jumps	No
Finite/infinite jump rate	Yes
Range of jump sizes	Yes
Distribution of returns	No
Heavy tails of increments	No
Independence of increments	No
Finite/infinite variation	Yes
Presence and volatility of diffusion component	Yes
Markov property	No
Absence of arbitrage	Yes

Example 9.3 Equivalent measure with infinite relative entropy

Suppose now that in the previous example $\alpha = 1$, $\beta_1 = 2$ and $\beta_2 = 1$. In this case, although $\mathbb{Q}$ and $\mathbb{P}$ are equivalent, the relative entropy of $\mathbb{Q}$ with respect to $\mathbb{P}$ is infinite. Indeed, on the negative half-axis $\frac{d\nu^Q}{d\nu^P} = e^{|x|}$ and the criterion (9.20) of absolute continuity is satisfied but $\frac{d\nu^Q}{d\nu^P} \ln\left(\frac{d\nu^Q}{d\nu^P}\right) d\nu^P = \frac{c}{|x|} dx$ and the second term in (9.29) diverges at infinity. □

9.7 Summary

We have discussed the notion of equivalent change of measure and its relation with absence of arbitrage and market completeness: in a given stochastic model, the existence of an equivalent martingale measure leads to absence of arbitrage while its uniqueness is related to market completeness. Moreover, each equivalent martingale measure gives a possible self-consistent pricing rule for contingent claims.

Using a particular change of measure — the Esscher transform — we showed that exponential-Lévy models are arbitrage-free: an equivalent martingale measure always exists. But they are also incomplete market models: the class of martingale measures equivalent is infinite. This may not seem to be a desirable property at first sight, because it means that after estimating a model from the time series we have many possible ways to price options. However, as discussed in the introductory chapter and in Section 9.2, market incompleteness is a realistic property that these models share with real markets.

Moreover as we will see in Chapter 13, the existence of many equivalent martingale measures leads to a great flexibility of exponential-Lévy models and their generalizations for fitting — "calibrating" — market prices of options. In the Black-Scholes model and, more generally, in a complete market model, once a model is estimated for historical returns, option prices are uniquely determined — and often they do not coincide with market prices of traded options. In exponential-Lévy model it is possible to find a martingale measure which is both equivalent to a given prior (estimated from historical data) and reproduces available market prices correctly. This is the basis of the calibration algorithm for Lévy processes, discussed in Chapter 13.

Since a given risk-neutral exponential-Lévy model is equivalent to many historical models with different parameters, the incompleteness property of Lévy models can also be interpreted in terms of *robustness* of option prices in these models with respect to specification of historical model [270]: agents with different views of the historical price process may still agree on option prices, because exponential-Lévy models with very different statistical properties (e.g., heavy/light tails) can still generate equivalent measures on scenarios. In the diffusion setting, on the contrary, if agents have different views about volatility then option prices that appear arbitrage-free to one of them, will seem arbitrageable to the other. However, even in exponential-Lévy models agents must agree on some basic qualitative properties of the price process such as the presence or absence of a diffusion component or the presence of jumps. Table 9.1 summarizes the distinction between pathwise properties of a process — which are the same in the historical returns and in a compatible risk-neutral model — and "statistical" properties, which may be different in the two models.

Further reading

The concepts of absence of arbitrage and market completeness and their characterization in terms of martingale measures were introduced by Harrison and Kreps [190] in a discrete time setting and by Harrison and Pliska [191, 192] in the continuous time setting. No-arbitrage theorems and relation with equivalent martingale measures are discussed at length in [109, 226, 90].

Measure transformations for Lévy processes are discussed in [345]. More general martingale measures for processes with independent increments are discussed in [176]. A general discussion of measure transformations for stochastic processes and random measures is given by Jacod & Shiryaev [215]. Absence of arbitrage and completeness for models with jumps is discussed in [22, 23, 24, 25]. The predictable representation property for processes with jumps is discussed by Chou and Meyer [91] for Lévy processesand by Yor and de Lazaro [385] for martingales with stationary increments. The predictable representation of random variables in terms of Brownian and compensated Poisson integrals given in Proposition 9.4 is discussed in [205], see also [241]. Other predictable representations for Lévy processes in terms of a sequence of jump martingales were introduced by Nualart and Schoutens in [310, 311]. A financial interpretation of the result of Nualart and Schoutens in terms of hedging with vanilla options is given by Balland in [21]. An example of a complete market model with jumps using the Azéma martingale is given by Dritschel and Protter [326].

Paul André Meyer

The French mathematician Paul André Meyer (1934–2003) is considered one of the pioneers of the modern theory of stochastic processes: his work opened the way for the extension of Itô's theory of stochastic integration beyond the Markov setting and laid the foundations of stochastic calculus for semimartingales.

Born in 1934, Meyer entered the Ecole Normale Supérieure to study mathematics in 1954 and obtained his doctorate in 1960. Apart from a short period in the Unites States where he met J. Doob, Meyer spent all his scientific career at Université Louis Pasteur in Strasbourg, where he founded the "Strasbourg school of probability." It is also in Strasbourg where he ran the "Séminaire de Probabilités," which led to the creation of a famous series of lecture notes published under the same name, in which many fundamental results on the theory of stochastic processes have appeared since 1967. These volumes constitute today a reference for researchers in probability and related fields: they contain, in addition to many research papers, numerous lecture notes by Paul André Meyer in which he reformulated and expanded recent results in probability in his personal style. A selection of fundamental articles which appeared in these lecture notes can be found in [132].

A crucial contribution of Meyer was to extend to a continuous time setting Doob's decomposition of any supermartingale into a martingale and an increasing process — today called the Doob-Meyer decomposition — which allowed to extend many results obtained by Itô in the case of Markov processes to a more general setting and led Courrège in France and Kunita and Watanabe [242] in Japan to extend Itô's stochastic integration theory to square integrable martingales. Meyer introduced the notion of predictable processes and, in his "Cours sur les intégrales stochastiques" [296], opened a new chapter in the theory of stochastic integrals, introducing the notion of semimartingale and extending the Kunita-Watanabe theory of stochastic integration for square integrable martingales to any semimartingale.

His work with Dellacherie, Doléans and other members of the Strasbourg school led to the development of the "general theory of stochastic processes," which culminated in the monumental multi-volume treatise Probabilities and potential [110] coauthored with Claude Dellacherie, which has become a reference for researchers in the theory of stochastic processes.

In addition to his work on stochastic integration and semimartingale theory, Meyer also authored many research papers on stochastic differential geometry. After the 1980s, Meyer focused his research efforts on the new field of quantum probability. But Meyer's curiosity and intellectual activities reached far beyond mathematics. He mastered several foreign languages, including Chinese and Sanskrit. His nonmathematical works include the translation into French of an ancient Indian literary classic.

Chapter 10

Pricing and hedging in incomplete markets

> The pricing of derivatives involves the construction of riskless hedges from traded securities.
>
> John Hull [202]

> The question is [...] whether the LTCM disaster was merely a unique and isolated event, a bad draw from nature's urn, or whether such disasters are the inevitable consequence of the Black-Scholes formula and the illusion it might give that all market participants can hedge away their risk at the same time.
>
> Merton Miller

While absence of arbitrage is both a reasonable property to assume in a real market and a generic property of many stochastic models, market completeness is neither a financially realistic nor a theoretically robust property. From a financial point of view, market completeness implies that options are redundant assets and the very existence of the options market becomes a mystery, if not a paradox in such models. From a theoretical viewpoint, market completeness is not a robust property. Indeed, in Chapter 9 we have seen that, given a complete market model, the addition of even a small (in statistical terms) jump risk destroys market completeness. Thus, in models with jumps market completeness is an exception rather than the rule.

In a complete market, there is only one arbitrage-free way to value an option: the value is defined as the cost of replicating it. In real markets, as well as in the models considered in this book, perfect hedges do not exist and options are not redundant: the notion of pricing by replication falls apart, not because continuous time trading is impossible in practice *but because there are risks that one cannot hedge even by continuous time trading.* Thus we are forced to reconsider hedging in the more realistic sense of approximating a target payoff with a trading strategy: one has to recognize that option hedging is a risky affair, specify a way to measure this risk and then try to minimize it. Different ways to measure risk thus lead to different approaches to hedging: superhedging, utility maximization and mean-variance hedging are among the approaches discussed in this chapter. Each of these hedging strategies has a cost, which can be computed in some cases. The value of the option will thus consist of two parts: the cost of the hedging strategy plus a risk premium, required by the option seller to cover her residual (unhedgeable) risk. We will deal here with the first component by studying various methods for hedging and their associated costs. Arbitrage pricing has nothing to say about the second component which depends on the preferences of investors and, in a competitive options market, this risk premium can be driven to zero, especially for vanilla options.

In this chapter we will discuss various approaches to pricing and hedging options in incomplete markets. Merton's approach [291], presented in Section 10.1, proposes to ignore risk premia for jumps; this assumption leads to a specific choice for pricing and hedging. The notion of superhedging, discussed in Section 10.2, is a preference-free approach to the hedging problem in incomplete markets and leads to bounds for prices. However, in most examples of models with jumps these bounds turn out to be too wide and the corresponding hedging strategies are buy and hold strategies. The (old) idea of expected utility maximization, combined with the possibility of dynamic trading, suggests choosing an optimal hedge by minimizing some measure of hedging error as measured by the expectation of a convex loss function. This leads to the notion of utility indifference price, discussed in Section 10.3. An important case, discussed in Section 10.4, is when the loss function is quadratic: the corresponding quadratic hedging problem has an explicit solution in the case of jump-diffusion and exponential-Lévy models. If one is simply interested in arbitrage-free pricing of options, one could also choose any (equivalent) martingale measure as a self-consistent pricing rule: as shown in Chapter 9, many choices are possible for this and we discuss some of them, such as the minimal entropy martingale measure, in Section 10.5. However the price given to an option in this way does not correspond anymore to the cost of a specific hedging strategy. From a financial point of view it is more appropriate to start by discussing hedging strategies and then derive a valuation for the option in terms of the cost of hedging, plus a risk premium.

All the above approaches deal with hedging options with the underlying asset. However the availability of liquid markets for "vanilla" options such as short term calls and puts on major indices, currencies and stocks allows to use such options as hedging instruments. Indeed, liquid calls and puts are commonly used for static hedging strategies of exotic options. This practice has important consequences for the choice of pricing and hedging methodologies, which are discussed in Section 10.6.

In the sequel, we consider a market described by a scenario space $(\Omega, \mathcal{F})$, asset prices $(S_t)_{t\in[0,T]}$ and an information flow $(\mathcal{F}_t)_{t\in[0,T]}$, which is the history of the assets. We take here $\mathcal{F} = \mathcal{F}_T$: terminal payoffs of all contingent claims will be expressed in terms of $(S_t, t \in [0,T])$. S_t will be one dimensional unless otherwise specified. Discounting is done using a numeraire S_t^0: for any portfolio with value V_t the discounted value is defined by $\hat{V}_t = V_t/S_t^0$ and $B(t,T) = S_t^0/S_T^0$ is called the discount factor. In all examples the numeraire is $S_t^0 = \exp(rt)$; the discount factor is then given by $B(t,T) = \exp[-r(T-t)]$. Throughout the chapter, $M(S)$ designates the set of probability measures $\mathbb{Q} \sim \mathbb{P}$ such that $\hat{S}_t$ is a $\mathbb{Q}$–martingale, $\mathcal{S}$ designates a set of admissible strategies — which contains all simple predictable strategies (see Definition 8.1) but may be larger — and

$$\mathbb{A} = \{V_0 + \int_0^T \phi_t dS_t,\ V_0 \in \mathbb{R},\ \phi \in \mathcal{S}\} \tag{10.1}$$

designates the set of terminal payoffs attainable by such strategies.

10.1 Merton's approach

The first application of jump processes in option pricing was introduced by Robert Merton in [291]. Merton considered the jump-diffusion model

$$\mathbb{P}: \qquad S_t = S_0 \exp[\mu t + \sigma W_t + \sum_{i=1}^{N_t} Y_i], \tag{10.2}$$

where W_t is a standard Wiener process, N_t is a Poisson process with intensity λ independent from W and $Y_i \sim N(m, \delta^2)$ are i.i.d. random variables independent from W, N. As observed in Chapter 9, such a model is incomplete: there are many possible choices for a *risk-neutral* measure, i.e., a measure $\mathbb{Q} \sim \mathbb{P}$ such that the discounted price $\hat{S}_t$ is a martingale. Merton proposed the following choice, obtained as in the Black-Scholes model by changing the drift of the Wiener process but leaving the other ingredients unchanged:

$$\mathbb{Q}_M: \qquad S_t = S_0 \exp[\mu^M t + \sigma W_t^M + \sum_{i=1}^{N_t} Y_i], \tag{10.3}$$

where W_t^M is a standard Wiener process, N_t, Y_i are as above, independent from W^M and μ^M is chosen such that $\hat{S}_t = S_t e^{-rt}$ is a martingale under $\mathbb{Q}^M$:

$$\mu^M = r - \frac{\sigma^2}{2} - \lambda\ E[e^{Y_i} - 1] = r - \frac{\sigma^2}{2} - \lambda[\exp(m + \frac{\delta^2}{2}) - 1]. \tag{10.4}$$

$\mathbb{Q}_M$ is the equivalent martingale measure obtained by shifting the drift of the Brownian motion but leaving the jump part unchanged. Merton justified the choice (10.3) by assuming that "jump risk" is diversifiable, therefore, no risk premium is attached to it: in mathematical terms, it means that the risk neutral properties of the jump component of S_t are (supposed to be) the same as its statistical properties. In particular, the distribution of jump times and jump sizes is unchanged. A European option with payoff $H(S_T)$ can then be priced according to:

$$C_t^M = e^{-r(T-t)} E^{\mathbb{Q}_M}[H(S_T)|\mathcal{F}_t]. \tag{10.5}$$

Since S_t is a Markov process under $\mathbb{Q}_M$, the option price C_t^M can be expressed as a deterministic function of t and S_t:

$$\begin{aligned} C_t^M = C^M(t, S_t) &= e^{-r(T-t)} E^{\mathbb{Q}_M}[(S_T - K)^+ | S_t = S] \\ &= e^{-r(T-t)} E[H(S e^{\mu^M (T-t) + \sigma W_{T-t}^M + \sum_{i=1}^{N_{T-t}} Y_i})]. \end{aligned} \tag{10.6}$$

By conditioning on the number of jumps N_t, we can express $C(t, S_t)$ as a weighted sum of Black-Scholes prices: denoting the time to maturity by $\tau = T - t$ we obtain:

$$\begin{aligned} C^M(t,S) &= e^{-r\tau} \sum_{n\geq 0} \mathbb{Q}_M(N_t = n) E^{\mathbb{Q}_M}[H(S \exp[\mu^M \tau + \sigma W_\tau^M + \overbrace{\sum_{i=1}^{n} Y_i}^{N(nm, n\delta^2)}])] \\ &= e^{-r\tau} \sum_{n\geq 0} \frac{e^{-\lambda\tau}(\lambda\tau)^n}{n!} E^{\mathbb{Q}_M}[H(S e^{nm + \frac{n\delta^2}{2} - \lambda \exp(m + \frac{\delta^2}{2}) + \lambda\tau} e^{r\tau - \sigma_n^2 \tau/2 + \sigma_n W_\tau})] \\ &= e^{-r\tau} \sum_{n\geq 0} \frac{e^{-\lambda\tau}(\lambda\tau)^n}{n!} C_H^{BS}(\tau, S_n; \sigma_n), \end{aligned} \tag{10.7}$$

where $\sigma_n^2 = \sigma^2 + n\delta^2/\tau$,

$$S_n = S \exp[nm + \frac{n\delta^2}{2} - \lambda \exp(m + \frac{\delta^2}{2}) + \lambda\tau]$$

and

$$C_H^{BS}(\tau, S; \sigma) = e^{-r\tau} E[H(S e^{(r - \frac{\sigma^2}{2})\tau + \sigma W_\tau})]$$

is the value of a European option with time to maturity τ and payoff H in a Black-Scholes model with volatility σ. This series expansion converges exponentially and can be used for numerical computation of prices in Merton's model. For call and put options $C^M(t, S)$ is a smooth, $C^{1,2}$ function so the Itô formula (Proposition 8.14) can be applied to $e^{-rt}C(t, S_t)$ between 0 and T. Since

$$\hat{C}_t^M = e^{-rt} C_t^M = E^{\mathbb{Q}_M}[e^{-rT}(S_T - K)^+ | \mathcal{F}_t] = E^{\mathbb{Q}_M}[\hat{C}_T^M | \mathcal{F}_t],$$

the discounted value $\hat{C}_t^M$ is a martingale under $\mathbb{Q}^M$, so $\hat{C}_t^M$ is equal to its martingale component:

$$\begin{aligned}\hat{C}_T^M - \hat{C}_0^M &= \hat{H}(S_T) - E^{\mathbb{Q}_M}[H(S_T)] \\ &= \int_0^T \frac{\partial C^M}{\partial S}(u, S_{u-})\hat{S}_{u-}\sigma dW_u^M \qquad (10.8)\\ &+ \int_0^t \int_{\mathbb{R}} [C^M(u, S_{u-}+z) - C^M(u, S_{u-})]\hat{S}_{u-}\tilde{J}_S(du\, dz),\end{aligned}$$

where $\tilde{J}_S$ is the compensated jump measure of S, which is the same under $\mathbb{P}$ and $\mathbb{Q}$. The hedging portfolio proposed by Merton is the self-financing strategy (ϕ_t^0, ϕ_t) given by

$$\phi_t = \frac{\partial C^M}{\partial S}(t, S_{t-}), \qquad \phi_t^0 = \phi_t S_t - \int_0^t \phi dS, \qquad (10.9)$$

which means that we choose to hedge only the risk represented by the diffusion part, i.e., the first term in (10.8). The initial cost of this hedging strategy is given by C_0^M defined above: the value of the option is defined as the cost of the proposed hedge (10.9). The (discounted) hedging error is equal to:

$$\begin{aligned}\hat{H} - e^{-rT}V_T(\phi) &= \hat{C}_T^M - \hat{C}_0^M - \int_0^T \frac{\partial C^M}{\partial S}(u, S_{u-})d\hat{S}_u \\ &= \int_0^t \int_{\mathbb{R}} [C^M(u, S_{u-}+z) - C^M(u, S_{u-}) - (1+z)\frac{\partial C^M}{\partial S}(u, S_{u-})]\hat{S}_{u-}\tilde{J}_S(du\, dz),\end{aligned} \qquad (10.10)$$

so this hedge has the effect of correcting for the average effect of jumps but leaves us otherwise completely exposed to jump risk.

This approach is justified if we assume that the investor holds a portfolio with many assets for which the diffusion components may be correlated, i.e., contain some common component — which is then identified as a "systemic risk" of the portfolio — but the jump components are independent (and thus diversifiable) across assets. Such a hypothesis would imply that in a large market a diversified portfolio such as the market index would not have jumps. Unfortunately this is not the case: market indexes such as the S&P 500 or the NIKKEI do exhibit large downward movements and in fact these large movements result from jumps highly correlated across index components, a market crash being the extreme example. We will see that other choices of hedging achieve a trade-off between jump risk and the risk in the diffusion part (if there is one). Finally, the assumption of diversifiability of jump risk is not justifiable if we are pricing index options: a jump in the index is not diversifiable.

These observations show that in models with jumps, contrarily to diffusion models, a pricing measure cannot be simply obtained by adjusting the "drift" coefficient μ: this choice means that we are not pricing the risk due to jumps and has implications that are not difficult to justify in terms of risk premia and hedging. This motivates us to explore other approaches to pricing and hedging in models with jumps.

10.2 Superhedging

If one cannot replicate a contingent claim H, a conservative approach to hedging is to look for a self-financing strategy ϕ such as to remain always on the safe side:

$$\mathbb{P}(\ V_T(\phi) = V_0 + \int_0^T \phi dS \geq H\ ) = 1. \tag{10.11}$$

Such a strategy is called a superhedging strategy for the claim H. For a self-financing superhedging strategy, the cost corresponds to initial capital V_0. The cost of the cheapest superhedging strategy is called the cost of superhedging:

$$\Pi^{\text{sup}}(H) = \inf\{V_0,\ \exists \phi \in \mathcal{S},\ \mathbb{P}(V_0 + \int_0^T \phi dS \geq H) = 1\}.$$

The superhedging cost is interpreted as the cost of eliminating all risk associated with the option. For an operator with a short position in the option — the option writer — it is the cost of hedging the option with probability 1. Obviously if the option writer is willing to take on some risk she will be able to partially hedge the option at a cheaper cost, thus the superhedging cost represents an upper bound for the selling price. Similarly, the cost of superhedging a short position in H, given by $-\Pi^{\text{sup}}(-H)$ gives a lower bound on the buying price. Note that Π^{sup} is not linear therefore in general $-\Pi^{\text{sup}}(-H) \neq \Pi^{\text{sup}}(H)$: the interval $[-\Pi^{\text{sup}}(-H), \Pi^{\text{sup}}(H)]$ therefore defines a price interval in which all prices must lie.

Computing the superhedging price involves the solution of the nontrivial optimization problem (10.14) but has the advantage of being preference-free: its definition does not involve any subjective risk aversion parameter nor does it involve any ad hoc choice of a martingale measure. The following result (see [240, Theorem 3.2.] and [145]) shows why: the cost of superhedging in an incomplete market corresponds to the most conservative risk neutral price, i.e., the supremum over *all* pricing rules compatible with $\mathbb{P}$:

PROPOSITION 10.1 Cost of superhedging (see Kramkov[240])
Consider a European option with a positive payoff H on an underlying asset

described by a semimartingale $(S_t)_{t\in[0,T]}$ *and assume that*

$$\sup_{\mathbb{Q}\in M(S)} E^{\mathbb{Q}}[H]. < \infty \tag{10.12}$$

Then the following duality relation holds:

$$\inf_{\phi\in\mathcal{S}}\{\hat{V}_t(\phi), \mathbb{P}(V_T(\phi) \geq H) = 1\} = \operatorname{ess\,sup}_{\mathbb{Q}\in M_a(S)} E^{\mathbb{Q}}[\hat{H}|\mathcal{F}_t]. \tag{10.13}$$

In particular, the cost of the cheapest superhedging strategy for H *is given by*

$$\Pi^{\text{sup}}(H) = \operatorname{ess\,sup}_{\mathbb{Q}\in M_a(S)} E^{\mathbb{Q}}[\hat{H}], \tag{10.14}$$

where $M_a(S)$ *the set of martingale measures absolutely continuous with respect to* $\mathbb{P}$.

This result, first obtained in [231] in the context of diffusion models, was generalized to discontinuous processes in [240, 145]. It means that, in terms of equivalent martingale measures, superhedging cost corresponds to the value of the option under the least favorable martingale measure. Kramkov [240] further shows that the supremum in (10.13) is attained: there exists a least favorable martingale measure $\mathbb{Q}^{\text{sup}}$ which corresponds to concentrating the probability on the "worst case scenarios" for H. However Π^{sup} defines a nonlinear valuation method and cannot be described in terms of a single martingale measure: the "least favorable martingale measure" depends on H. Proposition 10.1 allows in some cases to compute the cost of superhedging:

PROPOSITION 10.2 Superhedging in exponential-Lévy models
Consider a model defined by $S_t = S_0 \exp X_t$ *where* X *is a Lévy process on* $(\Omega, \mathcal{F}, (\mathcal{F}_t), \mathbb{P})$.

- *(Eberlein and Jacod [124]): if* X *is a Lévy process with infinite variation, no Brownian component, negative jumps of arbitrary size and Lévy measure verifying* ν *verifies:* $\int_0^1 \nu(dy) = +\infty$ *and* $\int_{-1}^0 \nu(dy) = +\infty$ *then the range of prices*

$$[\inf_{\mathbb{Q}\in M(S)} E^{\mathbb{Q}}[(S_T-K)^+], \sup_{\mathbb{Q}\in M(S)} E^{\mathbb{Q}}[(S_T-K)^+]]$$

  for a call option is given by

$$[(S_0e^{rT}-K)^+, S_0]. \tag{10.15}$$

- *(Bellamy and Jeanblanc [46]): if* X *is a jump-diffusion process with diffusion coefficient* σ *and compound Poisson jumps then the price range for a call option is given by:*

$$[C^{\text{BS}}(0, S_0; T, K; \sigma), S_0], \tag{10.16}$$

where $C^{BS}(t, S, T, K; \sigma)$ denotes the value of a call option in a Black-Scholes model with volatility σ.

Similar bounds hold for the price computed at any date t with the expectations replaced by conditional expectations and for other convex payoff structures[1] verifying some growth conditions [124, 46]. Models for which the above results are valid include all jump-diffusion models found in the literature as well as all exponential-Lévy models with infinite variation presented in Chapter 4. In fact Bellamy and Jeanblanc [46] show that a result analogous to (10.16) holds for jump-diffusions with time- and state-dependent coefficients:

$$\frac{dS_t}{S_{t-}} = b_t dt + \sigma(t, S_{t-})dW_t + a(t, S_{t-})dN_t,$$

where N is a Poisson process with time-dependent intensity $\lambda(t)$ and the coefficients σ, a are bounded Lipschitz functions. In this case, the lower bound in (10.16) is replaced by the option price computed in the diffusion model with local volatility function $\sigma(t, S_{t-})$.

The above result shows that even for call options the superhedging cost is too high (it is the maximum price allowed by absence of arbitrage) and the corresponding hedging strategies are too conservative. For a call option on an asset driven by a Wiener process and a Poisson process $S_t = S_0 \exp(\sigma W_t + a N_t)$ since for any martingale measure $\mathbb{Q}$, $E^{\mathbb{Q}}[(S_T - K)^+] \leq E^{\mathbb{Q}}[S_T] \leq S_0 \exp(rT)$ condition (10.12) is verified so by Proposition 10.1 the superhedging cost is given by S_0. So, however small the jump size, the cheapest superhedging strategy for a call option is actually a total hedge, requiring to buy one unit of asset (or forward contract) for each option sold! The same results hold for the pure jump models discussed in Proposition 10.2. It seems therefore that superhedging is not a great idea in such models: the option writer has to take on some risk in order to reduce hedging costs to a realistic level, which is indeed what happens in real markets! Therefore the hedging problem should be seen as an approximation problem: approximating — instead of perfectly hedging — a random payoff by the terminal value of a dynamic portfolio.

Note however that this example is based on stand-alone pricing and hedging of a derivative (call option, in this case). An options trader typically holds positions in many derivatives with different exposures in terms of sensitivity to the underlying. Under a linear pricing rule, a portfolio can be priced/hedged by computing prices/hedge ratios for each of its components. This is not true anymore for superhedging or other nonlinear hedging schemes: in this case nonlinearity saves the day and in fact superhedging a portfolio of derivatives is much less expensive than individually superhedging the options in the portfolio [15]. This remark suggests that superhedging is not as hopeless as it seems, provided it is applied to the aggregate position of a portfolio of derivatives.

[1] Note however that (10.14) was shown without any convexity hypothesis on the payoff in [240].

10.3 Utility maximization

The unrealistic results of the superhedging approach stem from the fact that it gives equal importance to hedging in all scenarios which can occur with nonzero probability, regardless of the actual loss in a given scenario. A more flexible approach involves weighting scenarios according to the losses incurred and minimizing this weighted average loss. This idea, which has a long tradition in the theory of choice under uncertainty, is formalized using the notion of expected utility[2]: an agent faced with an uncertain environment will choose among random future payoffs Z according to the criterion

$$\max_{Z} E^{\mathbb{P}}[U(Z)], \tag{10.17}$$

where $U : \mathbb{R} \to \mathbb{R}$ is a concave, increasing function called the utility function of the investor and $\mathbb{P}$ is a probability distribution which can be seen either as an "objective" description of future events [380] or as a subjective view held by the investor [347]. The concavity of U is related to the *risk aversion* of the agent. A typical example is the logarithmic utility function $U(x) = \ln \alpha x$. Another example is is the exponential utility function $U^{\alpha}(x) = 1 - \exp(-\alpha x)$ where $\alpha > 0$ determines the degree of risk aversion: a large α corresponds to a high degree of risk aversion. Exponential utilities are sometimes called "CARA" utilities.[3]

10.3.1 Certainty equivalent

Consider now an investor with a utility function U and an initial wealth x. In economic theory, a classical concept of value for an uncertain payoff H is the notion of *certainty equivalent* $c(x, H)$ defined as the sum of cash which, added to the initial wealth, results in the same level of expected utility:

$$U(x + c(x, H)) = E[U(x + H)] \Rightarrow c(x, H) = U^{-1}(E[U(x + H)]) - x.$$

The function $c(x, H)$ is interpreted as the compensation that a risk averse investor with utility U requires in order to assume the risk incurred by holding H. An investor who uses expected utility as a criterion is then indifferent between receiving the random payoff H or the lump sum $c(x, H)$.

The certainty equivalent is an example of a nonlinear valuation: in general, the certainty equivalent of $\lambda > 0$ units of the contract H is not obtained by multiplying by λ the value of one unit: $c(x, \lambda H) \neq \lambda c(x, H)$. Also, in general $c(x, H)$ depends on the initial wealth x held by the investor. In the case of an exponential utility, this dependence disappears.

---

[2] We refer to [147] for an introduction to this subject in relation with mathematical finance.

[3] "CARA" stands for "Constant Absolute Risk Aversion": that is four words instead of "exponential".

10.3.2 Utility indifference price

When H represents the payoff of an option, the certainty equivalent $c(x, H)$ corresponds to the (expected) utility obtained from holding a position in the option. But in a financial market where (dynamic) trading in the underlying is possible, an investor with a position in the option will not in general follow a buy and hold strategy: she will have the possibility to increase her utility/ reduce losses by trading in the underlying asset.

If the investor follows a self-financing strategy $(\phi_t)_{t\in[0,T]}$ during $[0, T]$ then her final wealth is given by

$$V_T = x + \int_0^T \phi_t dS_t. \tag{10.18}$$

A utility maximizing investor will therefore attempt to choose a trading strategy ϕ to optimize the utility of her final wealth:

$$u(x, 0) = \sup_{\phi\in\mathcal{S}} E^{\mathbb{P}}[U(x + \int_0^T \phi_t dS_t)]. \tag{10.19}$$

If the investor buys an option with terminal payoff H at price p and holds it until maturity T then the maximal utility obtained by trading in the underlying is

$$u(x - p, H) = \sup_{\phi\in\mathcal{S}} E^{\mathbb{P}}[U(x - p + H + \int_0^T \phi_t dS_t)]. \tag{10.20}$$

The utility indifference price, introduced by Hodges and Neuberger [197], is defined as the price $\pi_U(x, H)$ which equalizes these two quantities:

$$u(x, 0) = u(x - \pi_U(x, H), H). \tag{10.21}$$

Equation (10.21) is interpreted in the following way: an investor with initial wealth x and utility function U, trading in the underlying, will be indifferent between buying or not buying the option at price $\pi_U(x, H)$. The notion of utility indifference price extends the notion of certainty equivalent to a setting where uncertainty is taken into account.

Several remarks are noteworthy here. First, notice that indifference price is not linear:

$$\pi_U(x, \lambda H) \neq \lambda\pi_U(x, H) \quad \pi_U(x, H_1 + H_2) \neq \pi_U(x, H_1) + \pi_U(x, H_2).$$

Second, the utility indifference price depends in general on the initial wealth (more generally, on the initial portfolio) of the investor except for special utility functions such as $U_\alpha(x) = 1 - e^{-\alpha x}$. Third, buying and selling are not symmetric operations since the utility function weighs gains and losses in an asymmetric way: the utility indifference selling price defined as the price p which solves:

$$u(x,0) = u(x+p,-H), \tag{10.22}$$

which means that the selling price is given by $-\pi_U(x,-H)$ which, in general, is different from the buying price $\pi_U(x,H)$: this approach naturally leads to a pair of prices $\{\pi_U(x,H), -\pi_U(x,-H)\}$, which one is tempted to identify with bid/ask prices (see however the discussion in Section 10.3.4).

10.3.3 The case of exponential utility

The utility indifference price of an option depends in general on the initial wealth x of the investor. This means that two investors with the same utility functions but different initial wealths may not agree on the value of the option. However in the case of exponential utility functions the initial wealth cancels out in Equation (10.21) and one obtains an indifference price independent of initial wealth, which can then be used to define a (nonlinear) pricing rule $\Pi^\alpha(H)$. Indifference pricing with exponential utilities has been extensively studied and we summarize here some of its properties [44, 130, 107]:

PROPOSITION 10.3 Indifference prices for exponential utility
Let $\Pi_\alpha(H)$ be the utility indifference price for an exponential utility function $U^\alpha(x) = 1 - \exp(-\alpha x)$. Then:

1. *When the risk aversion parameter α tends to infinity the utility indifference price converges to the superhedging cost:*

$$\lim_{\alpha\to\infty} \Pi^\alpha(H) = \sup_{\mathbb{Q}\in M_e(S)} E^{\mathbb{Q}}[H]. \tag{10.23}$$

2. *When $\alpha \to 0$ the utility indifference price defines a linear pricing rule given by*

$$\lim_{\alpha\to 0} \Pi^\alpha(H) = E^{\mathbb{Q}^*}[H], \tag{10.24}$$

where $\mathbb{Q}^*$ is a martingale measure equivalent to $\mathbb{P}$ which minimizes the relative entropy with respect to $\mathbb{P}$:

$$\mathcal{E}(\mathbb{Q}^*|\mathbb{P}) = \inf_{\mathbb{Q}\in M^a(S)} \mathcal{E}(\mathbb{Q}|\mathbb{P}). \tag{10.25}$$

These results show that in the case of exponential utility the indifference price interpolates between the superhedging price (which is the maximal possible price in any model) and a *linear* pricing rule based on a martingale measure $\mathbb{Q}^*$ which minimizes the relative entropy with respect to $\mathbb{P}$. $\mathbb{Q}^*$ is called the minimal entropy martingale measure [154] and will be further discussed in Section 10.5.1

These results also show that the indifference price $\pi^\alpha(H)$ is not robust to changes in the risk aversion parameter α: since this parameter is a subjective, unobservable quantity, this should raise some doubts on the applicability of this method for pricing.

10.3.4 On the applicability of indifference pricing

Is utility maximization a realistic valuation method for pricing and hedging options?

Advocates of the use of utility functions in pricing and risk management often refer to the classic works of Von Neumann and Morgenstern [380] and Savage [347] to justify this approach. Von Neumann and Morgenstern [380] showed that, when facing uncertainty described by a probability distribution $\mathbb{P}$, a decision maker whose behavior obeys a set of axioms will behave as if she were maximizing expected utility $E^{\mathbb{P}}[U(X)]$ over possible portfolios X for some utility function U. Under additional axioms and without postulating the prior knowledge of any probability distribution, Savage showed that the decision maker behaves as if she were maximizing $E^{\mathbb{P}}[U(X)]$ over possible portfolios X for *some* utility function U and some (subjective) probability $\mathbb{P}$. The point is that, even if the "rational" decision maker fulfills these behavioral axioms, she does not necessarily know her (own) utility function U (nor, in the case of Savage, the probabilities $\mathbb{P}$).

One can sweep these questions aside by considering expected utility as a practical (normative) way to optimize portfolios. But then pricing and hedging by utility functions requires specifying U and $\mathbb{P}$. What is the "utility function" of an investment bank or an exotic derivatives trading desk? If the results of this analysis were robust to changes in U, one would not worry too much but as the example of exponential utility showed, even within a parametric class one can sweep a wide range of prices by changing the risk aversion parameter. As to $\mathbb{P}$, it requires the full specification of a joint statistical model for future market movements of all relevant assets, up to the maturity of the option. In fact a subtle point which may go unnoticed at first sight is that, given the formulation of the hedging problem as a portfolio optimization problem, even if one is concerned with pricing/hedging an option on as asset S^1, including or not including a second asset S^2 will affect in general the result of the optimization [350]: adding a newly listed stock to the universe of assets described by the joint statistical model $\mathbb{P}$ can potentially affect prices of options on blue-chip stocks such as IBM! In this respect, pricing by utility maximization is more similar to an portfolio allocation problem than to arbitrage pricing models.

Finally, as noted earlier, utility maximization leads to a nonlinear pricing rule. Nonlinear pricing may be acceptable for over the counter (OTC) structured products but for "vanilla" instruments, linear pricing is implicitly assumed by market participants. The only "utility" function which yields a linear pricing rule is $U(x) = -x^2$: the corresponding hedging approach is called quadratic hedging.

10.4 Quadratic hedging

Quadratic hedging can be defined as the choice of a hedging strategy which minimizes the hedging error in a mean square sense. Contrarily to the case of utility maximization, losses and gains are treated in a symmetric manner. This corresponds to the market practice of measuring risk in terms of "variance" and, as we will see, leads in some cases to explicitly computable hedging strategies.

The criterion to be minimized in a least squares sense can be either the hedging error at maturity or the "one step ahead hedging error," i.e., the hedging error, measured locally in time. The first notion leads to mean-variance hedging [69, 318] and the second notion leads to local risk minimization [149, 148, 355, 356]. In order to take expectations, a probability measure has to be specified. We will see that the two approaches — mean variance hedging and local risk minimization — are equivalent when the (discounted) price is a martingale measure but different otherwise.

10.4.1 Mean-variance hedging: martingale case

In mean variance hedging, we look for a self-financing strategy given by an initial capital V_0 and a portfolio (ϕ_t^0, ϕ_t) over the lifetime of the option which minimizes the terminal hedging error in a mean-square sense:

$$\inf_{\phi} E[\ |V_T(\phi) - H|^2] \text{ where } V_T(\phi) = V_0 + \int_0^T r\phi_t^0 dt + \int_0^T \phi_t dS_t. \quad (10.26)$$

The expectation is taken with respect to some probability which we have to specify. To begin, let us assume that we have chosen a pricing rule given by a risk neutral measure $\mathbb{Q}$ and the expectation in (10.26) is taken with respect to $\mathbb{Q}$. In particular, $\hat{S}_t$ is a martingale under $\mathbb{Q}$ and after discounting the problem above simplifies to:

$$\inf_{V_0,\phi} E^{\mathbb{Q}}|\epsilon(V_0,\phi)|^2, \quad \text{where} \quad \epsilon(V_0,\phi) = \hat{H} - \hat{V}_T = \hat{H} - V_0 - \int_0^T \phi_t d\hat{S}_t.$$

Assume now that H has finite variance — $H \in L^2(\Omega, \mathcal{F}, \mathbb{Q})$ — and $(\hat{S}_t)_{t\in[0,T]}$ is a square-integrable $\mathbb{Q}$-martingale. If we consider portfolios whose terminal values have finite variance:

$$\mathcal{S} = \{\phi \text{ caglad predictable and } E|\int_0^T \phi_t d\hat{S}_t|^2 < \infty \quad \} \tag{10.27}$$

then the set $\mathbb{A}$ of attainable payoffs defined by (10.1) is a closed subspace of the space of random variables with finite variance, denoted by $L^2(\Omega, \mathcal{F}, \mathbb{Q})$. Defining a scalar product between random variable as $(X,Y)_{L^2} := E[XY]$, this space becomes a Hilbert space: two random variables X, Y with finite variance are then said to be *orthogonal* if $E[XY] = 0$.

Rewriting the mean-variance hedging problem as

$$\inf_{V_0,\phi} E^{\mathbb{Q}}|\epsilon(V_0,\phi)|^2 = \inf_{A\in\mathbb{A}} ||\hat{H} - A||^2_{L^2(\mathbb{Q})}, \tag{10.28}$$

we see that the problem of minimizing the mean-square hedging error (10.26) can be interpreted as the problem of finding the orthogonal projection in $L^2(\mathbb{Q})$ of the (discounted) payoff $\hat{H}$ on the set of attainable claims $\mathbb{A}$. This orthogonal decomposition of a random variable with finite variance into a stochastic integral and an "orthogonal" component [242, 156] is called the Galtchouk-Kunita-Watanabe decomposition:

PROPOSITION 10.4 Galtchouk-Kunita-Watanabe decomposition
Let $(\hat{S}_t)_{t\in[0,T]}$ be a square-integrable martingale with respect to $\mathbb{Q}$. Any random variable $\hat{H}$ with finite variance depending on the history $(\mathcal{F}^S_t)_{t\in[0,T]}$ of $\hat{S}$ can be represented as the sum of a stochastic integral with respect to $\hat{S}$ and a random variable N orthogonal to $\mathbb{A}$: there exists a square integrable predictable strategy $(\phi^H_t)_{t\in[0,T]}$ such that, with probability 1

$$\hat{H} = E^{\mathbb{Q}}[\hat{H}] + \int_0^T \phi^H_t d\hat{S}_t + N^H, \tag{10.29}$$

where N^H is orthogonal to all stochastic integrals with respect to $\hat{S}$. Moreover, the martingale defined by $N^H_t = E^{\mathbb{Q}}[N^H|\mathcal{F}_t]$ is strongly *orthogonal to $\mathbb{A}$: for any square integrable predictable process $(\theta_t)_{t\in[0,T]}$, $N_t \int_0^t \theta dS$ is again a martingale.*

The integral $\int_0^T \phi^H_t dS_t$ is the orthogonal projection of the random variable H_T on the space of all stochastic integrals with respect to S. If a strategy that minimizes quadratic hedging error exists, it is given by ϕ^H: however as noted in Chapter 8, any predictable process ϕ^H cannot be interpreted as a trading strategy unless it is either a simple predictable process (piecewise constant in time) or can be approximated by such strategies (e.g., is caglad). In the next section we will compute ϕ^H in exponential-Lévy models and observe that it is indeed a caglad process. The random variable N^H represents the residual risk of the payoff H that cannot be hedged. Note that we started by looking for the orthogonal projection of the final payoff H on $\mathbb{A}$ but the decomposition (10.29) actually gives more: because of the martingale property of $\hat{S}$, for every t, the random variable $N_t^H = E^{\mathbb{Q}}[N^H|\mathcal{F}_t]$ is still orthogonal to payoffs of self-financing portfolios on $[0,t]$ so by optimizing the *global* hedging error (10.26) we obtain a strategy which is also *locally* risk minimizing. As we shall see below, this will not be true anymore if $\hat{S}$ is not a martingale.

In a complete market model $N^H = 0$ almost surely but we will now study an example where $N^H \neq 0$, resulting in a nonzero residual risk.

10.4.2 Mean-variance hedging in exponential-Lévy models

Although the quadratic hedging problem is "solved" by the Galtchouk-Kunita-Watanabe decomposition, from a practical point of view the problem is of course to compute the risk minimizing hedge ϕ_t^H which, in general, is not an easy task. Interesting models in this context are therefore models where an explicit expression (or an efficient computational procedure) is available for ϕ_t^H. We will now show that exponential-Lévy models introduced earlier fall into this category and give the explicit form for the risk minimizing hedge in this case.

Consider a risk-neutral model $(S_t)_{t\in[0,T]}$ given by $S_t = \exp(rt + X_t)$ where X_t is a Lévy process on $(\Omega, \mathcal{F}, \mathcal{F}_t, \mathbb{Q})$. Here $\mathcal{F}_t = \mathcal{F}_t^S = \mathcal{F}_t^X$ is the price history $(S_u, 0 \leq u \leq t)$. Using Proposition (3.13) S is a square integrable martingale if and only if

$$\int_{|y|\geq 1} e^{2y}\nu(dy) < \infty. \tag{10.30}$$

We will assume that this condition is verified; in particular X_t has finite variance. The characteristic function of X can then be expressed as:

$$E[\exp(iuX_t)] = \exp t[-\frac{\sigma^2 u^2}{2} + b_X t + \int \nu_X(dy)(e^{iuy} - 1 - iuy)]$$

with b_X chosen such that $\hat{S} = \exp X$ is a martingale. As explained in Section 8.4, $\hat{S}_t$ can also be written as a stochastic exponential of another Lévy process (Z_t):

$$d\hat{S}_t = \hat{S}_{t-}dZ_t,$$

where Z is a martingale with jumps > -1 and also a Lévy process with Lévy measure given by Proposition 8.22. Consider now a self-financing strategy $(\phi_t^0, \phi_t)_{t\in[0,T]}$. In order for quadratic hedging criteria to make sense, we must restrict ourselves to portfolios verifying (10.27) for which the terminal value has a well-defined variance. Using the isometry formulas (8.23) and (8.13) for stochastic integrals, (10.27) is equivalent to

$$E[\int_0^T |\phi_t \hat{S}_t|^2 dt + \int_0^T \int_{\mathbb{R}} dt\nu(dz) z^2 |\phi_t \hat{S}_t|^2] < \infty. \tag{10.31}$$

Denote by $L^2(S)$ the set of processes ϕ verifying (10.31). The terminal payoff of such a strategy is given by

$$G_T(\phi) = \int_0^T r\phi_t^0 dt + \int_0^T \phi_t S_{t-} dZ_u.$$

Since $\hat{S}_t$ is a martingale under $\mathbb{Q}$ and $\phi \in L^2(\hat{S})$ the discounted gain process, $\hat{G}_t(\phi) = \int_0^t \phi d\hat{S}$ is also a square integrable martingale given by the martingale part of the above expression:

$$\begin{aligned}\hat{G}_T(\phi) &= \int_0^T \phi_t S_{t-} \sigma dW_t + \int_0^T \int_{\mathbb{R}} \tilde{J}_X(dt\ dx) x \phi_t S_{t-} \\ &= \int_0^T \phi_t S_{t-} \sigma dW_t + \int_0^T \int_{\mathbb{R}} \tilde{J}_Z(dt\ dz) \phi_t S_{t-} (e^z - 1)\end{aligned}$$

using Proposition 8.22. The quadratic hedging problem can now be written as:

$$\inf_{\phi \in L^2(\hat{S})} E^{\mathbb{Q}} |\hat{G}_T(\phi) + V_0 - \hat{H}|^2. \tag{10.32}$$

First note that the expectation of the hedging error is equal to $V_0 - E^{\mathbb{Q}}[\hat{H}]$. Decomposing the above expression into

$$E^{\mathbb{Q}} |V_0 - E^{\mathbb{Q}}[\hat{H}]|^2 + \mathrm{Var}^{\mathbb{Q}}[\hat{G}_T(\phi) - \hat{H}],$$

we see that the optimal value for the initial capital is

$$V_0 = E^{\mathbb{Q}}[H(S_T)].$$

PROPOSITION 10.5 Quadratic hedge in exponential-Lévy models

Consider the risk neutral dynamics

$$\mathbb{Q}: \qquad d\hat{S}_t = \hat{S}_{t-} dZ_t, \tag{10.33}$$

where Z is a Lévy process with Lévy measure ν_Z and diffusion coefficient $\sigma > 0$. For a European option with payoff $H(S_T)$ where $H : \mathbb{R}^+ \to \mathbb{R}$ verifies

$$\exists K > 0, \qquad |H(x) - H(y)| \leq K|x - y|, \tag{10.34}$$

the risk minimizing hedge, solution of (10.32) amounts to holding a position in the underlying equal to $\phi_t = \Delta(t, S_{t-})$ *where:*

$$\Delta(t,S) = \frac{\sigma^2 \frac{\partial C}{\partial S}(t,S) + \frac{1}{S}\int \nu_Z(dy) z [C(t, S(1+z)) - C(t,S)]}{\sigma^2 + \int z^2 \nu_Z(dy)} \tag{10.35}$$

with $C(t,S) = e^{-r(T-t)} E^{\mathbb{Q}}[H(S_T)|S_t = S]$.

This result can be retrieved using the formalism for quadratic hedging in Markov models discussed in [69] or using stochastic flows and their derivatives as in [144]. We give here a direct proof using martingale properties of the stochastic integral.

PROOF Under the risk-neutral measure $\mathbb{Q}$, the discounted price $\hat{S}_t$ is a martingale. Consider now a self-financing trading strategy given by a nonanticipating caglad process (ϕ_t^0, ϕ_t) with $\phi \in L^2(\hat{S})$: the discounted value $(\hat{V}_t)$ of the portfolio is then a martingale whose terminal value is given by

$$\begin{aligned} \hat{V}_T &= \int_0^T \phi_t d\hat{S}_t = \int_0^T \phi_t \hat{S}_{t-} dZ_t \\ &= \int_0^T \phi_t \hat{S}_t \sigma dW_t + \int_0^T \int_{\mathbb{R}} \phi_t \hat{S}_t z \tilde{J}_Z(dt\ dz). \end{aligned} \tag{10.36}$$

Define now the function

$$C(t,S) = e^{-r(T-t)} E^{\mathbb{Q}}[H(S_T)|\mathcal{F}_t^S] = e^{-r(T-t)} E^{\mathbb{Q}}[H(S_T)|S_t = S]$$

and its discounted value by $\hat{C}(t,S) = e^{-rt} C(t,S)$. Note that $C(0,S_0) = E^{\mathbb{Q}}[H(S_T)]$ is the value of the option under the pricing rule $\mathbb{Q}$ while $C(T,S) = H(S_T)$ is the payoff of the option at maturity. By construction $\hat{C}(t,S_t) = e^{-rT} E^{\mathbb{Q}}[H(S_T)|\mathcal{F}_t^S]$ is a (square integrable) martingale. In Chapter 12 we will show that when $\sigma > 0$ (but this is not the only case), C is a smooth, $C^{1,2}$ function. Thus, the Itô formula can be applied to $\hat{C}(t,S_t) = e^{-rt} C(t,S_t)$ between 0 and t:

$$\begin{aligned} \hat{C}(t,S_t) - \hat{C}(0,S_0) &= \int_0^t \frac{\partial C}{\partial S}(u, S_{u-}) \hat{S}_{u-} \sigma dW_u \\ &+ \int_0^t \int_{\mathbb{R}} [C(u, S_{u-}(1+z)) - C(u, S_{u-})] \tilde{J}_Z(du\ dz) \\ &= \int_0^t \frac{\partial C}{\partial S}(u, S_{u-}) \hat{S}_{u-} \sigma dW_u \\ &+ \int_0^t \int_{\mathbb{R}} [C(u, S_{u-} e^x) - C(u, S_{u-})] \tilde{J}_X(du\ dx), \end{aligned} \tag{10.37}$$

where (X_t) is a Lévy process such that $\hat{S}_t = \exp(X_t)$ for all t (see Proposition 8.22). Since the payoff function H is Lipschitz, C is also Lipschitz with respect to the second variable:

$$C(t,x) - C(t,y) = e^{-r(T-t)}E[H(xe^{r(T-t)+X_{T-t}}) - H(ye^{r(T-t)+X_{T-t}})] \\ \leq K|x-y|E[e^{X_{T-t}}] = K|x-y|$$

since e^{X_t} is a martingale. Therefore the predictable random function $\psi(t,z) = [C(t,S_{t-}(1+z)) - C(t,S_{t-})]$ verifies

$$E\Big[\int_0^T dt \int_{\mathbb{R}} \nu_Z(dz)|\psi(t,z)|^2\Big] \\ = E\Big[\int_0^T dt \int_{\mathbb{R}} \nu_Z(dz)|C(t,S_{t-}(1+z)) - C(t,S_{t-})|^2\Big] \\ \leq E\Big[\int_0^T dt \int_{\mathbb{R}} z^2 S_{t-}^2 \nu(dz)\Big] < \infty$$

so from Proposition 8.7, the compensated Poisson integral in (10.37) is a square integrable martingale. Subtracting (10.36) from (10.37) we obtain the hedging error:

$$\epsilon(V_0,\phi) = \int_0^T [\phi_t S_{t-} - S_{t-}\frac{\partial C}{\partial S}(t,S_{t-})]\sigma dW_t \\ + \int_0^T dt \int_{\mathbb{R}} \tilde{J}_Z(dt\ dz)[\phi_t S_{t-} z - (C(t,S_{t-}(1+z)) - C(t,S_{t-}))],$$

where each stochastic integral is a well-defined, zero-mean random variable with finite variance. The isometry formulas (8.23) and (8.13) allow to compute the variance of the hedging error:

$$E|\epsilon(\phi)|^2 = E[\int_0^T dt \int_{\mathbb{R}} \nu_Z(dz)|C(t,S_{t-}(1+z)) - C(t,S_{t-}) - \hat{S}_{t-}\phi_t z|^2] \\ + E[\int_0^T \hat{S}_{t-}^2(\phi_t - \frac{\partial C}{\partial S}(t,S_{t-})\ )^2\sigma^2 dt]. \tag{10.38}$$

Notice now that the term under the integral in (10.38) is a positive process which is a quadratic function of ϕ_t with coefficients depending on (t,S_{t-}). The optimal (risk-minimizing) hedge is obtained by minimizing this expression with respect to ϕ_t: differentiating the quadratic expression we obtain the first order condition:

$$\hat{S}_{t-}^2\sigma^2(\phi_t - \frac{\partial C}{\partial S}(t,S_{t-})) + \\ \int_{\mathbb{R}} \nu_Z(dz)\hat{S}_{t-}z[\hat{S}_{t-}\phi_t z - C(t,S_{t-}(1+z)) - C(t,S_{t-})] = 0,$$

whose solution is given by (10.35). □

REMARK 10.1 Since the functions $H(S) = (S-K)^+$ and $H(S) = (K-S)^+$ verify the Lipschitz property $|H(x) - H(y)| \leq |x-y|$ these results hold in particular for call options, put options and any combination of these: straddles, strangles, butterfly spreads, etc. □

REMARK 10.2 In Chapter 12, we will see that the function $C(t, S)$ solves a partial integro-differential equation which can be solved using efficient numerical methods. The hedging strategy in (10.35) can thus be computed and used in order to set up a risk minimizing hedge. □

REMARK 10.3 In the exponential-Lévy formulation $S_t = S_0 \exp(rt + X_t)$, the optimal quadratic hedge can be expressed in terms of the Lévy measure ν_X of X as

$$\Delta(t,S) = \frac{\sigma^2 \frac{\partial C}{\partial S}(t,S) + \frac{1}{S}\int \nu_X(dx)(e^x-1)[C(t,Se^x) - C(t,S)]}{\sigma^2 + \int (e^x-1)^2 \nu_X(dx)}. \quad (10.39)$$

□

Proposition 10.5 gives an example where the Galtchouk-Kunita-Watanabe decomposition (10.29) can be computed explicitly. A similar formula is given for C^2 payoffs in the case of Markov processes with jumps in [144]; the more general case of convex/concave payoffs is treated in [213] but the formulas obtained are less explicit.

As a by-product of this computation we have obtained an expression for the residual risk of a hedging strategy (ϕ_t^0, ϕ_t):

$$R_T(\phi) = E[\int_0^T |\phi_t - \frac{\partial C}{\partial S}(t, S_{t-})|^2 S_{t-}^2 dt$$
$$+ \int_0^T dt \int_{\mathbb{R}} \nu(dz) |C(t, S_{t-}(1+z)) - C(t, S_{t-}) - z\, S_{t-}\phi_t|^2].$$

This allows us to examine whether there are any cases where the hedging error can be reduced to zero, i.e., where one can achieve a perfect hedge. A well-known case is when there are no jumps $\nu = 0$: the residual risk then reduces to

$$\epsilon(\phi) = E[\int_0^T (\phi_t S_{t-} - S_{t-}\frac{\partial C}{\partial S}(t, S_{t-}))^2 dt\ ]$$

and we retrieve the Black-Scholes delta hedge

$$\phi_t = \Delta^{BS}(t, S_t) = \frac{\partial C}{\partial S}(t, S_t),$$

which gives $\epsilon(\phi) = 0$ a.s. Another case is when $\sigma = 0$ and there is a single jump size $\nu = \delta_a$: $X_t = aN_t$ where N is a Poisson process. In this case

$$R_T(\phi) = E[\int_0^T dt S_{t-}^2 |C(t, S_{t-}(1+a)) - C(t, S_{t-}) - \phi_t|^2]$$

so by choosing

$$\phi_t = \frac{C(t, S_{t-}(1+a)) - C(t, S_{t-})}{S_{t-}a} \quad \text{and}$$

$$\phi_t^0 = e^{rt} S_t \phi_t - e^{rt} \int_0^t \phi_t dS_t, \tag{10.40}$$

we obtain a self-financing strategy (ϕ, ϕ^0) which is a replication strategy:

$$H(S_T) = V_0 + \int_0^T \frac{C(t, S_{t-}(1+a)) - C(t, S_{t-})}{S_{t-}a} dS_t + \int_0^T r\phi_t^0 dt.$$

These two cases are are the *only* cases where perfect hedging is possible: if there are at least two jump sizes a_1, a_2 then a perfect hedge should verify

$$\phi_t = \frac{C(t, S_{t-}(1+a_1)) - C(t, S_{t-})}{S_{t-}a_1} = \frac{C(t, S_{t-}(1+a_2)) - C(t, S_{t-})}{S_{t-}a_2},$$

which is impossible unless C is affine in S, i.e., for a forward contract. Also, if both jumps and the diffusion component are present, that is, $\nu \neq 0$ and $\sigma \neq 0$, then a perfect hedge should verify

$$\phi_t = \frac{C(t, S_{t-}(1+z)) - C(t, S_{t-})}{S_{t-}a} = \frac{\partial C}{\partial S}(t, S) \qquad \nu(dz) - a.e. \tag{10.41}$$

almost surely, which implies

$$\forall S > 0, \quad \frac{C(t, S(1+z)) - C(t, S)}{Sz} = \frac{\partial C}{\partial S}(t, S) \qquad \nu - a.e. \tag{10.42}$$

For convex functions C — calls and puts — this is not possible so calls and puts cannot be replicated. This gives a direct proof of the incompleteness of exponential-Lévy models: the only exponential-Lévy models which are complete are the Black-Scholes model and the exponential Poisson model $S_t = S_0 \exp(-\mu t + N_t)$, none of which are very realistic!

Note that in the case where jumps are present the risk minimizing hedge ratio is *not* given by the Merton strategy $\partial C/\partial S(t, S_{t-})$ discussed in Section 10.1: such a strategy is suboptimal and corresponds to minimizing only the diffusion term in (10.38), leaving the portfolio completely exposed to jump risk. In fact in the case where $\sigma = 0$ and Z is a compensated Poisson process, there exists a perfect hedge as observed above but it is not equal to the Merton hedge $\phi_t^M = \partial C/\partial S(t, S_{t-})$. This point is neglected in several papers (see, e.g., [57]) where, by analogy to the Black-Scholes delta hedge it is simply assumed that the hedge ratio is given by the derivative with respect to the underlying. The quadratic hedge achieves a mean-variance trade-off between the risk due to the diffusion part and the jump risk.

10.4.3 Global vs. local risk minimization (*)

In the quadratic hedging approach described above, we used as a starting point a risk-neutral measure $\mathbb{Q}$ and chose a hedging strategy ϕ in order to minimize the mean square hedging error $E^{\mathbb{Q}}|H - V_T(\phi)|^2$. However using the *risk-neutral* variance $E^{\mathbb{Q}}|\epsilon(\phi)|^2$ of the hedging error as a criterion for measuring risk is not very natural: $\mathbb{Q}$ represents a pricing rule and not a statistical description of market events, so the profit and loss (P&L) of a portfolio may have a large variance while its "risk neutral" variance can be small.

A natural generalization is therefore to try to repeat the same analysis under a statistical model $\mathbb{P}$: choose a self-financing strategy ϕ with initial capital V_0 such as to minimize

$$E^{\mathbb{P}}\left[(V_0 + \int_0^T \phi dS - H)^2\right].$$

This approach is called mean-variance hedging [69, 318, 357, 53]. The solution defines a linear pricing rule $H \mapsto V_0$ which, as any linear arbitrage-free pricing rule, can be represented as:

$$V_0 = E^{\mathbb{Q}_{MV}}[\hat{H}]$$

for some martingale measure $\mathbb{Q}_{MV} \sim \mathbb{P}$, called the variance optimal martingale measure. Unfortunately the analysis is not as easy as in the martingale case and explicit solutions are difficult to obtain for models with jumps, see [318, 53, 200]. Furthermore, as in the case of utility maximization, this approach requires estimating a statistical model for the historical evolution of the underlying and in particular, estimating its expected return, which turns out to be quite difficult.

Another solution is to choose first a risk neutral measure $\mathbb{Q} \sim \mathbb{P}$ and then proceed as in (10.29), by decomposing the claim into a hedgeable part, represented by a self-financing strategy, and a residual risk, orthogonal to (the martingale) $\hat{S}$ under $\mathbb{Q}$. But this decomposition depends on the choice of $\mathbb{Q}$: different choices will lead to different pricing rules and hedging strategies. In particular the notion of orthogonality is not invariant under change of measure: this is unpleasant because when we interpret variables orthogonal to $\hat{S}$ as unhedgeable risks, of course we mean *objectively* unhedgeable and not with respect to some ad hoc choice of martingale measure $\mathbb{Q}$.

A solution proposed by Föllmer and Schweizer [148] was precisely to look for a martingale measure $\mathbb{Q}^{FS} \sim \mathbb{P}$ which respects orthogonality under $\mathbb{P}$. If $\hat{S}_t = M_t + A_t$ where M is the martingale component of S under $\mathbb{P}$, any martingale (N_t) which is strongly orthogonal to (M_t) under $\mathbb{P}$ should remain a martingale orthogonal to $\hat{S}$ under $\mathbb{Q}^{FS}$. Such a choice of $\mathbb{Q}^{FS}$ is called a "minimal martingale measure". It is minimal in the sense that it changes $\hat{S}_t$ into a martingale while perturbing other statistical quantities in a minimal manner and in particular preserving orthogonality relations. If such a measure exists then, writing the Galtchouk-Kunita-Watanabe decomposition (10.29) of $\hat{H}$ under $\mathbb{Q}^{FS}$ we obtain

$$\hat{H} = E^{FS}[\hat{S}] + \int_0^T \phi_t^{FS} d\hat{S}_t + N^H \ a.s., \tag{10.43}$$

where $\Pi^{FS}[H] \in \mathbb{R}$ and N^H is orthogonal to the gains $\int_0^T \theta dS$ of all portfolios $\theta \in L^2(\hat{S})$. Since $\mathbb{Q}^{FS}$ preserves orthogonality, N^H is also orthogonal to the martingale component of $\hat{S}$ under $\mathbb{P}$. So (10.43) provides now a decomposition under the real-world probability $\mathbb{P}$ of the payoff into the sum of

- an initial capital $\Pi^{FS}[H] = E^{\mathbb{Q}^{FS}}[H]$.
- the gain $\int_0^t \phi dS$ of a strategy ϕ^{FS}.
- a martingale $N_t = E^{\mathbb{P}}[N^H|\mathcal{F}_t]$ which is orthogonal to the risk (martingale component) of S_t under $\mathbb{P}$.

The decomposition (10.43), when it exists, is called the Föllmer-Schweizer decomposition and is discussed in [148, 355, 356, 357, 200] and the hedging strategy ϕ_t^H is called a "locally risk minimizing" strategy for H.[4] Note however that it is *not* a self-financing strategy in general: while for a self-financing strategy the cost process is a constant, this strategy has a cost process given by

$$N_t = E^{\mathbb{P}}[N^H|\mathcal{F}_t].$$

However N_t is a martingale with mean zero: the locally risk minimizing strategy ϕ_t^H is self-financing *on average*. A procedure for finding the locally risk minimizing hedge is therefore the following:

- Find the (dynamics of S_t under the) minimal martingale measure $\mathbb{Q}^{FS}$.
- Perform an orthogonal decomposition (10.29) of H into a stochastic integral with respect to $\hat{S}_t$ and a component orthogonal to $\hat{S}_t$:

  $$H = E^{\mathbb{Q}^{FS}}[H] + \int_0^T \phi_t^H dS_t + N_T.$$

  The process ϕ^H is then the locally risk minimizing hedging strategy for H with a cost process given by $N_t = E^{\mathbb{P}}[N_T|\mathcal{F}_t]$.
- The initial capital of the quadratic hedging strategy is given by $E^{\mathbb{Q}^{FS}}[H]$.

[4] This terminology is justified in [355, 336] where ϕ^H is characterized as minimizing a quantity which is interpreted as local quadratic risk.

If $\hat{S}_t$ is an exponential-Lévy process under $\mathbb{Q}^{FS}$, Proposition 10.5 can be used to compute the hedge ratio ϕ^H. Unfortunately, the Föllmer-Schweizer decomposition and the minimal martingale measure $\mathbb{Q}^{FS}$ do not always exist when S is a process with jumps [357].

Let us see how this machinery works in the case of the jump-diffusion model:

$$\frac{dS_t}{S_{t-}} = dZ_t, \qquad Z_t = \mu t + \sigma W_t + \sum_{i=1}^{N_t} Y_i,$$
$$E[N_t] = \lambda t, \qquad Y_i \overset{\text{i.i.d.}}{\sim} F, \qquad E[Y_i] = m, \qquad \text{Var}(Y_i) = \delta^2.$$

In this case, Zhang [390] shows that a minimal martingale measure exists if and only if

$$-1 \leq \eta = \frac{\mu + \lambda m - r}{\sigma^2 + \lambda\left(\delta^2 + m^2\right)} \leq 0.$$

This assumption means that the risk premium in the asset returns should be *negative.*[5] When this condition is verified, the minimal martingale measure $\mathbb{Q}^{FS}$ is given by

$$\frac{d\mathbb{Q}^{FS}}{d\mathbb{P}} = \exp[-\sigma\eta W_T + \lambda\eta m T - \frac{\sigma^2\eta^2}{2}T]\prod_{j=1}^{N_t}(1-\eta U_j)^{N_T}. \tag{10.44}$$

From the results of Chapter 9, the risk-neutral dynamics of the asset under $\mathbb{Q}^{FS}$ can be expressed as

$$\frac{dS_t}{S_{t-}} = r dt + dU_t,$$
$$U_t = \lambda[\eta(m^2 + \delta^2) - m]t + \sigma W_t^{'} + \sum_{i=1}^{N_t^{'}} \Delta U_i, \tag{10.45}$$

where under $\mathbb{Q}^{FS}$

- W_t' is a standard Wiener process.
- N_t' is a Poisson process with intensity $\lambda' = \lambda(1 - \eta m)$.
- The jump sizes (ΔU_i) are i.i.d. with distribution F_U where

$$dF_U = \frac{1-\eta x}{1-\eta m} dF(x).$$

[5] At the time of the publication of this book, this may seem a realistic assumption!

Thus the minimal martingale model is still a jump-diffusion of exponential-Lévy type. Combining this with the results of the preceding section, we obtain an expression for the locally risk minimizing hedge:

PROPOSITION 10.6 Locally risk minimizing hedge in a jump-diffusion model
Consider the jump-diffusion model

$$\frac{dS_t}{S_{t-}} = dZ_t, \quad \text{where} \quad Z_t = \mu t + \sigma W_t + \sum_{i=1}^{N_t} Y_i$$

with $Y_i \sim F$ i.i.d. and N a Poisson process with intensity λ. Then the locally risk minimizing hedge for a European option with payoff $H(S_T)$ verifying (10.34) is given by $\phi_t = \Delta(t, S_{t-})$ where:

$$\Delta(t,S) = \frac{\sigma^2 \frac{\partial C}{\partial S}(t,S) + \frac{\lambda}{S}\int F(dy)y(1-\eta y)[C(t,S(1+y)) - C(t,S)]}{\sigma^2 + \lambda \int y^2(1-\eta y)F(dy)}$$

with $C(t,S) = e^{-r(T-t)}E^{FS}[H(S_T)|S_t = S]$ is the expected discounted payoff taken with respect to (10.45).

When $\hat{S}_t$ is a martingale ($\eta = 0$) local risk minimization and mean-variance hedging lead to the same result, given by the Galtchouk-Kunita-Watanabe decomposition described above. In general, they are not equivalent and lead to different hedging strategies and hedging costs: they correspond to different choices $\mathbb{Q}_{MV} \neq \mathbb{Q}_{FS}$ for the risk-neutral probabilities. Mean-variance hedging is more intuitive: the hedging strategies are self-financing so the initial capital really corresponds to the cost of the hedge and the objective function is the mean-square fluctuation of the P&L. Local risk minimization is mathematically more tractable but does not have a neatly expressible objective function and involves strategies which are not self financing; thus the interpretation of the initial capital as a "cost" of hedging is not justified. Also, as shown by the jump-diffusion example above, the existence of the minimal martingale measure only holds under restrictive assumptions on model parameters. A numerical comparison of these two approaches has been done in the case of diffusion-based stochastic volatility models in [225, Chapter 14]; for the case of models with jumps a systematic study remains to be done.

Both approaches require a good knowledge of the objective probabilities $\mathbb{P}$ of future market scenarios. In particular, we need to specify the drift part, i.e., the expected (future) rate of return on the asset S_t, which is a difficult task from an econometric point of view. On the other hand, as we will see in Chapter 13, a lot of cross-sectional data is available which gives information on the risk-neutral probabilities $\mathbb{Q}$ in the form of option prices: a feasible way to use quadratic hedging (though not the way originally proposed in

[148, 355]) is then to infer a pricing measure $\mathbb{Q}$ by "calibrating" an exp-Lévy model to market prices of options[6] and then perform a quadratic hedge under the martingale measure $\mathbb{Q}$, which can then be done explicitly using Proposition 10.5.

10.5 "Optimal" martingale measures

By the fundamental theorem of asset pricing, choosing an arbitrage-free pricing method is basically equivalent to choosing a martingale measure $\mathbb{Q} \sim \mathbb{P}$. In the literature, this is typically done by solving an optimization problem. A widely studied family of objective functions for choosing probability measures consists of criteria which can be expressed in the form

$$J_f(\mathbb{Q}) = E^{\mathbb{P}}\left[f(\frac{d\mathbb{Q}}{d\mathbb{P}})\right], \tag{10.46}$$

where $f : [0, \infty[\to \mathbb{R}$ is some strictly convex function. $J_f(\mathbb{Q})$ can be seen as measuring a deviation from the prior $\mathbb{P}$. Examples are the Kullback-Leibler distance, also called *relative entropy*:

$$H(\mathbb{Q}, \mathbb{P}) = E^{\mathbb{P}}\left[\frac{d\mathbb{Q}}{d\mathbb{P}} \ln \frac{d\mathbb{Q}}{d\mathbb{P}}\right] \tag{10.47}$$

or the quadratic distance:

$$E\left[\left(\frac{d\mathbb{Q}}{d\mathbb{P}}\right)^2\right]. \tag{10.48}$$

In some cases, the minimal martingale measure and the variance optimal martingale measure introduced in the quadratic hedging problem can also be expressed as "optimal" with respect to some f [357]. We discuss here an example, the minimal entropy martingale measure, already encountered in Section 10.3.

10.5.1 Minimal entropy martingale measure

An important example of the deviation measures described above is the relative entropy, previously introduced in Section 9.6, which corresponds to choosing $f(x) = x \ln x$ in (10.46):

$$\mathcal{E}(\mathbb{Q}, \mathbb{P}) = E^{\mathbb{Q}}\left[\ln \frac{d\mathbb{Q}}{d\mathbb{P}}\right] = E^{\mathbb{P}}\left[\frac{d\mathbb{Q}}{d\mathbb{P}} \ln \frac{d\mathbb{Q}}{d\mathbb{P}}\right].$$

---

[6] See Chapter 13.

Given a stochastic model $(S_t)_{t\in[0,T]}$ the minimal entropy martingale model is defined as a martingale $(S^*_t)_{t\in[0,T]}$ such that the law $\mathbb{Q}^*$ of S^* minimizes the relative entropy with respect to $\mathbb{P}^S$ among all martingale processes on $(\Omega, \mathcal{F}, (\mathcal{F}_t), \mathbb{P})$: it minimizes the relative entropy under the constraint of being a martingale. Or, in the "change of measure" language, one can fix the reference probability measure $\mathbb{P}$ and define the minimal entropy martingale measure as the measure on the path space $D([0,T])$ which is the solution of

$$\inf_{\mathbb{Q}\in M^a(S)} \mathcal{E}(\mathbb{Q}|\mathbb{P}). \tag{10.49}$$

The minimal entropy martingale model has an information theoretic interpretation: minimizing relative entropy corresponds to choosing a martingale measure by adding the least amount of information to the prior model. Finally, we observed in Section 10.3 that option prices computed with the minimal entropy martingale measure are related to the utility indifference price $\Pi_\alpha(H)$ for exponential utility functions [299, 44, 107, 130]:

$$\Pi_\alpha(H) \underset{\alpha\to 0}{\to} E^{\mathbb{Q}^*}[H].$$

Given a stochastic model, the corresponding minimal entropy martingale does not always exist. But in exponential-Lévy models one is able to give an analytic criterion for the existence of the minimal entropy martingale measure and compute it explicitly in cases where it exists. Furthermore, the minimal entropy martingale is also given by an exponential-Lévy model. The following result is given in [299] (see also [88, 170, 298, 136]):

PROPOSITION 10.7 Minimal entropy measure in exp-Lévy model

If $S_t = S_0\exp(rt + X_t)$ where $(X_t)_{t\in[0,T]}$ is a Lévy process with Lévy triplet (σ^2, ν, b). If there exists a solution $\beta\in\mathbb{R}$ to the equation:

$$b + (\beta + \frac{1}{2})\sigma^2 + \int_{-1}^{+1} \nu(dx)[(e^x - 1)e^{\beta(e^x-1)} - x]$$
$$+ \int_{|x|>1} e^x - 1)e^{\beta(e^x-1)}\nu(dx) = 0 \tag{10.50}$$

then the minimal entropy martingale S^*_t is also an exponential-Lévy process $S^*_t = S_0\exp(rt + X^*_t)$ where $(Z_t)_{t\in[0,T]}$ is a Lévy process with Lévy triplet (σ^2, ν^*, b^*) given by:

$$b^* = b + \beta\sigma^2 + \int_{-1}^{+1} \nu(dx)[e^{\beta(e^x-1)}x], \tag{10.51}$$

$$\nu^*(dx) = \exp[\beta(e^x - 1)]\nu(dx). \tag{10.52}$$

A general feature of the minimal entropy martingale is that its statistical properties closely resemble the original process (the prior) so the specification

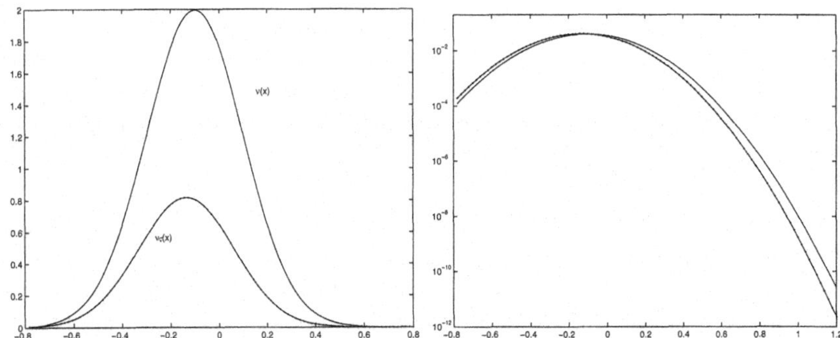

FIGURE 10.1: Left: Lévy density $\nu^*(x)$ (dotted lower curve) of minimal entropy martingale in the Merton model described in Example 10.2, compared with the initial Lévy density $\nu(x)$ (solid line). Right: Jump size densities corresponding to these Lévy densities.

of the prior is quite important. We now give two examples where the minimal entropy martingale model is computed from an exponential-Lévy model and its properties are compared to the initial model:

Example 10.1 Black-Scholes with Poisson jumps
Consider a jump-diffusion model where $S_t = S_0 \exp[bt+\sigma W_t+aN_t]$ where N is a Poisson process with intensity $\lambda = 1$, $b = +10\%, \sigma = 30\%$ and the jump size can take the single, negative value of $a = -10\%$. Equation (10.50) then has a unique solution, numerically found to be $\beta \simeq -1.5$. The minimal entropy martingale model is again of the same type: $S_t^* = S_0 \exp[b^*t + \sigma W_t + aN_t^*]$ where the risk neutral jump intensity is now $\lambda^* = \exp[\beta(e^a - 1)] \simeq 1.15$. □

Example 10.2 Merton model
Assume the underlying asset follows Merton's jump-diffusion model with log-normal price jumps $S_t = S_0 \exp[bt + \sigma W_t + \sum_{j=1}^{N_t} Y_i]$ where N is a Poisson process with intensity $\lambda = 1$, $b = +10\%, \sigma = 30\%$ and the jump sizes $Y_i \sim N(-0.1, \delta^2)$ with $\delta = 0.1$. Equation (10.50) is numerically solved to obtain $\beta \simeq -2.73$. The minimal entropy martingale model is again a jump-diffusion model with compound Poisson jumps: $S_t^* = S_0 \exp[b^*t + \sigma W_t + \sum_{j=1}^{N_t^*} Y_i^*]$ where a (risk neutral) Lévy density, now given by

$$\nu^*(x) = \frac{\lambda^*}{\sqrt{2\pi}\delta} \exp[-\frac{(x-a)^2}{2} + \beta(e^x - 1)].$$

The initial (historical) Lévy measure and the minimal entropy risk-neutral Lévy measure are depicted in Figure 10.1: the minimal entropy Lévy measure lies below the initial one but, as observed on the right hand figure, the jump size densities are similar and the main effect is a decrease in the jump intensity: $\lambda^* = \exp[\beta(e^x - 1)] \simeq 0.4 < \lambda = 1$. Since for small x, $e^x - 1 \simeq x$ the main change in the center of the distribution is a negative shift. However since $\beta < 0$ the term $\exp[\beta(e^x - 1)]$ strongly damps the right tail of the Lévy density. □

These examples and others studied in [299] show that for many models of interest, the parameter β is found to be *negative* so the right tail of the Lévy measure is strongly damped by a factor $\exp[\beta(e^x - 1)]$: the risk neutral process given by the minimal entropy martingale has. Also, this damping is asymmetric: if $\beta < 0$ then large positive jumps will be nearly irrelevant for pricing options, while large negative jumps will still contribute. This is of course, quite reasonable from a financial point of view and it is interesting that such a property stems from the abstract principle of relative entropy minimization.

REMARK 10.4 Relation with the Esscher transform The minimal entropy martingale may also be characterized as the Esscher transform of the Lévy process Z_t where $dS_t = S_{t-}(rdt + dZ_t)$, see [88, 299]. This remark lends some justification to the (otherwise ad hoc) use of Esscher transforms for pricing. □

10.5.2 Other martingale measures

Obviously relative entropy is not the only criterion one can minimize to choose a martingale measure and the same exercise can be done with other criteria, as in (10.46). But are these "optimal" martingale measures of any relevance? Unless the martingale measure is a by-product of a hedging approach, the price given by such martingale measures is not related to the cost of a hedging strategy therefore the meaning of such "prices" is not clear. In the minimal entropy case, we do have an interpretation in terms of exponential hedging in the limit of small risk aversion but the procedure does not allow to compute any hedging strategy associated to the price.

10.6 Hedging with options and model calibration

In incomplete markets, options are not redundant assets; therefore, if options are available as hedging instruments they *can* and *should* be used to improve hedging performance. Thirty years after the opening of the first organized options market, options have become liquidly traded securities in their own right and "vanilla" call and put options on indices, exchange rates and major stocks are traded in a way similar to their underlying stocks. Thus, liquidly traded options are available as instruments for hedging more complex, exotic or illiquid options and are commonly used as such by risk managers. While the lack of liquidity in the options market prevents in practice from using dynamic hedges involving options, options are commonly used for static hedging: call options are frequently used for dealing with "vega" (volatility) or "gamma" (convexity) exposures and for hedging barrier options [78, 4]. In fact as shown by Dupire [122] *any* European option can be perfectly hedged by an (infinite) static portfolio of European calls so the notions of market completeness and attainable claims must be revised when options are allowed as hedging instruments. More realistically, it is well known that *even* in diffusion models many exotic options such as binary or lookback options cannot be delta hedged in practice because of the large deltas implied by these options: hedging with options is the *only* feasible alternative in these cases. In the context of jump-diffusion models, Andersen and Andreasen [7] show that adding static positions in a few vanilla options can considerably decrease the dispersion of hedging error for a sample portfolio of derivatives.

While hedging with options is a common practice in risk management and was considered as early as 1976 by Ross [339], this point seems to be largely ignored in the recent literature on pricing and hedging in incomplete markets which is focused on a detailed formal analysis of utility-based hedging using only the underlying asset, a topic of secondary interest when options are available; see however [201, 15].

What are the implications of hedging with options for the choice of a pricing rule? Consider a contingent claim H and assume that we have as hedging instruments a set of benchmark options with prices $C_i^*, i = 1 \dots n$ and terminal payoffs $H_i, i = 1 \dots n$. A static hedge of H is a portfolio composed from the options $H_i, i = 1 \dots n$ and the numeraire, in order to match as closely as possible the terminal payoff of H:

$$H = V_0 + \sum_{i=1}^{n} x_i H_i + \int_0^T \phi dS + \epsilon, \tag{10.53}$$

where ϵ is an hedging error representing the nonhedgeable risk. Typically H_i are payoffs of call or put options and are not possible to replicate using the underlying so adding them to the hedge portfolio increases the span of hedgeable claims and reduces residual risk.

Consider now a pricing rule $\mathbb{Q}$. Assume that $E^{\mathbb{Q}}[\epsilon] = 0$ (otherwise $E^{\mathbb{Q}}[\epsilon]$ can be added to V_0). Then the claim H is valued under $\mathbb{Q}$ as:

$$e^{-rT} E^{\mathbb{Q}}[H] = V_0 + \sum_{i=1}^{n} x_i e^{-rT} E^{\mathbb{Q}}[H_i] \tag{10.54}$$

since the stochastic integral term, being a $\mathbb{Q}$-martingale, has zero expectation. On the other hand, the cost of setting up the hedging portfolio (10.53) is given by

$$V_0 + \sum_{i=1}^{n} x_i C_i^*. \tag{10.55}$$

So the value of the claim given by the pricing rule $\mathbb{Q}$ corresponds to the cost the hedging portfolio (10.53) if the model prices of the benchmark options H_i correspond to their market prices C_i^*:

$$\forall i = 1 \dots n, \qquad e^{-rT} E^{\mathbb{Q}}[H_i] = C_i^*. \tag{10.56}$$

Condition (10.56) is called a calibration condition: a pricing rule verifying (10.56) is said to be calibrated to the option prices $C_i^*, i = 1 \dots n$. This condition is necessary to guarantee the coherence between model prices and the cost of hedging with portfolios such as (10.53): if the model is not calibrated then the model price for a claim H may have no relation with the effective cost of hedging it using the available options H_i.

If a pricing rule $\mathbb{Q}$ is specified in an ad hoc way, the calibration conditions (10.56) will not be verified: one way to ensure them is to incorporate it as a constraint in the choice of the pricing measure $\mathbb{Q}$. We will return to this point in Chapter 13 where we will present some methods for incorporating calibration constraints into the choice of a pricing model.

All the hedging methods above can be reconsidered by allowing hedging strategies of the form (10.53). The superhedging approach to this problem was examined by Avellaneda and Paras [15] in the framework of uncertain volatility models: unlike superhedging with the underlying alone, the incorporation of calibration constraints leads to much more realistic results. A utility maximization approach to this problem has been recently considered by Hugonnier and Kramkov [201]. We delay further discussion of these issues to Chapter 13 where model calibration is discussed in more detail.

10.7 Summary

In complete market models such as the Black-Scholes model, probability does not really matter: the "objective" evolution of the asset is only there to define the set of "impossible" events and serves to specify the class of equivalent measures. Thus, two statistical models $\mathbb{P}_1 \sim \mathbb{P}_2$ with equivalent measures lead to the same option prices in a complete market setting.

This is not true anymore in incomplete markets: probabilities matter and model specification has to be taken seriously since it will affect hedging decisions. This situation is more realistic but also more challenging and calls for an integrated approach between option pricing methods and statistical modelling.

In incomplete markets, not only does probability matter but attitudes to risk also matter: utility based methods explicitly incorporate these into the hedging problem via utility functions. The difficulty of specifying firmwide utility functions and the lack of robustness of the result with respect to this specification has prompted research into preference-free methods for pricing and hedging: superhedging and quadratic hedging are among them.

While these methods are focused on hedging with the underlying asset, common practice is to use liquid call/put options to hedge exotic options. The introduction of options as hedging instruments improves hedging performance and can be conjugated with the hedging approaches described above. At the pricing level, hedging with options leads to calibration constraints on pricing models which have to be respected in order to keep consistency between model prices and the cost of hedging.

Table 10.1 summarizes the different valuation methods and their associated hedging strategies. Among the different valuation methods, (local or global) quadratic hedging and minimal entropy martingale measures yield linear pricing rules, which are therefore representable by a risk-neutral pricing rule. As noted above, the associated risk-neutral process is still an exponential-Lévy process. Thus, the class of exponential Lévy models can also provide suitable candidates for risk-neutral dynamics.

Further reading

Option pricing in incomplete markets has been an active research topic in mathematical finance in the last decade. An exposition of these ideas in a discrete time setting may be found in [147]. Incomplete markets in continuous time, where uncertainty is represented by Brownian motion, are discussed in [230].

Superhedging was introduced by El Karoui and Quenez [231] and has initiated new mathematical results on optional decompositions [231, 240, 145]. The price ranges given in Proposition 10.2 are computed in [124, 46].

Utility indifference pricing was introduced by Hodges and Neuberger [197]. The case of exponential utility has been studied by many authors; see [230] and [130] for the Brownian case and [107, 369, 44] for more general processes allowing for jumps.

Quadratic hedging under a risk neutral measure was introduced by Föllmer and Sondermann [149] using the Kunita-Watanabe decomposition [242]. Mean-variance hedging for Markov models was introduced in [69]. Explicit computations of minimal variance hedging strategies in the case of Markov processes are given in [144, 213]. The Föllmer-Schweizer decomposition was introduced in [148] for continuous processes; generalizations can be found in

TABLE 10.1: Comparison of different pricing methods in incomplete markets

Pricing method	**Hedging method**	**Risk criterion**
Maximize over EMMs: $\sup_{\mathbb{Q}\in M(S)} E^{\mathbb{Q}}[H]$	Superhedging	Perfect hedge: $\mathbb{P}(V_T(\phi) \geq H) = 1$
Utility indifference price	Utility maximization	Expected utility of terminal wealth: $-E^{\mathbb{P}}[U(V_0 + \int_0^T \phi dS - H)]$
Minimal martingale measure: $E^{\mathbb{Q}^{FS}}[H]$	Leads to non self-financing hedges	Local quadratic risk minimization
Mean-variance optimal measure: $E^{\mathbb{Q}^{MV}}[H]$	Mean-variance hedging	Mean square hedging error: $E^{\mathbb{P}}\|V_0 + \int_0^T \phi dS - H\|^2$
Minimal entropy martingale measure $E^{\mathbb{Q}^*}[H]$	Exponential hedging with risk aversion $\to 0$	Limit $\alpha \to 0$ of: $E^{\mathbb{P}}[\exp -\alpha(V_0+\int_0^T \phi dS-H)]$
Risk-neutral model calibrated to options H_i: $E^{\mathbb{Q}}[H]$ with $E^{\mathbb{Q}}[H_i] = C_i^*$	Static hedging with benchmark options H_i	Mean square static hedging error: $E^{\mathbb{Q}}\|V_0-H+\sum_i \phi_i(H_i-C_i^*)\|^2$

[356] and [108]. Local risk minimization in jump-diffusion models was studied in [390, 179, 178]. Formulas involving Fourier integrals in the case of exp-Lévy models are given in [200], see also [67]. A review of quadratic hedging approaches is given in [357].

Minimal entropy martingale measures are discussed in [154] in an abstract setting and studied in [130] in the case of a model driven by Brownian motion and by [299, 298, 88, 170, 136] in the case of exponential-Lévy models. Other criteria of "optimality" for martingale measures have been discussed in the literature: Hellinger distance [204], general f-distance [171].

Static hedging with options is discussed in [78, 4]. Optimal martingale measures consistent with a set of observed option prices are discussed in an abstract framework in [172] and in the framework of exp-Lévy models in [97]. Other references to model calibration are given in Chapter 13.

Another approach to hedging in incomplete markets, not discussed here, is based on the minimization of expected shortfall: this approach was introduced by Föllmer and Leukert [146] and has been applied to jump-diffusion models in [304].

Some confusion surrounds the notion of dynamic hedging and market completeness in the academic literature, where market completeness is sometimes considered to be a "desirable" feature of a model and incompleteness is assumed to result only from transaction costs or trading constraints. An interesting discussion of the dynamic hedging paradigm from the point of view of an option trader is given by Taleb [372].

Chapter 11

Risk-neutral modelling with exponential Lévy processes

Consider an arbitrage-free market where asset prices are modelled by a stochastic process $(S_t)_{t\in[0,T]}$, $\mathcal{F}_t$ represents the history of the asset S and $\hat{S}_t = e^{-rt}S_t$ the discounted value of the asset. In Chapter 9 we saw that in such a market, the prices of all instruments may be computed as discounted expectations of their terminal payoffs with respect to some martingale measure $\mathbb{Q}$ such that:

$$\hat{S}_t = E^{\mathbb{Q}}[\hat{S}_T|\mathcal{F}_t].$$

Under the pricing rule given by $\mathbb{Q}$ the value $\Pi_t(H_T)$ of an option with payoff H_T on an underlying asset S_t is given by:

$$\Pi_t(H_T) = e^{-r(T-t)}E^{\mathbb{Q}}[H_T|\mathcal{F}_t]. \tag{11.1}$$

Specifying an option pricing model is therefore equivalent to specifying the law of $(S_t)_{t\geq 0}$ under $\mathbb{Q}$, also called the "risk-neutral" or "risk-adjusted" dynamics of S.

In the Black-Scholes model, the risk-neutral dynamics of an asset price was described by the exponential of a Brownian motion with drift:

$$S_t = S_0\exp(B_t^0), \text{ where } B_t^0 = (r-\sigma^2/2)t + \sigma W_t. \tag{11.2}$$

Applying the Itô formula we obtain:

$$\frac{dS_t}{S_t} = rdt + \sigma dW_t = dB_t^1, \text{ where } B_t^1 = rt + \sigma W_t. \tag{11.3}$$

Therefore there are two ways to define the risk neutral dynamics in the Black-Scholes model using a Brownian motion with drift: by taking the exponential as in (11.2) or by taking the stochastic exponential as in (11.3).

A tractable class of risk neutral models with jumps generalizing the Black-Scholes model can be obtained by replacing the Brownian motion with drift by a Lévy process. One way to do this is to make the replacement in (11.2):

$$S_t = S_0\exp(rt + X_t). \tag{11.4}$$

We will call such a model an *exponential-Lévy model.* By Propositions 3.18 and 8.20, in order to guarantee that $e^{-rt}S_t$ is indeed a martingale we have to impose the additional restrictions on the characteristic triplet (σ^2, ν, γ) of X:

$$\int_{|x|\geq 1} e^x \nu(dx) < +\infty \qquad \text{and}$$

$$\gamma + \frac{\sigma^2}{2} + \int (e^y - 1 - y 1_{|y|\leq 1})\nu(dy) = 0. \tag{11.5}$$

$(X_t)_{t\geq 0}$ is then a Lévy process such that $E^{\mathbb{Q}}[e^{X_t}] = 1$ for all t.

Another approach is to replace B_t^1 by a Lévy process Z_t in (11.3):

$$dS_t = rS_{t-}dt + S_{t-}dZ_t. \tag{11.6}$$

S_t then corresponds to the *stochastic exponential* of Z, defined in Section 8.4. Then by Proposition 8.23, $e^{-rt}S_t$ is a martingale if and only if the Lévy process Z_t is a martingale which is verified in turn if $E[Z_1] = 0$.

As shown in Proposition 8.22, these two constructions lead to the same class of processes — any risk-neutral exponential Lévy model can also be expressed as a stochastic exponential of another Lévy process Z — so we will use the first approach (ordinary exponentials), pointing out equivalent formulations in terms of Z when necessary.

Exponential-Lévy models offer analytically tractable examples of positive jump processes. The availability of closed-form expressions for characteristic function of Lévy processes allows to use Fourier transform methods for option pricing. Also, the Markov property of the price enables us to express option prices as solutions of partial integro-differential equations. Finally, the flexibility of being able to choose the Lévy measure makes calibrating the model to market prices of options and reproducing implied volatility skews/smiles possible.

As shown in Chapter 9, the equivalence class of a Lévy process contains a wide range of jump processes with various types of dependence and non-stationarity in their increments. This means that many econometric models with complex properties are compatible with the use of an exponential Lévy model as a pricing rule.

Let us stress that specifying risk-neutral dynamics does not mean that we are working under the hypothesis that "investors are risk neutral" or anything of the sort. It simply means that all contingent claims are priced in a coherent, arbitrage-free way and that the pricing rule is linear: we are specifying a model for the pricing rule $\mathbb{Q}$.

Examples of risk-neutral exponential-Lévy models can be constructed from the Lévy processes from Chapter 4 by exponentiating them and imposing the martingale condition (11.5) on the triplet. We will use the name of a Lévy process to denote the corresponding exponential-Lévy model, e.g., variance gamma model stands for the risk-neutral exponential-Lévy model obtained by exponentiating the variance gamma process.

Having specified the risk neutral dynamics, we can now proceed to price different types of options using (11.1). The analytic properties of Lévy processes combined with those of the exponential function will allow us to derive efficient pricing procedures for various types of options. Section 11.1 discusses European options and Fourier transform methods for their valuation. Section 11.2 addresses the pricing of forward start options. Section 11.3 discusses barrier options and American options are treated in Section 11.4. Finally, Section 11.5 discusses multi-asset contracts.

11.1 European options in exp-Lévy models

A European call option on an asset S with maturity date T and strike price K is defined as a contingent claim that gives its holder the right (but not the obligation) to buy the asset at date T for a fixed price K. Since the holder can immediately sell the asset at its prevailing price recovering the positive part of the difference $S_T - K$, the option can be seen as an asset that pays to its holder the payoff $H(S_T) = (S_T - K)^+$ at date T. In the same way a put option is defined as an option with payoff $(K - S_T)^+$. These two types of options are called vanilla options because of their popularity in financial markets.

More generally a European option with maturity T and payoff function $H(.)$ is an option which pays the holder $H(S_T)$ at the maturity date T. If H is convex, denoting its left derivative by H' and its second derivative in the sense of distributions (which, by the way, is a positive Radon measure) by $\rho(.)$, we have

$$H(S) = H(0) + H'(0)S + \int_0^\infty \rho(dK)(S - K)^+ \tag{11.7}$$

so any European payoff can be represents as a (possibly infinite) superposition of call payoffs:

$$H(S_T) = H(0) + H'(0)S_T + \int_0^\infty \rho(dK)(S_T - K)^+. \tag{11.8}$$

This fact is due to the particular payoff structure of the call option and shows that, aside from being popular, call options can be regarded as fundamental building blocks for synthesizing more complex payoffs. Moreover, most popular European payoff structures (straddles, butterfly spreads, etc.) are linear combinations of a *finite* number of call/put payoffs.

The price of a call option may be expressed as the risk-neutral conditional expectation of the payoff:

$$C_t(T, K) = e^{-r(T-t)} E^Q[(S_T - K)^+ | \mathcal{F}_t] \tag{11.9}$$

and, due to the Markov property of S_t, it is a function of the characteristics of the option and the current value of the underlying: $C_t(T,K) = C(t, S_t; T, K)$. The quantity $(S_t - K)^+$ is usually called the *intrinsic* value of the option and $C_t(T,K) - (S_t - K)^+$ its *time* value. A call option is said to be "at the money" if $S_t = K$, "at the money forward" if $S_t = Ke^{r(T-t)}$, "out of the money" if $S_t < K$, "out of the money forward" if $S_t < Ke^{r(T-t)}$. For out of the money options, the intrinsic value is zero so C_t is equal to the time value.

If the price of a put option is denoted by $P_t(T,K)$, static arbitrage arguments lead to the call-put parity relation:

$$C_t(T,K) - P_t(T,K) = S_t - e^{-r(T-t)}K.$$

11.1.1 Call options

In an exponential-Lévy model, the expression (11.9) may be simplified further. By stationarity and independence of increments, the conditional expectation in (11.9) may be written as an expectation of the process at time $\tau = T - t$:

$$\begin{aligned} C(t,S,T = t+\tau, K) &= e^{-r\tau}E[(S_T - K)^+ | S_t = S] \\ &= e^{-r\tau}E[(Se^{r\tau + X_\tau} - K)^+] = Ke^{-r\tau}E(e^{x+X_\tau} - 1)^+, \end{aligned} \tag{11.10}$$

where x is the log forward moneyness defined by $x = \ln(S/K) + r\tau$. For options that are at the money forward $x = 0$. We see that similarly to the Black-Scholes model, in all exp-Lévy models call option price depends on the time remaining until maturity but not on the actual date and the maturity date and is a homogeneous function of order 1 of S and K. Defining the relative forward option price in terms of the relative variables (x, τ):

$$u(\tau, x) = \frac{e^{r\tau}C(t, S; T = t+\tau, K)}{K}, \tag{11.11}$$

we conclude that the entire structure of option prices in exponential-Lévy models — which a priori has four degrees of freedom — is parametrized by only two variables:

$$u(\tau, x) = E[(e^{x+X_\tau} - 1)^+].$$

This is a consequence of temporal and spatial homogeneity of Lévy processes. $u(\tau, .)$ can also be written as a convolution product: $u(\tau, .) = \rho_\tau * h$, where ρ_τ is the transition density of the Lévy process. Thus, if the process has smooth transition densities, $u(\tau, .)$ will be smooth, even if the payoff function h is not.

11.1.2 Implied volatility

In the Black-Scholes model $\nu = 0$ and call option prices are uniquely given by the Black-Scholes formula:

$$C^{BS}(S_t, K, \tau, \sigma) = S_t N(d_1) - Ke^{-r\tau} N(d_2) \tag{11.12}$$
$$\text{with} \quad d_1 = \frac{x + \tau\sigma^2/2}{\sigma\sqrt{\tau}} \quad \text{and} \quad d_2 = \frac{x - \tau\sigma^2/2}{\sigma\sqrt{\tau}},$$

where x is the log forward moneyness defined above, $\tau = T - t$ and

$$N(u) \equiv (2\pi)^{-1/2} \int_{-\infty}^{u} \exp(-\frac{z^2}{2}) dz$$

is the Gaussian CDF. If all other parameters are fixed, (11.12) is an increasing continuous function of σ, mapping $]0, \infty[$ into $](S_t - Ke^{-r\tau})^+, S_t[$. The latter interval is the maximal interval allowed by arbitrage bounds on call option prices. Therefore given the market price $C_t^*(T, K)$ of a call option one can always invert (11.12) and find the value of volatility parameter which, when substituted into the Black-Scholes formula, gives the correct option price:

$$\exists! \quad \Sigma_t(T, K) > 0, \qquad C^{BS}(S_t, K, \tau, \Sigma_t(T, K)) = C_t^*(K, T). \tag{11.13}$$

This value is called the (Black-Scholes) implied volatility of the option. For fixed (T, K), the implied volatility $\Sigma_t(T, K)$ is in general a stochastic process and, for fixed t, its value depends on the characteristics of the option such as the maturity T and the strike level K: the function $\Sigma_t : (T, K) \to \Sigma_t(T, K)$ is called the *implied volatility surface* at date t. As noted in the introductory chapter, the implied volatility surface is the representation adopted by the market for prices of call/put options. Using the *moneyness* $m = K/S_t$ of the option, one can also represent the implied volatility surface as a function of moneyness and time to maturity: $I_t(\tau, m) = \Sigma_t(t + \tau, mS(t))$. This representation is convenient since there is usually a range of moneyness around $m = 1$ for which the options are most liquid and therefore the empirical data are most readily available. Typical examples of implied volatility surfaces observed in the market were shown in Figures 1.6 and 1.7 and their properties are described in Section 1.2: implied volatility patterns are typically *skewed*: out of the money calls have lower implied volatilities than in the money ones. This effect becomes less important as the time to maturity grows. As becomes clear from Figures 11.1, 11.2 and 11.3, exponential-Lévy models can easily accommodate this feature.

In general, the implied volatility surface $I_t(\tau, m)$ computed from a model varies in time: it may depend not only on the maturity of options but also on the current date or the spot price, making the simulation of the P&L of option portfolios tedious. However in exponential-Levy models the evolution in time of implied volatilities is particularly simple, as shown by the following proposition:

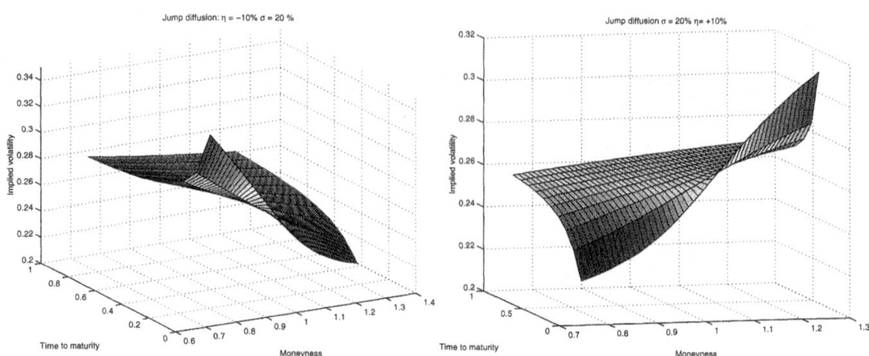

FIGURE 11.1: Profile of the implied volatility surface as a function of time to maturity and moneyness for a jump diffusion model with $\sigma = 20\%$ and Poisson jumps with size $\eta = -10\%$ (left) and $\eta = +10\%$ (right) and annual intensity $\lambda = 1$.

PROPOSITION 11.1 Floating smile property
When the risk neutral dynamics is given by an exponential-Lévy process, the implied volatility for a given moneyness level $m = K/S_t$ *and time to maturity* τ *does not depend on time*

$$\forall t \geq 0, I_t(\tau, m) = I_0(\tau, m). \tag{11.14}$$

PROOF Equation (11.10) entails that in an exponential Lévy model

$$\frac{C_t(S_t, K, T)}{S_t} = me^{-r\tau} E(m^{-1}e^{r\tau + X_\tau} - 1)^+ = g(\tau, m),$$

that is, the ratio of option price to the underlying only depends on the moneyness and time to maturity. The same is true for the Black-Scholes model because it is a particular case of the above:

$$\frac{C^{BS}(t, S_t, T, K, \sigma)}{S_t} = g_{BS}(\tau, m, \sigma).$$

The implied volatility $I_t(\tau, m)$ is defined by solving the equation

$$C^{BS}(t, S_t, T, K, \sigma) = S_t g(\tau, m) \iff g_{BS}(\tau, m, I_t(\tau, m)) = g(\tau, m).$$

Since each side depends only on (τ, m) and not on t one concludes that the implied volatility for a given moneyness m and time to maturity τ does not evolve in time:

$$\forall t \geq 0, I_t(\tau, m) = I_0(\tau, m).$$

□

Exponential-Lévy models are therefore "sticky moneyness" models in the vocabulary introduced by E. Derman [111]. Note however that the implied volatility for a given strike level is not constant in time: it evolves stochastically according to the rule:

$$\Sigma_t(T,K) = I_0(\frac{K}{S_t}, T-t).$$

Since the implied volatility surface I_t does not vary with t, it is enough to study it for $t = 0$. This has been done for various models in the literature; here are some salient features of implied volatility surfaces in exponential-Lévy models:

1. Skew/smile features: a negatively skewed jump distribution gives rise to a skew (decreasing feature with respect to moneyness) in implied volatility. Similarly, a strong variance of jumps leads to a curvature (smile pattern) in the implied volatility: Figure 11.2 shows the implied volatility patterns observed in the Merton jump diffusion model with an annual variance of jumps $\delta = 1$.

2. Short term skew: contrarily to diffusion models which produce little skew for short maturities, exponential-Lévy models (and more generally, models with jumps in the price) lead to a strong short term skew. Figure 11.1 shows that the addition of Poisson jumps to Black-Scholes model creates a strong skew for short maturities.

3. Flattening of the skew/smile with maturity: for a Lévy process with finite variance, the central limit theorem shows that when the maturity T is large, the distribution of $(X_T - E[X_T)])/\sqrt{T}$ becomes approximately Gaussian. The result of this so-called aggregational normality is that for long maturities prices of options will be closer to Black-Scholes prices and the implied volatility smile will become flat as $T \to \infty$. While this phenomenon is also observed in actual market prices, the effect is more pronounced in exponential-Lévy models. Figure 11.3 clearly illustrates this feature for the variance gamma model. This shortcoming and solutions for it will be discussed in more detail in Chapters 13, 14 and 15.

   However, there are exceptions to this rule: Carr and Wu [84] give an example of an exponential-Lévy models where the risk neutral dynamics of an asset is modelled as the stochastic exponential of an α-stable Lévy process with maximum negative skewness. In this model the return distribution of the underlying index has infinite second moment: as a result, the central limit theorem no longer applies and the volatility smile does not flatten out as in finite variance exp-Lévy models. Nevertheless, the extreme asymmetry of the Lévy measure guarantees that all moments of the index level itself and thus option prices at all maturities are well defined: in particular, (11.5) continues to hold.

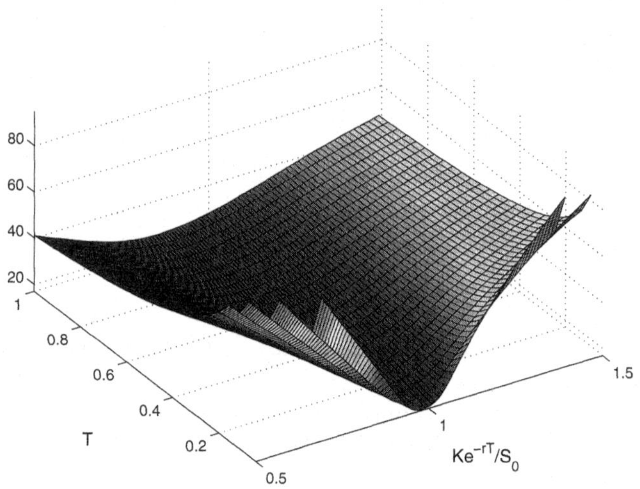

FIGURE 11.2: Profile of the implied volatility surface as a function of time to maturity and moneyness for the Merton jump-diffusion model with $\sigma = 15\%$, $\delta = 1$ and $\lambda = 0.1$. Option prices were computed using the series expansion in Section 10.1.

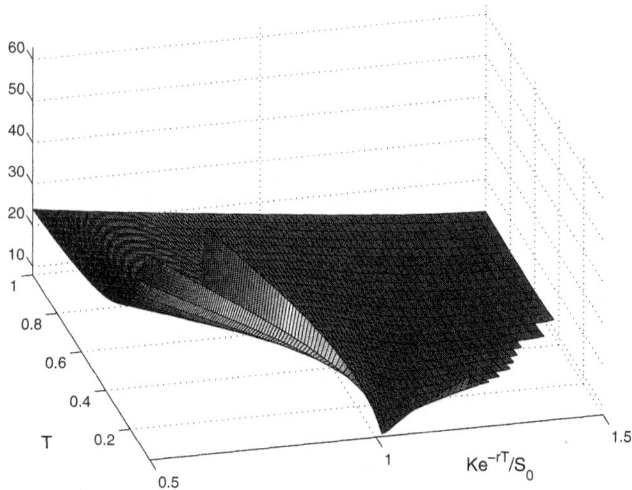

FIGURE 11.3: Implied volatility surface as a function of time to maturity and moneyness for the variance gamma model using parameters taken from [271]. Option prices were computed using the Carr-Madan Fourier transform. Note the flattening of the skew with maturity.

11.1.3 Fourier transform methods for option pricing

Contrary to the classical Black-Scholes case, in exponential-Lévy models there are no explicit formulae for call option prices, because the probability density of a Lévy process is typically not known in closed form. However, the characteristic function of this density can be expressed in terms of elementary functions for the majority of Lévy processes discussed in the literature. This has led to the development of Fourier-based option pricing methods for exponential-Lévy models. In these methods, one needs to evaluate one Fourier transform numerically but since they simultaneously give option prices for a range of strikes and the Fourier transform can be efficiently computed using the FFT algorithm, the overall complexity of the algorithm per option price is comparable to that of evaluating the Black-Scholes formula.

We will describe two Fourier-based methods for option pricing in exp-Lévy models. The first method, due to Carr and Madan [83] is somewhat easier to implement but has lower convergence rates. The second one, described by Lewis [262] converges faster but requires one intelligent decision which makes it more delicate to produce a robust automatic implementation. Recall the definition of the Fourier transform of a function f:

$$\mathbf{F}f(v) = \int_{-\infty}^{\infty} e^{ixv} f(x)dx$$

Usually v is real but it can also be taken to be a complex number. The inverse Fourier transform is given by:

$$\mathbf{F}^{-1}f(x) = \frac{1}{2\pi}\int_{-\infty}^{\infty} e^{-ixv} f(v)dx$$

For $f \in L^2(\mathbb{R})$, $\mathbf{F}^{-1}\mathbf{F}f = f$, but this inversion formula holds in other cases as well. In what follows we denote by $k = \ln K$ the log strike and assume without loss of generality that $t = 0$.

Method of Carr and Madan [83] In this section we set $S_0 = 1$, i.e., at time 0 all prices are expressed in units of the underlying. An assumption necessary in this method is that the stock price have a moment of order $1+\alpha$ for some $\alpha > 0$:

(H1) $$\exists\, \alpha > 0 : \quad \int_{-\infty}^{\infty} \rho_T(s)e^{(1+\alpha)s}ds < \infty,$$

where ρ_T is the risk-neutral density of X_T. In terms of the Lévy density it is equivalent to the condition

$$\exists\, \alpha > 0 \qquad \int_{|y|\geq 1} \nu(dy)e^{(1+\alpha)y} < \infty. \tag{11.15}$$

This hypothesis can be satisfied in all models discussed in Chapter 4 by putting a constraint on the exponential decay parameter for positive jumps (negative jumps do not affect it). In order to compute the price of a call option

$$C(k) = e^{-rT}E[(e^{rT+X_T} - e^k)^+]$$

we would like to express its Fourier transform in strike in terms of the characteristic function $\Phi_T(v)$ of X_T and then find the prices for a range of strikes by Fourier inversion. However we cannot do this directly because $C(k)$ is not integrable (it tends to a positive constant as $k \to -\infty$). The key idea of the method is to instead compute the Fourier transform of the (modified) time value of the option, that is, the function

$$z_T(k) = e^{-rT}E[(e^{rT+X_T} - e^k)^+] - (1 - e^{k-rT})^+. \tag{11.17}$$

Let $\zeta_T(v)$ denote the Fourier transform of the time value:

$$\zeta_T(v) = \mathbf{F}z_T(v) = \int_{-\infty}^{+\infty} e^{ivk} z_T(k)dk. \tag{11.18}$$

It can be expressed in terms of characteristic function of X_T in the following way. First, we note that since the discounted price process is a martingale, we can write

$$z_T(k) = e^{-rT}\int_{-\infty}^{\infty} \rho_T(x)dx(e^{rT+x} - e^k)(1_{k\leq x+rT} - 1_{k\leq rT}).$$

Condition (H1) enables us to compute $\zeta_T(v)$ by interchanging integrals:

$$\begin{aligned}
\zeta_T(v) &= e^{-rT}\int_{-\infty}^{\infty} dk \int_{-\infty}^{\infty} dx e^{ivk}\rho_T(x)(e^{rT+x} - e^k)(1_{k\leq x+rT} - 1_{k\leq rT}) \\
&= e^{-rT}\int_{-\infty}^{\infty} \rho_T(x)dx \int_{x+rT}^{rT} e^{ivk}(e^k - e^{rT+x})dk \\
&= \int_{-\infty}^{\infty} \rho_T(x)dx \left\{ \frac{e^{ivrT}(1-e^x)}{iv+1} - \frac{e^{x+ivrT}}{iv(iv+1)} + \frac{e^{(iv+1)x+ivrT}}{iv(iv+1)} \right\}
\end{aligned}$$

The first term in braces disappears due to martingale condition and, after computing the other two, we conclude that

$$\zeta_T(v) = e^{ivrT}\frac{\Phi_T(v-i) - 1}{iv(1+iv)} \tag{11.19}$$

The martingale condition guarantees that the numerator is equal to zero for $v = 0$. Under the condition (H1), we see that the numerator becomes an analytic function and the fraction has a finite limit for $v \to 0$. Option prices can now be found by inverting the Fourier transform:

$$z_T(k) = \frac{1}{2\pi}\int_{-\infty}^{+\infty} e^{-ivk}\zeta_T(v)dv \tag{11.20}$$

Note that in this method we need the condition (H1) to derive the formulae but we do not need the exact value of α to do the computations, which makes the method easier to implement. The price to pay for this is a slower convergence of the algorithm: since typically $\Phi_T(z) \to 0$ as $\Re z \to \infty$, $\zeta_T(v)$ will behave like $|v|^{-2}$ at infinity which means that the truncation error in the numerical evaluation of integral (11.20) will be large. The reason of such a slow convergence is that the time value (11.17) is not smooth; therefore its Fourier transform does not decay sufficiently fast at infinity. For most models of Chapter 4 the convergence can be dramatically improved by replacing the time value with a smooth function of strike. Namely, instead of subtracting the intrinsic value of the option (which is non-differentiable) from its price, we suggest to subtract the Black-Scholes call price with suitable volatility (which is a smooth function). The resulting function will be both integrable and smooth. Denote

$$\tilde{z}_T(k) = e^{-rT} E[(e^{rT+X_T} - e^k)^+] - C^{\sigma}_{BS}(k),$$

where $C^{\sigma}_{BS}(k)$ is the Black-Scholes price of a call option with volatility σ and log-strike k for the same underlying value and the same interest rate. By a reasoning similar to the one used above, it can be shown that the Fourier transform of $\tilde{z}_T(k)$, denoted by $\tilde{\zeta}_T(v)$, satisfies

$$\tilde{\zeta}_T(v) = e^{ivrT} \frac{\Phi_T(v-i) - \Phi^{\sigma}_T(v-i)}{iv(1+iv)}, \tag{11.21}$$

where $\Phi^{\sigma}_T(v) = \exp(-\frac{\sigma^2 T}{2}(v^2 + iv))$. Since for most models of Chapter 4 (more precisely, for all models except variance gamma) the characteristic function decays faster than every power of its argument at infinity, this means that the expression (11.21) will also decay faster than every power of v as $\Re v \to \infty$, and the integral in the inverse Fourier transform will converge very fast. This is true for *every* $\sigma > 0$ but some choices, or course, are better than others. Figure 11.4 shows the behavior of $|\tilde{\zeta}_T|$ for different values of σ compared to the behavior of $|\zeta_T|$ in the framework of Merton jump-diffusion model with volatility 0.2, jump intensity equal to 5 and jump parameters $\mu = -0.1$ and $\delta = 0.1$ for the time horizon $T = 0.5$. The convergence of $\tilde{\zeta}_T$ to zero is clearly very fast (faster than exponential) for all values of σ and it is particularly good for $\sigma = 0.3575$, the value of σ for which $\tilde{\zeta}(0) = 0$.

Method of Lewis [262] We present this method from a different angle than the previous one, in order to show how arbitrary payoff structures (and not just vanilla calls) can be priced. Since for an arbitrary payoff the notion of strike is not defined, we will show how to price options for a range of different initial values of the underlying. Let $s = \ln S_0$ denote the logarithm of current stock value and f be the payoff function of the option. The price of this option

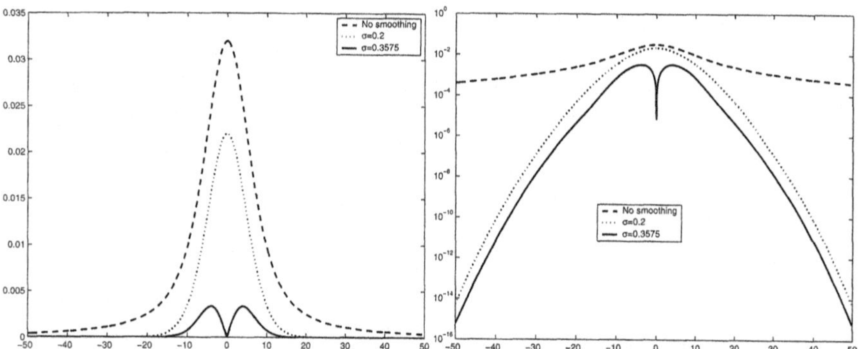

FIGURE 11.4: Convergence of Fourier transform of option's time value to zero in Merton model — see page 363. Left graph: linear scale; right graph: logarithmic scale.

is

$$C(s) = e^{-rT} \int_{-\infty}^{\infty} f(e^{s+x+rT}) \rho_T(x) dx.$$

In this method, instead of subtracting something from the call price to obtain an integrable function, one computes the Fourier transform for complex values of the argument: the Fourier transform is defined, as usual, by

$$\mathbf{F}g(z) = \int_{-\infty}^{\infty} e^{iuz} g(u) du$$

but z may now be a complex number.

For $a, b \in \mathbb{R}$ we say that $g(u)$ is Fourier integrable in a strip (a, b) if $\int_{-\infty}^{\infty} e^{-au}|g(u)|du < \infty$ and $\int_{-\infty}^{\infty} e^{-bu}|g(u)|du < \infty$. In this case $\mathbf{F}g(z)$ exists and is analytic for all z such that $a < \Im z < b$. Moreover, within this strip the generalized Fourier transform may be inverted by integrating along a straight line parallel to the real axis (see [262]):

$$g(x) = \frac{1}{2\pi} \int_{iw-\infty}^{iw+\infty} e^{-izx} \mathbf{F}g(z) dz \tag{11.22}$$

with $a < w < b$. To proceed, we need an additional hypothesis ($\bar{S}$ denotes the complex conjugate set of S):

(H2) $\rho_T(x)$ is Fourier integrable in some strip S_1,
$f^*(x) \equiv f(e^{x+rT})$ is Fourier integrable in some strip S_2 and the intersection of $\bar{S}_1$ with S_2 is nonempty: $S = \bar{S}_1 \cap S_2 \neq \emptyset$.

Using this hypothesis we can now compute the generalized Fourier transform of $C(s)$ by interchanging integrals: for every $z \in S$

$$\int_{-\infty}^{\infty} e^{izs} C(s) ds = e^{-rT} \int_{-\infty}^{\infty} \rho_T(x) dx \int_{-\infty}^{\infty} e^{izs} f(e^{s+x+rT}) ds$$
$$= e^{-rT} \int_{-\infty}^{\infty} e^{-izx} \rho_T(x) dx \int_{-\infty}^{\infty} f(e^{y+rT}) e^{izy} dy.$$

Finally we obtain

$$\mathbf{F}C(z) = e^{-rT} \Phi_T(-z) \mathbf{F} f^*(z) \quad \forall z \in S.$$

Option prices can now be computed from (11.22) for a range of initial values using the FFT algorithm (see below).

Application to call options The payoff function of a European call option is Fourier integrable in the region $\Im z > 1$, where its generalized Fourier transform can be computed explicitly:

$$\mathbf{F} f^*(z) = \int_{-\infty}^{\infty} e^{iyz} (e^{y+rT} - e^k)^+ dy = \frac{e^{k+iz(k-rT)}}{iz(iz+1)}.$$

The hypothesis (H2) now requires that $\rho_T(x)$ be integrable in a strip (a, b) with $a < -1$. Since $\rho_T(x)$ is a probability density, 0 belongs to its strip of integrability which means that $b \geq 0$. Therefore, in this setting the hypothesis (H2) is equivalent to (H1).

Finally, the generalized Fourier transform of call option price takes the form

$$\mathbf{F}C(z) = \frac{\Phi_T(-z) e^{(1+iz)(k-rT)}}{iz(iz+1)}, \quad 1 < \Im z < 1 + \alpha.$$

The option price can be computed using the inversion formula (11.22), which simplifies to

$$C(x) = \frac{\exp(wx + (1-w)(k-rT))}{2\pi} \int_{-\infty}^{\infty} \frac{e^{iu(k-rT-x)} \Phi_T(-iw-u) du}{(iu-w)(1+iu-w)}$$

for some $w \in (1, 1+\alpha)$. The integral in this formula is much easier to approximate at infinity than the one in (11.20) because the integrand decays exponentially (due to the presence of characteristic function). However, the price to pay for this is having to choose w. This choice is a delicate issue because choosing big w leads to slower decay rates at infinity and bigger truncation errors and when w is close to one, the denominator diverges and the discretization error becomes large. For models with exponentially decaying tails of Lévy measure, w cannot be chosen a priori and must be adjusted depending on the model parameters.

Computing Fourier transforms In order to implement the algorithms above, one needs to numerically compute Fourier transform in an efficient manner. This can be done using the discrete Fourier transform

$$F_n = \sum_{k=0}^{N-1} f_k e^{-2\pi ink/N}, \quad n = 0 \dots N-1.$$

To compute $F_0, \dots, F_{N-1}$, one needs a priori N^2 operations, but when N is a power of 2, an algorithm due to Cooley and Tukey [99] and known as the *fast Fourier transform* (FFT) reduces the computational complexity to $O(N \ln N)$ operations, see [321]. Subroutines implementing the FFT algorithm are available in most high-level scientific computation environments. A C language library called FFTW can be downloaded (under GPL license) from www.fftw.org.

Suppose that we would like to approximate the inverse Fourier transform of a function $f(x)$ with a discrete Fourier transform. The integral must then be truncated and discretized as follows:

$$\int_{-\infty}^{\infty} e^{-iux} f(x)dx \approx \int_{-A/2}^{A/2} e^{-iux} f(x)dx \approx \frac{A}{N} \sum_{k=0}^{N-1} w_k f(x_k) e^{-iux_k},$$

where $x_k = -A/2 + k\Delta$, $\Delta = A/(N-1)$ is the discretization step and w_k are weights corresponding to the chosen integration rule (for instance, for the trapezoidal rule $w_0 = w_{N-1} = 1/2$ and all other weights are equal to 1). Now, setting $u_n = \frac{2\pi n}{N\Delta}$ we see that the sum in the last term becomes a discrete Fourier transform:

$$\mathbf{F}f(u_n) \approx \frac{A}{N} e^{iuA/2} \sum_{k=0}^{N-1} w_k f(x_k) e^{-2\pi ink/N}$$

Therefore, the FFT algorithm allows to compute $\mathbf{F}f(u)$ at the points $u_n = \frac{2\pi n}{N\Delta}$. Notice that the grid step d in the Fourier space is related to the initial grid step Δ:

$$d\Delta = \frac{2\pi}{N}$$

This means that if we want to compute option prices on a fine grid of strikes, and at the same time keep the discretization error low, we must use a large number of points. Another limitation of the FFT method is that the grid must always be uniform and the grid size a power of 2. The functions that one has to integrate are typically irregular at the money and smooth elsewhere but increasing the resolution close to the money without doing so in other regions is not possible. These remarks show that the use of FFT is only justified when one needs to price a large number of options with the same maturity (let us say, more than 10) — for pricing a single option adaptive variable-step integration algorithms perform much better.

When one only needs to price a single option, Boyarchenko and Levendorskiĭ [71] suggest to transform the contour of integration in the complex plane (using Cauchy theorem) to achieve a better convergence. Their method, called "integration along cut" is applicable to tempered stable model with $\alpha \leq 1$ and in many other cases and performs especially well for options close to maturity. However, when one only needs to price a single option, the speed is not really an issue since on modern computers *all methods* can do this computation quickly. Speed becomes an issue when one repeatedly needs to price a large number of options (as in calibration or in scenario simulation for portfolios). If one is interested in pricing several options with the same maturity, FFT is hard to beat.

11.2 Forward start options

A forward start (or delayed) option is an option with some contractual feature, typically the strike price, that will be determined at some future date before expiration, called the fixing date. A typical example of forward start option is a stock option. When an employee begins to work, the company may promise that he or she will receive call options on the company's stock at some future date. The strike of these options will be such that the options are at the money or, for instance, 10% out of the money at the fixing date of the strike. Due to the homogeneous nature of exponential-Lévy models, forward start options are easy to price in this framework.

The simplest type of a forward start option is a forward start call, starting at T_1 and expiring at T_2, with strike equal to a fixed proportion m of the stock price at T_1 (often $m = 1$: the option is at the money on the date of issue). The payoff of such an option at T_2 is therefore

$$H = (S_{T_2} - mS_{T_1})^+. \tag{11.23}$$

At date T_1 the value of this option is equal to that of a European call with maturity T_2 and strike mS_{T_1}. Denoting the price of a forward start option by P_t we obtain

$$P_{T_1} = S_{T_1} e^{-r(T_2-T_1)} E\{(e^{r(T_2-T_1)+X_{T_2-T_1}} - m)^+\}.$$

Notice that the expectation is not conditional: the result is deterministic. Therefore the value of option at T_1 is equal to S_{T_1} times a constant. This option is then simply equivalent to a certain number of forward contracts and its price at t is given by

$$P_t = e^{-r(T_1-t)}E\{S_{T_1}|\mathcal{F}_t\}e^{-r(T_2-T_1)}E\{(e^{r(T_2-T_1)+X_{T_2-T_1}} - m)^+\}$$
$$= S_t e^{-r(T_2-T_1)}E\{(e^{r(T_2-T_1)+X_{T_2-T_1}} - m)^+\}. \quad (11.24)$$

Hence, *the price of a forward start call starting at T_1 and expiring at T_2 is equal to the price of a regular call with time to maturity $T_2 - T_1$ and strike mS_t.* The implied volatility of a forward start call, or the forward implied volatility, is defined as the volatility σ such that the Black-Scholes price of a call with moneyness m, time to maturity $T_2 - T_1$ and volatility σ is equal to the price of this forward start call. The pricing formula for forward start calls (11.24) leads to the following result.

PROPOSITION 11.2 Forward smiles in exp-Lévy models
In an exponential-Lévy model, the forward smile is stationary: the implied volatility of a forward start call $I^F_{T_1}(T_2 - T_1, m)$ only depends on $T_2 - T_1$ and not on the fixing date T_1. In particular, it is equal to today's implied volatility for the maturity $T_2 - T_1$.

$$I^F_{T_1}(T_2 - T_1, m) = I_0(T_2 - T_1, m)$$

Therefore in an exponential-Lévy model, the implied volatility smile for forward start options can be deduced from the current implied volatility surfaces. By contrast in a stochastic volatility or local volatility model the forward smile has to computed numerically.

11.3 Barrier options

Barrier options are options in which the right of the holder to buy or sell the stock at a specific price at the end of the contract is conditioned on the stock crossing (or not crossing) a certain barrier before the expiry date. For example, the up-and-out call option has the following payoff at maturity

$$(S_T - K)^+ 1_{\max_{0\le t\le T} S_t < B}.$$

If the stock price crosses the barrier B, this option becomes worthless (sometimes the holder may receive a fixed amount of money called *rebate*). For an up-and-out option without rebate to have a nonzero value, the strike K must be lower than H. An up-and-out call may be interesting for the buyer because while offering the same guarantees in normal market conditions (if the barrier is not crossed), it can be purchased at a considerably lower price.

The payoff of a barrier option may be written using the notion of *exit time*, introduced in Section 2.4.3. Define the exit time T_B of the barrier by

$$T_B = \inf\{t > 0, S_t > B\}. \tag{11.25}$$

T_B is a nonanticipating random time: it is the hitting time of an open interval. If S_t is continuous, then $S_{T_B} = B$: the barrier cannot be crossed without being hit. But for a jump process this may no longer be true: S_t can jump over the barrier and the value $S_{T_B} - B$ is a random variable called the *overshoot*. If S_t has a nonzero probability of *hitting* the barrier, i.e., if

$$\mathbb{Q}(T_{\{B\}} < \infty) \neq 0 \text{ where } T_{\{B\}} = \inf\{t > 0, S_t = B\},$$

the random time associated to crossing *or hitting* the barrier[1]

$$\tau_B = \inf\{t > 0, S_t \geq B\} \tag{11.26}$$

is not necessarily the same as T_B: the process may jump up to B and then jump down again without crossing the barrier so τ_B can be strictly smaller than T_B! However, the probability of such an event is zero if there is a nontrivial diffusion component [48]: for jump-diffusion processes $\tau_B = T_B$ almost surely. Points which can never be reached, i.e., for which $T_{\{B\}} = \infty$ a.s. for any starting point S_0, are called regular points. The difference between the two random times T_B and τ_B depends on the regularity of B and is related to the *local time* of the process at B. The regularity of B is also linked to the smoothness at the barrier of the barrier option value as a function of the underlying. These properties may be quite delicate to study for a jump process and this is just the tip of an iceberg called *potential theory*, see [49], [345, Chapter 8] for the case of Lévy processes. In the case of pure jump processes, Kesten [234] (see also [345, Theorems 43.13 and 43.21]) gives sufficient conditions on the Lévy process X such that $T_B = \tau_B$ a.s. These conditions are verified in particular for all Lévy processes with a non-zero diffusion component and tempered stable processes (see [70]). In the sequel we will limit ourselves to the case where $T_B = \tau_B$.

Three different approaches are available for pricing barrier options in exponential-Lévy models. The first one is based on Wiener-Hopf factorization identities. In principle it applies to all Lévy processes and allows to treat barrier and lookback options. However, in most cases of practical interest the Wiener-Hopf factors are not known in closed form and computing option prices requires integration in several dimensions. We present nevertheless the principles behind this approach.

---

[1] Although this is not obvious, τ_B can also shown to be a nonanticipating random time (stopping time), see [345, Theorem 40.13].

Wiener-Hopf factorization The term Wiener-Hopf factorization refers to a number of identities (see [345, Chapter 9] for an overview) that are helpful in evaluating Fourier transforms of quantities related to maximum and minimum of a Lévy process, which explains why they are useful for barrier option pricing. Here we only give the necessary definitions. In the following $(X_t)_{t\geq 0}$ denotes a Lévy process on $\mathbb{R}$ with Lévy triplet (σ, ν, γ) and characteristic exponent $\psi(u)$. Denote by ρ_t the distribution of X_t. We also introduce the maximum and minimum processes associated with (X_t):

$$M_t = \max_{0\leq s\leq t} X_s \qquad N_t = \min_{0\leq s\leq t} X_s$$

The next proposition defines the Wiener-Hopf factors, associated to the process X_t. The proof can be found in [345, Theorem 45.2].

PROPOSITION 11.3
Let $q > 0$. There exists a unique pair of characteristic functions $\Phi_q^+(z)$ and $\Phi_q^-(z)$ of infinitely divisible distributions having drift 0 and supported on $[0, \infty)$ and $(-\infty, 0]$ respectively such that

$$\frac{q}{q - \psi(z)} = \Phi_q^+(z)\Phi_q^-(z), \quad z \in \mathbb{R}.$$

These functions have the following representations

$$\Phi_q^+(z) = \exp\left\{\int_0^\infty t^{-1}e^{-qt}dt \int_0^\infty (e^{izx} - 1)\rho_t(dx)\right\}, \tag{11.27}$$

$$\Phi_q^-(z) = \exp\left\{\int_0^\infty t^{-1}e^{-qt}dt \int_{-\infty}^0 (e^{izx} - 1)\rho_t(dx)\right\}. \tag{11.28}$$

The function $\Phi_q^+(z)$ can be continuously extended to a bounded analytic function without zeros on the upper half plane and $\Phi_q^-(z)$ can be similarly extended to the lower half plane.

Example 11.1 Wiener-Hopf factors for Brownian motion with drift
If $(X_t)_{t\geq 0}$ is a Brownian motion with drift, $\psi(u) = -\frac{1}{2}\sigma^2 z^2 + i\gamma z$ and the Wiener-Hopf factors can be computed explicitly:

$$\frac{q}{q - \psi(z)} = \frac{q}{q + \sigma^2 z^2/2 - i\gamma z} = \frac{\lambda_+}{\lambda_+ + iz}\frac{\lambda_-}{\lambda_- - iz}, \tag{11.29}$$

where $\lambda_+ = \frac{\gamma}{\sigma^2} + \frac{\sqrt{\gamma^2+2\sigma^2 q}}{\sigma^2}$ and $\lambda_- = -\frac{\gamma}{\sigma^2} + \frac{\sqrt{\gamma^2+2\sigma^2 q}}{\sigma^2}$. The first factor in (11.29) is the characteristic function of exponential distribution with parameter λ_+ and the second factor is the characteristic function of the dual of the exponential distribution (exponential distribution on the negative half axis) with parameter λ_-. □

This example can be generalized to Lévy processes having only negative (or only positive) jumps. If (X_t) has only negative jumps,

$$\Phi_q^+(\xi) = \frac{\lambda_q}{\lambda_q - i\xi} \tag{11.30}$$

where $-i\lambda_q$ is the (unique) root of $q + \psi(z)$ (see [71, Section 2.2]).

However, in the general case these factors are not known explicitly and must be evaluated numerically. Computations using Equations (11.27) and (11.28) are not very efficient because they involve the probability distribution of X_t which is usually not available in closed form. Boyarchenko and Levendorskiĭ [71, Section 3.6] give a more efficient expression which is valid for tempered stable, normal inverse Gaussian and several other examples of Chapter 4):

$$\Phi_q^+(z) = \exp\left\{ \frac{z}{2\pi i} \int_{-\infty+iw}^{+\infty+iw} \frac{\ln(q+\psi(\xi))}{\xi(z-\xi)} d\xi \right\}$$

with some $w < 0$ such that $\Phi_q^+(z)$ is analytic in the half plane $\Im z > w$. Note however that this integral must be computed for all values of z and q and it cannot be approximated using the FFT method.

The next theorem [345, Theorem 45.7] shows how Wiener-Hopf factors may be used for computing quantities related to the maximum of a Lévy process.

THEOREM 11.1 Wiener-Hopf factorization

The Laplace transform in t of the joint characteristic function of $(X_t, M_t - X_t)$ is given by:

$$q \int_0^{+\infty} e^{-qt} E[e^{ixM_t + iy(X_t - M_t)}] dt = \Phi_q^+(x)\Phi_q^-(y) \tag{11.31}$$

for any $q > 0, x, y \in \mathbb{R}$.

In probabilistic terms, this result can be interpreted as follows: the distribution of M evaluated at an independent exponential random time is independent of of $M - X$ evaluated at the same exponential random time.

Application to barrier option pricing Suppose for simplicity that $S = 1$ and the interest rate is zero. Then the price of an up-and-out call option is

$$C(T,k,b) = E\{(e^{X_T} - e^k)^+ 1_{M_t < b}\} = \int_{\mathbb{R}^2} (e^x - e^k)^+ 1_{y<b} p_T(x,y) dx dy,$$

where b is the log-barrier, k the log strike and $p_T(x,y)$ denotes the joint density of X_T and M_T. For sufficiently regular Lévy processes we can compute by interchanging integrals the generalized Fourier transform in log-strike and log-barrier of this option price:

$$\int_{\mathbb{R}^2} e^{iuk+ivb} C(T,k,b) dk db = \frac{\mathbf{F}p_T(u-i,v)}{uv(1+iu)}, \quad \Im u < 0, \quad \Im v > 0,$$

where $\mathbf{F}p$ is the characteristic function of (X_T, M_T). The last step is to apply the Laplace transform to both sides and use the factorization identity (11.31):

$$q \int_{\mathbb{R}^2} e^{iuk+ivb} dk db \int_0^\infty e^{-qT} C(T,k,b) dT = \frac{\Phi_q^+(v+u-i)\Phi_q^-(u-i)}{uv(1+iu)}.$$

Computing barrier option prices for a range of strikes, barrier values and maturities requires inverting the Fourier transform (using Equation (11.22)) and the Laplace transform. If Fourier inversion cannot be done analytically, it may be performed using the FFT algorithm. However, the inversion of a Laplace transform is a well-known ill-posed problem: although numerical methods exist, they are known to be unstable. This pricing method is more or less feasible if Wiener-Hopf factors are known in closed form, like in the case of Lévy processes with no positive jumps; in other cases it is too computationally expensive and we recommend using the two other methods available: Monte Carlo simulation or numerical solution of partial integro-differential equations.

A special case where closed form expressions can be obtained for the Laplace transform of barrier, double barrier and lookback options is the Kou jump diffusion model described in Section 4.3, see [239, 360, 359]. However, the pricing formulae for European options given in [238] involve special functions whose computation is delicate, so we recommend using numerical methods such as FFT.

Monte Carlo methods for barrier options Monte Carlo methods perform well for pricing barrier options in a jump-diffusion framework because one can control the behavior of the Lévy process between the jump times: between two jumps the log-price follows a Brownian bridge process. This remark can be used to devise efficient pricing algorithms for barrier options [292]. We refer the reader to Example 6.1 and Algorithm 6.4 in Chapter 6, which explain in detail how to price an up-and-out call in a jump-diffusion model by the Monte Carlo method. Moreover, in this case the price of vanilla call or put (computed using Fourier transform methods of this chapter) may be used as control variate to reduce the error of Monte Carlo method.

In infinite activity setting the Monte Carlo methods for barrier options are less precise and more time consuming because, in order to see if the underlying has crossed the barrier or not, one has to simulate the entire trajectory of the process.

PIDE methods for barrier options The third and maybe the most general method to price barrier options in exponential-Lévy models is by solving corresponding partial integro-differential equations. In this framework, barrier options are easier to price than European ones, because to price European options one must impose artificial boundary conditions whereas for barrier options the boundary condition is natural: the price is either zero or equal to the rebate on the barrier and beyond. PIDE methods for barrier options are discussed in Chapter 12.

11.4 American options

American options differ from their European counterparts by giving their holder the right to exercise at any date prior to maturity and not only at the expiry date. The holder is thus free to adopt any strategy that binds the exercise decision to the past values of stock price: the exercise date may be any nonanticipating random time τ. The value of an American put at date t is therefore given by the highest value obtained by optimizing over exercise strategies:

$$P_t = \text{ess}\ \sup_{\tau \in \mathcal{T}(t,T)} E[e^{-r(\tau-t)}(K - S_\tau)^+ | \mathcal{F}_t],$$

where and $\mathcal{T}(t,T)$ denotes the set of nonanticipating exercise times τ satisfying $t \leq \tau \leq T$. The essential supremum over an uncountable set of random variables is defined as the smallest random variable dominating all those in the set. As T also belongs to $\mathcal{T}$, it is clear that the price of an American put is always greater or equal to the price of its European counterpart: the difference between the price of an American option and the European option with the same characteristics is called the *early exercise premium*. The price of an American call may be computed using a similar formula, however, as shown by Merton [290] using arbitrage arguments, it is not optimal to exercise an American call before maturity if the underlying stock pays no dividends: the value of an American call on an asset paying no dividends is equal to the value of the European call with same strike and maturity.

General properties of American option values have been studied by [47] in the framework of optimal stopping problems. American options in models with jumps have been studied in [316, 390] in the finite activity case and in [250] for some infinite activity models. In an exponential-Lévy model, using the strong Markov property of S_t one can show that

$$P_t = \text{ess}\ \sup_{\tau \in \mathcal{T}(t,T)} E[e^{-r(\tau-t)}(K - S_\tau)^+ | S_t = S] = P(t, S_t),$$
$$\text{where} \quad P(t,S) = \text{ess}\ \sup_{\tau \in \mathcal{T}(t,T)} E[(Ke^{-r(\tau-t)} - Se^{X_{\tau-t}})^+]. \quad (11.32)$$

The following properties are shown in [317, Proposition 2.1.] in the case of a jump-diffusion model:

- $P(t,S)$ is decreasing and convex with respect to the underlying S for every $t \in [0,T]$.
- $P(t,S)$ is increasing with respect to t.

As in the Black-Scholes model, exponential-Lévy models do not give closed-form expressions for American options. Also, evaluation of (11.32) by Monte Carlo simulation is not straightforward: it involves the computation of a *conditional* expectation. The least squares Monte Carlo method [266, 86] allows to approximate conditional expectations in a Monte Carlo framework. An overview of this and other Monte Carlo based methods for American option pricing can be found in [165, Chapter 8]

Boyarchenko and Levendorskiĭ [71] give methods using Wiener-Hopf factorization for computing prices of *perpetual* American options, i.e., for $T = \infty$. See also [250] and the discussion in Section 12.5. Alternative characterizations of American options using partial integro-differential equations are discussed in Chapter 12.

11.5 Multi-asset options

A *basket* option is an option whose payoff depends on the value of several underlying assets. The simplest example is a call on a portfolio of assets $S^{(1)} \dots S^{(d)}$, whose value is given by

$$C_t(T,K) = e^{-r(T-t)} E\left(\sum_{i=1}^{d} \alpha_i S_T^{(i)} - K\right)^+ . \qquad (11.33)$$

If the stocks are modelled by the exponentials of Lévy processes, the sum $\sum_{i=1}^{d} \alpha_i S_t^{(i)}$ will no longer be a Lévy process, and the Fourier methods of Section 11.1.3 cannot be used directly. Since the complexity of PIDE methods grows exponentially with dimension, the only remaining solution is the Monte Carlo method. Fortunately, for multivariate models of Chapter 5, efficient simulation methods are available (see Section 6.6). Moreover, in exponential-Lévy models one can construct efficient variance reduction algorithms that allow to achieve acceptable precision already with several hundred trajectories of Monte Carlo. In this section we discuss these variance reduction methods and illustrate their application on a numerical example.

We consider the following problem:

$$C(K) = e^{-rT} E\left(\sum_{i=1}^{d} \alpha_i e^{rT + X_T^{(i)}} - K\right)^+ ,$$

where $(X^{(1)} \dots X^{(d)})$ is a d-dimensional Lévy process and $\alpha_1, \dots, \alpha_d$ are positive weights that sum up to one. Problem (11.33) can be reduced to this form by renormalizing the strike. At first glance, this problem seems difficult to solve by Monte Carlo methods because the payoff function is unbounded and may have very high variance. However, on one hand, due to the convexity of payoff we always have

$$e^{-rT}E\left(\sum_{i=1}^{d}\alpha_i e^{rT+X_T^{(i)}}-K\right)^+ \leq \sum_{i=1}^{d}\alpha_i e^{-rT}E\left(e^{rT+X_T^{(i)}}-K\right)^+ \equiv C^+(K)$$

On the other hand, using the well-known identity $ta+(1-t)b \geq a^t b^{1-t}$ for a and b positive and $0 \leq t \leq 1$, we can write:

$$e^{-rT}E\left(\sum_{i=1}^{d}\alpha_i e^{rT+X_T^{(i)}}-K\right)^+$$
$$\geq e^{-rT}E\left(\exp\left\{rT+\sum_{i=1}^{d}\alpha_i X_T^{(i)}\right\}-K\right)^+ \equiv C^-(K)$$

We have found two bounds for basket option price: $C^-(K) \leq C(K) \leq C^+(K)$. Notice that $C^+(K)$ is a linear combination of calls on individual assets of the basket and $C^-(K)$ is a call on a fictitious asset (which is by the way not a martingale), described by the Lévy process $\sum_{i=1}^d \alpha_i X^{(i)}$ whose characteristics can often be computed from those of the d-dimensional process. Therefore, both $C^+(K)$ and $C^-(K)$ can be evaluated either analytically or by Fourier methods and used as control variates for the computation of $C(K)$.

Variance reduction with control variates Suppose that we need to compute the expectation of a random variable Y by Monte Carlo method. Find another random variable X (called control variate) whose expectation may be computed analytically and which is highly correlated with Y. Denote the sample means over N i.i.d. realizations of X and Y by $\bar{X}$ and $\bar{Y}$: $\bar{X} \equiv \frac{1}{N}\sum_{i=1}^N X_i$. The control variate estimator of Y is then given by

$$\hat{Y}(b) = \bar{Y}_i + b(EX_i - \bar{X}_i) \tag{11.34}$$

for any fixed b. It is now straightforward to see that the variance of $\hat{Y}(b)$ is minimal at $b^* = \frac{\operatorname{Cov}(Y,X)}{\operatorname{Var} X}$ where it is equal to $\frac{1}{N}\operatorname{Var} Y(1-\rho_{XY}^2)$, with ρ_{XY} denoting the correlation coefficient between X and Y. The variance reduction factor compared to plain Monte Carlo is equal to $1-\rho_{XY}^2$ which means that a strong reduction of variance may be achieved if X and Y are highly correlated. The optimal value of b cannot typically be computed explicitly (if we knew how to compute $\operatorname{Cov}(Y,X)$ we would probably also be able to compute EY without simulation). However, in most cases it can safely be estimated from the same sample:

$$\hat{b}^* = \frac{\sum_{i=1}^N (Y_i-\bar{Y})(X_i-\bar{X})}{\sum_{i=1}^N (X_i-\bar{X})^2} \tag{11.35}$$

Using the estimated value of b^* introduces some bias into the estimator of Y but for sufficiently large samples this bias is small compared to the Monte Carlo error (see [165, Chapter 4] for a detailed discussion of small sample issues).

Equation (11.34) can be generalized to the case when several control variates $X^{(1)}, \dots, X^{(m)}$ are available. Now

$$\hat{Y}(b) = \bar{Y}_i + \sum_{k=1}^{m} b_k (EX_i^{(k)} - \bar{X}_i^{(k)}) \tag{11.36}$$

and the optimal vector $\mathbf{b}^*$ is given by

$$\mathbf{b}^* = \Sigma_X^{-1} \Sigma_{XY},$$

where Σ_X is the covariance matrix of control variates $X^{(1)}, \dots, X^{(m)}$ and $[\Sigma_{XY}]_k = \mathrm{Cov}(Y, X^{(k)})$. For details on the use of control variates and other methods of variance reduction in Monte Carlo simulations we refer the reader to [165].

Let us now return to the problem of basket option pricing in Lévy models. Let $(X_j^{(1)} \dots X_j^{(d)})$ with $j = 1 \dots N$ be independent random vectors having the distribution of $(X_T^{(1)} \dots X_T^{(d)})$. We write

$$C_j(K) = e^{-rT} \left(\sum_{i=1}^{d} \alpha_i e^{rT + X_j^{(i)}} - K \right)^+$$

$$C_j^+(K) = \sum_{i=1}^{d} \alpha_i e^{-rT} \left(e^{rT + X_j^{(i)}} - K \right)^+$$

$$C_j^-(K) = e^{-rT} \left(\exp\left\{ rT + \sum_{i=1}^{d} \alpha_i X_j^{(i)} \right\} - K \right)^+$$

Clearly, both $C_j^+(K)$ and $C_j^-(K)$ are highly correlated with $C_j(K)$ and can be used to construct efficient Monte Carlo estimators, based either on one of the control variates or on both of them:

$$\begin{aligned} \hat{C}_1(K) &= \bar{C}(K) + \hat{b}_1 (C^+(K) - \bar{C}^+(K)), \\ \hat{C}_2(K) &= \bar{C}(K) + \hat{b}_2 (C^-(K) - \bar{C}^-(K)), \\ \hat{C}_3(K) &= \bar{C}(K) + \hat{b}_{31} (C^+(K) - \bar{C}^+(K)) + \hat{b}_{32} (C^-(K) - \bar{C}^-(K)). \end{aligned} \tag{11.37}$$

Here, b_1, b_2 and (b_{31}, b_{32}) are the in-sample approximations of optimal parameters for the corresponding estimators — see Equation (11.35).

Example 11.2 Basket call in Merton's model

Consider the problem of pricing a basket call (11.33) on two equally weighted assets with maturity 1 year and strike equal to 1. Suppose that the first asset

TABLE 11.1: Control variate estimators of a basket option price under the Merton jump-diffusion model; see Example 11.2.

Estimator	$\hat{C}_1$	$\hat{C}_2$	$\hat{C}_3$	**Plain MC**
Price	0.1396	0.1371	0.1378	0.1436
Std. dev. (of price)	1.6×10^{-3}	9.4×10^{-4}	5.9×10^{-4}	6.6×10^{-3}
Implied vol.	0.2928	0.2863	0.2881	0.3035
Optimal param. $\hat{b}^*$	0.9826	1.0442	0.3364 0.7145	—

$X^{(1)}$ is described by Merton's jump-diffusion model (see Chapter 4) with volatility $\sigma_1 = 0.3$, jump intensity $\lambda_1 = 5$, mean jump size $\mu = -0.1$ and jump size dispersion $\Delta = 0.1$ and that the second asset has $\sigma_2 = 0.2$, $\lambda_2 = 3$ and the same jump size distribution. Assume further that the jump parts of the assets are independent, their Gaussian parts have correlation $\rho = 0.5$, $S = 1$ for both assets and the interest rate is $r = 0.05$.

In this case $\frac{1}{2}(X^{(1)} + X^{(2)})$ also follows the same model with $\sigma = 0.218$, $\lambda = 8$, $\mu = 0.05$ and $\Delta = -0.05$. Analytic computations yield the bounds: $C^+ = 0.1677$ and $C^- = 0.1234$. These bounds correspond to implied volatility bounds of 36.7% and 25%: they already give us some information about the price of basket.

Table 11.1 shows the price of the basket option and its Black-Scholes implied volatility computed using the three estimators (11.37) and the plain Monte Carlo estimator along with corresponding standard deviations. The prices were evaluated using 1000 Monte Carlo runs. It is clear that the estimator that uses both control variates outperforms both $\hat{C}_1$ and $\hat{C}_2$ and achieves a reduction of standard deviation by a factor greater than ten compared to the plain Monte Carlo. This means that to compute the price with the same precision with plain MC one needs one hundred times more runs than with the control variate estimator. The "true" price, computed with a great number of Monte Carlo runs, is equal to 0.1377, which means that the bias in $\hat{C}_3$, produced by the in-sample estimation of optimal parameter b_3, is insignificant.

Figure 11.5 further compares the behavior of the four estimators for different number of Monte Carlo runs. Notice that with only 100 evaluations the implied volatility error is already less than one per cent for the estimator $\hat{C}_3$ and that at 500 runs this estimator practically attains its limiting value. □

Efficient Monte Carlo simulation of American basket options (with a reasonable number of underlyings) has become possible in the recent years using the Carrière-Longstaff-Schwartz least squares algorithm [266, 86], which can also be applied to exponential-Lévy models using the simulation methods described in Chapter 6.

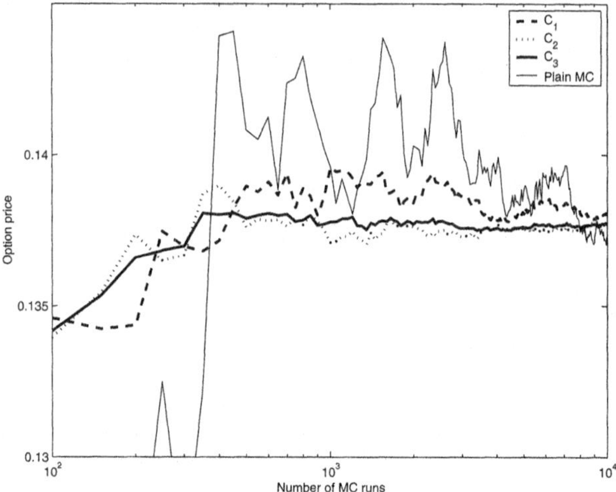

FIGURE 11.5: Comparison of the three improved Monte Carlo estimators of the basket call price to the plain Monte Carlo estimator. The price interval between 0.135 and 0.14 corresponds roughly to one and a half per cent of implied volatility difference.

Further reading

Various examples of risk neutral exponential-Lévy models have been studied in the literature: Merton's jump-diffusion model [291], the variance gamma model [271, 270, 272], the normal inverse Gaussian model [30, 342], the hyperbolic model [125, 319, 123], generalized hyperbolic models [123], the tempered stable model [288, 71, 81], the Kou model [238] and the Meixner model [352].

Fourier transform methods for European option pricing were introduced by Carr and Madan [83], see also [261, 393]. The recent book [71] contains an in-depth discussion of the Fourier approach to the pricing of options in exponential-Lévy models using tools from complex analysis and pseudo-differential operators for a class of exp-Lévy models including tempered stable, normal inverse Gaussian and some others.

Kou and Wang [239] obtain closed formulas for barrier options and Sepp [360, 359] obtains expressions for double barrier and lookback options in the Kou jump-diffusion model [238] using special functions. Barrier options in exponential-Lévy models are discussed in [70, 386]. The Wiener-Hopf factorization is simpler to understand in the case of random walks: a good discussion is given in [334, Chapter 7]. Wiener-Hopf factorization for Lévy processes is discussed in [345, Chapter 9] and [49, Chapter VI]. For Lévy processeswith positive jumps a summary of results is given in [244]. The use of Wiener-Hopf factorizations for pricing exotic options in exp-Lévy models is discussed in [386] using probabilistic methods.

Pricing of American options on infinite-activity Lévy-driven assets is a current research topic. American options on underlying assets with jumps of finite activity have been studied in [317, 316, 390, 391]. Pham [316] discusses the relation between the optimal stopping problem and the integro-differential free boundary problem. Zhang [390] and Gukhal [182] give a decomposition of the American option price into the corresponding European option price and the early exercise premium and discusses the structure of early exercise premium. Gerber and Shiu [163] discuss perpetual American options in compound Poisson models; perpetual American options in exp-Lévy models are discussed in [72] using Wiener-Hopf techniques. A popular Monte Carlo procedure for pricing American options is the least-squares Monte Carlo method described in [266, 86]. A survey of Monte Carlo methods for American options may be found in [165]. Pricing methods based on the numerical solution of partial integro-differential equations will be presented in Chapter 12.

Chapter 12

Integro-differential equations and numerical methods

> An approximate answer to the right problem is worth a good deal more than an exact answer to an approximate problem.
>
> John Tukey

> At that time, the notion of partial differential equations was very, very strange on Wall Street.
>
> Robert C. Merton, *Derivative Strategies*, March 1998, p. 32.

In the Black-Scholes model and, more generally, in option pricing models where the risk neutral dynamics can be described by a diffusion process:

$$\frac{dS_t}{S_t} = rdt + \sigma(t, S_t)dW_t,$$

the value $C(t, S_t)$ of a European or barrier option can be computed by solving a parabolic partial differential equation:

$$\frac{\partial C}{\partial t}(t, S) + rS\frac{\partial C}{\partial S} + \frac{\sigma^2(t, S)S^2}{2}\frac{\partial^2 C}{\partial S^2} - rC(t, S) = 0$$

with boundary conditions depending on the type of option considered. Many numerical methods exist for solving such partial differential equations and have given rise to efficient methods for option pricing in diffusion models.

A similar result holds when the risk-neutral dynamics is given by an exp-Lévy model or a jump-diffusion model: the value of a European or barrier option is given by $C(t, S_t)$ where C solves a *second-order partial integro-*

differential equation (PIDE):

$$\frac{\partial C}{\partial t}(t,S) + rS\frac{\partial C}{\partial S} + \frac{\sigma^2 S^2}{2}\frac{\partial^2 C}{\partial S^2} - rC(t,S)$$
$$+ \int \nu(dy)[C(t,Se^y) - C(t,S) - S(e^y - 1)\frac{\partial C}{\partial S}(t,S)] = 0$$

with boundary conditions depending on the payoff structure of the option. The new element is the integral term in the equation, due to the presence of jumps. This is a *nonlocal* term: it depends on the whole solution $C(t,.)$ and not only on its behavior at the point S. Its presence leads to new theoretical and numerical issues making PIDE less easy to solve than partial differential equations. In the recent years, option pricing in models with jumps has provided an impetus for the study of numerical methods for parabolic equations and free boundary problems for such integro-differential operators.

While parabolic integro-differential equations have been repeatedly used in the literature on option pricing in models with jumps, their derivation and the discussion of their properties are often omitted. This chapter serves as an introduction to this topic: we will show how to derive the integro-differential equations verified by various types of options (Section 12.1), discuss some properties of these equations and different concepts of solution (Section 12.2) and present various numerical methods for solving them: multinomial trees (Section 12.3), finite difference schemes (Section 12.4), the analytic method of lines (Section 12.5) and Galerkin schemes (Section 12.6). The advantages and drawbacks of these methods are summarized in Section 12.7.

Most statements are given for the time-homogeneous case but similar results hold when coefficients depend on time: this case is discussed further in Chapter 14.

12.1 Partial integro-differential equations for computing option prices

Consider a market where the risk-neutral dynamics of the asset is given by an exponential-Lévy model:

$$S_t = S_0 \exp(rt + X_t), \tag{12.1}$$

where X is a Lévy process with characteristic triplet (σ^2, ν, γ) under some risk-neutral measure $\mathbb{Q}$ such that $\hat{S}_t = e^{-rt}S_t = e^{X_t}$ is a martingale. We will assume that

$$\int_{|y|\geq 1} e^{2y}\nu(dy) < \infty. \tag{12.2}$$

This condition is equivalent to the existence of a second moment of the price process and allows to use L^2 methods in the proofs below. The risk-neutral dynamics of S_t is then given by

$$S_t = S_0 + \int_0^t rS_{u-}du + \int_0^t S_{u-}\sigma dW_u + \int_0^t \int_{-\infty}^{\infty} (e^x - 1)S_{u-}\tilde{J}_X(du\ dx), \quad (12.3)$$

where $\tilde{J}_X$ denotes the compensated jump measure of the Lévy process X (see Proposition 8.20) and $\hat{S}_t = \exp X_t$ is a square-integrable martingale:

$$\frac{d\hat{S}_t}{\hat{S}_{t-}} = \sigma dW_t + \int_{-\infty}^{\infty} (e^x - 1)\tilde{J}_X(dt\ dx), \qquad \sup_{t\in[0,T]} E[\hat{S}_t^2] < \infty.$$

The value of a European option with terminal payoff $H(S_T)$ is defined as a discounted conditional expectation of its terminal payoff under risk-neutral probability $\mathbb{Q}$: $C_t = E[e^{-r(T-t)}H(S_T)|\mathcal{F}_t]$. From the Markov property, $C_t = C(t, S)$ where

$$C(t, S) = E[e^{-r(T-t)}H(S_T)|S_t = S]. \quad (12.4)$$

Introducing the change of variable $\tau = T - t$, $x = \ln(S/K) + r\tau$ where K is an arbitrary constant and defining: $h(x) = H(Ke^x)/K$ and $u(\tau, x) = e^{r(T-t)}C(t, S)/K$, we can rewrite the above equation as

$$u(\tau, x) = E[h(x + X_\tau)]. \quad (12.5)$$

If h is in the domain of the infinitesimal generator then we can differentiate with respect to τ to obtain the following integro-differential equation:

$$\frac{\partial u}{\partial \tau} = L^X u \quad \text{on} \quad]0, T] \times \mathbb{R}, \qquad u(0, x) = h(x), \qquad x \in \mathbb{R}, \quad (12.6)$$

where L is the infinitesimal generator[1] of X:

$$L^X f(x) = \gamma\frac{\partial f}{\partial x} + \frac{\sigma^2}{2}\frac{\partial^2 f}{\partial x^2} + \int \nu(dy)[f(x + y) - f(x) - y1_{\{|y|<1\}}\frac{\partial f}{\partial x}(x)].$$

Similarly, if u is smooth then using a change of variable we obtain a similar equation for $C(t, S)$:

$$\frac{\partial C}{\partial t}(t, S) + rS\frac{\partial C}{\partial S} + \frac{\sigma^2 S^2}{2}\frac{\partial^2 C}{\partial S^2} - rC(t, S) + \int \nu(dy)[C(t, Se^y) - C(t, S) - S(e^y - 1)\frac{\partial C}{\partial S}(t, S)] = 0. \quad (12.7)$$

[1]See Proposition (3.16) for a definition of the infinitesimal generator.

These equations are similar to the Black-Scholes partial differential equation, except that a new integral term appears in the equation.

However, the above reasoning is heuristic: the payoff function h is usually not in the domain of L^X and in fact it is usually not even differentiable. For example $h(x) = (1-e^x)^+$ for a put option and $h(x) = 1_{x \geq x_0}$ for a binary option. For barrier options, the presence of boundary conditions may lead to additional complications.

We will nevertheless show now that under some conditions, the value of European options, barrier options and American options in this model can be expressed as solutions of second order partial integro-differential equations (PIDEs) such as the one above, with appropriate boundary conditions. When these conditions are not always verified, it will lead us to consider, in Section 12.2.4, the notion of viscosity solution; we will then show that under more general conditions, values of European or barrier options can be expressed as viscosity solutions of appropriate PIDEs.

12.1.1 European options

Consider a European option with maturity T and payoff $H(S_T)$. The payoff function H is assumed to verify:

$$|H(x) - H(y)| \leq c|x - y| \tag{12.8}$$

for some $c > 0$. This condition is of course verified by call and put options with $c = 1$. The value C_t of such an option is given by $C_t = C(t, S_t)$ where

$$C(t,S) = e^{-r(T-t)}E[H(S_T)|S_t = S] = e^{-r(T-t)}E[H(Se^{r(T-t)+X_{T-t}})].$$

PROPOSITION 12.1 Backward PIDE for European options
Consider the exponential-Lévy model $S_t = S_0 \exp(rt + X_t)$ where X is a Lévy process verifying (12.2). If either $\sigma > 0$ or

$$\exists \beta \in]0,2[, \qquad \liminf_{\epsilon \downarrow 0} \epsilon^{-\beta} \int_{-\epsilon}^{\epsilon} |x|^2 d\nu(x) > 0 \tag{12.9}$$

then the value of a European option with terminal payoff $H(S_T)$ is given by $C(t,S)$ where:

$$C : [0,T] \times]0,\infty[\to \mathbb{R},$$
$$(t,S) \mapsto C(t,S) = E[H(S_T)|S_t = S]$$

is continuous on $[0,T] \times [0,\infty[$, $C^{1,2}$ on $]0,T[\times]0,\infty[$ and verifies the partial integro-differential equation:

$$\frac{\partial C}{\partial t}(t,S) + rS\frac{\partial C}{\partial S}(t,S) + \frac{\sigma^2 S^2}{2}\frac{\partial^2 C}{\partial S^2}(t,S) - rC(t,S)$$
$$+ \int \nu(dy)[C(t,Se^y) - C(t,S) - S(e^y - 1)\frac{\partial C}{\partial S}(t,S)] = 0 \tag{12.10}$$

on $[0,T[\times]0,\infty[$ *with the terminal condition:*

$$\forall S > 0 \qquad C(T,S) = H(S). \tag{12.11}$$

REMARK 12.1 The regularity condition (12.9) will be used to show smoothness with respect to the underlying, which is needed to derive the PIDE. Conditions (12.8) and (12.2) simplify the proof but can be weakened. □

PROOF The proof involves, as in the Black-Scholes case, applying the Itô formula to the martingale $\hat{C}(t,S_t) = e^{-rt}C(t,S_t)$, identifying the drift component and setting it to zero.

First we need to show that $\hat{C}(t,S)$ is a smooth function. Let us start by showing that it is a differentiable function of t. To simplify the notation suppose for the moment that interest rates are zero. The option price is a differentiable function of time if and only if its time value is differentiable. The Fourier transform of the time value is given by formula (11.19):

$$\zeta_\tau(v) = \frac{\phi_\tau(v-i)-1}{iv(1+iv)}.$$

Its derivative can be easily computed via the Lévy-Khinchin formula:

$$\frac{\partial\zeta_\tau(v)}{\partial\tau} = \frac{\psi(v-i)\phi_\tau(v-i)}{iv(1+iv)}.$$

Under the martingale condition for $\exp X_t$, $\psi(-i) = \psi'(-i) = 0$ so $\psi(v-i) = O(|v|^2)$. Therefore at infinity the denominator compensates the growth of $\psi(v-i)$ and $\frac{\partial\zeta_\tau(v)}{\partial\tau}$ is integrable if $\phi_\tau(v-i)$ is. Under the condition (12.9) Re $\psi_\tau(v) \leq -|u|^{2-\beta}$ and also, under the martingale condition, Re $\psi_\tau(v-i) \leq -|u|^{2-\beta}$. Therefore $\phi_\tau(v-i)$ decays faster than any negative power of $|v|$ at infinity and is in any case integrable. Finally using the dominated convergence theorem it follows that $\hat{C}(t,S_t)$ is a differentiable function of t.

The condition (12.9) also allows to conclude that X_t has a C^∞ density with all derivatives vanishing at infinity (see Proposition 3.12 or [345, Proposition 28.3]) and as we observed in Chapter 11, $C(t,S)$ is then a smooth function of S.

By construction, $\hat{C}_t = E[e^{-rT}H|\mathcal{F}_t]$ is a martingale. Applying the Itô formula to $\hat{C}_t = e^{-rt}C(t, S_t)$ and using Equation (12.3) we obtain:

$$\begin{aligned}
d\hat{C}_t &= e^{-rt}[-rC_t + \frac{\partial C}{\partial t}(t, S_{t-}) + \frac{\sigma^2 S_t^2}{2}\frac{\partial^2 C}{\partial S^2}(t, S_{t-})dt + \frac{\partial C}{\partial S}(t, S_{t-})dS_t\ ] \\
&+ e^{-rt}[C(t, S_{t-}e^{\Delta X_t}) - C(t, S_{t-}) - S_{t-}(e^{\Delta X_t} - 1)\frac{\partial C}{\partial S}(t, S_{t-})] \\
&= a(t)dt + dM_t, \quad \text{where} \qquad (12.12) \\
a(t) &= e^{-rt}\left[-rC + \frac{\partial C}{\partial t} + \frac{\sigma^2 S_{t-}^2}{2}\frac{\partial^2 C}{\partial S^2} + rS_{t-}\frac{\partial C}{\partial S}\right](t, S_{t-}) \\
&+ \int_{-\infty}^{\infty} \nu(dx)e^{-rt}[C(t, S_{t-}e^x) - C(t, S_{t-}) - S_{t-}(e^x - 1)\frac{\partial C}{\partial S}(t, S_{t-})], \\
dM_t &= e^{-rt}\left\{\frac{\partial C}{\partial S}(t, S_{t-})\sigma S_{t-}dW_t + \int_{\mathbb{R}}[C(t, S_{t-}e^x) - C(t, S_{t-})]\tilde{J}_X(dt\ dx)\right\}.
\end{aligned}$$

Let us now show that M_t is a martingale. Since the payoff function H is Lipschitz, C is also Lipschitz with respect to the second variable:

$$\begin{aligned}
&C(t, x) - C(t, y) \\
&\quad = e^{-r(T-t)}\{E[H(xe^{r(T-t)+X_T-X_t})] - E[H(ye^{r(T-t)+X_T-X_t})]\} \\
&\quad \leq c|x - y|E[e^{X_{T-t}}] = c|x - y|, \qquad (12.13)
\end{aligned}$$

since e^{X_t} is a martingale. Therefore the predictable random function $\psi(t, x) = C(t, S_{t-}e^x) - C(t, S_{t-})$ verifies:

$$\begin{aligned}
E[\int_0^T dt \int_{\mathbb{R}} \nu(dx)|\psi(t, x)|^2] &= E[\int_0^T dt \int_{\mathbb{R}} \nu(dx)|C(t, S_{t-}e^x) - C(t, S_{t-})|^2] \\
&\leq E[\int_0^T dt \int_{\mathbb{R}} c^2(e^{2x} + 1)S_{t-}^2\nu(dx)] \\
\text{using (12.2)} \qquad &\leq c^2 \int_{\mathbb{R}}(e^{2x} + 1)\nu(dx)\ E[\int_0^T S_{t-}^2 dt] < \infty,
\end{aligned}$$

so from Proposition 8.8, the compensated Poisson integral

$$\int_0^t \int_{-\infty}^{\infty} e^{-rt}[C(t, S_{t-}e^x) - C(t, S_{t-})]\ \tilde{J}_X(dt\ dx)$$

is a square-integrable martingale. Also, since C is Lipschitz,

$$||\frac{\partial C}{\partial S}(t, .)||_{L^\infty} \leq c \text{ so } E[\int_0^T S_{t-}^2|\frac{\partial C}{\partial S}(t, S_{t-})|dt] \leq c^2 E[\int_0^T S_{t-}^2 dt] < \infty$$

so, by Proposition 8.6, $\int_0^t \sigma S_t \frac{\partial C}{\partial S}(t, S_{t-})dW_t$ is also a square-integrable martingale. Therefore, M_t is a square-integrable martingale. $\hat{C}_t - M_t$ is thus a (square-integrable) martingale; but $\hat{C}_t - M_t = \int_0^t a(t)dt$ is also a *continuous* process with finite variation. It is therefore a continuous martingale with finite variation hence from Remark 8.2 we must have $a(t) = 0$ $\mathbb{Q}$-almost surely which yields the PIDE (12.10). □

Introducing the integro-differential operator:

$$L^S g(x) = rxg'(x) + \frac{\sigma^2 x^2}{2} g''(x) + \int \nu(dy)[g(xe^y) - g(x) - x(e^y - 1)g'(x)],$$

which is well defined, under hypothesis (12.2), for any C^2 function g with bounded first derivative, the equation for European options can be written as:

$$\frac{\partial C}{\partial t}(t, S) + L^S C(t, S) - rC(t, S) = 0 \quad \text{on} \quad [0, T[\times]0, \infty[$$
$$\text{with the terminal condition} \quad C(T, S) = H(S).$$

In general the three terms under the integral cannot be separated otherwise the integral does not converge. In the case where the process is of finite variation, i.e., $\sigma = 0$ and $\int_{-1}^{1} |y|\nu(dy) < \infty$ the derivative term under the integral can be absorbed into the first order derivative term and the equation reduces to a first order PIDE:

$$\frac{\partial C}{\partial t}(t, S) + (r + c)S\frac{\partial C}{\partial S}(t, S) - rC(t, S) + \int \nu(dy)[C(t, Se^y) - C(t, S)] = 0,$$
$$\text{where} \quad c = \int \nu(dy)(1 - e^y).$$

REMARK 12.2 "Boundary conditions" As we will observe in Section 12.2.1, the terminal value (Cauchy) problem (12.10)–(12.11) is well posed and has a unique solution continuous on $[0, T[\times[0, \infty[$. By substituting $S = 0$ in (12.10) it is seen that any such solution verifies $C(t, 0) = H(0) \exp[-r(T - t)]$ for $t \in [0, T]$. In the case of the call option, this gives $C(t, 0) = 0$ for $t \in [0, T]$. As in the case of diffusion models, this is a *property* of solutions and not a condition imposed on the solution. Some authors incorrectly refer to such behavior at the boundary of the domain as a "boundary condition." In fact the terminal condition (payoff) alone determines uniquely the solution and one does *not* require the specification of boundary conditions to have a unique solution. The uniqueness of the solution can be seen in financial terms as a consequence of absence of arbitrage: two European options with the same terminal payoff should have the same price so the terminal condition should determine the solution in a unique manner. In other words, any solution verifies $C(t, 0) = H(0) \exp[-r(T - t)]$ automatically so if we try to impose a boundary condition at $S = 0$ it is either redundant or impossible to satisfy! This is due to the fact that the coefficients of the integro-differential operator in (12.10) vanish at $S = 0$. □

Since various changes of variable will be useful in the sequel, we will give different forms of the PIDE (12.10). Define

$$f(t,y) = e^{r(T-t)}C(t, S_0 e^y), \qquad y = \ln\frac{S}{S_0}. \tag{12.14}$$

Then f has the following probabilistic representation:

$$f(t,y) = E[h(X_{T-t} + r(T-t) + y)],$$

where $h(z) = H(S_0 e^z)$. Using the change of variable formula to compute the partial derivatives of f, we obtain a PIDE with constant coefficients in terms of this new parameterization:

COROLLARY 12.1 PIDE in log-price

Under the conditions (12.2) and (12.9) the forward value of the option $f: [0,T]\times\mathbb{R}\mapsto\mathbb{R}$ given by (12.14) is a $C^{1,2}$ solution of

$$\forall (t,y)\in[0,T[\times\mathbb{R},\quad \frac{\partial f}{\partial t}(t,y) + \frac{\sigma^2}{2}\frac{\partial^2 f}{\partial y^2}(t,y) + (r-\frac{\sigma^2}{2})\frac{\partial f}{\partial y}(t,y)]$$
$$+\int \nu(dz)[f(t,y+z) - f(t,y) - (e^z-1)\frac{\partial f}{\partial y}(t,y)] = 0.$$
$$\forall x\in\mathbb{R},\quad f(T,x) = h(x).$$

For call (or put) options one can make a further simplification by using the homogeneity of the payoff function. Define the log forward moneyness x and the time to maturity τ by

$$\tau = T - t, \qquad x = \ln(S/K) + r\tau. \tag{12.15}$$

Rewriting the payoff and the option value in terms of these new variables:

$$u(\tau,x) = \frac{e^{r\tau}}{K}C(T-\tau, Ke^{x-r\tau}), \qquad h(x) = (e^x - 1)^+, \tag{12.16}$$

we obtain a simple expression for the forward value of the option $u(\tau,x)$:

$$u(\tau,x) = E[h(X_\tau + x)] = P_\tau h(x), \tag{12.17}$$

where P_τ is the semi-group or evolution operator of the process X. Computing the derivatives of u by the change of variable formula we obtain the following equation for u:

COROLLARY 12.2 PIDE in log-moneyness coordinates
Under the conditions (12.2) and (12.9) the function

$$u : [0,T] \times \mathbb{R} \to \mathbb{R},$$

defined by (12.15)–(12.16) is continuous on $[0,T] \times [0,\infty[$, $C^{1,2}$ *on* $]0,T] \times]0,\infty[$ *and verifies the partial integro-differential equation:*

$$\frac{\partial u}{\partial \tau}(\tau,x) = \frac{\sigma^2}{2}[\frac{\partial^2 u}{\partial x^2}(\tau,x) - \frac{\partial u}{\partial x}(\tau,x)]$$
$$+ \int \nu(dy)[u(\tau,x+y) - u(\tau,x) - (e^y - 1)\frac{\partial u}{\partial x}(\tau,x)]$$

with the initial condition: $\forall x \in \mathbb{R}, \quad u(0,x) = (e^x - 1)^+$.

12.1.2 Barrier options

Barrier options were discussed in Chapter 11. Consider for instance an up-and-out call option with maturity T, strike K and (upper) barrier $B > K$. The terminal payoff is given by

$$H_T = (S_T - K)^+ \text{ if } \max_{t\in[0,T]} S_t < B,$$
$$= 0 \text{ if } \max_{t\in[0,T]} S_t \geq B.$$

The value of the barrier option at time t can be expressed as discounted expectation of the terminal payoff: $C_t = e^{-r(T-t)}E[H_T|\mathcal{F}_t]$. By construction, $e^{-rt}C_t$ is a martingale. Defining $H(S) = (S-K)^+ 1_{S<B}$ and the first exit time from $]0,B[$

$$\tau_B = \inf\{s \geq t, X_s \geq B\},$$

we can introduce a deterministic function

$$C(t,S) = e^{-r(T-t)} \; E[H(S_{T\wedge\tau_B})|S_t = S]. \tag{12.18}$$

Due to the strong Markov property of S_t, C_t coincides with $C(t,S_t)$ before the barrier is crossed, i.e., for $t \leq \tau_B$. Therefore,

$$C_{t\wedge\tau_B} = C(t\wedge\tau_B, S_{t\wedge\tau_B}) \tag{12.19}$$

for all $t \leq T$. Note that for $t \geq \tau_B$, C_t and $C(t,S_t)$ may be different: if the barrier has already been crossed C_t stays equal to zero , but $C(t,S_t)$ may become positive again if the stock returns to the region below the barrier.

By analogy with the case of European options, one can derive a pricing equation in the following way. Assume that the function $C(t,S) : [0,T]\times]0,\infty[\rightarrow]0,\infty[$ is $C^{1,2}$, i.e., smooth enough for the Itô formula to be applied. Applying the Itô formula to $\hat{C}_t = \exp(-rt)C(t,S_t)$ between $t=0$ and $t = T\wedge\tau_B$ we obtain, as in (12.12):

$$d\hat{C}_t = a(t)dt + dM_t \quad \text{where}$$
$$dM_t = e^{-rt}\frac{\partial C}{\partial S}(t,S_{t-})\sigma S_{t-}dW_t + \int_{\mathbb{R}} e^{-rt}[C(t,S_{t-}e^x) - C(t,S_{t-})]\tilde{J}_X(dt\, dx)$$
$$\text{and} \quad a(t) = e^{-rt}[\frac{\partial C}{\partial t}(t,S_{t-}) - rC(t,S_{t-}) + L^S C(t,S_{t-})].$$

From (12.19) we know that $\hat{C}_{t\wedge\tau_B}$ is a martingale. Therefore, its drift part must be equal to zero which means that $\int_0^{t\wedge\tau_B} a(t)dt = 0$ for all $t \leq T$. This implies that with probability 1, $a(t) = 0$ almost everywhere on $[0,\tau_B]$:

$$\frac{\partial C}{\partial t}(t,S_{t-}) + L^S C(t,S_{t-}) - rC(t,S_{t-}) = 0,$$

which should be verified for almost every $t \in]0,\tau_B[$ with probability 1 so:

$$\frac{\partial C}{\partial t}(t,S) + L^S C(t,S) = rC(t,S) \qquad \forall (t,S) \in [0,T[\times]0,B[. \quad (12.20)$$

For these arguments to work, we need the smoothness of C which is not obvious, especially at the barrier $S = B$. The above reasoning can be rigorously justified if $\sigma > 0$ (see [48, Chapter 3] and [337]):

PROPOSITION 12.2 PIDE for barrier options

Let $C(t,S)$ be the value of a up-and-out call option defined by (12.18) in an exponential-Lévy model with diffusion coefficient $\sigma > 0$ and Lévy measure ν. Then $(t,S) \rightarrow C(t,S;T,K)$ verifies

$$\forall (t,S) \in [0,T[\times]0,B[,$$
$$\frac{\partial C}{\partial t}(t,S) + rS\frac{\partial C}{\partial S}(t,S) + \frac{\sigma^2 S^2}{2}\frac{\partial^2 C}{\partial S^2}(t,S) - rC(t,S)$$
$$+ \int \nu(dy)[C(t,Se^y) - C(t,S) - S(e^y - 1)\frac{\partial C}{\partial S}(t,S)] = 0. \quad (12.21)$$
$$\forall S \in]0,B[, \quad C(t=T,S) = (S-K)^+. \quad (12.22)$$
$$\forall S \geq B, \quad \forall t \in [0,T], \quad C(t,S) = 0. \quad (12.23)$$

The main difference between this equation and the analogous PDEs for diffusion models is in the "boundary condition": (12.23) not only specifies the behavior of the solution *at* the barrier $S = B$ but also *beyond* the barrier $(S > B)$. This is necessary because of the nonlocal nature of the operator L^S:

to compute the integral term we need the function $C(t,.)$ on $]0,\infty[$ and (12.23) extends the function beyond the barrier by zero. In the case of a rebate, the function would be replaced by the value of the rebate in the knock-out region $S > B$.

The case $\sigma > 0$ is not the only case where the barrier price verifies (12.21): more generally, if $C(.,.)$ defined by (12.18) can be shown to be $C^{1,2}$ (or simply $C^{1,1}$ in the case of finite variation models) then C is a solution of the boundary value problem above. Since

$$C(t,S) = e^{-r(T-t)} \int q^B_{t,T}(u) H(u) du,$$

where $q^B_{t,T}$ is the (risk-neutral) transition density of the process $S_{t\wedge\tau_B}$ stopped at the first exit from the barrier, the smoothness of $(t,S) \mapsto C(t,S)$ is linked to the smoothness of this transition density. In analytical terms $G(t,T,S,u) = e^{-r(T-t)} q^B_{t,T}(u)$ is nothing but the *Green function* for the operator L^S with boundary conditions (12.23). Smoothness results for Green functions associated to integro-differential operators are studied using an analytical approach in [158, 157] in the case of a diffusion coefficient bounded away from zero. In the case of pure jump models where $\sigma = 0$ smoothness is not obvious: the behavior of the option price at the barrier depends on the local behavior of the Lévy process at the barrier. Some smoothness results may be obtained using Fourier methods [71] or by probabilistic methods [58] but the conditions for such results to hold are not easy to verify in general.

Defining as above

$$f(t,y) = e^{r(T-t)} C(t, S_0 e^y), \quad y = \ln \frac{S}{S_0}, \quad b = \ln \frac{B}{S_0}, \tag{12.24}$$

we obtain a PIDE with constant coefficients in terms of f:

$$\forall (t,y) \in [0,T[\times]-\infty,b[, \quad \frac{\partial f}{\partial t}(t,y) + \frac{\sigma^2}{2}\frac{\partial^2 f}{\partial y^2}(t,y) + (r - \frac{\sigma^2}{2})\frac{\partial f}{\partial y}(t,y)]$$
$$+ \int \nu(dz)[f(t,y+z) - f(t,y) - (e^z - 1)\frac{\partial f}{\partial y}(t,y)] = 0. \tag{12.25}$$
$$\forall t \in [0,T[, \ \forall y \geq b, \quad f(t,y) = 0.$$
$$\forall y \in \mathbb{R}, \quad f(T,y) = (S_0 e^y - K)^+ 1_{y<b}. \tag{12.26}$$

Similarly, the function $u : [0,T] \times \mathbb{R} \to \mathbb{R}$ defined as in (12.16) is a $C^{1,2}$ solution of the partial integro-differential equation:

$$\frac{\partial u}{\partial \tau}(\tau,x) = \frac{\sigma^2}{2}[\frac{\partial^2 u}{\partial x^2}(\tau,x) - \frac{\partial u}{\partial x}(\tau,x)]$$
$$+ \int \nu(dz)[u(\tau,x+z) - u(\tau,x) - (e^z - 1)\frac{\partial u}{\partial x}(\tau,x)].$$
$$\forall x \in \mathbb{R}, \quad u(0,x) = (e^x - 1)^+ 1_{x < \ln B - \ln K + r\tau}.$$
$$\forall \tau \in [0,T], \ \forall x \geq \ln \frac{B}{K} + r\tau, \quad u(\tau,x) = 0.$$

Notice that in this last parameterization the boundary is not constant anymore but an affine function of time. PIDEs for values of other types of barrier options may be derived in a similar fashion, using in-out/call-put parity relations [70, 383].

12.1.3 American options

As discussed in Section 11.4, the value of an American put is given by the supremum over all nonanticipating exercise strategies, i.e., the set $\mathcal{T}(t,T)$ of nonanticipating random times τ with $t \leq \tau \leq T$, of the value of the payoff at exercise:

$$P_t = \operatorname*{ess\,sup}_{\tau\in\mathcal{T}(t,T)} E[e^{-r(\tau-t)}(K-S_\tau)^+|S_t=S] = P(t,S_t),$$

$$\text{where} \qquad P(t,S) = \sup_{\tau\in\mathcal{T}(t,T)} E[(Ke^{-r(\tau-t)} - Se^{X_{\tau-t}})^+]. \tag{12.27}$$

Results shown in [48] in the case $\sigma > 0$ allow to characterize the function $P(t,S)$ in terms of the solution of a system of partial integro-differential inequalities:

PROPOSITION 12.3 Linear complementarity problem
Let $P: [0,T[\times]0,\infty[$ be the unique bounded continuous solution of

$$\begin{aligned}
&\frac{\partial P}{\partial t}(t,S) + L^S P(t,S) - rP(t,S) \leq 0,\\
&P(t,S) - (K-S)^+ \geq 0,\\
&\left\{\frac{\partial P}{\partial t}(t,S) + L^S P(t,S) - rP(t,S)\right\}\{P(t,S)-(K-S)^+\} = 0 \qquad (12.28)\\
&\text{on } [0,T]\times]0,\infty[\quad \text{and} \quad \forall S \in]0,\infty[, \quad P(T,S) = (K-S)^+,
\end{aligned}$$

where $\sigma > 0$ and

$$L^S P(t,S) = rS\frac{\partial P}{\partial S} + \frac{\sigma^2 S^2}{2}\frac{\partial^2 P}{\partial S^2} + \int \nu(dy)[P(t,Se^y) - P(t,S) - S(e^y-1)\frac{\partial C}{\partial S}(t,S)],$$

then the value of the American put option is given by:

$$P_t = P(t,S_t). \tag{12.29}$$

If P were a smooth $C^{1,2}$ function then the above proposition could be proved by applying the Itô formula.[2] However, even in the case where $\sigma > 0$, the

[2] A heuristic proof along these lines is given in [245].

solution $P(.,.)$ is known not to be $C^{1,2}$ so we cannot proceed in this way. A proper proof of this result involves the dynamic programming principle, see [48, 316, 317]. The linear complementarity problem above can also be written in the form of a nonlinear Hamilton-Jacobi-Bellman equation on $[0,T] \times \mathbb{R}$:

$$\min\{-\frac{\partial P}{\partial t}(t,S) - L^S P(t,S) + rP(t,S), P(t,S) - (K-S)^+\} = 0,$$
$$P(T,S) = (K-S)^+.$$

The strong Markov property of (S_t) implies that the exercise decision at any time t only depends on t and S_t. Therefore, for each t there exists an exercise region $\mathcal{E}_t$ such that it is optimal to exercise the option if S_t is in $\mathcal{E}_t$ and to continue holding it otherwise. In an exponential-Lévy model the value $P(t,S)$ of an American put is convex and nonincreasing in S and satisfies

$$\left|P(t,S_1) - P(t,S_2)\right| \leq |S_1 - S_2|.$$

This entails that the exercise region has the simple form $\{S : S \leq b(t)\}$: there exists an *exercise boundary* $\{(t,b(t)), t \in [0,T]\}$ such that it is optimal to exercise if S_t is below the boundary and to continue otherwise. The domain $[0,T] \times \mathbb{R}^+$ of the put option price $P(t,S)$ is divided into the *exercise region*

$$\mathcal{E} = \{(t,S) : x \leq b(t)\} = \{(t,S) : P(t,S) = (K-S)^+\}$$

and the *continuation region*

$$\mathcal{C} = \{(t,S) : S > b(t)\} = \{(t,x) : P(t,S) > (K-S)^+\}.$$

REMARK 12.3 Smooth pasting In the case of a jump-diffusion model with $\sigma > 0$ and finite jump intensity, it is shown in [390, 316] that the value $P : (t,S) \mapsto P(t,S)$ of the American put option is continuously differentiable with respect to the underlying on $[0,T[\times]0,\infty[$ and in particular the derivative is continuous across exercise boundary:

$$\forall t \in [0,T[, \qquad \frac{\partial P}{\partial S}(t',S) \underset{(t',S)\to(t,b(t))}{\to} -1. \tag{12.30}$$

Relation (12.30) is called the smooth pasting condition. However this condition does not always hold when $\sigma = 0$, i.e., for pure jump models. Boyarchenko and Levendorskii [72, Theorem 7.2] show that the smooth pasting condition fails for tempered stable processes with finite variation, i.e., with $\alpha_\pm < 1$. These authors also give a sufficient condition for smooth pasting to hold [72, Theorem 7.1] but this condition is given in terms of Wiener-Hopf factors and is not easy to verify in a given model. In the case of models with jumps of a given sign (i.e., only positive or only negative jumps) necessary and sufficient conditions for smooth pasting to hold have been given by Avram et al. [16]:

if the process can jump over the exercise boundary then the derivative is continuous at the boundary if and only if the process is of infinite variation. But it is not clear whether this result carries over to processes with both negative and positive jumps. □

Under an additional condition, Pham [316] gives the following characterization for the couple $(P(.,.), b(.))$:

PROPOSITION 12.4 Free boundary problem
Assume that $\sigma > 0$, the intensity of jumps is finite and $r - \int (e^y - 1)\nu(dy) \geq 0$. Then (P, b) is the unique pair of continuous functions verifying

$$\begin{aligned}
&P : [0, T[\to [0, \infty[\text{ nonincreasing, convex,} && 0 \leq b(t) \leq K,\\
&\frac{\partial P}{\partial t} + L^S P(t,S) - rP(t,S) = 0, && \text{while } S > b(t),\\
&\lim_{S \downarrow b(t)} P(t,S) = K - b(t), && \forall t \in [0, T[,\\
&\lim_{S \downarrow b(t)} \frac{\partial P}{\partial S}(t,S) = -1, && \forall t \in [0, T[,\\
&P(T,S) = (K-S)^+, && \forall S \in [0, \infty[,\\
&P(t,S) > (K-S)^+, && \text{if } S > b(t),\\
&P(t,S) = (K-S)^+, && \text{if } S \leq b(t).
\end{aligned}$$

This formulation, where the unknowns are both the pricing function $f(.,.)$ and the boundary b, is called a free boundary problem. There are few results available on free boundary problems for integro-differential operators, see, e.g., [157, 151].

12.2 Second order integro-differential equations

All the equations obtained above involve operators of type:

$$\begin{aligned}
Lf(x) &= a_1(x)\frac{\partial f}{\partial x} + a_2(x)\frac{\partial^2 f}{\partial x^2}\\
&+ \int \nu(dy)[f(x + a_0(x,y)) - f(x) - a_0(x,y)1_{\{|a_0(x,y)|<1\}}\frac{\partial f}{\partial x}(x)],
\end{aligned}$$

which are all incarnations under different changes of variable of the infinitesimal generator L^X of the Lévy process

$$L^X f(x) = \gamma\frac{\partial f}{\partial x} + \frac{\sigma^2}{2}\frac{\partial^2 f}{\partial x^2} + \int \nu(dy)[f(x+y) - f(x) - y1_{\{|y|<1\}}\frac{\partial f}{\partial x}(x)].$$

Since we are concerned here with risk-neutral dynamics, the relation (11.5) is verified so L^X takes the form:

$$L^X f(x) = \frac{\sigma^2}{2}(\frac{\partial^2 f}{\partial x^2} - \frac{\partial f}{\partial x}) + \int \nu(dy)[f(x+y) - f(x) - (e^y - 1)\frac{\partial f}{\partial x}(x)],$$

which, under condition (11.5), is well defined for any Lipschitz function $f \in C^2(\mathbb{R})$. In the Black-Scholes model, $X_t = -\sigma^2 t/2 + \sigma W_t$ and L^X reduces to a second-order differential operator

$$L^X f(x) = \frac{\sigma^2}{2}(\frac{\partial^2 f}{\partial x^2} - \frac{\partial f}{\partial x}), \tag{12.31}$$

which is well-defined for any $f \in C^2(\mathbb{R})$. When the Lévy process is of infinite variation, i.e., if $\sigma > 0$ or $\int_{-1}^{1} |x|\nu(dx) = \infty$, f is required to be twice differentiable in order for the derivatives and the integral term to make sense: L^X is then a second-order operator (even if $\sigma = 0$). When $\sigma = 0$ and $\int |x|\nu(dx) < \infty$ (i.e., the process has finite variation) the operator can be rewritten as

$$L^X.f(x) = b\frac{\partial f}{\partial x} + \int \nu(dy)[f(x+y) - f(x)],$$

where f is now only required to be C^1: L^X is a first order operator. Finally, if X is a compound Poisson process (without drift!) L is an operator of order zero and f can need not even be differentiable:

$$L^X f(x) = \int \nu(dy)[f(x+y) - f(x)],$$

L^X is thus of "order zero." But smoothness is not the only condition required on f to define Lf: f (and its derivatives up to the order of the operator) have to be integrable with respect to ν for the integral term to be defined, which means that f and its derivatives can increase at infinity but their growth is determined by the decay rate of the Lévy density. For example if

$$\int_{|x|\geq 1} |x|^p \nu(dx) < \infty$$

then $L\varphi$ can be defined for $\varphi \in C^2([0,T]\times\mathbb{R})$ which grows slower than $1+|x|^p$: $\exists C > 0, |\varphi(t,x)| \leq C(1+|x|^p)$:

$$L\varphi(x) = A\varphi(x) + \int_{|y|\leq 1} \nu(dy)[\varphi(x+y) - \varphi(x) - y\frac{\partial \varphi}{\partial x}] \tag{12.32}$$

$$+ \int_{|y|>1} \nu(dy)[\varphi(x+y) - \varphi(x)], \tag{12.33}$$

where A is a second-order differential operator. The terms in (12.32) are well defined for $\varphi \in C^2([0,T] \times \mathbb{R})$ since for $|y| \leq 1$

$$|\varphi(\tau, x+y) - \varphi(\tau, x) - y\frac{\partial \varphi}{\partial x}(\tau, x)| \leq y^2 \sup_{B(x,1)} |\varphi''(\tau, .)|,$$

while the term in (12.33) is well defined because of the conditions on the decay of ν. In particular $L^X\varphi$ is well-defined for $\varphi \in C_0^2(\mathbb{R})$ but one can allow φ to grow at a faster rate by imposing a faster decay at infinity of the Lévy measure. In exponential-Lévy models, since the right tail of ν decays exponentially this allows for functions which grow at an exponential rate when $x \to +\infty$. This is useful since the value of a call option grows linearly in $S = S_0 e^x$ therefore exponentially in x.

An important property, that the integro-differential operator L^X shares with the more well known second order differential operators, is the (positive) *maximum principle* [65, 159]: for $\varphi \in C_0^2(\mathbb{R})$

$$\varphi(x_0) = \sup_{x\in\mathbb{R}} \varphi(x) \geq 0 \Rightarrow L^X\varphi(x_0) \leq 0. \tag{12.34}$$

12.2.1 Nonlocality and its consequences

The main difference between (12.31) and (12.31) is the *nonlocal* character of (12.31): whereas $L^0 f(x)$ only involves the value and the derivatives of f at x, i.e., the local behavior of f at x, in order to evaluate $L^X f(x)$ we need to know f not only in a neighborhood of x but for all values which the process can jump to from x: $\{x+y, y \in \text{supp}(\nu)\}$. Since in the majority of the models considered here (except models with Poisson jumps) jumps are unbounded, this means that values of f at all points $y \in \mathbb{R}$ are used in computing $L^X f(x)$.

For considering boundary value problems as in the case of barrier options, we need to define $L^X f$ for functions f defined on a bounded interval $[a, b]$. The nonlocality alluded to above has the following practical consequence: if we want to define $L^X f$ for a function f defined on a bounded interval $[a, b]$, the integral term in expression (12.31) becomes meaningless unless we *extend* f by assigning values to f outside $[a, b]$. Obviously there are many ways to extend a function beyond its domain of definition; each of these extensions will lead to a different value for $L^X f(x)$ (even when $x \in [a, b]$). One way of extending functions beyond $[a, b]$ will be to choose some function $h : \mathbb{R} \to \mathbb{R}$ and set $u(x) = h(x)$ for $x \notin]a, b[$. Note that by doing so we may destroy the regularity of the function f at the boundary. This has a probabilistic interpretation in terms of the process X stopped at its first exit from $[a, b]$, which was already encountered when discussing barrier options. Smooth extensions are possible but do not have this probabilistic interpretation.

12.2.2 The Fourier view: pseudo-differential operators

For readers tired of seeing the ∂ sign let us make a digression in Fourier space. Recall the definition of the Fourier transform:

$$\mathbf{F}f(z) = \int_{-\infty}^{\infty} e^{ixz} f(x)dx,$$

where $i = \sqrt{-1}$ and its inverse:

$$\mathbf{F}^{-1}f(z) = \frac{1}{2\pi}\int_{-\infty}^{\infty} e^{-ixz} f(x)dx.$$

One of the important properties of the Fourier transform is that it maps the derivative ∂_x into a multiplication operator:

$$\mathbf{F}f' = iz\mathbf{F}f(z).$$

More generally, any differential operator $Lf = \sum_{j=1}^{n} a_j f^{(j)}$ with constant coefficients is mapped into multiplication by a polynomial:

$$\mathbf{F}\sum_{j=1}^{n} a_j f^{(j)} = \sum_{j=1}^{n} a_j (iz)^j \mathbf{F}f(z)$$

for any smooth f in the domain of L such that the corresponding Fourier integrals are well defined. The polynomial $\psi_L(z) = \sum_{j=1}^{n} a_j (iz)^j$ is called the Fourier symbol of the operator L. So, if you prefer computing fast Fourier transforms to computing derivatives, one way of computing Lf is to compute the Fourier transform $\mathbf{F}f$, multiply by the symbol of L and then invert the Fourier transform:

$$Lf = \mathbf{F}^{-1}[\psi_L(z)\mathbf{F}f(z)]. \tag{12.35}$$

Thus, any differential operator L with constant coefficients corresponds to multiplication by a polynomial ψ_L in Fourier space. What happens if we replace polynomials by more general functions ψ_L? Well, the operator L can still be defined by (12.35) as long as Fourier inversion can be done on $\psi_L(z)\mathbf{F}f(z)$ which leaves a wide range of choices for ψ_L: analytic functions are among the most studied ones. When ψ_L is not a polynomial, the operator L defined by (12.35) is not a differential operator anymore: it is called a *pseudo-differential* operator, with symbol ψ_L. L is sometimes denoted $L = \psi_L(\partial)$. Consider now the integro-differential operator encountered in the preceding section:

$$L^X f(x) = \gamma\frac{\partial f}{\partial x} + \frac{\sigma^2}{2}\frac{\partial^2 f}{\partial x^2} + \int \nu(dy)[f(x+y) - f(x) - y1_{\{|y|<1\}}\frac{\partial f}{\partial x}(x)].$$

Taking Fourier transforms yields:

$$\mathbf{F}L^X f(z) = \{i\gamma z - \frac{z^2\sigma^2}{2} + \int \nu(dy)[e^{izy} - 1 - izy1_{\{|y|<1\}}]\}\mathbf{F}f(z),$$

where we recognize the Lévy-Khinchin representation of the characteristic exponent ψ of X:

$$\mathbf{F}(L^X f)(z) = \psi_X(z)\mathbf{F}f(z), \quad \text{i.e.,} \quad L^X = \psi_X(\partial). \tag{12.36}$$

So, like Molière's Monsieur Jourdain who spoke in prose without knowing it, we have been manipulating pseudo-differential operators without knowing it: the infinitesimal generator L^X of a Lévy process X is a pseudo-differential operator whose symbol is given by the characteristic exponent ψ_X of X.

This remark opens the way to the application of many results known for pseudo-differential operators to the study of partial integro-differential equations and Lévy processes. Though we will not employ such techniques in this volume, curious readers are referred to [137] for a systematic study of pseudo-differential operators and related boundary value problems, [209] for a brief account of their relation with Lévy processes and Markov processes and [71] for applications to option pricing in exp-Lévy models.

REMARK 12.4 Fractional order A more intrinsic definition of the order of the operator L^X, which is justified by the expression of its Fourier symbol, can be given by examining the rate of growth of the Lévy measure at zero [137]: when $\sigma = 0$ and $\gamma = 0$, the order of L can be defined as the smallest $\alpha \geq 0$ such that

$$\int_{|x|\leq 1} |x|^\alpha \nu(dx) < \infty. \tag{12.37}$$

Notice that since $\nu(.)$ is a Lévy measure $\alpha \leq 2$. This definition coincides with the stability index α for a (tempered) stable processes, is smaller than one for processes with finite variation and equal to zero for compound Poisson processes. When $\sigma > 0$, the order is equal to 2, because a second derivative is present in the operator, and if $\sigma = 0$ but $\gamma \neq 0$, the order is equal to $\max(1, \alpha)$ because a first derivative is present. □

12.2.3 Classical solutions and Feynman-Kac representations

Now let us get back to real space. Consider an integro-differential operator L as above. A smooth function $f : [0, T] \times \mathbb{R} \mapsto \mathbb{R}$ is said to be a "classical" solution of the associated boundary value problem if u is continuously differentiable in t, $u(t, .) : \mathbb{R} \mapsto \mathbb{R}$ belongs to the domain of definition of L for each $t \in [0, T]$, verifies the boundary conditions and

$$\frac{\partial u}{\partial \tau}(\tau, x) = Lu.$$

The requirement that $u(t, .) : \mathbb{R} \mapsto \mathbb{R}$ belongs to the domain of definition of L ensures that the derivatives in Lu are well-defined continuous functions and $Lu(t, x)$ is defined pointwise. This imposes that $u \in C^{1,2}([0, T] \times \mathbb{R})$; if there is no diffusion component and the process is of finite variation, one can simply require the function to be $C^{1,1}$.

Existence and uniqueness of classical solutions are discussed in [157, 48] for the case $\sigma > 0$ and [297, 77] for the case $\sigma = 0$. A precise formulation of such results requires defining the appropriate function spaces in which existence and uniqueness holds and involves in particular specifying the behavior of solutions at infinity. Since we are only interested in the pointwise values of solutions, namely option prices, we will refrain from going into such details. In the next section we will see a notion of solution which avoids specifying various function spaces in order to discuss existence and uniqueness of solutions.

More interesting for our purpose is the link with option values (defined via risk neutral expectations of terminal payoffs) which requires a probabilistic representation of solutions of PIDEs. In the case of second order parabolic PDEs the well-known Feynman-Kac formula gives a probabilistic representation of the solution in terms of an associated Markovian diffusion process. Similar results are available in the integro-differential case [48, Theorems 3.3 - 8.1.].

Let $\gamma : [0,T] \to \mathbb{R}$ be a bounded (L^∞) function and $\sigma : [0,T] \to \mathbb{R}^+$ be a positive bounded function. J_X denotes a Poisson random measure on $[0,T]\times\mathbb{R}$ with intensity $\mu(dy\ dt) = \nu(dy)\ dt$ with ν a Lévy measure and $\tilde{J}_X$ the compensated version of J_X: $\tilde{J}_X(A) = J_X(A) - \int_A dt\ \nu(dy)$. For a given $t \in [0,T[, x\in\mathbb{R}$ define the jump process $(X_s^{t,x})_{s\in[t,T]}$ by

$$X_s^{t,x} = x + \int_t^s \gamma(u)du + \int_t^s \sigma(u)dW_u \qquad (12.38)$$
$$+ \int_t^s \int_{|y|\geq 1} yJ_X(du\ dy) + \int_t^s \int_{|y|\leq 1} y\tilde{J}_X(du\ dy).$$

$X_s^{t,x}$ is the position at time $s > t$ of a jump process starting in x at time t and having drift $\gamma(.)$, a time-dependent volatility $\sigma(.)$ and a jump component described by a (pure jump) Lévy process with Lévy measure ν. If $\sigma(t) = \sigma$ and $\gamma(t) = \gamma$ then $X_s^{t,x} = x + X_{s-t}$ where X is a Lévy process with triplet (σ^2, ν, γ) and (12.38) is simply the Lévy-Itô decomposition.

PROPOSITION 12.5 Feynman-Kac representation: whole space
Consider a bounded function $h \in L^\infty(\mathbb{R})$. If

$$\exists\ \overline{a},\ \underline{a} > 0, \forall t \in [0,T],\quad \overline{a} \geq \sigma(t) \geq \underline{a} \qquad (12.39)$$

then the Cauchy problem

$$\frac{\partial f}{\partial t}(t,x) + \frac{\sigma^2(t)}{2}\frac{\partial^2 f}{\partial x^2}(t,x) + \gamma(t)\frac{\partial f}{\partial x}(t,x)$$
$$+ \int \nu(dy)[f(t,x+y) - f(t,x) - y1_{|y|\leq 1}\frac{\partial f}{\partial x}(t,x)] = 0,$$
$$\forall x\in\mathbb{R},\quad f(T,x) = h(x), \qquad (12.40)$$

has a unique solution given by

$$f(t,x) = E[h(X_T^{t,x})],$$

where $X^{t,x}$ is the the process given by (12.38).

Of course, this representation does not come as a surprise since we derived the PIDE verified by option prices starting from their probabilistic representation in the first place. But the assertion above means that there is no other solution to the PIDE, which could not be concluded from our derivation.

A similar representation holds in the case of bounded domains [48, Theorem 8.2.]: define $\tau(t,x)$ as the first exit time from the domain $[a,b]$ of the process $(X_s^{t,x})_{t\in[0,t]}$ defined in (12.38):

$$\tau(t,x) = \inf\{s \geq t,\ X_s^{t,x} \notin]a,b[\}. \tag{12.41}$$

PROPOSITION 12.6 Feynman-Kac representation: bounded domain

Consider a bounded function $h \in L^\infty(\mathbb{R})$. If

$$\exists \overline{a}, \underline{a} > 0,\ \overline{a} \geq \sigma(t) \geq \underline{a}$$

then the boundary value problem

$$\begin{aligned}
&\frac{\partial f}{\partial t}(t,x) + \frac{\sigma^2(t)}{2}\frac{\partial^2 f}{\partial x^2}(t,x) + \gamma(t)\frac{\partial f}{\partial x}(t,x)\\
&\quad + \int \nu(dy)[f(t,x+y) - f(t,x) - y1_{|y|\leq 1}\frac{\partial f}{\partial x}(t,x)] = 0 \text{ on } [0,T[\times]a,b[,\\
&\forall x \in [a,b],\quad f(T,x) = h(x),\\
&\forall t \in [0,T],\quad \forall x \notin]a,b[,\quad f(t,x) = 0,
\end{aligned} \tag{12.42}$$

has a unique (classical) solution given by:

$$f(t,x) = E[h(X_T^{t,x})1_{T<\tau(t,x)}],$$

where $X^{t,x}$ is the the process given by (12.38).

REMARK 12.5 The Feynman-Kac formula for jump processes in bounded domains, shown in [48] for zero boundary conditions, is generalized by Rong [337] to time-dependent boundary conditions $f(t,x) = g(t,x)$ for $x \notin O$ where g is only differentiable in the sense of distributions and $\partial g/\partial x(t,.)$, $\partial g/\partial t(t,.)$, $\partial^2 g/\partial x^2(t,.) \in L^p(\mathbb{R})$. □

An important property of solutions, which holds both for Cauchy problems and boundary value problems is the *comparison principle*: if f_1 and f_2 are two solutions then

$$\forall x \in O,\ f_1(T,x) \leq f_2(T,x) \Rightarrow \forall t \in [0,T],\ \forall x \in O,\ f_1(t,x) \leq f_2(t,x).$$

TABLE 12.1: Correspondence between analytical and probabilistic properties

PIDE	**Jump process $X^{t,x}$**	**Financial interpretation**
x	Starting point of process $X^{t,x}$	Log moneyness
Integro-differential operator	Infinitesimal generator	
Terminal condition		Payoff at maturity
Solution of PIDE $f(t,x)$	$E[h(X_T^{t,x})]$	Value of option with payoff h
Fundamental solution of Cauchy problem	Transition density $\rho_t(x)$ of X_t	$e^{-r(T-t)} \times$ Gamma of call option
Zero boundary condition for $x \geq b$	Stopping at first exit from b	Knock-out barrier b
Green function with zero boundary condition for $x \geq b$	Density of stopped process	$e^{-r(T-t)} \times$ Gamma of up-and-out call
Smoothing property	Smoothness of density	Price is smoother than payoff
Comparison principle	$H \geq 0 \Rightarrow E[H] \geq 0$	Static arbitrage relations

This property is synonymous with the positiveness of the density of the process. Another property is the parabolic smoothing effect: under condition (12.39) for any bounded terminal condition $h \in L^\infty$, the solution $f(t,.) : x \to f(t,x)$ at time $t < T$ is smoother in x than the terminal condition [159, 158]. From the probabilistic point of view this corresponds to the smoothness of the density of the (stopped) process. Table 12.1 summarizes the correspondence between the analytic and probabilistic properties discussed above. Note that these results hold under the condition (12.39) that the volatility σ is bounded away from zero, which guarantees that the solutions are smooth. For a pure jump process where $\sigma = 0$ the regularity of the solutions is more difficult to obtain and requires the use of advanced techniques [58, 137, 77, 56]. In addition, in the case of American options, solutions are known *not* to be regular even in the case of diffusion models. We will now see an alternative notion of solution which does away with smoothness requirements.

12.2.4 Viscosity solutions (*)

As noted above, in order for the option price to be a classical solution of the pricing equation, we need to show that it is continuously differentiable up to some degree (once in t, twice in x) which is not always easy in models with jumps. To obtain existence and uniqueness of classical solutions we need to specify a function space in which the solution lies and this requires in turn knowing a priori the regularity of the solution solving the equation. The notion of *viscosity solution* gives an intrinsic definition of a solution which is local in nature and does not impose a priori the existence of derivatives: continuity will be enough.

Consider first a smooth ($C^{1,2}$) solution of

$$\frac{\partial u}{\partial \tau} = L^X u, \tag{12.43}$$

$$u(0,x) = h(x), \tag{12.44}$$

and take a smooth test function $\varphi \in C^2([0,T] \times \mathbb{R})$ such that $u(\tau_0, x_0) = \varphi(\tau_0, x_0)$ and (τ_0, x_0) is a global maximum of the function $u - \varphi$:

$$\forall (\tau, x) \in [0,T] \times \mathbb{R}, u(\tau,x) - \varphi(\tau,x) \leq u(\tau_0,x_0) - \varphi(\tau_0,x_0) = 0.$$

Writing the first and second order conditions for the maximum at (τ_0, x_0) yields:

$$\frac{\partial u}{\partial x}(\tau_0,x_0) = \frac{\partial \varphi}{\partial x}(\tau_0,x_0), \quad \frac{\partial u}{\partial \tau}(\tau_0,x_0) = \frac{\partial \varphi}{\partial \tau}(\tau_0,x_0),$$
$$\frac{\partial^2 u}{\partial x}(\tau_0,x_0) \leq \frac{\partial^2 \varphi}{\partial x^2}(\tau_0,x_0).$$

Such a test function φ therefore captures the first order behavior of u at (τ_0, x_0) exactly and its second order behavior up to an inequality. Assume now that $L^X \varphi$ is well defined (we will give sufficient conditions on φ below). Substituting in L^X gives

$$\frac{\partial \varphi}{\partial \tau}(\tau_0,x_0) - L^X \varphi(\tau_0,x_0) \leq \frac{\partial u}{\partial \tau}(\tau_0,x_0) - L^X u(\tau_0,x_0) = 0. \tag{12.45}$$

Similarly, if $\varphi \in C^2([0,T] \times \mathbb{R})$ such that $u(\tau_0,x_0) - \varphi(\tau_0,x_0) = 0$ is a global *minimum* of $u - \varphi$ on $[0,T] \times \mathbb{R}$ then

$$\frac{\partial \varphi}{\partial \tau}(\tau_0,x_0) - L^X \varphi(\tau_0,x_0) \geq \frac{\partial u}{\partial \tau}(\tau_0,x_0) - L^X u(\tau_0,x_0) = 0. \tag{12.46}$$

Conversely, if we require (12.45)–(12.46) to hold for all test functions $\varphi \in C^2([0,T] \times \mathbb{R})$ we recover (12.43) using the fact that L^X verifies a maximum principle: in fact, this way of characterizing solutions can be extended to any operator L (even nonlinear ones) verifying the maximum principle. But in (12.45) - (12.46) the derivatives only involve the test function φ so the inequalities involving φ make sense even if u is not differentiable, but simply continuous (or even semicontinuous, see below). This simple yet powerful idea leads to the notion of *viscosity solutions*, introduced by M. Crandall and P.L. Lions for first order PDEs and by P.L. Lions for second order PDEs [103]. Using test functions in the manner above allows to make sense of nonsmooth solutions for nonlinear or degenerate PDEs [103]. The notion of viscosity solution was generalized to integro-differential equations by Soner [366] and Sayah [348] for first order operators and by Alvarez and Tourin [5], Barles, Buckdahn and Pardoux [28] and Pham [317] for problems involving a second order operator.

Let us now give a proper definition of a viscosity solution. A locally bounded function $v : [0,T] \times \mathbb{R} \to \mathbb{R}$ is said to be upper-semicontinuous if

$$(t_k, x_k) \to (\tau, x) \text{ implies } v(\tau, x) \geq \limsup_{k\to\infty} u(t_k, x_k) := \lim_{k_0\to\infty} \sup_{k\geq k_0} u(t_k, x_k)$$

and lower-semicontinuous if

$$(t_k, x_k) \to (\tau, x) \text{ implies } v(\tau, x) \leq \liminf_{k\to\infty} u(t_k, x_k) := \lim_{k_0\to\infty} \inf_{k\geq k_0} u(t_k, x_k).$$

If v is both upper- and lower-semicontinuous then it is continuous. Denote by USC (respectively LSC) the class of upper semicontinuous (respectively lower semicontinuous) functions $v :]0,T] \times \mathbb{R} \to \mathbb{R}$ and by $C_p([0,T] \times \mathbb{R})$ the set of measurable functions on $[0,T] \times \mathbb{R}$ with polynomial growth of degree p:

$$\varphi \in C_p([0,T] \times \mathbb{R}) \iff \exists C > 0, |\varphi(t,x)| \leq C(1 + |x|^p). \tag{12.47}$$

Under a polynomial decay condition on the tails of the Lévy density, $L\varphi$ can be defined for $\varphi \in C^2([0,T] \times \mathbb{R}) \cap C_p([0,T] \times \mathbb{R})$:

$$L\varphi(x) = A\varphi(x) + \int_{|y|\leq 1} \nu(dy) \left[\varphi(x+y) - \varphi(x) - y\frac{\partial \varphi}{\partial x}\right] \tag{12.48}$$

$$+ \int_{|y|>1} \nu(dy)[\varphi(x+y) - \varphi(x)], \tag{12.49}$$

where A is a second-order differential operator. The terms in (12.48) are well defined for $\varphi \in C^2([0,T] \times \mathbb{R})$ since for $|y| \leq 1$,

$$|\varphi(\tau, x+y) - \varphi(\tau, x) - y\frac{\partial \varphi}{\partial x}(\tau, x)| \leq y^2 \sup_{B(x,1)} |\varphi''(\tau, .)|,$$

while the term in (12.49) is well defined for $\varphi \in C_p([0,T] \times \mathbb{R})$ if

$$\int_{|y|\geq 1} |x|^p \nu(dx) < +\infty. \tag{12.50}$$

This condition is equivalent to the existence of a moment of order p for the Lévy process X_t. We will assume in the sequel that it holds for $p = 2$.

Consider the following parabolic integro-differential equation on $[0,T] \times \mathbb{R}$ and $O = (a,b) \subset \mathbb{R}$ an open interval:

$$\frac{\partial u}{\partial \tau} = Lu, \quad \text{on } (0,T] \times O, \tag{12.51}$$

$$u(0,x) = h(x), \quad x \in O; \quad u(\tau,x) = g(\tau,x), \quad x \notin O, \tag{12.52}$$

where $g \in C_p([0,T] \times \mathbb{R} - O)$ is a given continuous function with $g(0,x) = h(x)$, defining the boundary conditions outside O. Denote by $\partial O = \{a,b\}$ the boundary of O and by $\overline{O} = [a,b]$ its closure.

DEFINITION 12.1 Viscosity solution *$v \in USC$ is a viscosity subsolution of (12.51)–(12.52) if for any $(\tau,x) \in [0,T] \times \mathbb{R}$ and any (test function) $\varphi \in C^2([0,T] \times \mathbb{R}) \cap C_2([0,T] \times \mathbb{R})$ such that $v - \varphi$ has a global maximum at (τ,x):*

$$\forall (t,y) \in [0,T] \times \mathbb{R}, \quad v(\tau,x) - \varphi(\tau,x) \geq v(t,y) - \varphi(t,y),$$

the following is satisfied:

$$\frac{\partial \varphi}{\partial \tau}(\tau,x) - L\varphi(\tau,x) \leq 0, \quad \text{for} \quad (\tau,x) \in]0,T] \times O, \tag{12.53}$$

$$\min\{\frac{\partial \varphi}{\partial \tau}(\tau,x) - L\varphi(\tau,x),\ v(\tau,x) - h(x)\} \leq 0, \quad \text{for} \quad \tau = 0,\ x \in \overline{O},$$

$$\min\{\frac{\partial \varphi}{\partial \tau}(\tau,x) - L\varphi(\tau,x),\ v(\tau,x) - g(\tau,x)\} \leq 0, \quad \text{for} \quad (\tau,x) \in]0,T] \times \partial O$$

$$\text{and} \quad v(\tau,x) \leq g(\tau,x) \quad \text{for} \quad x \notin O. \tag{12.54}$$

$v \in LSC$ is a viscosity supersolution of (12.51)–(12.52) if for any $(\tau,x) \in [0,T] \times \mathbb{R}$ and any (test function) $\varphi \in C^2([0,T] \times \mathbb{R}) \cap C_2([0,T] \times \mathbb{R})$ such that $v - \varphi$ has a global minimum at (τ,x):

$$\forall (t,y) \in [0,T] \times \mathbb{R}, \quad v(\tau,x) - \varphi(\tau,x) \leq v(t,y) - \varphi(t,y),$$

the following is satisfied:

$$\frac{\partial \varphi}{\partial \tau}(\tau,x) - L\varphi(\tau,x) \geq 0, \quad \text{for} \quad (\tau,x) \in]0,T] \times O, \tag{12.55}$$

$$\max\{\frac{\partial \varphi}{\partial \tau}(\tau,x) - L\varphi(\tau,x),\ v(\tau,x) - h(x)\} \geq 0, \quad \text{for} \quad \tau = 0,\ x \in \overline{O},$$

$$\max\{\frac{\partial \varphi}{\partial \tau}(\tau,x) - L\varphi(\tau,x),\ v(\tau,x) - g(\tau,x)\} \geq 0, \quad \text{for} \quad (\tau,x) \in]0,T] \times \partial O$$

$$\text{and} \quad v(\tau,x) \geq g(\tau,x) \quad \text{for} \quad x \notin O. \tag{12.56}$$

A function $v \in C_2([0,T] \times \mathbb{R})$ is called a viscosity solution of (12.51)–(12.52) if it is both a subsolution and a supersolution. v is then continuous on $]0,T] \times \mathbb{R}$ and verifies (12.47) with $p = 2$.

Since L verifies a maximum principle, one can show that a classical solution $u \in C^{1,2}([0,T] \times \mathbb{R}) \cap C_2([0,T] \times \mathbb{R})$ is also a viscosity solution. However, since the definition above only involves applying derivatives to the test functions φ, a viscosity solution need not be smooth: it is simply required to be continuous on $]0,T] \times \mathbb{R}$. Note that a subsolution/supersolution need not be continuous so the initial condition is verified in a generalized sense.

REMARK 12.6 In the case of second-order PDEs (with no integral term) $u - \varphi$ can be assumed to have a *local* minimum/maximum and the resulting definition is unchanged. This is not necessarily true here due to the nonlocality of the integral operator. □

Several variations on this definition can be found in the articles cited above. First, one can restrict the maximum/minimum of $u - \varphi$ to be equal to zero:

LEMMA 12.1
$v \in USC$ is a viscosity subsolution of (12.51-12.52) if and only if for any $(\tau, x) \in [0,T] \times \mathbb{R}$ and any $\varphi \in C^2([0,T] \times \mathbb{R}) \cap C_2([0,T] \times \mathbb{R})$, if $v(\tau, x) = \varphi(\tau, x)$ and $v(\tau, x) \leq \varphi(\tau, x)$ on $[0,T] \times \mathbb{R}$ then φ verifies (12.53)–(12.54).

PROOF Clearly the definition of a subsolution implies the property above. Inversely, let $\varphi \in C^2([0,T] \times \mathbb{R}) \cap C_2([0,T] \times \mathbb{R})$ be such that $v - \varphi$ has a global maximum at (τ, x). We can modify φ by adding a constant:

$$\psi(t,y) = \varphi(t,y) + u(\tau,x) - \varphi(\tau,x).$$

ψ then satisfies the following properties:

$$\left(\frac{\partial \psi}{\partial \tau} - L\psi\right)(t,y) = \left(\frac{\partial \varphi}{\partial \tau} - L\varphi\right)(t,y), \quad \forall (t,y) \in [0,T] \times \mathbb{R}, \tag{12.57}$$

$$\psi(\tau,x) = u(\tau,x), \tag{12.58}$$

and (τ, x) is a global maximum point of $u - \psi$ so

$$\forall (t,y) \in [0,T] \times \mathbb{R}, \quad u(t,y) \leq \psi(t,y) \leq 0. \tag{12.59}$$

□

A similar result holds for supersolutions. As shown in [28], one can also replace “maximum” and “minimum” by “strict maximum” and “strict minimum” in Definition 12.1. Finally, one can require the test functions to be $C^{1,2}$

or C^∞ instead of C^2. The growth condition at infinity $\varphi \in C_p$ on test functions is essential for $L\varphi$ to make sense. It may be replaced by other growth conditions under stronger hypotheses on the decay of the Lévy density.

Defining

$$F\left(\frac{\partial^2\varphi}{\partial x^2}(\tau,x), \frac{\partial\varphi}{\partial x}(\tau,x), \varphi, v(\tau,x), \tau, x\right) = \frac{\partial\varphi}{\partial\tau}(\tau,x) - L^X\varphi(\tau,x),$$

the Equations (12.45)–(12.46) can be rewritten as:

$$\begin{aligned} F\left(\frac{\partial^2\varphi}{\partial x^2}(\tau,x), \frac{\partial\varphi}{\partial x}(\tau,x), \varphi, u(\tau,x), \tau, x\right) &\geq 0, \\ \text{if} \quad u(\tau,x) - \varphi(\tau,x) = \max_{[0,T]\times\mathbb{R}} u - \varphi; \\ F\left(\frac{\partial^2\varphi}{\partial x^2}(\tau,x), \frac{\partial\varphi}{\partial x}(\tau,x), \varphi, u(\tau,x), \tau, x\right) &\leq 0 \\ \text{if} \quad u(\tau,x) - \varphi(\tau,x) = \min_{[0,T]\times\mathbb{R}} u - \varphi. \end{aligned}$$

If the equation is verified on a bounded domain, e.g., $O = [0,T]\times]a,b[$ with the condition $u(\tau,x) = h(x)$ outside the domain as for barrier options, then F should be replaced by:

$$\begin{aligned} F(\frac{\partial^2\varphi}{\partial x^2}(\tau,x), \frac{\partial\varphi}{\partial x}(\tau,x), \varphi, v(\tau,x), \tau, x) &= \frac{\partial\varphi}{\partial\tau}(\tau,x) - L^X\varphi(\tau,x) \text{ for } (\tau,x)\in O \\ &= v(\tau,x) - h(x) \text{ for } (\tau,x)\notin \overline{O} \end{aligned}$$

with the additional condition that *on* the boundary $x = a, b$

$$\max\{v(\tau,x) - h(x), \frac{\partial\varphi}{\partial\tau}(\tau,x) - L^X\varphi(\tau,x)\} \geq 0 \quad \text{for subsolutions,}$$

$$\min\{v(\tau,x) - h(x), \frac{\partial\varphi}{\partial\tau}(\tau,x) - L^X\varphi(\tau,x)\} \leq 0 \quad \text{for supersolutions.}$$

In this form, the definition can be generalized to nonlinear equations: F may be nonlinear with respect to all variables as long as it is decreasing with respect to the first variable $\partial^2\varphi/\partial x^2$.

REMARK 12.7 Boundary conditions We noted above that, for classical solutions, "boundary" conditions have to be imposed on $\mathbb{R} - O$ and not only on the boundary $\partial O = \{a,b\}$. Since the nonlocal integral term in (12.53)–(12.54) only involves the test function φ and not the solution itself so one can be led to think that behavior of u beyond the boundary is not involved in the definition (see remark in [317, Sec. 5.1.]). Note however that the test functions have verify $\varphi \leq v$ (respectively $\varphi \geq v$) on $[0,T]\times\mathbb{R}$ and not only on $[0,T]\times O$, which requires specifying u outside O. □

While the class of viscosity solutions is sufficiently large to allow for existence of solutions in the cases where smooth solution do not exist, it is sufficiently small to obtain uniqueness, using *comparison principles*. Comparison principles for integro-differential equations were obtained by Soner [366], Sayah [348], Alvarez and Tourin [5]; the case of a general Lévy measure is treated in [317, 220]:

PROPOSITION 12.7 Comparison principle [5, 317, 220]
If $u \in USC$ is a subsolution with initial condition u_0 and $v \in LSC$ is a supersolution of (12.43) with initial condition v_0 then

$$\forall x \in \mathbb{R},\; u_0(x) \leq v_0(x) \Rightarrow u(\tau, x) \leq v(\tau, x) \quad \text{on} \quad]0, T] \times \mathbb{R}. \tag{12.60}$$

In financial terms, comparison principles simply translate into arbitrage inequalities: if the terminal payoff of a European options dominates the terminal payoff of another one, then their values should verify the same inequality. These results can be used to show uniqueness of viscosity solutions. Existence and uniqueness of viscosity solutions for such parabolic integro-differential equations are discussed in [5] in the case where ν is a finite measure and in [28] and [317] for general Lévy measures. Growth conditions other than $u \in C_2$ can be considered (see, e.g., [5, 28]) with additional conditions on the Lévy measure ν. An important advantage with respect to the classical theory [48, 157] is that we do not require that the diffusion part be nondegenerate: σ is allowed to be zero. Another important property of viscosity solutions is their stability under various limits, which is important when studying the convergence of numerical schemes.

In the case of European and barrier options, the following result [98] shows that the option value in an exponential-Lévy model can be characterized as the unique continuous viscosity solution of the backward PIDE (12.51-12.52):

PROPOSITION 12.8 Option prices as viscosity solutions [98]
Assume the Lévy measure $\nu(.)$ verifies the decay condition (12.47) with $p = 2$. *Then:*

- *Let $u(\tau, x)$ be the (forward) value of a European option defined by (12.5). If the payoff function H verifies a Lipschitz condition 12.8 and $h(x) = H(S_0 e^x)$ has quadratic growth at infinity, u is the (unique) viscosity solution of the Cauchy problem (12.6).*

- *The forward value of a knockout (double) barrier call or put option $f(t, y)$ defined by (12.24) is the (unique) viscosity solution of (12.25-12.26).*

Note that we do not require that the diffusion component to be nonzero, nor do we need growth conditions on the intensity of small jumps as in Proposition 12.1. A similar result for American options is shown in [317, Theorem 3.1]:

PROPOSITION 12.9 American option prices as viscosity solutions

Consider an exp-Lévy model verifying the hypotheses (12.2) and (11.5). Then the value of the American put

$$P : (t, S) \mapsto P(t, S) = \sup_{t \leq \tau \leq T} E[e^{-r(\tau - t)}(K - Se^{r(\tau-t)+X_{\tau-t}})^+]$$

is the unique continuous viscosity solution of:

$$\min\{-\frac{\partial P}{\partial t}(t, S) - L^S P(t, S) + rP(t, S), P(t, S) - (K - S)^+\} = 0,$$
$$P(T, S) = (K - S)^+$$

on $[0, T] \times]0, \infty[$.

As we will see later, these results provide an appropriate framework for studying the convergence of finite difference schemes, regardless of whether the smoothness of the option price as a function of the underlying is known.

12.3 Trees and Markov chain methods

The simplest numerical method in the case of the Black-Scholes model is obtained by approximating the continuous time diffusion process by a discrete time Markov chain: the Cox-Ross-Rubinstein binomial tree. Since the binomial tree process converges weakly[3] to the Black-Scholes process, from the definition of weak convergence the value of any European option with a continuous bounded payoff computed by the binomial tree will converge to its value in the limiting Black-Scholes model.

This idea can be generalized to many other stochastic models, including Markov models with jumps, using the methods described by Kushner and Dupuis [243]: given a continuous time stochastic model (for instance, a jump-diffusion model), we construct a discrete time Markov chain $S^{\Delta t, \Delta x}$ — typically, a multinomial tree — with transition probabilities specified such that when $\Delta t \to 0$, the process $S^{\Delta t, \Delta x}$, defined by interpolating the Markov chain, converges weakly to S. Sufficient conditions for weak convergence to jump diffusions in terms of transition probabilities of the Markov chain can be found in

[3]See 2.3.3 for a definition of weak convergence.

[243]. Under these conditions, option prices and hedge ratios computed with the discrete model $S^{\Delta t,\Delta x}$ will provide approximations for option prices/hedge ratios in the continuous-time model defined by S [322, 269].

12.3.1 Multinomial trees

A family of Markov chain approximations is provided by multinomial trees, studied by Amin [6] in the context of jump-diffusion models. Given a continuous time exp-Lévy/jump-diffusion model $S_t = S_0 \exp X_t$ with jumps one can construct the discrete time Markov chain by setting

$$S_{t+\Delta t}^{\Delta t,\Delta x} = s^{n+1} = S_t^{\Delta t,\Delta x} \exp[\epsilon^n] = s^n \exp[\epsilon^n], \qquad t = n\Delta t, \quad (12.61)$$

where ϵ^n is an i.i.d. family of random variables taking a finite number k of values, called the "branching number" of the tree. The method is defined by the choice of the k values (the nodes) and the transition probabilities. The computationally tractable case is that of a recombining trees: the values taken by ϵ^n are multiples of a given step size Δx. In order to represent the asymmetry of jumps one should allow the transitions to be asymmetric:

$$\epsilon^n \in \{-k_1\Delta x, \ldots, -\Delta x, 0, \Delta x, \ldots, +k_2\Delta x\}$$

with $k = k_1 + k_2 + 1$. Then the paths of the Markov chain $\ln s^n$ fall on a lattice with step size $(\Delta t, \Delta x)$ hence the name of "lattice method." We will denote quantities on the lattice by A_j^n where n denotes the time index of a node and j the price index. The value of the asset at node (n, j) is given by S_j^n and can evolve to k states in the next step given by

$$S_{j+i}^{n+1} = S_j^n \exp[i\Delta x], \qquad i = -k_1, \ldots 0, \ldots k_2.$$

The values of k_1, k_2 and the transition probabilities are chosen to obtain absence of arbitrage and such that the transition density of the Markov chain s^n approximates the transition density of S_t. Absence of arbitrage is obtained by imposing that the discounted Markov chain $e^{-nr\Delta t} s_n$ is a martingale. If $q_i = \mathbb{Q}(\epsilon^n = i\Delta x)$ are the transition probabilities this means:

$$\sum_{i=-k_1}^{k_2} q_i \exp(i\Delta x) = \exp(r\Delta t), \quad (12.62)$$

where r is the discount rate. This can also be interpreted as moment matching condition for the first moment of $\exp(X_{t+\Delta t} - X_t)$. Matching other moments of $\exp(X_{t+\Delta t} - X_t)$ can give extra conditions to determine the transition probabilities:

$$\sum_{i=-k_1}^{k_2} q_i \exp(ji\Delta x) = \exp(r\Delta t) E[\exp(j\epsilon^n)]. \quad (12.63)$$

In most models based on Lévy processes or affine jump-diffusion models, the characteristic function is known analytically so the moment conditions are known and the transition probabilities can be determined by solving a linear system. Moment matching is not the only way to specify transition probabilities but is widely used in practice. Since moments are precisely the derivatives of the characteristic function at zero, exponential moment matching is equivalent to matching the Taylor expansion of the characteristic function of the returns $S_{t+\Delta t}/S_t$ at zero. However to match accurately the behavior of large jumps, i.e., the tails of the risk neutral density one needs to use large values of k which increases correspondingly the complexity of the scheme.

The value of a European option can then be computed by backward induction: starting from the payoff (the final node $N\Delta t = T$) we go backwards and at each step/each node in the tree we compute the value of the option as the discounted expectation of the values on the branches

$$C_j^n = e^{-r\Delta t} \sum_{i=-k_1}^{k_2} q_i C_{j+i}^{n+1}. \tag{12.64}$$

In the case of an American put option, this backward step is replaced by an optimization step where the risk neutral expectation is compared to the payoff from exercise:

$$P_j^n = \max\{(K - S_j^n)^+,\ e^{-r\Delta t} \sum_{i=-k_1}^{k_2} q_i P_{j+i}^{n+1}\}. \tag{12.65}$$

This is one of the main advantages of tree methods: pricing American options is as easy as pricing European ones since conditional expectations are simple to compute on a tree.

If the transition probabilities are chosen such that the (interpolated) Markov chain converges weakly to S as $(\Delta x, \Delta t) \to 0$ convergence of European option prices follows for put options (which have bounded continuous payoffs) and, by put-call parity, for call options. Convergence of American option prices to their continuous-time value is by no means an obvious consequence of weak convergence, but can be shown to hold [243]: more generally the value function of an optimal stopping or dynamic portfolio optimization problems in the multinomial tree converges to the value function of the analogous problem in the limiting exp-Lévy model [243, 322].

12.3.2 Multinomial trees as finite difference schemes

Multinomial trees are usually introduced as we have done here, in terms of an approximation of the risk neutral process $(S_t)_{t\in[0,T]}$ by a discrete time Markov chain $(s^n)_{n=1..N}$ with finite state space and their construction does not involve the PIDEs discussed above. Since they give rise to a lattice in (t, S)-space, they can also be interpreted as explicit finite difference schemes for the associated PIDE, constructed on this lattice. A multinomial tree with k branches corresponds to an explicit finite difference scheme on a uniform grid for the log-price, using k neighboring points to approximate the terms in the PIDE. Therefore, their convergence and properties can also be studied in the framework of finite difference schemes and they suffer from the well known drawbacks of explicit schemes, namely restrictive stability conditions on the size of the time step. But two properties distinguish a multinomial tree from a general explicit finite difference scheme.

First, the transition matrix in a multinomial tree (or more generally, any Markov chain approximation method) has *positive* entries corresponding to the risk neutral transition probabilities of the underlying Markov chain. This guarantees that a positive terminal condition/payoff will lead to a positive price and inequalities between payoffs will lead to inequalities between prices. In mathematical terms it means the scheme verifies a discrete version of the comparison principle: it is a *monotone* (or positive) scheme. In financial terms, it means that arbitrage inequalities will be respected by the approximation.

Second, the one-step transition probabilities are chosen such that the discounted value of the approximating Markov chain is a martingale. Because of the way backward induction is done, the martingale property is also inherited by prices of all European options computed in the tree. This guarantees that the prices obtained are arbitrage-free, not only when $\Delta t, \Delta x \to 0$, but also for any finite $\Delta t, \Delta x$. In particular, put-call parity will be respected exactly and not only up to first order in Δt. This property does not hold for arbitrary finite difference schemes, but can be imposed at the price of extra complications: it means that certain functions (here the function $C(t, S) = \exp(-rt)S$) have to be integrated exactly by the scheme. Of course, both of these properties are interesting from a financial point of view.

12.4 Finite difference methods: theory and implementation

Finite difference methods are approximation schemes for partial differential equations based on replacing derivatives by finite differences in the pricing equation. In the case of PIDEs we have an additional integral term which can be discretized by replacing it with Riemann sums.

There are three main steps in the construction of a finite difference scheme:

- Localization: if the PIDE is initially given on an unbounded domain, we reduce it to a bounded one and introduce "artificial" boundary conditions. This induces an approximation error usually called localization error. In the PIDE case, the domain of integration in the integral term also has to be localized, which gives another error term. Both terms have to be estimated and controlled by an appropriate choice of localization.

- Approximation of small jumps: when the Lévy measure diverges at zero, the contribution of this singularity to the integral term can be approximated by a second derivative term. This corresponds to approximating small jumps of a Lévy process by a Brownian motion as in Chapter 6.
- Discretization in space: the spatial domain is replaced by a discrete grid and the operator L is replaced by a matrix M, acting on the vector $U^n = (u(n\Delta t, j\Delta x), j = 0 \dots N)$ representing the solution on the grid points:
$$Lu(n\Delta t, j\Delta x) \simeq (MU^n)_j.$$
- Discretization in time: the time derivative is replaced by a finite difference. There are several choices, leading to different *time stepping* schemes.

The result is a linear system to be solved iteratively at each time step, starting from the payoff function. Finally, after all these approximations, we need to examine under what conditions the solutions of the numerical scheme do converge to the solutions of the PIDE when the discretization parameters go to zero and try to obtain estimates of these errors. We will now detail this step for the PIDE:

$$\frac{\partial u}{\partial \tau} = L^X u, \qquad u(0,x) = h(x), \tag{12.66}$$

to be solved on $[0,T] \times O$ where either $O = \mathbb{R}$ (European option) or $O =]a,b[$ is an open interval (barrier options) with appropriate boundary conditions imposed outside of O.

12.4.1 Localization to a bounded domain

Since numerical computations can only be performed on finite domains, the first step is to reduce the PIDE to a bounded domain. For a double barrier knock-out option, this is already done: the PIDE is solved between the barriers and the boundary conditions are naturally given by the terms of the contract which is worth zero outside the boundary. For a European option (or a single barrier option) we localize the problem by replacing O with $]-A, A[$:

$$\begin{aligned} \frac{\partial u_A}{\partial \tau} &= L^X u_A, \qquad]0,T]\times]-A,A[, \\ u_A(0,x) &= h(x), \qquad x \in]-A,A[. \end{aligned} \tag{12.67}$$

In order for the resulting problem to be a well posed one, should add *artificial* boundary conditions for u_A at all points outside of $]-A,A[$ and not only at $\pm A$. Since the option price $u(\tau, x)$ behaves asymptotically like the payoff function h for large $|x|$, a sensible choice is to set:

$$u_A(\tau, x) = h(x) \quad \forall x \notin]-A, A[. \tag{12.68}$$

This boundary condition has a probabilistic representation in terms of the process $X_t + x$ stopped at first exit from the domain $]-A, A[$. Using this probabilistic interpretation the following error estimate is shown in [98] for bounded payoff functions:

PROPOSITION 12.10 Bound on localization error *[98]*
If $\|h\|_{L^\infty} < \infty$, and $\exists\, \alpha > 0$, $\int_{|x|>1} \nu(dx) e^{\alpha|x|} < +\infty$ then

$$|u(\tau, x) - u_A(\tau, x)| \leq 2C_{\tau,\alpha} \|h\|_\infty \exp[-\alpha(A - |x|)], \tag{12.69}$$

where $C_{\tau,\alpha} = Ee^{\alpha M_\tau}$, $M_\tau = \sup_{t\in[0,\tau]} |X_t|$.

Similarly, the integral term in L^X has to be truncated at some finite upper/lower bounds B_r, B_l: this corresponds to truncating the large jumps of the Lévy process, i.e., replacing it by a new process $\tilde{X}_\tau$ characterized by the Lévy triplet $(\tilde{\gamma}, \sigma, \nu \mathbf{1}_{x\in[B_l,B_r]})$, where $\tilde{\gamma}$ is determined by the martingale condition:

$$\tilde{\gamma} = -\frac{\sigma^2}{2} - \int_{B_l}^{B_r} (e^y - 1 - y 1_{|y|\leq 1}) \nu(dy).$$

Denoting by $\tilde{u}(\tau, x) = E[h(x + \tilde{X}_\tau)]$ the solution of the PIDE with the integral term truncated as above, the following error estimate is given in [98]:

PROPOSITION 12.11 Truncation of large jumps *[98]*
Let $h(x)$ be an almost everywhere differentiable function with $\|h'\|_{L_\infty} < \infty$. Assume $\exists \alpha_r, \alpha_l > 0$, such that

$$\int_1^\infty e^{(1+\alpha_r)y} \nu(dy) < \infty \quad \text{and} \quad \int_{-\infty}^{-1} |y| e^{\alpha_l |y|} \nu(dy) < \infty. \tag{12.70}$$

Then the error due to the truncation of large jumps can be estimated by

$$|u(\tau, x) - \tilde{u}(\tau, x)| \leq \|h'\|_{L_\infty} \tau (C_1 e^{-\alpha_l |B_l|} + C_2 e^{-\alpha_r |B_r|}). \tag{12.71}$$

Therefore, when the Lévy density decays exponentially, both the truncation of the domain and the truncation of large jumps lead to errors that are exponentially small in the truncation parameters. We will assume in the sequel that both approximations have been done and focus on the numerical solution of the localized problem.

12.4.2 Discretization in space

The next step is to replace the domain $[-A, A]$ by a discrete grid: consider a uniform grid on $[0, T] \times [-A, A]$:

$$\tau_n = n\Delta t, \qquad n = 0 \dots M, \qquad \Delta t = T/M,$$
$$x_i = -A + i\Delta x, \qquad i = 0 \dots N, \qquad \Delta x = 2A/N.$$

Let $\{u_i^n\}$ be the solution on discretized grid, extended by zero outside of $[-A, A]$.

Consider first the finite activity case where $\nu(\mathbb{R}) = \lambda < +\infty$. Then the integro-differential operator can be written as

$$Lu = \frac{\sigma^2}{2}\frac{\partial^2 u}{\partial x^2} - \left(\frac{\sigma^2}{2} + \alpha\right)\frac{\partial u}{\partial x} + \int_{B_l}^{B_r} \nu(dy)u(\tau, x + y) - \lambda u, \quad (12.72)$$

where $\alpha = \int_{B_l}^{B_r}(e^y - 1)\nu(\mathrm{d}y)$. To approximate the integral terms one can use the trapezoidal quadrature rule with the same grid resolution Δx. Let K_l, K_r be such that $[B_l, B_r] \subset [(K_l - 1/2)\Delta x, (K_r + 1/2)\Delta x]$. Then:

$$\int_{B_l}^{B_r} \nu(dy)u(\tau, x_i + y) \simeq \sum_{j=K_l}^{K_r} \nu_j u_{i+j}, \quad \lambda \simeq \hat{\lambda} = \sum_{j=K_l}^{K_r} \nu_j,$$
$$\alpha \simeq \hat{\alpha} = \sum_{j=K_l}^{K_r} (e^{y_j} - 1)\nu_j, \qquad \text{where } \nu_j = \int_{(j-1/2)\Delta x}^{(j+1/2)\Delta x} \nu(\mathrm{d}y). \quad (12.73)$$

As noted above since we need the solution on $[-A + B_l, A + B_r]$ to compute the integral term, the actual computational grid extends from $i = K_l$ to $i = N + K_r$ but $u_i^n = h(x_i)$ for $i \notin [0, N]$. Alternatively, Andersen and Andreasen [7] propose to compute the jump integral using a Fourier transform: this methods is well suited when boundary conditions are periodic (for example, when the function is extended by zero) but if this is not the case (for example when the payoff function is used as a boundary condition) it may lead to oscillating error terms.

The space derivatives can be discretized using finite differences:

$$\left(\frac{\partial^2 u}{\partial x^2}\right)_i \simeq \frac{u_{i+1} - 2u_i + u_{i-1}}{(\Delta x)^2}, \quad (12.74)$$

$$\left(\frac{\partial u}{\partial x}\right)_i \simeq \begin{cases} \frac{u_{i+1} - u_i}{\Delta x}, & \text{if} \quad \sigma^2/2 + \hat{\alpha} < 0 \\ \frac{u_i - u_{i-1}}{\Delta x}, & \text{if} \quad \sigma^2/2 + \hat{\alpha} \geq 0. \end{cases} \quad (12.75)$$

The choice of approximation for the first-order derivative is determined by stability requirements (see below).

12.4.3 An explicit-implicit finite difference method

Denoting by D and J the matrices representing respectively the discretizations of the differential and the integral parts of L, the explicit scheme is given by:

$$\frac{u^{n+1}-u^n}{\Delta t} = Du^n + Ju^n \quad \Rightarrow \quad u^{n+1} = [I+\Delta t(D+J)]u^n. \quad (12.76)$$

In this scheme, each step consists in multiplying by the matrix $I+\Delta t(D+J)$. However, the stability of the scheme imposes stringent conditions on the time step: a sufficient condition for stability is

$$\Delta t \leq \inf\{\frac{1}{\hat{\lambda}}, \frac{\Delta x^2}{\sigma^2}\}.$$

While the first term is not very constraining given usual values of λ, the second term (also present in the diffusion case) forces to use a very small time step, therefore increasing computation time.

Another choice is the implicit scheme

$$\frac{u^{n+1}-u^n}{\Delta t} = Du^{n+1} + Ju^{n+1} \quad \Rightarrow \quad [I-\Delta t(D+J)]u^{n+1} = u^n. \quad (12.77)$$

This scheme does not suffer from the stringent stability condition on the step size Δt but now we have to solve a linear system at each iteration.

In the case of diffusion models, $J=0$ and the matrix $I-\Delta tD$ is tridiagonal: the resulting linear system is thus very easy to solve. In the present case D is still tridiagonal but J is a dense matrix: in general all terms are nonzero and the solution of the system by a linear solver requires $O(N^2)$ operations. More generally one can use any combination of the above schemes, known as θ-scheme:

$$\frac{u^{n+1}-u^n}{\Delta t} = \theta(Du^n + Ju^n) + (1-\theta)[Du^{n+1} + Ju^{n+1}]. \quad (12.78)$$

For $\theta = 0$ we recover the explicit scheme but the complexity for $\theta \neq 0$ is the same as for the implicit scheme.

This discussion on computational complexity at each step vs. the number of steps shows that if $J=0$ then an implicit scheme is a good choice while if $D=0$, an explicit scheme should be chosen. When both terms are present one can use *operator splitting*: the integral term is computed using the solution computed at the preceding iteration, while the differential term is treated implicitly, leading to the following *explicit-implicit* time stepping scheme [98]:

$$\frac{u^{n+1}-u^n}{\Delta t} = Du^{n+1} + Ju^n, \quad \tau_n = n\Delta t, \quad n = 0\dots M.$$

This leads to the tridiagonal system

$$(I-\Delta tD)u^{n+1} = (I+\Delta tJ)u^n = Au^n, \quad (12.79)$$

which has the same complexity as in the Black-Scholes case.

ALGORITHM 12.1 Explicit-implicit scheme for PIDE
Initialization:

$$u_i^0 = h(x_i), \quad \forall i.$$

For $\mathbf{n = 0, \dots, M-1}$: solve

$$(I - \Delta t D)u^{n+1} = (I + \Delta t J)u^n = Au^n. \tag{12.80}$$

Impose boundary conditions: replace

$$u_i^{n+1} = h(x_i), \quad for\, i \notin \{0, \dots, N\},$$

where

$$(Du^{n+1})_i = \frac{\sigma^2}{2}\frac{u_{i+1}^{n+1} - 2u_i^{n+1} + u_{i-1}^{n+1}}{(\Delta x)^2} - \left(\frac{\sigma^2}{2} + \hat{\alpha}\right)\frac{u_{i+1}^{n+1} - u_i^{n+1}}{\Delta x},$$

$$(Ju^n)_i = \sum_{j=-K}^{K} \nu_j u_{i+j}^n - \hat{\lambda} u_i^n,$$

$$\hat{\alpha} = \sum_{j=K_l}^{K_r} (e^{y_j} - 1)\nu_j, \qquad \nu_j = \int_{(j-1/2)\Delta x}^{(j+1/2)\Delta x} \nu(dy).$$

Two key properties of a finite difference scheme are its *consistence* with the continuous equation and its *stability*. The numerical scheme is said to be (locally) consistent with the PIDE (12.67) if the discretized operator converges to its continuous version when applied to any test function $v \in C^\infty([0,T]\times\mathbb{R})$: we require that

$$\left| \frac{v_i^{n+1} - v_i^n}{\Delta t} - (Dv)_i^{n+1} - (Jv)_i^n - \left(\frac{\partial v}{\partial \tau} - Lv\right)(\tau_n, x_i)\right| = r_i^n(\Delta t, \Delta x) \to 0$$

at all points in the computational domain when $(\Delta t, \Delta x) \to 0$. Consistency means the discrete equation approximates the continuous equation (which does not ensure that the *solution* approximates the continuous one).

The scheme is said to be *stable* if for a bounded initial condition, the solution u_i^n is uniformly bounded at all points of the grid, independently from Δt, Δx:

$$\exists C > 0,\ \forall\ \Delta t > 0,\ \Delta x > 0,\ i \in \{0..N\},\ n \in \{0, \dots, M\}: \quad |u_i^n| \le C.$$

Stability ensures that the numerical solution at a given point (i.e., the option value for a given date/underlying) does not blow up when $(\Delta t, \Delta x) \to 0$.

These properties hold for the explicit-implicit scheme presented above:

PROPOSITION 12.12 Consistence, stability and monotonicity[98]

If the time step is smaller than $1/\hat{\lambda}$*:*

$$\Delta t \leq 1/\hat{\lambda}, \tag{12.81}$$

the explicit-implicit scheme given by Algorithm 12.1 is stable and consistent with the PIDE (12.67) as $(\Delta t, \Delta x) \to 0$*. Moreover it is a* monotone scheme*:*

$$u^0 \geq v^0 \Rightarrow \quad \forall n \geq 1, \quad u^n \geq v^n.$$

REMARK 12.8 Other notions of stability Various definitions for the stability of numerical schemes can be found in the literature: for example, the "von Neumann stability" studied for a similar scheme in [7] represents u^n as a discrete Fourier transform and requires the corresponding Fourier coefficients to remain bounded under iteration of the scheme. This is essentially equivalent to stability in L^2 norm with periodic boundary conditions and simply controls a global error in the least square sense but does not control the error on the value of a given option. Moreover it does not allow to capture the effect of boundary conditions when they are nonperiodic, which is the case here. □

When the Lévy measure is singular near zero, the above scheme cannot be applied directly. In Chapter 6 we observed that any infinite activity Lévy process can be approximated by a compound Poisson process by truncating the Lévy measure near zero, i.e., jumps smaller than ε and that the approximation can be further improved by replacing the small jump by a Brownian motion (see Section 6.4): X_t is approximated by X_t^ε where $\tilde{X}_t^\varepsilon$ has Lévy triplet $(\gamma(\varepsilon), \sigma^2 + \sigma^2(\varepsilon), \nu 1_{|x|\geq\varepsilon})$ with

$$\sigma^2(\varepsilon) = \int_{-\varepsilon}^{\varepsilon} y^2 \nu(\mathrm{d}y).$$

One could choose $\gamma(\varepsilon) = \gamma$ but this violates the martingale condition for $\exp X_t^\varepsilon$ so put-call parity is violated. If we require in addition that the approximating finite activity model is risk neutral we obtain:

$$\gamma(\varepsilon) = -\frac{\sigma^2 + \sigma^2(\varepsilon)}{2} - \int_{|x|\geq\varepsilon} (e^y - 1 - y 1_{|y|\leq 1}) \nu(dy).$$

Since X_t^ε is a jump-diffusion process with compound Poisson jumps, the scheme (12.1) can be applied to compute the solution $u^\varepsilon(\tau, x)$ of

$$\frac{\partial u^\varepsilon}{\partial \tau} = L_\varepsilon u^\varepsilon, \quad u^\varepsilon(0,x) = h(x), \quad \text{where}$$
$$L^\varepsilon f = \frac{\sigma^2 + \sigma^2(\varepsilon)}{2}\frac{\partial^2 f}{\partial x^2} - \left(\frac{\sigma^2+\sigma^2(\varepsilon)}{2} + \alpha(\varepsilon)\right)\frac{\partial f}{\partial x}$$
$$+ \int_{|y|\geq\varepsilon} \nu(dy) f(x+y) - \lambda(\varepsilon) f(x) \quad \text{with}$$
$$\sigma^2(\varepsilon) = \int_{-\varepsilon}^{\varepsilon} y^2 \nu(\mathrm{d}y), \quad \alpha(\varepsilon) = \int_{|y|\geq\varepsilon} (e^y - 1)\nu(\mathrm{d}y), \quad \lambda(\varepsilon) = \int_{|y|\geq\varepsilon} \nu(\mathrm{d}y).$$

Using the method in (6.2) one can show the following bound for the truncation error:

$$|u(\tau, x) - u^\varepsilon(\tau, x)| \leq K\|h'\|_\infty \frac{\int_{-\varepsilon}^{\varepsilon} |x|^3 \nu(\mathrm{d}x)}{\sigma^2(\varepsilon)}.$$

As seen above, this scheme is stable and consistent if $\lambda(\varepsilon)\Delta t < 1$, i.e., the time step is smaller than the average interval between jumps. However for barrier options the convergence may be oscillatory as in the case of diffusion models [395] and it may be better to use nonuniform meshes at the boundary. The finite difference scheme discussed above can be adapted to the case of American options by treating the early exercise feature using a penalty method, see [114].

12.4.4 Convergence

In the usual approach to the convergence of finite difference schemes for PDEs, consistency and stability ensure convergence under regularity assumptions on the solution. This approach is not feasible here because, as discussed in Section 12.2.4, solutions may be nonsmooth and higher order derivatives may not exist. For example, in the variance gamma model the value of a call option is C^1 in the price variable but not C^2.

This is where viscosity solutions come to the rescue: in the case of second-order parabolic PDEs verifying a comparison principle as in Proposition (12.7), Barles and Souganidis [29] show that, for elliptic/parabolic PDEs, any finite difference scheme which is locally consistent, stable and monotone converges uniformly on each compact subset of $[0,T] \times \mathbb{R}$ to the unique continuous viscosity solution, even when solutions are not smooth.

This result can be extended to the PIDEs considered here [98]:

PROPOSITION 12.13 Convergence of explicit-implicit scheme
The solution of the explicit-implicit finite difference scheme described above converges uniformly on each compact subset of $]0,T] \times \mathbb{R}$ *to the unique viscosity solution of (12.51)–(12.52).*

PROOF We give here the main steps of the method, see [98] for details. The scheme 12.1 verifies the properties of local consistency, stability and monotonicity. Now define

$$\underline{u}(\tau, x) = \liminf_{(\Delta t, \Delta x) \to 0\ (n\Delta t, j\Delta x) \to (\tau, x)} u_j^n,$$
$$\overline{u}(\tau, x) = \limsup_{(\Delta t, \Delta x) \to 0\ (n\Delta t, j\Delta x) \to (\tau, x)} u_j^n.$$

The stability of the scheme implies that the limits defining $\underline{u}$ and $\overline{u}$ are finite bounded functions and by definition $\underline{u} \leq \overline{u}$. The monotonicity and consistency of the scheme then implies that $\overline{u} \in USC$ is a subsolution[4] and $\underline{u} \in LSC$ a supersolution of (12.66).
Using the comparison principle for semicontinuous solutions (Proposition 12.7), we can therefore conclude that $\overline{u} \leq \underline{u}$. So $\overline{u} = \underline{u} = u$ is a continuous viscosity solution. Using the continuity of the limit u one can then show, along the lines of [29], that the convergence to u is uniform on all compact subsets of $]0, T] \times \mathbb{R}$ using a variation on the proof of Dini's theorem. □

Readers used to convergence results for finite difference schemes should note that the requirements on the scheme are quite weak: we have only used *local consistency* whereas the usual conditions for convergence to classical solutions (Lax theorem) require *global* consistency, i.e., the convergence in (12.81) must not be pointwise but with respect to some global norm, which requires knowing in advance in what function space the solutions live. This shows the flexibility of the notion of viscosity solution. The price to pay is the loss of information about the order of convergence. Of course in cases where smoothness of the solution can be shown by other means (for example if $\sigma > 0$) one can discuss the order of convergence using classical methods and higher order schemes can be used.

The numerical examples shown in Figures 12.1–12.2, taken from [98], illustrate the performance of the scheme when compared to Carr and Madan's FFT method (described in Section 11.1) in the case of a put option $h(x) = (1-e^x)^+$. The errors are computed in terms of Black-Scholes implied volatility:

$$\varepsilon(\tau, x) = |\Sigma^{\text{PIDE}}(\tau, x) - \Sigma^{\text{FFT}}(\tau, x)| \qquad \text{in} \quad \%.$$

The computations were done in variance gamma models with Lévy density

$$\nu(x) = a\frac{\exp(-\eta_\pm |x|)}{|x|}$$

and two sets of parameters $a = 6.25, \eta_- = 14.4$, $\eta_+ = 60.2$ (VG1) and $a = 0.5$, $\eta_- = 2.7$, $\eta_+ = 5.9$ (VG2) and a Merton model with Gaussian jumps in

[4] Note that $\overline{u}, \underline{u}$ are defined as pointwise limits and cannot be assumed to be continuous in general: this shows why we cannot restrict a priori subsolutions or supersolutions in Definition 12.1 to be continuous. However, semicontinuity is preserved in the limit.

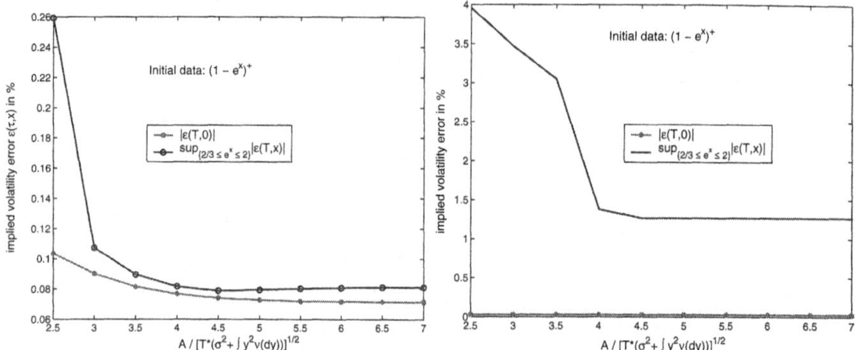

FIGURE 12.1: Influence of domain size on localization error for the explicit-implicit finite difference scheme. Left: Put option in Merton jump-diffusion model. Right: Put option in variance gamma model.

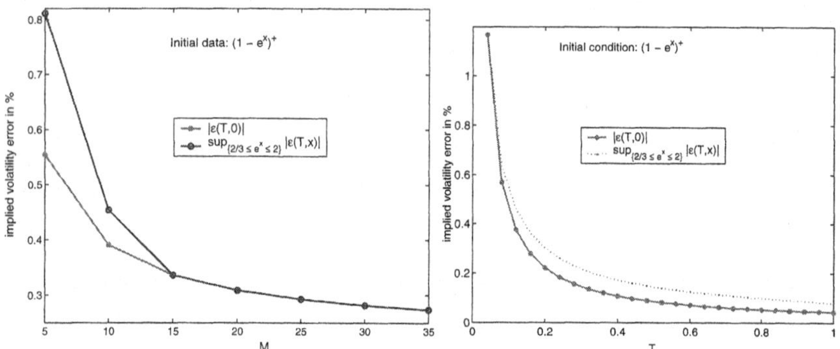

FIGURE 12.2: Numerical precision for a put option in the Merton model. Left: Influence of number of time steps M. $\Delta x = 0.05$, $\Delta t = T/M$. Right: Decrease of error with maturity.

log-price with Lévy density

$$\nu(x) = 0.1 \frac{e^{-x^2/2}}{\sqrt{2\pi}} \quad \text{and volatility } \sigma = 15\%$$

for a put of maturity of $T = 1$ year. When the FFT method is available it is more efficient but the finite difference scheme described here can be used in a much more general context: in presence of barriers and in cases where the characteristic function is not known in closed form.

The localization error is shown in Figure 12.1: domain size A is represented in terms of its ratio to the standard deviation of X_T. An acceptable level is obtained as soon as this ratio is of order $\simeq 5$.

We observe that a nonsmooth initial condition leads to a lack of precision for small T. This phenomenon, which is not specific to models with jumps,

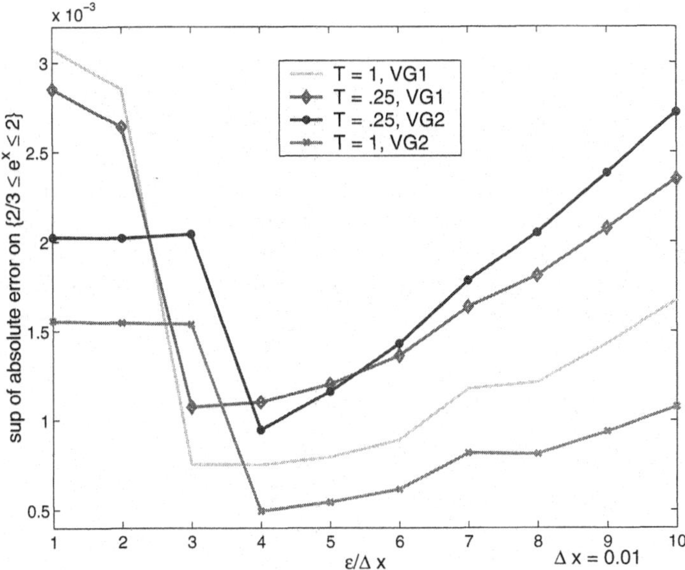

FIGURE 12.3: Influence of truncation of small jumps on numerical error in various variance gamma models. Put option.

can be overcome using an irregular time-stepping scheme which exploits the smoothness in time of the solution. Here the optimal time stepping scheme is obtained by using logarithmically spaced time steps [353].

In the case of infinite activity models an additional parameter which influences the solution is the truncation parameter ε for the small jumps. Choosing this parameter as small as possible is not necessarily a good idea: Figure 12.3 shows that, for a given $\Delta x > 0$ the minimal error is obtained for a finite ε which in this case is *larger* than Δx. The optimal choice of ε is not universal depends on the growth of the Lévy density near zero, see [98].

12.5 Analytic method of lines

For American options, the finite difference methods outlined above cannot be used directly since the exercise boundary is not known and has to be computed iteratively at the same time as the price.

A method which has been used by several authors [79, 293, 294, 71] both for the Black-Scholes model and for models with jumps is the *method of lines*: the idea here is to discretize time derivatives only, obtaining a differential-difference equation in x:

$$\frac{f^{n+1}(x) - f^n(x)}{\Delta t} + Lf^n(x) - rf^n(x) = 0, \quad x > b^n, \tag{12.82}$$

$$f^n(x) = K - S_0 e^y, \quad x \leq b^n. \tag{12.83}$$

This equation must be solved iteratively, going backwards from the maturity: each iteration step involves solving the integro-differential equation above in the x variable. In some special cases, this can be done analytically. In [71, Chapter 5] an expression of the solution is given based on the Wiener-Hopf factors ϕ_q^+, ϕ_q^- introduced in Chapter 11: denoting the Fourier transform of g by $\mathbf{F}g$:

$$f^n(x) = K - S_0 e^x + (1 + \frac{r}{\Delta t})^{-1} g^n(x), \tag{12.84}$$

$$\text{where} \quad \mathbf{F}g^n = \phi_q^- \ \mathbf{F}(1_{[b^n,\infty[} w^n) \tag{12.85}$$

$$\text{and} \quad \mathbf{F}w^n = \phi_q^+ \ \mathbf{F}[f^{n+1}(x) + S_0 e^x - (1 + r\Delta t)K]. \tag{12.86}$$

Applying this method requires, of course, knowing the Wiener-Hopf factors and evaluating several Fourier transforms at each iteration, one for computing w^n from f^{n+1}, one for computing g^n from w^n and finally a Fourier inversion for computing f^n from g^n. Examples where this method can be used are given in [250].

This method can also be viewed as a special case of the Galerkin method, outlined in the next section.

12.6 Galerkin methods

While finite difference methods represent the solution by its values on a discrete grid, Galerkin methods are based on the representation of the solution u using a basis of functions:

$$u(\tau, x) = \sum_{i \geq 1} a_i(\tau) e_i(x),$$

which is then approximated by restricting to a finite number of basis functions:

$$u_N(\tau, x) = \sum_{i=1}^{N} a_i(\tau) e_i(x). \tag{12.87}$$

Advantages of this representation include the ability to directly estimate the derivatives (Greeks) of the option value — if the basis functions e_i have derivatives known in closed form — and to compute values of the solution at points not necessarily belonging to a uniform grid.

12.6.1 Variational formulation and the Galerkin method

Consider the PIDE in the time-to-maturity/log moneyness variables (τ, x):

$$\frac{\partial u}{\partial \tau} = Lu, \tag{12.88}$$

where $L = L^X$ is the integro-differential operator

$$Lu = \frac{\sigma^2}{2}[\frac{\partial^2 u}{\partial x^2}(x) - \frac{\partial u}{\partial x}(x)] + \int \nu(dy)[u(x+y) - u(x) - (e^y - 1)\frac{\partial u}{\partial x}(\tau, x)].$$

The Galerkin method is based on a *variational* formulation of the PIDE [48]. First we choose a suitable Hilbert space H of functions of x which contains the solutions of our equation and a complete basis of H, that is, a sequence $(e_i)_{i\geq 1}$ of linearly independent elements of H such that

$$\forall i \geq 1, \langle e_i, f\rangle_H = 0 \quad \Rightarrow \quad f = 0. \tag{12.89}$$

Here $\langle .,. \rangle_H$ denotes the scalar product in H. A possible choice for H is a weighted L^2 space

$$\langle f, g\rangle_{L^2(w)} = \int f(x)g(x)w(x)dx,$$

where w is a weight function which fixes the growth at infinity of the solutions or associated Sobolev spaces [48, 219]. In view of the remarks in Section 12.2.1, an exponential weight $w(x) = \exp(a|x|)$ is a natural choice and allows for exponentially growing solutions (call options in particular); this choice has been used in the variational formulation of the Black-Scholes equation in [219] and for PIDEs in the jump-diffusion case in [390]. If the equation has boundary conditions, these can be incorporated into the choice of H and of the basis functions (e_i). If the solution u is known to be smooth enough such that $\frac{\partial u}{\partial \tau} \in H$ and $Lu \in H$ then the equation (12.88) is equivalent to

$$\forall i \geq 1, \quad \langle \frac{\partial u}{\partial \tau}, e_i\rangle_{L^2(w)} = \langle Lu, e_i\rangle_{L^2(w)}. \tag{12.90}$$

But in general $Lu \notin L^2(w)$ so this formulation is restrictive. A more flexible approach is to interpret the scalar products as duality products: instead of an L^2 space one can choose $e_i, u \in H$ where $H = H^1(w)$, a weighted Sobolev space of order 1, such that $Lu \in H^*$ and $\langle Lu, e_i\rangle$ is well defined as a duality product [48]:

$$\forall i \geq 1, \qquad \langle \frac{\partial u}{\partial \tau}, e_i\rangle = \langle Lu, e_i\rangle. \tag{12.91}$$

Note that if u, v are both in $L^2(w)$ then $\langle u, v\rangle = \langle u, v\rangle_{L^2(w)}$ so this formulation includes the preceding one but it more flexible.

Equation (12.91) is called a variational formulation of the PIDE and allows for solutions which are less regular than classical solutions: a solution $u \in H = H^1(w)$ is not necessarily $C^{1,2}$ and the derivatives of u which appear in $Lu \in H^*$ are defined in the sense of distributions. Note however that, unlike the case of classical solutions, it is not straightforward to obtain a probabilistic representation for such generalized solutions since the usual Itô formula does not apply for $u \in H^1(w)$. Approximation arguments must be used in this case to obtain such probabilistic representations [48, 337].

The formulation (12.91) can be used to obtain a numerical solution in the following way. As in the case of the finite difference method, we begin by localizing the problem to a bounded domain:

$$\frac{\partial u_A}{\partial \tau} = L^X u_A, \quad (0,T]\times]-A,A[, \tag{12.92}$$
$$u_A(\tau,x) = h(x), \quad x \notin]-A,A[, \qquad u_A(0,x) = h(x), \quad x \in]-A,A[.$$

To repeat the procedure above we have to choose a space H of functions on $]-A,A[$ and a basis (e_i) of H. A first guess would be the Sobolev space $H^1_0(\,]-A,A[\,)$ with boundary conditions imposed at $\pm A$ but to compute the nonlocal integral term in $\langle Lu, e_i\rangle$, we need to know u outside of $]-A,A[$ and extending by zero is a crude approximation. A solution proposed in [287] is to replace u by $U = u_A - h$, which tends exponentially to zero at $\pm\infty$ and can therefore be extended by zero beyond the domain with an exponentially small error. This gives rise to a source term $F = L^X h$ in the equation:

$$\frac{\partial U}{\partial \tau} - L^X U = L^X h = F, \quad]0,T]\times]-A,A[, \tag{12.93}$$
$$U(\tau,x) = 0, \quad x \notin]-A,A[, \qquad U(0,x) = 0, \quad x \in]-A,A[.$$

This equation can now be written in the variational form above using a basis of $H = \{f \in H^1(\mathbb{R}), f(x) = 0 \text{ for } |x| \geq A\}$: the "discretization" then consists in projecting onto a *finite* number of basis elements $e_i, i = 1\dots N$:

$$\forall i = 1\dots N, \quad \langle \frac{\partial U}{\partial \tau}, e_i\rangle = \langle LU, e_i\rangle \tag{12.94}$$

and looking for a solution U_N in the form (12.87) verifying:

$$\forall i = 1\dots N, \quad \langle \frac{\partial U_N}{\partial \tau}, e_i\rangle = \langle LU_N,, e_i\rangle + \langle F, e_i\rangle$$
$$\iff \quad \sum_{j=1}^{n} K_{ij}\frac{d}{d\tau}a_i(\tau) = \sum_{j=1}^{N} L_{ij}a_i(\tau) + F_i,$$
$$\text{where} \quad A_{ij} = \langle Le_j, e_i\rangle, \qquad K_{ij} = \langle e_j, e_i\rangle = \int_{-A}^{A} e_i(x)e_j(x)dx. \tag{12.95}$$

This is now an initial value problem for a N-dimensional *ordinary differential equation* which can be written in matrix form as:

$$K\frac{d}{d\tau}v_N + Av_N = F_N, \quad v_N = (a_i(\tau), i = 1\dots N). \tag{12.96}$$

This equation can now be iteratively solved by time discretization, using one of the time stepping schemes described in Section 12.4.3: this may be an explicit, implicit or θ-scheme. Representing by v_N^n the vector of components of $U_N(n\Delta t)$ on the basis $(e_i, i = 1 \dots N)$, we obtain:

$$K\frac{v_N^{n+1} - v_N^n}{\Delta t} = \theta A v_N^{n+1} + (1-\theta) A v_N^n + F_N.$$

Unlike the case of a finite difference method, the resulting solution is not only defined on a discrete grid but is a continuous function $U_N^n(\tau, x) = \sum_{i=1}^N v_{N,i}^n e_i(x)$ defined at all points of the domain. U_N^n converges[5] in H to the solution U as N, the dimension of the approximation, is increased [287].

12.6.2 Choice of a basis

Different choices of basis are possible in the Galerkin method and lead to numerical methods of varying efficiency. When L is a differential operator, a common choice is a finite element basis, using "hat" functions associated to a (possibly irregular) mesh x_j:

$$\begin{aligned} e_j(x) &= \frac{x - x_{j-1}}{x_j - x_{j-1}} \quad \text{for } x \in [x_{j-1}, x_j], \\ &= \frac{x_{j+1} - x}{x_{j+1} - x_j} \quad \text{for } x \in [x_j, x_{j+1}], \\ &= 0, \quad x \notin [x_{j-1}, x_{j+1}]. \end{aligned}$$

This grid can be defined in log-price but also in the price directly since we do not require it to be regularly spaced. This choice yields piecewise linear approximations for u and a piecewise constant approximation defined almost everywhere for its derivative. In fact, using the above basis on a regular grid yields a finite difference scheme. The main difference with the finite difference method is the capability of using an irregular grid which is dense at points where a higher accuracy is required: for example, when solving the forward equation one might be interested in having more strikes closer to the money. Generalizations using higher order (piecewise quadratic or cubic) functions can be constructed along the same lines. These choices are interesting when dealing with differential operators since their locality leads to a tridiagonal matrix $A = [A_{ij}]$. In this case the integral operator leads to a dense matrix so finite elements are not particular appealing.

Another choice is to use complex exponentials

$$e_j(x) = \exp\left(\frac{2j\pi x}{A}\right),$$

[5]Note that unlike the convergence of finite difference schemes to viscosity solutions which is pointwise, the convergence should be interpreted here in the sense of a global L^2 or Sobolev norm.

where A is the domain size. The representation of the function in this basis is nothing else but the discrete Fourier transform! In particular, if we choose a regular grid with size 2^n and solve the equation in space variable by performing FFT at each time step, this corresponds to the method of lines outlined in Section 12.5. This representation is well suited when the solution is known to be smooth with periodic boundary conditions. In cases where the solution is not smooth (for instance, at the barrier for barrier options) a well-known feature of Fourier transform — called the Gibbs phenomenon — is that the Fourier series exhibits oscillatory behavior at such points. When the resolution is increased the convergence is not polynomial and the implementation of nonperiodic boundary conditions can be problematic.

The above approaches do not yield "sparse" representations of the operator L: they give rise to a dense matrix with a lot of nonzero elements which means a lot of computation time will go into the solution of the linear system at each time step if non-explicit schemes are used. The use of a *wavelet* basis can help to remove these difficulties: a wavelet basis is generated from a function $\psi(x)$ called the mother wavelet by dilating and translating it by powers of 2:

$$e_{j,k}(x) = 2^{-nj/2}\psi(2^{-j}x - k). \tag{12.97}$$

ψ can be chosen such that $e_{j,k}$ have compact support, are orthogonal to polynomials of a given degree and form an orthogonal basis of $L^2(\mathbb{R})$; other choices are also possible/interesting [106]. Compact support and exact representation of polynomials of a fixed degree allow accurate representation of regions with strong gradients such as price regions close to the barrier for barrier options or the neighborhood of the exercise boundary for American options. Fast algorithms, with performance comparable to FFT, are available for computing wavelet coefficients [106].

It is possible [104, 52] to construct wavelet bases adapted to a given (integro-differential) operator in the sense that functions at different refinement levels are orthogonal with respect to the bilinear form induced by the operator. With this choice of basis, the matrix is "as sparse as possible." The basic idea is to use an appropriate choice of wavelet to obtain a lot of near zero terms in the matrix and then approximate the dense matrix of the discrete integral operator by dropping small terms so as to end up with a sparse matrix. A Galerkin method for PIDEs using a wavelet basis has been suggested by Matache et al. [287]. To compute the matrix A in (12.95) the Lévy density has to be known in analytic form, which is the case in most parametric exp-Lévy models. In the case of time-inhomogeneous models described in Chapter 14 the computation has to be done at each time step which makes the method less easy to use. Also, when the Lévy measure is empirically calibrated to market data as in Chapter 13, an additional quadrature is necessary to compute the matrix A.

Finally, note that the variational formulation extends to the case of American options [219, 48] and Galerkin methods can also be applied in this case [286].

12.7 A comparison of numerical methods for PIDEs

The different numerical methods described above yield different representations of the option price: in tree methods the solution is represented by its values on the nodes of a tree, in finite difference methods the solution is represented by its values on a (typically regular) grid and in Galerkin methods the function C is represented by its coefficients on a basis of functions. Therefore Galerkin methods come with a natural interpolation scheme makes things easier if one is interested in computing the solution at points not belonging to the initial grid.

As already pointed out above, multinomial trees are special cases of explicit finite difference schemes. Also, finite difference schemes can be seen as special cases of finite element methods which are themselves special cases of the Galerkin approach. Thus, in terms of generality we have:

Trees $\subset$ Finite difference methods $\subset$ Finite elements $\subset$ Galerkin methods

The Galerkin method is therefore the most general and special choices of basis lead to the other methods including the finite difference/FFT method.

However, more general does not mean simpler to implement: the complexity of implementation increases as we go from trees to general Galerkin methods. While the tree or finite difference methods can be implemented by any quant with reasonable programming knowledge, finite elements and Galerkin methods require the use of specialized toolboxes to be implemented in a reasonable time. Also, finite difference programs use less memory than Galerkin methods since there is no overhead for managing grids, at the expense of not easily being able to refine the grid. This explains why finite element and Galerkin methods have had a limited scope of application in computational finance until now, at least for European or barrier options. On the other hand, Galerkin methods become interesting when there are delicate boundary conditions/irregular boundaries involved as in the case of American options or convertible bonds. Aside from these two cases, finite difference methods such as the one proposed in Section 12.4 seem to be the best choice.

A good property of a numerical scheme is monotonicity: in financial terms it means that if h, g are two payoff functions and $h \geq g$ then the prices $u^n(h), u^n(g)$ computed by the scheme also verify $u^n(h) \geq u^n(g)$ at each point. Monotonicity guarantees that the prices computed using the scheme respect arbitrage inequalities. In computational terms, monotonicity means that the discretized equation verifies a comparison principle just as in the continuous equation. Multinomial trees, as well as the finite difference schemes presented in Section 12.4 are monotone schemes while Galerkin/finite element methods are not monotone in general: prices computed from nonmonotone schemes are only arbitrage-free up to a numerical error.

TABLE 12.2: Numerical methods for partial integro-differential equations

Method	Advantages	Drawbacks	Ref.
Multinomial tree/lattice	Monotonicity, ease of implementation	Inaccurate representation of jumps, slow convergence	[6]
Analytic method of lines	Fast when feasible	Needs closed form Wiener-Hopf factor.	[79, 71]
Explicit/Implicit finite difference scheme	Fast, simple to implement, monotone, handles barriers efficiently.		[98, 114]
Crank-Nicholson/FFT	Fast	Boundary conditions not handled efficiently.	[7]
Finite elements	Extends to American options	Dense matrix: slow computation.	
Wavelet-Galerkin	Extends to American options	Implementation is difficult.	[287]

Galerkin methods are particularly useful when a robust approximation is sought to solve partial differential equations on an inhomogeneous mesh. This is the case for example when solving *forward* PIDEs [82] in order to price call/put options for a whole range of strikes and maturities at once. Also, they can be extended to the case of American options.

A major advantage of PIDE methods for valuation is their computational efficiency for single asset options and the ability to handle barriers, early exercise and obtain the sensitivities (Greeks) as a by-product of the computation. But, as in the case of diffusion models, PIDE methods become inefficient as soon as we have more than two underlying assets/factors. The only feasible method for pricing American/barrier options in high dimensional problems such as the pricing of basket options is the Monte Carlo method, discussed in Chapters 6 and 11.

Table 12.2 further compares various numerical methods for PIDE, discussed in this chapter, and sums up their relative advantages and drawbacks.

Further reading

The literature on parabolic integro-differential equations, especially regarding numerical methods and viscosity solutions, is rapidly growing.

The link between PIDEs and risk neutral valuation is the tip of an iceberg called *potential theory* which explores the deep connection between partial differential equations and Markov processes. Potential theory was first developed in the context of diffusion processes (Markov processes without jumps) and later extended to Markov processes with jumps [110]. A balanced overview of the relation between integro-differential operators and Markov processes with jumps is given by Jacob [208], see also [370].

Smoothness and integrability of densities of Lévy processes are studied using Fourier techniques in [345]. These results have been generalized to Markov processes with jumps using the tools of Malliavin calculus by Bismut [58]; see also [309] and [248].

Elliptic and parabolic PIDEs of the type studied in this chapter were systematically studied by Bensoussan and Lions [48] and Garroni and Menaldi [157] under the hypothesis of a diffusion component which is nonzero and bounded from below ("ellipticity"). Properties of the fundamental solutions (Green functions) for such equations are studied in [158]. Feynman-Kac representations are discussed under this hypothesis in [48, Chapter 3], see also [337]. Integro-differential equations with boundary conditions are studied in [77] using pseudo-differential operator methods and in [337] using a probabilistic approach.

An introduction to viscosity solutions of PDEs is given in [103] and [143]. The relevance of viscosity solutions for numerical methods in finance is discussed in [27] in the case of diffusion models. Our discussion of convergence of numerical schemes to viscosity solutions follows the method introduced by Barles and Souganidis [29] for nonlinear PDEs. Viscosity solutions for PIDEs where studied by Soner [366], Alvarez and Tourin [5], Sayah [348], Barles et al. [28] and Pham [317]. The case of general Lévy measures is only treated in Barles, Buckdahn and Pardoux [28] and Pham [317]: while the hypotheses in [28] exclude the case of exponential-Lévy models treated here, Pham [317] gives a general framework applicable to European and American options in all the models treated in this book.

The relation between optimal stopping and the free boundary problem for PIDEs is discussed in [48, 157] in the nondegenerate case and [348, 317] in the general case where the absence of a diffusion component is allowed. American options on underlying assets with jumps and the relation between the optimal stopping problem and the integro-differential free boundary problem have been studied in the case of a jump-diffusion with compound Poisson jumps by Zhang [392, 390, 389] using a variational inequality approach and by Pham [316] using a viscosity solution approach. Variational methods for American options was introduced in [219] for the Black-Scholes equation and extended by Zhang [392, 390] to jump-diffusion models with finite activity and finite exponential moments.

In the context of option pricing models with jumps, PIDEs for option prices were considered in [7] for models with compound Poisson jumps and in [22, 329] for exp-Lévy models; see also [98, 114, 376, 383].

Multinomial tree methods for option pricing in jump-diffusion models were proposed by Amin [6]. Weak convergence of approximation schemes is studied by Kushner and Dupuis [243]; financial examples in relation to incomplete market models and models with jumps can be found in the book Prigent [322]. Andersen and Andreasen [7] use an operator splitting approach combined with a fast Fourier transform (FFT) evaluation of the jump integral to price European options in a jump-diffusion model with finite jump intensity, where the diffusion terms involve nonconstant local volatilities. Finite difference methods for models with finite and infinite activity are studied by Cont and Voltchkova [98]. The method of lines was used in [79, 294] for American option valuation; Meyer [293] uses the method of lines to value American options where the number of jump sizes is finite. Galerkin methods for parabolic PDEs are reviewed in [378]. Wavelets have been used in many contexts for the discretization of differential and integral operators see, e.g., [169, 104]. Wavelet Galerkin methods for pricing options in exp-Lévy models have been proposed by Matache et al. [287, 286].

Chapter 13

Inverse problems and model calibration

> If you want to know the value of a security, use the price of another security that is as similar to it as possible. All the rest is modelling.
>
> Emmanuel Derman, *The boys guide to pricing and hedging*

As discussed in Chapter 10, exponential-Lévy models, jump-diffusion models and most other models encountered in this book correspond to incomplete markets: perfect hedges do not exist, option prices cannot be uniquely identified from the underlying price process by arbitrage arguments alone. In fact, as shown by Eberlein and Jacod [124] and Bellamy and Jeanblanc [46] (see Proposition 10.2), in most models with jumps different choices of (equivalent) martingale measure allow to generate any given price for a given European option: the range of possible prices obtained by picking various equivalent martingale measures is the maximal interval allowed by static arbitrage arguments.[1] These arguments illustrate the main difficulty of using the historical approach for identification of exponential-Lévy models: *due to the fact that the market is incomplete, knowledge of the historical price process alone does not allow to compute option prices* in a unique way.

On the other hand, we observed in Chapter 9 that in any arbitrage-free market the prices of securities can be represented as discounted conditional expectations with respect to a *risk-neutral* measure $\mathbb{Q}$ under which discounted asset prices are martingales. In incomplete market models, $\mathbb{Q}$ bears only a weak relation to the time series behavior described by $\mathbb{P}$: it cannot be identified from $\mathbb{P}$ but only inherits some qualitative properties such as the presence of jumps, infinite or finite activity, infinite or finite variation,... from the historical price process (see Table 9.1). Furthermore, for an underlying on which options are traded in an organized market, option prices quoted in the

[1] However this does not mean that any given prices for several options can be simultaneously reproduced by such a model.

market provide an extra source of information for selecting an appropriate pricing model. Therefore a natural approach, known as "implied" or risk-neutral modelling, is to model directly the risk-neutral dynamics of the asset by choosing a pricing measure $\mathbb{Q}$ respecting the qualitative properties of the asset price.

However an ad hoc choice of equivalent martingale measure — for instance, starting from a Lévy process estimated from time series of log-prices and performing an Esscher transform as in Proposition 9.9 — does not give values consistent with the market prices of traded options, when they are available. As argued in Section 10.6, consistency with market prices of options is an important constraint on a pricing model.

In fact, as argued in Section 9.1, when option prices are quoted on the market, a market-consistent pricing model $\mathbb{Q}$ can not be obtained only by an econometric analysis of the time series of the underlying but by looking at prices of contingent claims today ($t = 0$). In a market where options are traded on the market, their prices are available and can be used as source of information for selecting $\mathbb{Q}$. Choosing a risk-neutral model such as to reproduce the prices of traded options is known as *model calibration*: given market prices $(C_i)_{i\in I}$ at $t = 0$ for a set of benchmark options (typically call options with different strikes K_i and maturities T_i), one looks for a risk-neutral model $\mathbb{Q}$ which correctly prices these options:

$$\forall i \in I,\ C_i = e^{-rT} E^{\mathbb{Q}}[(S_{T_i} - K_i)^+].$$

The logic is the following: one calibrates a risk-neutral model to a set of observed market prices of options and then uses this model to price exotic, illiquid or OTC options and to compute hedge ratios.

While the pricing problem is concerned with computing values of option given model parameters, here we are interested in backing out parameters describing risk-neutral dynamics from observed option prices: model calibration is thus the *inverse problem* associated to the pricing problem. But this inverse problem is ill-posed: there may be many pricing models which generate the same prices for the benchmark options thus the solution of the inverse problem is not necessarily unique. Another problem is, of course, the computation of a solution for the inverse problem, for which efficient and stable algorithms are needed.

If we add to calibration constraints the requirement that our risk-neutral model belongs to a certain prespecified class of models, such as exponential-Lévy of jump-diffusion models, then there is no guarantee that a solution exists: one may not be able to find a model in the specified class which *exactly* reproduces market prices of options. Of course market prices are given up to bid-ask spreads so exact calibration is not meaningful, so a more feasible interpretation of the calibration problem is to achieve the *best approximation* of market prices of options with a given model class. If the approximation is interpreted in a least squares sense, this leads to a nonlinear least squares problem, discussed in Section 13.2.

In this chapter we will study these problems in the context of exponential-Lévy models. Consider a risk-neutral exponential-Lévy model defined by its characteristic triplet $(\sigma(\theta), \nu(\theta), \gamma(\theta))$ where θ is some set of parameters in the parametric case or simply the characteristic triplet itself in the nonparametric case. To calibrate this model to option prices one must solve the following problem:

Calibration Problem 1 *Given an exponential-Lévy model* $(\sigma(\theta), \nu(\theta), \gamma(\theta))$ *and observed prices* C_i *of call options for maturities* T_i *and strikes* K_i, $i \in I$, *find* θ *such that the discounted asset price* $S_t \exp(-rt)$ *is a martingale and the observed option prices are given by their discounted risk-neutral expectations:*

$$\forall i \in I, \; C_i = e^{-rT} E^{\theta}[(S_{T_i} - K_i)^+].$$

where E^{θ} *denotes the expectation computed in the exponential-Lévy model with triplet* $(\sigma(\theta), \nu(\theta), \gamma(\theta))$.

The solution of this problem then gives an *implied Lévy measure* describing the jumps of the risk-neutral process. Despite its simple form, Problem 1 presents formidable difficulties related to the fact that it is an ill-posed inverse problem: there may exist no solution at all or an infinite number of solutions. Even if we use an additional criterion to choose one solution from many, the dependence of the solution on input prices may be discontinuous, which results in numerical instability of calibration algorithm.

We will argue that in order to identify a risk-neutral model correctly, one should take into account the information both from the historical time series and from prices of traded options. To achieve this, one can restrict the choice of pricing rules to the class of martingale measures equivalent to a prior model, either resulting from historical estimation or otherwise specified according to the views held by a risk manager: the calibration procedure then updates the prior according to the information in market prices of options. This approach is discussed in Section 13.1.

As we will argue, even when restricting to exponential-Lévy models the resulting minimization problem is ill-posed because there may be many Lévy triplets reproducing option prices equally well and the solution may depend on the input data in a discontinuous way. In Section 13.3, using the notion of relative entropy, we present a regularization method that removes both of these problems. The role of relative entropy is two-fold: it ensures that the calibrated Lévy measure is equivalent to the prespecified prior and it regularizes the ill-posed inverse problem.

13.1 Integrating prior views and option prices

We argued in Chapter 9 that if a market is arbitrage-free, it must be possible to express option prices as discounted expectation of payoffs under some martingale probability *equivalent to the historical one.* Therefore, if the historical price process has been estimated statistically from the time series of the underlying, the calibration must be restricted to martingale measures *equivalent* to the estimated historical probability. Borrowing the terminology from Bayesian statistics we use the word *prior* to denote this historical probability. The calibration problem now takes the following form:

Calibration Problem 2 (Calibrating with a prior) *Given a prior model $(\sigma_0, \nu_0, \gamma_0)$ and the observed prices C_i of call options for maturities T_i and strikes K_i, $i \in I$, find an exponential-Lévy model $(\sigma(\theta), \nu(\theta), \gamma(\theta))$ such that the discounted asset price $S_t \exp(-rt)$ is a martingale, the probability measure $\mathbb{Q}_\theta$ generated by $(\sigma(\theta), \nu(\theta), \gamma(\theta))$ is equivalent to the prior $\mathbb{P}_0$ and the observed option prices are given by discounted risk-neutral expectations under $\mathbb{Q}_\theta$:*

$$\forall i \in I,\ C_i = e^{-rT} E^\theta[(S_{T_i} - K_i)^+ | S_0 = S].$$

Using the absolute continuity conditions of Proposition 9.8, all Lévy processes can be split into equivalence classes. If the prior was chosen from one such class, the calibrated measure must also lay in this class. If one wants to calibrate a model from a given parametric family, equivalence imposes some constraints on the parameters.

If the prior is of jump-diffusion type with finite jump intensity a nonzero diffusion component, the class of equivalent models includes all jump-diffusion models with compound Poisson jumps that have the same diffusion component and a Lévy measure that is equivalent to that of prior. For example, if the prior Lévy measure has a density that is everywhere positive, the calibrated measure must also have an everywhere positive density. When the diffusion component is present, we can obtain an equivalent martingale measure simply by changing the drift, therefore no additional constraints are imposed on the Lévy measure and it can be freely changed during calibration.

When the prior is an infinite-activity Lévy process, stricter conditions are imposed on the behavior of calibrated Lévy measure near zero by the integrability condition (9.20). Let ν be the prior Lévy measure and ν' be the calibrated one. Define $\phi(x) = \log \frac{d\nu'}{d\nu}(x)$ as in Chapter 9.

If ν has a density with stable-like singularity at zero, that is, $\nu(x) = \frac{f(x)}{|x|^{1+\alpha}}$ with f positive and finite, the condition (9.20) is satisfied if $|\phi(x)| \leq C|x|^{\alpha/2+\varepsilon}$ in a neighborhood of zero for some $\varepsilon > 0$, which means that ν' will also have a density with stable-like behavior near zero.

If ϕ is finite and satisfies $|\phi(x)| \leq C|x|$ in a neighborhood of zero, then the integral (9.20) is finite for every Lévy measure ν. Below we give some examples of equivalence classes for some parametric models.

Example 13.1 Tempered stable process
Let us suppose that the prior is a tempered stable process with center γ_c (we use the representation (4.29), that is, we do not truncate the large jumps) and Lévy measure ν of the form

$$\nu(x) = \frac{c_-}{|x|^{1+\alpha}} e^{-\lambda_-|x|} 1_{x<0} + \frac{c_+}{x^{1+\alpha}} e^{-\lambda_+ x} 1_{x>0}.$$

We would like to calibrate a Lévy process from the same family to option prices. In this case, the absolute continuity condition requires that the parameters of the calibrated process satisfy $\alpha' = \alpha$, $c'_+ = c_+$, $c'_- = c_-$ (see Example 9.1) and the condition (9.21):

$$\gamma'_c - \gamma_c = \Gamma(1-\alpha)\left\{c_+(\lambda'_+)^{\alpha-1} - c_+(\lambda_+)^{\alpha-1} + c_-(\lambda'_-)^{\alpha-1} - c_-(\lambda_-)^{\alpha-1}\right\}.$$

Finally, since the calibrated measure must be risk-neutral, it must satisfy

$$\gamma'_c = -\int_{-\infty}^{\infty} (e^x - 1 - x)\nu'(dx),$$

which entails

$$-\gamma'_c = \Gamma(-\alpha)\lambda'_+ c_+ \left\{\left(1 - \frac{1}{\lambda'_+}\right)^{\alpha} - 1 + \frac{\alpha}{\lambda'_+}\right\} + \Gamma(-\alpha)\lambda'_- c_- \left\{\left(1 + \frac{1}{\lambda'_-}\right)^{\alpha} - 1 - \frac{\alpha}{\lambda'_-}\right\}.$$

This gives us five equations for six unknown parameters in the pricing model. Therefore, only one of them actually needs to be calibrated, i.e., one can fix arbitrarily only one decay rate or only the mean of the process. □

Example 13.2 Kou jump-diffusion model
When the prior corresponds to a double exponential model with characteristic triplet (σ, ν, γ) where ν has a density

$$\nu(x) = c_+ e^{-\lambda_+ x} 1_{x>0} + c_- e^{-\lambda_-|x|} 1_{x<0}, \tag{13.1}$$

the only condition imposed by absolute continuity is that $\sigma' = \sigma$. By adjusting γ' we can fulfil the risk-neutrality condition, which leaves us four free parameters to calibrate: as in every jump-diffusion model, all parameters of the Lévy measure are free. □

These examples show, that although Kou's double exponential model and the tempered stable model have the same number of parameters, the former one has a much greater potential of reproducing option prices because it has four free parameters when the prior is fixed whereas in the tempered stable process all parameters but one are fixed by the equivalence to the prior and the risk-neutrality.

Choice of prior measure Ideally, parameters of the prior should be estimated from historical data. If such data is not available, but the calibration to options on the same asset is carried out regularly, say, every day, then a good choice for the prior of the day n is the calibrated measure of the day $n-1$. Indeed, a model which is good enough to calibrate a cross section of option prices may not be rich enough to represent correctly the dynamics of the option prices across time: as a result, calibrating the model to current option prices may result in time varying parameters. In order to avoid arbitrage these different models should generate the same set of scenarios, i.e., generate equivalent measures. Choosing at each re-calibration the preceding pricing model as a prior automatically enforces this property: all risk neutral probabilities thus obtained by successive updating will remain absolutely continuous with respect to the first one. This choice of prior thus enforces the consistency of calibration over time. What about the initial choice of the prior? If no historical data is available, its only role is to regularize the calibration at the first iteration and to stabilize the results of our model choice. Its exact form is thus not very important; it will be corrected at later stages. One can therefore take any reasonable parametric model — for instance, the Merton jump-diffusion model — with parameters that may or may not correspond to typical values of option prices.

13.2 Nonlinear least squares

In order to obtain a practical solution to the calibration problem, many authors have resorted to minimizing the in-sample quadratic pricing error (see for example [7, 41]):

$$\mathcal{S}(\theta) = \sum_{i=1}^{N} \omega_i |C^\theta(T_i, K_i) - C_i|^2, \tag{13.2}$$

the optimization being usually done by a gradient-based method. More precisely, in the least squares method one must solve the following calibration problem:

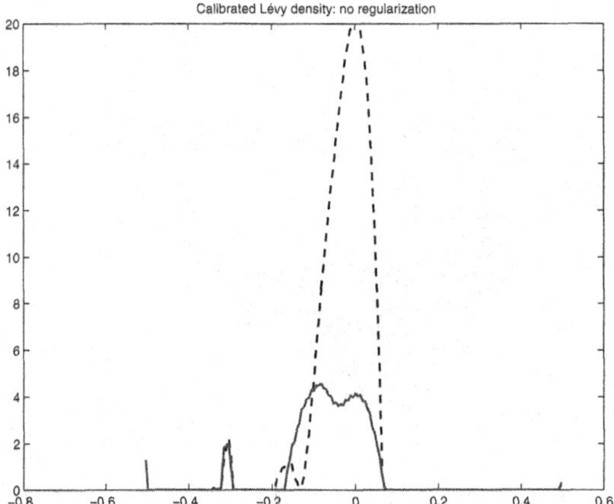

FIGURE 13.1: Lévy measure calibrated to DAX option prices via a nonlinear least squares method. The starting measure for both graphs is a Merton's model (Gaussian jumps); the jump intensity is initialized to 1 for the solid curve and to 5 for the dashed one.

Calibration Problem 3 (Least-squares calibration) *Given an exp-Lévy model $(\sigma(\theta), \nu(\theta), \gamma(\theta))$ and the observed prices C_i of call options for maturities T_i and strikes K_i, $i \in I$, find*

$$\theta^* = \arg \min_{\mathbb{Q}_\theta \in \mathcal{Q}} \sum_{i=1}^{N} \omega_i |C^\theta(T_i, K_i) - C_i|^2,$$

where C^θ denotes the call option price computed for the exponential-Lévy model with triplet $(\sigma(\theta), \nu(\theta), \gamma(\theta))$ and $\mathcal{Q}$ is the set of martingale measures.

Prior can be introduced into this problem by restricting the minimization to the set of martingale measures, equivalent to the prior. The choice of weights ω_i is addressed later in this section.

While, unlike Problem 1, here one can always find a solution, at least if the minimization is restricted to a compact set, the objective functional is nonconvex so a gradient descent method may not succeed in locating the minimum. Given that the number of calibration constraints (option prices) is finite (and not very large), there may be many Lévy triplets reproducing call prices with equal precision, which means that pricing error can have many local minima or, more typically, the error landscape will have flat regions in which the error has a low sensitivity to variations in model parameters (see below). As a result, the calibrated Lévy measure is very sensitive not only to the input prices but also to the numerical starting point in the minimization algorithm [97]. Figure 13.1 shows an example of this instability: the two graphs represent the result of a nonlinear least squares minimization where

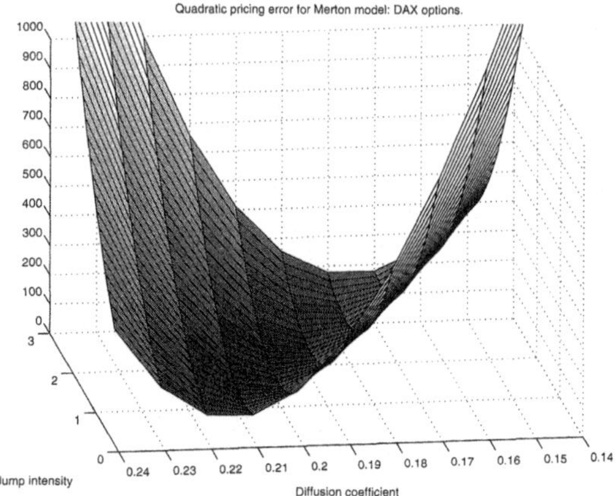

FIGURE 13.2: Quadratic pricing error as a function of model parameters, Merton model, DAX options.

the variable θ is the vector of discretized values of ν on a grid. In both cases the same option prices are used, the only difference being the starting points of the optimization routines. In the first case (dashed line), we used Merton's model with intensity $\lambda = 5$ and in the second case (solid line), it was Merton's model with intensity $\lambda = 1$. As can be seen in Figure 13.1, the results of the minimization are totally different!

One may think that in a parametric model with few parameters one will not encounter this problem of multiple minima as long as there are (many) more options than parameters. However, even if the minimum is unique, the problem remains ill-posed, because the solution will depend on the input data in a nonsmooth way, as illustrated by the following empirical example. Figure 13.2 represents the quadratic pricing error (13.2) for the Merton's model on a data set of DAX index options, as a function of the diffusion coefficient σ and the jump intensity λ, other parameters remaining fixed. It can be observed that if one increases the jump intensity while decreasing the diffusion volatility in a suitable manner, the calibration error stays almost at the same level, leading to a flat direction in the error landscape. Therefore, a small change in the input data or in the initial value of the minimization algorithm can move the solution a long way along the valley's bottom. Even in this simple parametric case one therefore needs some regularization or preconditioning to solve the calibration problem in a stable way. In fact the number of parameters is much less important from a numerical point of view than the *convexity* and well-conditionedness of the objective function.

Choice of weights in the objective functional The relative weights ω_i of option prices in the functional to be minimized should reflect our confidence in individual data points, which is determined by the liquidity of a given option.

This can be assessed from the bid-ask spreads:

$$\omega_i = \frac{1}{|C_i^{\text{bid}} - C_i^{\text{ask}}|^2}.$$

However, the bid and ask prices are not always available from option price data bases. On the other hand, it is known that at least for the options that are not too far from the money, the bid-ask spread is of order of tens of basis points of implied volatility ($\approx 1\%$). This means that in order to have errors proportional to bid-ask spreads, one must minimize the differences of implied volatilities and not those of the option prices. However, this method involves many computational difficulties (numerical inversion of the Black-Scholes formula at each minimization step). A feasible solution to this problem is to minimize the squared differences of option prices weighted by squared Black-Scholes "vegas"[2] evaluated at the implied volatilities of the market option prices, in view of the following approximation.

$$\sum_{i=1}^{N}(I(C^\theta(T_i,K_i)) - I_i)^2 \approx \sum_{i=1}^{N}\left(\frac{\partial I}{\partial C}(I_i)(C^\theta(T_i,K_i) - C_i)\right)^2$$
$$= \sum_{i=1}^{N}\frac{(C^\theta(T_i,K_i) - C_i)^2}{\text{Vega}^2(I_i)},$$

where $I(.)$ denotes the Black-Scholes implied volatility as a function of option price and I_i denotes the market implied volatilities.

13.3 Regularization using relative entropy

The discussion of preceding section has made it clear that reformulating the calibration problem as a nonlinear least squares problem does not resolve the uniqueness and stability issues, even if the calibration is restricted to the class of measures, equivalent to the prior: the inverse problem remains ill-posed. To obtain a unique solution in a stable manner we must introduce a regularization method [133]. One way to achieve this is to add to the least-squares criterion (13.2) a *penalization* term and minimize the functional

$$\mathcal{J}(\theta) = \sum_{i=1}^{N}\omega_i|C^\theta(T_i,K_i) - C_i|^2 + \alpha F(\mathbb{Q}_\theta,\mathbb{P}_0) \tag{13.3}$$

[2]Vega is the common name of the first derivative of Black-Scholes price with respect to volatility.

over the set of parameters θ that correspond to martingale measures equivalent to the prior. The first role of the penalization term F is to ensure uniqueness of solution. Indeed, if F is convex in θ then for α large enough the entire functional $\mathcal{J}(\theta)$ will be convex and the solution will be unique. Empirical studies [97] show that for DAX option prices and nonparametric calibration of Lévy measure uniqueness of solution is achieved already for α small enough for calibration quality to be acceptable. The second role of F is to ensure stability of solution. As a measure of closeness of the model $\mathbb{Q}_\theta$ to the prior $\mathbb{P}_0$, it penalizes the models that are too far from the prior. In other words, here we do not trust the data completely: to a model which reproduces option prices exactly but is very far from the prior we prefer a model which has slightly worse calibration quality but is more similar to the prior. This allows to obtain a more stable solution of the calibration problem while remaining within the same error bounds fixed by the bid-ask fork.

Many choices are possible for the penalization term (see [133]) but in this setting it is particularly convenient to use the *relative entropy* or *Kullback Leibler distance* $\mathcal{E}(\mathbb{Q}_\theta|\mathbb{P}_0)$ of the the pricing measure $\mathbb{Q}_\theta$ with respect to the prior model $\mathbb{P}_0$ (see Section 9.6).

The relative entropy has several interesting properties which make it a popular choice as a regularization criterion [133]. As explained in Section 9.6 it is a convex functional of the Lévy measure, with a unique minimum corresponding to the prior Lévy measure. The same is true in the parametric case, if the Lévy measure is a convex function of the parameters. In other parametric models, the relative entropy may no longer be a globally convex function of the parameter vector, but if the parameterization is sufficiently well-behaved, that is, if the Lévy measure depends on the parameters in a smooth way and different parameter sets correspond to different Lévy measures, then the relative entropy will still have its global minimum at the prior and will be convex in some neighborhood of this minimum. This means that one can still use it as a penalization term. Figure 13.3 plots the relative entropy for the double exponential model as a function of parameters c_+ and λ_+ the other ones being fixed. Note the nice convex profile around the global minimum (dark region).

An important property of the relative entropy which makes it particularly convenient for calibration is that it preserves absolute continuity of the calibrated measure $\mathbb{Q}_\theta$ with respect to the prior: if the Lévy measure approaches zero at some point where the prior Lévy measure is nonzero, the gradient of the relative entropy term becomes arbitrarily large and pushes it away from zero. Using relative entropy as penalty function therefore guarantees that the solution will be an equivalent martingale measure.

From the point of view of information theory, minimizing relative entropy with respect to some prior measure corresponds to adding the least possible amount of information to the prior in order to correctly reproduce observed option prices. Finally, the relative entropy of $\mathbb{Q}_\theta$ with respect to $\mathbb{P}_0$ is a simple function of the parameter vector θ: in the nonparametric case it is given by (9.28) and (9.29) by a one-dimensional integration and for most parametric

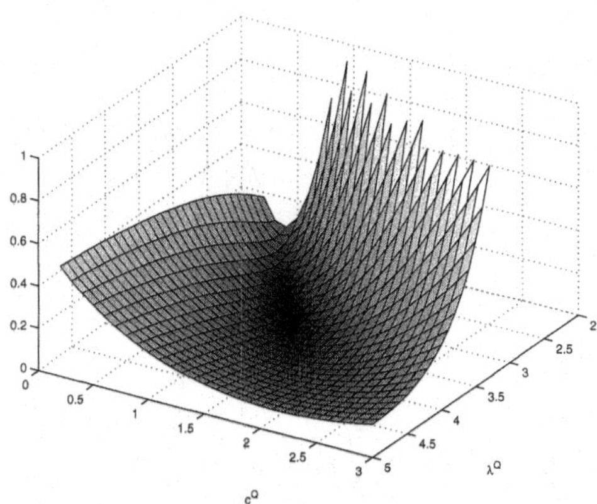

FIGURE 13.3: Relative entropy in Kou's double exponential model as a function of parameters c^Q and λ^Q of the positive part of Lévy measure.

models it is explicitly computable.

Example 13.3 Relative entropy for double exponential model
If both $\mathbb{P}$ and $\mathbb{Q}$ are of double exponential type with Lévy measures of the form (13.1), the relative entropy can be easily computed from Equation (9.28):

$$T^{-1}\mathcal{E}(\mathbb{Q}|\mathbb{P}) = \frac{1}{2\sigma^2}(b^Q - b^P)^2 + \frac{c_+^Q}{\lambda_+^Q}\ln\frac{c_+^Q}{c_+^P} - c_+^Q\frac{\lambda_+^Q - \lambda_+^P}{(\lambda_+^Q)^2} + \frac{c_+^P}{\lambda_+^P} - \frac{c_+^Q}{\lambda_+^Q}$$
$$+ \frac{c_-^Q}{\lambda_-^Q}\ln\frac{c_-^Q}{c_-^P} - c_-^Q\frac{\lambda_-^Q - \lambda_-^P}{(\lambda_-^Q)^2} + \frac{c_-^P}{\lambda_-^P} - \frac{c_-^Q}{\lambda_-^Q},$$

where b^Q and b^P are drifts of the two processes with respect to zero truncation function. □

The calibration problem now takes the form:

Calibration Problem 4 (Least squares calibration regularized by relative entropy)
Given a prior exponential-Lévy model $\mathbb{P}_0$ with characteristics $(\gamma_0, \sigma_0, \nu_0)$, find a parameter vector θ which minimizes

$$\mathcal{J}(\theta) = \sum_{i=1}^{N} \omega_i |C^\theta(T_i, K_i) - C_i|^2 + \alpha H(\theta), \tag{13.4}$$

where $H(\theta) = \mathcal{E}(\mathbb{Q}_\theta|\mathbb{P}_0)$, relative entropy of the model $(\gamma(\theta), \sigma(\theta), \nu(\theta))$ with respect to the prior.

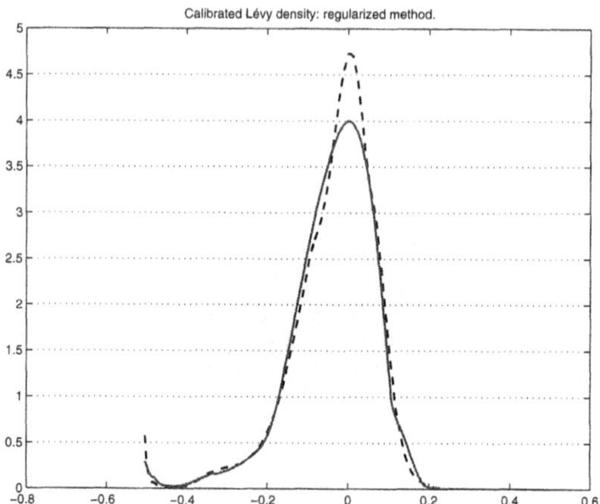

FIGURE 13.4: Lévy measure implied from DAX option prices by solving Problem 4. The starting measure for both graphs is a Merton's model (Gaussian jumps); the jump intensity is initialized to 1 for the solid curve and to 5 for the dashed one.

In the discretized nonparametric case where θ corresponds to the entire Lévy measure discretized on a grid, it was shown in [97] that Problem 4 always has a finite solution that depends continuously on the input prices and converges to the least squares solution with minimal entropy when α goes to zero. Moreover, for α large enough this solution is unique. Figure 13.4 shows the effect of regularization on the calibration problem represented in Figure 13.1. The calibrated measure is now little sensitive to the starting values of minimization algorithm: for two very different initializers the calibrated measures are quite similar. Notice also the smoothing effect of regularization on calibrated measures.

Choice of the regularization parameter The functional (13.4) consists of two parts: the relative entropy functional, which is convex (at least locally) in its argument θ and the quadratic pricing error which measures the precision of calibration. The coefficient α, called the *regularization parameter* defines the relative importance of the two terms: it characterizes the trade-off between prior knowledge of the Lévy measure and the information contained in option prices. If α is large enough, the convexity properties of the entropy functional stabilize the solution and when $\alpha \to 0$, we recover the nonlinear least squares criterion (13.2). Therefore the correct choice of α is important: it cannot be fixed in advance but its "optimal" value depends on the data at hand and the level of error present in it.

One way to achieve this trade-off is by using the Morozov discrepancy principle [300, 134]. Assume that bid and ask price data are available. This means that we can measure the error on inputs by

$$\varepsilon_0 = ||C^{\text{bid}} - C^{\text{ask}}|| \equiv \sqrt{\sum_i \omega_i |C_i^{\text{bid}} - C_i^{\text{ask}}|^2}.$$

Since the "true" price lies somewhere between the bid price and the ask price and not exactly in the middle, it is useless to calibrate the midmarket prices exactly: we only need to calibrate them with the precision ε_0. On the other hand, by increasing α we improve the stability, so the best possible stability for this precision is achieved by

$$\alpha^* = \sup\{\alpha : ||C^{\theta(\alpha)} - C^{\text{mm}}|| \leq \varepsilon_0\}, \tag{13.5}$$

where $\theta(\alpha)$ denotes the parameter vector found with a given value of α and C^{mm} is the vector of midmarket prices. In practice $||C^{\theta(\alpha)} - C^{\text{mm}}||$ is an increasing function of α so the solution can be found by Newton's or dichotomy method with a few low-precision runs of the minimization routine.

What if the bid and ask quotes are not available? In this case a possible solution is to use the model error as a proxy for the market error. To do so, we first minimize the quadratic pricing error (13.2) without regularization. The minimal least squares error level $\hat{\varepsilon}_0$ achieved by solving this optimization problem can now be interpreted as a measure of model error: if $\hat{\varepsilon}_0 = 0$ then it means that perfect calibration can be achieved but this almost never happens: typically $\hat{\varepsilon}_0 > 0$ represents the "distance" of market to the model and gives an a priori level of quadratic pricing error that one cannot really hope to improve upon while staying in the same class of models. Note that here we only need to find the minimal value of (13.2) and not to locate its minimum so a rough estimate is sufficient and the presence of "flat" directions is not a problem.

The discrepancy principle in this setting consists in authorizing a maximum error ε_0 which is of the same order of magnitude as $\hat{\varepsilon}_0$ but slightly greater (we must sacrifice some precision to gain stability). Then the optimal α can be found as before by solving Equation (13.5). In practice taking $\varepsilon_0 = \delta\hat{\varepsilon}_0$ with δ between 1.1 and 1.5 produces satisfactory results.

13.4 Numerical implementation

We will now show how the general ideas of the preceding sections can be applied to the specific problem of nonparametric calibration of the (discretized) Lévy measure when the prior is a jump-diffusion model with compound Poisson jumps and a nonzero diffusion component. We represent the calibrated Lévy measure ν by discretizing it on a grid $(x_i,\ i = 0 \dots N)$ where

$x_i = x_0 + i\Delta x$:

$$\nu = \sum_{i=1}^{N} \nu_i \delta(x - x_i). \tag{13.6}$$

The grid must be uniform in order to use the FFT algorithm for option pricing. This means that we effectively allow a fixed (but large) number of jump sizes and calibrate the intensities of these jumps. The Lévy process is then represented as a weighted sum of independent standard Poisson processes with different intensities. The prior Lévy measure must also be discretized on the same grid, for example, using the formula

$$\nu_i^P = \int_{x_i - \Delta x}^{x_i + \Delta x} \nu^P(dx)$$

for the points $x_1, \dots, x_{N-1}$ and integrating up to infinity or minus infinity for the points x_0 and x_N. The calibrated Lévy measure will of course be equivalent to the discretized prior and not to the original one, but if the grid is sufficiently fine, one can construct a good approximation of the continuous case. There are four main steps in the numerical solution of the calibration problem.

- Choice of the weights assigned to each option in the objective function.
- Choice of the prior measure $\mathbb{P}_0$.
- Choice of the regularization parameter α.
- Minimization of the functional (13.4) for given α and $\mathbb{P}_0$.

Since the first three steps have already been discussed, in this section we concentrate on the last one. The minimum of functional (13.4) is searched in the space of discretized Lévy measures using a gradient descent method: a possible choice is the L-BFGS-B routine written by Zhu, Byrd and Nocedal [75]. Therefore, to find the minimum we must be able to compute the functional itself and its gradient. To compute the functional, we first evaluate the characteristic function of the calibrated Lévy process for a given maturity T using the Lévy-Khinchin formula:

$$\phi_T(v) = \exp T\{-\frac{1}{2}\sigma^2 v^2 + ib(\nu)v + \sum_{i=0}^{N}(e^{ivx_i} - 1)\nu_i\}, \tag{13.7}$$

where the drift b can be computed from ν using the martingale condition (it will again be a finite sum). Notice that when the jump part is a sum of Poisson processes the characteristic function can be evaluated exactly: this is not an approximation. Using the Fast Fourier transform, the characteristic function can be evaluated simultaneously at a large number of points. The second step is to evaluate option prices using formulae (11.18) and (11.19) of Chapter 11. We recall these formulae from Chapter 11:

$$\zeta_T(v) = \frac{e^{-rT}\phi_T(v-i) - e^{ivrT}}{iv(1+iv)},$$
$$z_T(k) = \frac{1}{2\pi}\int_{-\infty}^{+\infty} e^{-ivk}\zeta_T(v)dv. \tag{13.8}$$

Here $z_T(k) = e^{-rT}E[(e^{s_T} - e^k)^+] - (1-e^{k-rT})^+$ is the time value of the option. Note that this time we need to *approximate* the Fourier integral (13.8) by a finite sum in order to use the Fast Fourier transform. Notice that using only two Fourier transforms, we are able to compute option prices for an arbitrary number of strikes and a single maturity. This means that computing the functional requires twice as many Fourier transforms as there are maturities in the price sheet and this number is usually quite small. The entropy part of the objective functional is easy to compute since it only contains one integral which becomes a finite sum in our setting.

The essential step that we are going to discuss now is the computation of the gradient of the functional to be minimized with respect to the discretized Lévy measure ν_i. If we were to compute this gradient numerically, the complexity would increase by a factor equal to the number of grid points. Fortunately, some preliminary thought allows to develop a method to compute the gradient of the option prices with only a two-fold increase of complexity compared to computing prices alone. Due to this optimization, the execution time of the program changes on average from several hours to about a minute on a standard PC. We will now show how to compute the derivative of option price with respect to a discretized variable ν_i.

First observe that the derivative of option price with respect to ν_i is equal to the derivative of z_T. Direct computation shows that

$$\frac{\partial\zeta_T(v)}{\partial\nu_j} = \frac{Te^{-rT}\phi_T(v-i)}{iv(1+iv)}\left\{iv(1-e^{x_j}) + e^{x_j}(e^{ivx_j}-1)\right\} =$$
$$T(1-e^{x_j})e^{-rT}\frac{\phi_T(v-i)}{1+iv} + Te^{x_j}e^{ivx_j}\zeta_T(v) - Te^{x_j}\zeta_T(v) + Te^{x_j}e^{ivrT}\frac{e^{ivx_j-1}}{iv(1+iv)}.$$

Computing the Fourier transform term by term yields:

$$\begin{aligned}\frac{\partial z_T(k)}{\partial\nu_j} &= T(1-e^{x_j})e^{-rT}\frac{1}{2\pi}\int_{-\infty}^{\infty} e^{ivk}\frac{\phi_T(v-i)}{1+iv} \\ &\quad + Te^{x_j}\{z_T(k-x_j) + (1-e^{k-x_j-rT})^+ - z_T(k) - (1-e^{k-rT})^+\} \\ &= T(1-e^{x_j})e^{-rT}\frac{1}{2\pi}\int_{-\infty}^{\infty} e^{ivk}\frac{\phi_T(v-i)}{1+iv} + Te^{x_j}\{C_T(k-x_j) - C_T(k)\}.\end{aligned}$$

Therefore, the gradient may be represented in terms of the option price and one auxiliary function. Since we are using FFT to compute option prices for the whole price sheet, we already know these prices for the whole range of strikes. As the auxiliary function will also be computed using the FFT algorithm, the computational time will only increase by a factor of two. The computation of the gradient of entropy term of the objective functional is straightforward and we do not dwell on it [97].

Here is the final calibration algorithm as implemented in the numerical examples of the next section.

1. Choose the prior. Since we did not use the historical price data, we chose the prior by calibrating a simple auxiliary jump-diffusion model (Merton model) to option prices to obtain an estimate of volatility σ_0 and a candidate for the prior Lévy measure ν_0.

2. Compute the implied volatilities and vegas corresponding to market option prices and choose the weights ω_i.

3. Fix $\sigma = \sigma_0$ and run least squares ($\alpha = 0$) using gradient descent method with low precision to get estimate of "model error"

$$\hat{\varepsilon}_0^2 = \inf_{\nu} \sum_{i=1}^{N} \omega_i |C_i^{\sigma,\nu} - C_i|^2.$$

4. Compute the optimal regularization parameter α^* to achieve trade-off between precision and stability using the a posteriori method described at the end of Section 13.3:

$$\varepsilon^2(\alpha^*) = \sum_{i=1}^{N} \omega_i |C_i^{\sigma,\nu(\alpha^*)} - C_i^*|^2 \simeq \hat{\varepsilon}_0^2. \qquad (13.9)$$

   The optimal α^* is found by running the gradient descent method (BFGS) several times with low precision.

5. Solve variational problem for $\mathcal{J}(\nu)$ with α^* by gradient-based method (BFGS) with high precision.

13.5 Numerical results

In order to assess the performance of the algorithms described above, we will first examine their performance on artificial data sets of "option prices" simulated from a given exponential-Lévy model. Then, we will apply the regularized algorithm to empirical data sets of index options and examine the implied Lévy measures thus obtained.

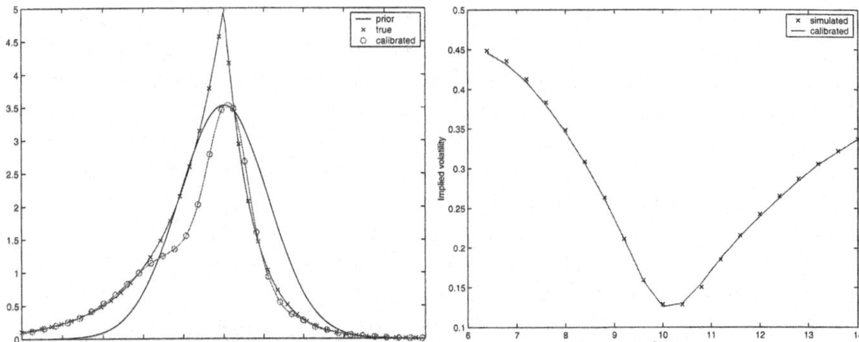

FIGURE 13.5: Left: Implied Lévy measure calibrated to option prices simulated from Kou's jump-diffusion model with $\sigma_0 = 10\%$. σ has been calibrated in a separate step ($\sigma = 10.5\%$). Right: Calibration accuracy. The solid line corresponds to calibrated implied volatilities and crosses correspond to the true ones.

13.5.1 Tests on simulated data

Figure 13.5 shows the results of nonparametric calibration of Lévy measure to simulated option prices. A data set of option prices was generated using Kou's jump-diffusion model (with double exponential jump size distribution) with a diffusion part $\sigma_0 = 10\%$ and a continuous Lévy density

$$\nu(x) = c(e^{-\lambda_+ x}1_{x>0} + e^{-\lambda_-|x|}1_{x<0}). \tag{13.10}$$

The density was taken asymmetric with the left tail heavier than the right one ($\lambda_- = 1/0.07$ and $\lambda_+ = 1/0.13$) and the constant c was chosen to have the jump intensity equal to 1. The fast Fourier pricer was used to generate prices of 21 options with maturity 5 weeks and equidistant strikes ranging from 6 to 14 (the spot being at 10). The calibration algorithm with Merton's model as prior was then applied to this simulated price sheet. The right graph in Figure 13.5 compares the implied volatilities of calibrated option prices (solid line) to the true (simulated) implied volatilities (crosses). With only 21 options the accuracy of calibration is already quite satisfactory. The left graph compares the nonparametric reconstruction of the Lévy density to the true double exponential density. The main features of the true density are successfully reconstructed with the nonparametric approach. The only region in which we observe a detectable error is near zero. This is due to the fact that the volatility of the prior was automatically fixed by the algorithm at a level slightly greater than the true one. The redundancy of small jumps and diffusion component is well known in the context of statistical estimation on time series [45, 276]. Here we retrieve another version of this redundancy in a context of calibration to a cross-sectional data set of options.

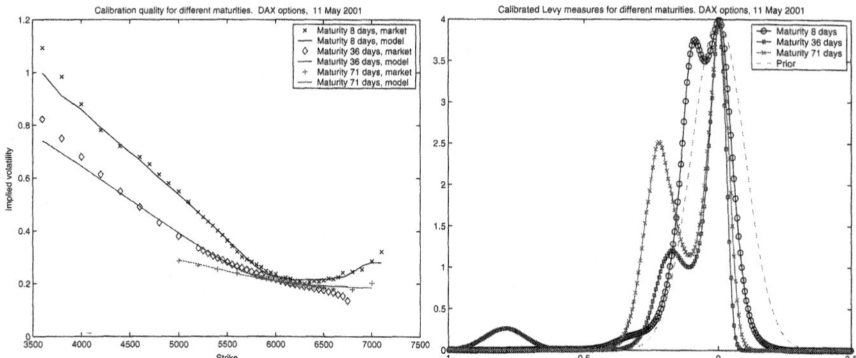

FIGURE 13.6: Left: Market implied volatilities for DAX options compared with model implied volatilities. Each maturity has been calibrated separately. Right: Implied Lévy measures.

13.5.2 Empirical result: single maturity

Figure 13.6 presents the result of calibration of Lévy measures to options on DAX index traded on the German stock exchange. Here each maturity has been calibrated separately. The left graph shows the calibration accuracy for three different maturities: clearly the model allows to reproduce a single maturity quite well. The right graph shows the corresponding implied Lévy measures which share the following features:

- The calibrated measures are strongly asymmetric; the left tail is much heavier than the right one.
- The calibrated measures are bimodal with one mode at zero and another one corresponding to large negative jump, whose magnitude becomes larger as the maturity increases.
- The intensities of calibrated measures are quite small; note however that these are risk-neutral intensities and that jump intensity of a compound Poisson process is not preserved under equivalent measure changes.

Figure 13.7 shows the Lévy measures calibrated to options of roughly the same maturity but at different calendar dates. The exact form of the density changes but its qualitative behavior remains the same.

13.5.3 Empirical results: several maturities

We have seen that time homogeneous jump-diffusion (exponential-Lévy) models can calibrate an implied volatility smile for a single maturity quite accurately. Unfortunately, due to the independence and stationarity of their increments these models perform poorly when calibrating several maturities

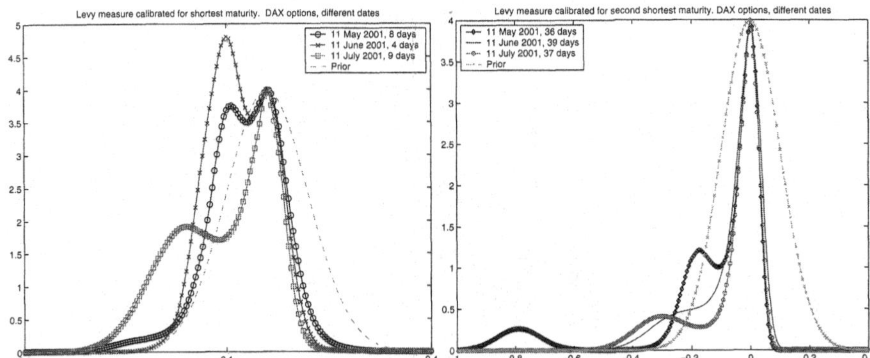

FIGURE 13.7: Stability of calibration over calendar dates. Implied Lévy measures have been calibrated at different dates for shortest (left) and second shortest (right) maturity.

at the same time: the implied volatility smile in exponential-Lévy models flattens too quickly with maturity. Figure 13.8 depicts a market implied volatility surface for DAX options. This surface flattens out as the maturity grows but even for the last maturity it is still quite skewed. In the left graph of Figure 13.9 the implied volatilities have been computed for an exponential-Lévy model calibrated to the first market maturity. The implied volatility smile generated by the model flattens out too rapidly compared to the market data. In the right graph the model was calibrated to the last maturity and we observe the same effect working in the opposite direction: this time the smile is too skewed and ATM volatility is too low for the first maturity. These remarks suggest that to reproduce the term structure of implied volatility correctly we have to relax either stationarity or independence of increments. Additive processes, which are time inhomogeneous extensions of Lévy processes, are discussed in Chapter 14. Stochastic volatility models with jumps, where increments are no longer independent, are discussed in Chapter 15.

Further reading

Empirical evidence in option prices for the presence of jumps is discussed in [40, 43, 85] and [97]. Nonlinear least squares methods are used to calibrate jump-diffusion models in [7] and stochastic volatility models with jumps in [41]. Nonparametric calibration methods for jump-diffusion models based on relative entropy minimization were introduced in [97]. The ill-posed nature of model calibration problems has been progressively recognized in various contexts. Ill-posed inverse problems are an active topic of research in applied

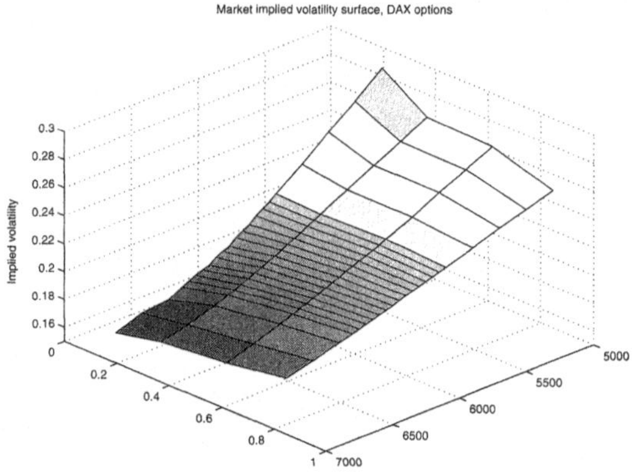

FIGURE 13.8: Market implied volatility surface: DAX options, 2001.

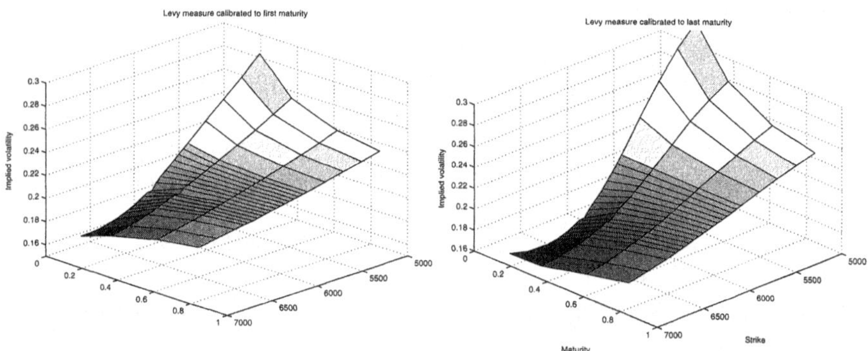

FIGURE 13.9: Implied volatilities for all maturities were computed, using the implied Lévy measure calibrated to the first maturity (left graph) and the last maturity (right graph).

mathematics. Regularization methods for ill-posed inverse problems are discussed in [133, 134]. The use of (relative) entropy for model selection in finance was initiated by Stutzer [370] and Avellaneda [13, 12, 14] in the framework of discrete state spaces. Calibration of jump-diffusion and exponential-Lévy models by relative entropy minimization is discussed in [97]. An alternative approach to model calibration which does not make use of regularization methods is proposed by Cont and BenHamida in [93] using random search methods. An updated list of references on calibration methods for option pricing models may be found at:

http://www.cmap.polytechnique.fr/~rama/dea/

Part IV

Beyond Lévy processes

Chapter 14

Time inhomogeneous jump processes

> The sciences do not try to explain, they hardly even try to interpret, they mainly make models. By a model is meant a mathematical construct which, with the addition of certain verbal interpretations, describes observed phenomena. The justification of such a mathematical construct is solely and precisely that it is expected to work.
>
> John von Neumann

While Lévy processes offer nice features in terms of analytical tractability, the constraints of independence *and* stationarity of their increments prove to be rather restrictive. On one hand, in Chapter 7 we have seen that the stationarity of increments of Lévy processes leads to rigid scaling properties for marginal distributions of returns, which are not observed in empirical time series of returns. On the other hand from the point of view of risk neutral modelling, we have observed in Chapter 13 that exponential-Lévy models allow to calibrate to implied volatility patterns for a single maturity but fail to reproduce option prices correctly over a range of different maturities. Both problems are due to the fact that exponential-Lévy models do not allow for time inhomogeneity. In this chapter we discuss the class of exponential additive models, which represent the risk-neutral dynamics of an asset as $S_t = S_0 \exp X_t$, where X is a process with independent but not stationary increments, called an additive process. We will see that this generalization allows to take into account *deterministic* time inhomogeneities: the parameters describing local behavior will now be time dependent but nonrandom. The main advantage of this approach is that it allows to preserve almost all the tractability of Lévy processes while enabling us to reproduce the whole range of option prices across strikes and maturities.

14.1 Additive processes

Additive processes are obtained from Lévy processes by relaxing the condition of stationarity of increments.

DEFINITION 14.1 Additive process *A stochastic process* $(X_t)_{t\geq 0}$ *on* $\mathbb{R}^d$ *is called an* additive process *if it is cadlag, satisfies* $X_0 = 0$ *and has the following properties:*

1. *Independent increments: for every increasing sequence of times* $t_0 \dots t_n$*, the random variables* $X_{t_0}, X_{t_1} - X_{t_0}, \dots, X_{t_n} - X_{t_{n-1}}$ *are independent.*
2. *Stochastic continuity:* $\forall \varepsilon > 0,\ \mathbb{P}[|X_{t+h} - X_t| \geq \varepsilon] \underset{h\to 0}{\to} 0.$

Before discussing generalizations of the Lévy-Khinchin formula for additive processes, we give some simple examples illustrating this notion.

Example 14.1 Brownian motion with time dependent volatility
Let $(W_t)_{t\geq 0}$ be a standard Brownian motion on $\mathbb{R}$, $\sigma(t) : \mathbb{R}^+ \to \mathbb{R}^+$ be a measurable function such that $\int_0^t \sigma^2(s)ds < \infty$ for all $t > 0$ and $b(t) : \mathbb{R}^+ \to \mathbb{R}$ be a continuous function. Then the process

$$X_t = b(t) + \int_0^t \sigma(s)dW_s \tag{14.1}$$

is an additive process. In fact, all one-dimensional continuous additive processes have this form. □

Example 14.2 Cox process with deterministic intensity
Let $\lambda(t) : \mathbb{R}^+ \to \mathbb{R}^+$ be a positive measurable function such that $\Lambda(t) = \int_0^t \lambda(s)ds < \infty$ for all t and let M be a Poisson random measure on $\mathbb{R}^+$ with intensity measure $\mu(A) = \int_A \lambda(s)ds$ for all $A \in \mathcal{B}(\mathbb{R}^+)$. Then the process $(X_t)_{t\geq 0}$ defined path by path via

$$X_t(\omega) = \int_0^t M(\omega, ds) \tag{14.2}$$

is an additive process. It can also be represented as a time changed Poisson process: if $(N_t)_{t\geq 0}$ is a standard Poisson process and

$$X_t'(\omega) = N_{\Lambda(t)}(\omega) \tag{14.3}$$

then $(X_t')_{t\geq 0}$ and $(X_t)_{t\geq 0}$ have the same law. The independent increments property now follows from properties of Poisson random measures and continuity of probability is a consequence of the continuity of time change. This process can be seen as a Poisson process with time dependent intensity $\lambda(t)$: the probability of having a jump between t and $t+\Delta$ is given by $\lambda\Delta + o(\Delta)$.. It is an example of a Cox process, which is a generalization of the Poisson process allowing for *stochastic* intensity [235]. $\Lambda(t)$ is sometimes called the hazard function and $\lambda(t)$ the hazard rate. □

The sum of two independent additive processes is obviously an additive process. Adding the two examples above we get a model frequently used in option pricing applications:

Example 14.3 Time inhomogeneous jump-diffusion
Given positive functions $\sigma : \mathbb{R}^+ \to \mathbb{R}^+$, $\lambda : \mathbb{R}^+ \to \mathbb{R}^+$ as above and a sequence of independent random variables (Y_i) with distribution F the process defined by

$$X_t = \int_0^t \sigma(s)dW_s + \sum_{i=1}^{N_{\Lambda(t)}} Y_i \tag{14.4}$$

is an additive process. □

More generally, in the same way that Lévy processes are superpositions of a Brownian motion and independent (compensated) Poisson processes (see Chapter 3), we will see that any additive processes can be represented as a superposition of independent processes of type (14.2) and (14.1).

Example 14.4 Lévy process with deterministic volatility
Extending Example 14.1, we now consider Lévy processes with time dependent volatility. Consider a continuous function $\sigma(t) : \mathbb{R}^+ \to \mathbb{R}^+$. Let $(L_t)_{t\geq 0}$ be a Lévy process on $\mathbb{R}$. Then

$$X_t = \int_0^t \sigma(s)dL_s \tag{14.5}$$

is an additive process: the independent increments property is straightforward, and to show that X is continuous in probability, one way is to decompose L into a sum of a compound Poisson process incorporating big jumps and the residual process having only jumps of size smaller than one. Then the integral with respect to compound Poisson part can be easily shown to be continuous in probability and the rest can be shown to be L^2-continuous hence also continuous in probability. □

Example 14.5 Time changed Lévy process
In Example 14.2, we have constructed an additive process by applying a continuous deterministic time change to a standard Poisson process. This method

can be generalized to arbitrary Lévy process. Let $(L_t)_{t\geq 0}$ be a Lévy process on $\mathbb{R}^d$ and let $v(t) : \mathbb{R}^+ \to \mathbb{R}^+$ be a continuous increasing function such that $v(0) = 0$. Then the process $(X_t)_{t\geq 0}$ defined path by path via

$$X_t(\omega) = L_{v(t)}(\omega) \tag{14.6}$$

is an additive process. This follows from independent increments property of L and the continuity of time change. □

The last three examples are already sufficiently rich and can be used to construct flexible parametric models, based on additive processes, for option pricing. The exact form of the volatility function or the time change can then be adjusted to reproduce the term structure of option prices while the driving Lévy process L can be chosen to calibrate the smile (see below).

In all the above examples X_t has an infinitely divisible distribution for all t. This is a general property of additive processes [345, Theorem 9.1.]: for any t, the characteristic function of X_t has a Lévy-Khinchin representation:

$$E[\exp(iu.X_t)] = \exp \psi_t(u) \qquad \text{with} \tag{14.7}$$
$$\psi_t(u) = -\frac{1}{2}u.A_t u + iu.\Gamma_t + \int_{\mathbb{R}^d} \mu_t(dx)(e^{iu.x} - 1 - iu.x1_{|x|\leq 1}),$$

where, as usual, A_t is a positive definite $d \times d$ matrix and μ_t is a Lévy measure, that is, a positive measure on $\mathbb{R}^d$ satisfying $\int_{\mathbb{R}^d}(|x|^2 \wedge 1)\mu_t(dx) < \infty$. Note that, unlike the case of Lévy processes, $\psi_t(u)$ is not linear in t anymore. In a model where $X_t = \ln(S_t/S_0)$ describes the log-return, the triplet $(A_t, \mu_t, \Gamma(t))$ completely describes the distribution of the asset at time t, as viewed from $t = 0$: we will call $(A_t, \mu_t, \Gamma(t))$ the spot characteristics of X.

Similarly, for any $t > s$, $X_t - X_s$ is a sum of infinitely divisible random variables therefore infinitely divisible. Independence of increments implies the following relation between the characteristic functions of X_s, X_t and $X_t - X_s$:

$$\phi_{X_s}(u)\phi_{X_t - X_s}(u) = \phi_{X_t}(u). \tag{14.8}$$

The characteristic function of $X_t - X_s$ can therefore be written as:

$$E[\exp(iu.(X_t - X_s))] = \exp \psi_s^t(u) \qquad \text{where} \tag{14.9}$$
$$\psi_s^t(u) = -\frac{1}{2}u.(A_t - A_s)u + iu.(\Gamma_t - \Gamma_s)$$
$$+ \int_{\mathbb{R}^d} (\mu_t(dx) - \mu_s(dx))(e^{iu.x} - 1 - iu.x1_{|x|\leq 1}).$$

If $A_t - A_s$ is a positive matrix and $\mu_t - \mu_s$ a Lévy measure (therefore a positive measure), the above equation is none other than the Lévy-Khinchin representation of $X_t - X_s$, which has therefore characteristic triplet $(A_t - A_s, \mu_t - \mu_s, \Gamma_t - \Gamma_s)$. The triplet $(A_t - A_s, \mu_t - \mu_s, \Gamma_t - \Gamma_s)$ completely describes the distribution of the increments on $[s, t]$ and we will call it the forward Lévy triplet associated to the dates (s, t). The positiveness of $A_t - A_s$, $\mu_t - \mu_s$ for each $t \geq s$, which was used to define the forward characteristics, means that the spot volatility A_t and the spot Lévy measure μ_t should increase with t.

Conversely, when does a set of spot characteristics, $(A_t, \mu_t, \Gamma(t))_{t\geq 0}$, specify an additive process? The following result shows that the positiveness of forward characteristics plus a continuity condition on spot characteristics suffice:

THEOREM 14.1 (see Sato [345], Theorems 9.1–9.8)

Let $(X_t)_{t\geq 0}$ be an additive process on $\mathbb{R}^d$. Then X_t has infinitely divisible distribution for all t. The law of $(X_t)_{t\geq 0}$ is uniquely determined by its spot characteristics $(A_t, \mu_t, \Gamma_t)_{t\geq 0}$:

$$E[\exp(iu.X_t)] = \exp \psi_t(u) \qquad \textit{where} \tag{14.10}$$
$$\psi_t(u) = -\frac{1}{2}u.A_t u + iu.\Gamma_t + \int_{\mathbb{R}^d} \nu_t(dx)(e^{iu.x} - 1 - iu.x1_{|x|\leq 1}).$$

The spot characteristics $(A_t, \mu_t, \Gamma_t)_{t\geq 0}$ satisfy the following conditions

1. *For all t, A_t is a positive definite $d \times d$ matrix and μ_t is a positive measure on $\mathbb{R}^d$ satisfying $\mu_t(\{0\}) = 0$ and $\int_{\mathbb{R}^d}(|x|^2 \wedge 1)\mu_t(dx) < \infty$.*
2. *Positiveness: $A_0 = 0$, $\nu_0 = 0$, $\gamma_0 = 0$ and for all s, t such that $s \leq t$, $A_t - A_s$ is a positive definite matrix and $\mu_t(B) \geq \mu_s(B)$ for all measurable sets $B \in \mathcal{B}(\mathbb{R}^d)$.*
3. *Continuity: if $s \to t$ then $A_s \to A_t$, $\Gamma_s \to \Gamma_t$ and $\mu_s(B) \to \mu_t(B)$ for all $B \in \mathcal{B}(\mathbb{R}^d)$ such that $B \subset \{x : |x| \geq \varepsilon\}$ for some $\varepsilon > 0$.*

Conversely, for family of triplets $(A_t, \mu_t, \Gamma_t)_{t\geq 0}$ satisfying the conditions (1), (2) and (3) above there exists an additive process $(X_t)_{t\geq 0}$ with $(A_t, \mu_t, \Gamma_t)_{t\geq 0}$ as spot characteristics.

Examples of spot characteristics $(A_t, \mu_t, \Gamma_t)_{t\in[0,T]}$ verifying these conditions can be constructed in the following way: consider

- A continuous, matrix valued function $\sigma : [0, T] \to M_{d\times n}(\mathbb{R})$ such that $\sigma(t)$ is symmetric and verifies $\int_0^T \sigma^2(t)dt < \infty$.
- A family $(\nu_t)_{t\in[0,T]}$ of Lévy measures verifying

$$\int_0^T dt \int (1 \wedge |x|^2)\nu_t(dx) < \infty.$$

- A deterministic function with finite variation (e.g., a piecewise continuous function) $\gamma : [0, T] \to \mathbb{R}$.

Then the spot characteristics $(A_t, \mu_t, \Gamma_t)_{t\in[0,T]}$ defined by

$$A_t = \int_0^t \sigma^2(s)ds, \tag{14.11}$$

$$\mu_t(B) = \int_0^t \nu_s(B)ds \quad \forall B \in \mathcal{B}(\mathbb{R}^d), \tag{14.12}$$

$$\Gamma(t) = \int_0^t \gamma(s)ds, \tag{14.13}$$

satisfy the conditions (1), (2) and (3) and therefore define a unique additive process $(X_t)_{t\in[0,T]}$ with spot characteristics $(A_t, \mu_t, \Gamma_t)_{t\in[0,T]}$. $(\sigma^2(t), \nu_t, \gamma(t))$ are called *local* characteristics of the additive process.

Example 14.6

In Example 14.4, assume that the jumps sizes Y_i verify $|Y_i| \leq 1$. Then the local characteristics are given by

$$(\sigma^2(t), \lambda(t)F, \lambda(t)E[Y_1]).$$

□

Not all additive processes can be parameterized in this way, but we will assume this parameterization in terms of local characteristics in the rest of this chapter. In particular, the assumptions above on the local characteristics imply that X is a semimartingale which will allow us to apply the Itô formula (see below).

REMARK 14.1 Let $(X_t)_{t\geq 0}$ be a Lévy process with characteristic triplet (A, ν, γ). It is also an additive process, with spot characteristics given by $(A_t = tA, \mu_t = t\nu, \Gamma_t = t\gamma)_{t\geq 0}$, forward characteristics on $[s,t]$ given by $((t-s)A, (t-s)\nu, (t-s)\gamma)$ and local characteristics given by (A, ν, γ). □

Sample paths of additive processes The local characteristics of an additive process enable to describe the structure of its sample paths: the positions and sizes of jumps of $(X_t)_{t\in[0,T]}$ are described by a Poisson random measure on $[0,T] \times \mathbb{R}^d$

$$J_X = \sum_{t\in[0,T]} \delta_{(t,\Delta X_t)} \tag{14.14}$$

with (time inhomogeneous) intensity given by $\nu_t(dx)dt$:

$$E[J_X([t_1,t_2] \times A)] = \mu_T([t_1,t_2] \times A) = \int_{t_1}^{t_2} \nu_t(A)dt. \tag{14.15}$$

The compensated version $\tilde{J}_X$ can therefore be defined by: $\tilde{J}_X(dt\ dx) = J_X(dt\ dx) - \nu_t(dx)dt$. In terms of this Poisson random measure, the additive process X_t satisfy a generalized version of Lévy-Itô decomposition [345] of which we give the one-dimensional version:

$$X_t = \Gamma_t + \int_0^t \sigma(s)dW_s + \int_{s\in(0,t],|x|\geq 1} xJ_X(ds \times dx) \\ + \int_{s\in(0,t],0<|x|<1} x\tilde{J}_X(ds \times dx), \quad (14.16)$$

where $(W_t)_{t\geq 0}$ is a standard Wiener process, J_X is a Poisson random measure on $[0,T]\times\mathbb{R}$ with intensity measure $\nu_t(dx)dt$ and the terms in the decomposition are independent. This decomposition shows that X is a semimartingale. Using the results of Section 8.2, we can compute its quadratic variation $[X,X]$:

$$\begin{aligned}[X,X]_t &= \int_0^t \sigma(s)^2 ds + \int_{s\in(0,t],|x|\geq 1} x^2 J_X(ds\times dx) \\ &= \int_0^t \sigma(s)^2 ds + \sum_{s\in(0,t]} |\Delta X_s|^2.\end{aligned}$$

Additive processes as Markov processes Due to the independence of their increments, additive processes are spatially (but not temporally) homogeneous Markov processes.

Consider an additive process $(X_t)_{t\in[0,T]}$ described by the local characteristics $(\sigma^2(t),\nu_t,\gamma(t))$. The family of transition operators $(P_{s,t})_{0\leq s\leq t}$ defined by:

$$P_{s,t}f(x) \equiv E[f(X_t)|X_s = x] = E[f(X_t - X_s + x)]$$

is described by the forward characteristics of X. Let $\mathcal{C}_0$ be the set of continuous functions, vanishing at infinity and $\mathcal{C}_0^2$ be the set of functions f such that f, f' and f'' are in $\mathcal{C}_0$. By analogy with the case of Lévy processes, one can define an infinitesimal generator:

$$L_t f(x) = \frac{1}{2}\sigma^2(t)\frac{\partial^2 f}{\partial x^2} + \gamma(t)\frac{\partial f}{\partial x} + \int_{\mathbb{R}} \nu_t(dy)[f(x+y) - f(x) - y\frac{\partial f}{\partial x}(x)1_{\{|y|\leq 1\}}]$$

for $f\in\mathcal{C}_0^2$. Since X is a semimartingale, we can apply the Itô formula (Proposition 8.19) to $f(X_t)$:

$$f(X_t) = f(0) + \int_0^t L_s f(X_s)ds + M_t, \quad (14.17)$$

where

$$M_t = \int_0^t f'(X_s)dW_s + \int_{\mathbb{R}\times(0,t]} \{f(X_{s-}+x) - f(X_{s-})\}\{J_X(dx\times ds) - \nu_s(dx)ds\} \quad (14.18)$$

is a martingale for every $f \in \mathcal{C}_0^2$. Formula (14.17) implies that

$$E\{f(X_t)\} = f(0) + E\left\{\int_0^t L_s f(X_s)ds\right\}. \quad (14.19)$$

This is already the key result that we need to price exotic options. It can also be used to prove that

$$L_t f(x) = \lim_{s\downarrow 0} \frac{P_{t,t+s}f(x) - f(x)}{s} \quad (14.20)$$

for every $f \in \mathcal{C}_0^2$.

14.2 Exponential additive models

In exponential additive models, the risk-neutral dynamics of an asset price is given by $S_t = \exp(rt + X_t)$, where $(X_t)_{t\geq 0}$ is an additive process such that $(e^{X_t})_{t\geq 0}$ is a martingale. By Proposition 3.17, the characteristic function of X must then satisfy $\phi_t(-i) = 1$ which implies the following conditions on the spot characteristics:

$$\forall t, \int_{|y|>1} \mu_t(dy)e^y < \infty \qquad \text{and}$$
$$\Gamma_t = -\frac{A_t}{2} - \int (e^y - 1 - y1_{|y|\leq 1})\mu_t(dy). \quad (14.21)$$

Equivalently, the discounted price $(S_t e^{-rt}) = (\exp(X_t))_{t\geq 0}$ is a martingale if and only if the local characteristics satisfy the following conditions:

$$\int_{|y|>1} \nu_t(dy)e^y \quad \text{is integrable in } t \text{ and} \quad (14.22)$$
$$\gamma_t = -\frac{\sigma_t^2}{2} - \int (e^y - 1 - y1_{|y|\leq 1})\nu_t(dy),$$

where the last inequality holds dt-almost everywhere.

14.2.1 Option pricing in risk-neutral exp-additive models

European options The price at time $t = 0$ of a European call with maturity T, and strike K is

$$C_0(S,T,K) = e^{-rT} E[(Se^{rT+X_T} - K)^+]. \tag{14.23}$$

The price is determined by the spot characteristics (A_T, ν_T, γ_T). Since X_T is infinitely divisible with triplet (A_T, ν_T, γ_T), we come to the conclusion:

PROPOSITION 14.1
The price of a European option with maturity T in an exp-additive model with spot characteristics $(A_t, \mu_t, \Gamma_t)_{t\geq 0}$ is equal to the price of the same option in an exp-Lévy model with characteristic triplet $(A_T/T, \mu_T/T, \Gamma_T/T)$.

The pricing methods and algorithms developed for exp-Lévy models (e.g., Fourier methods discussed in Section 11.1) can therefore be used without modification for exp-additive models.

Forward start options A forward start call[1] starting at T_1 and maturing at T_2 has payoff at expiry given by

$$H = (S_{T_2} - mS_{T_1})^+ = (Se^{rT_2+X_{T_2}} - mSe^{rT_1+X_{T_1}})^+$$
$$= Se^{rT_1+X_{T_1}}(e^{r(T_2-T_1)+X_{T_2}-X_{T_1}} - m)^+,$$

where m is the proportion of S_{T_1} at which the strike of the option is fixed at date T_1. Using the independent increments property of X we can compute the price of this option at date 0:

$$P_0 = e^{-rT_2}E[H] = Se^{-r(T_2-T_1)}E[e^{r(T_2-T_1)+X_{T_2}-X_{T_1}} - m)^+]. \tag{14.24}$$

The price of a forward start option is thus determined by the *forward* characteristics $(A_{T_2} - A_{T_1}, \mu_{T_2} - \mu T_1, \Gamma_{T_2} - \Gamma_{T_1})$. More precisely, we have the following result:

[1]See Section 11.2.

PROPOSITION 14.2 Forward start options in exp-additive models
The price of a forward start call starting at T_1 and expiring at T_2 in an exp-additive model with spot characteristics $(A_t, \mu_t, \Gamma_t)_{t\geq 0}$ is equal to the price of a European call with the same moneyness and maturity $T_2 - T_1$ in an exp-Lévy model with characteristic triplet

$$\left(\frac{A_{T_2} - A_{T_1}}{T_2 - T_1}, \frac{\mu_{T_2} - \mu_{T_1}}{T_2 - T_1}, \frac{\Gamma_{T_2} - \Gamma_{T_1}}{T_2 - T_1}\right). \tag{14.25}$$

Note that today's price of this forward start option is also equal to the future price (at T_1) of a European option with maturity T_2. This implies that the future smile is completely determined by today's forward smile hence also by today's smile. This sheds light on an important drawback of exponential additive models: having observed the smile today for all strikes and maturities, we know option prices at all future dates. This means that the evolution of smile as a whole is completely deterministic and does not depend on any market parameters. Therefore, influence of new information on the dynamics of smile cannot be incorporated.

Path dependent options Since path dependent options are not only sensitive to the terminal distribution of the process but to the whole path, their pricing in exp-additive models cannot be completely reduced to the exp-Lévy case but the key methods are not much different. The value of path-dependent option with maturity T depends on the *local* characteristics $(\sigma(t), \nu_t, \gamma(t))_{t\in[0,T]}$ and not only on spot characteristics for the maturity T. As an example consider the pricing of an up-and-out option with payoff at maturity

$$(S_T - K)^+ 1_{\sup_{[0,T]} X_t < B}. \tag{14.26}$$

Then the payoff may be represented as

$$\begin{aligned} &H(X_{T\wedge T_b}) \quad \text{where} \quad T_b = \inf\{t : S_t \geq B\} \\ &\text{and} \quad H(S) = (S-K)^+ 1_{S>B}. \end{aligned} \tag{14.27}$$

Following the lines of Proposition 12.2 we can show that the value $UOC(t)$ is given by $UOC(t) = C(t, S_t)$ where $C : [0,T] \times [0, \infty[$ is a solution of the following boundary value problem involving the local characteristics: ν.

$$\forall (t,S) \in [0,T]\times]0,\infty[,$$
$$\frac{\partial C}{\partial t}(t,S) + rS\frac{\partial C}{\partial S}(t,S) + \frac{\sigma^2(t)S^2}{2}\frac{\partial^2 C}{\partial S^2}(t,S) - rC(t,S)$$
$$+ \int \nu_t(dy)[C(t,Se^y) - C(t,S) - S(e^y-1)\frac{\partial C}{\partial S}(t,S)] = 0, \quad (14.28)$$
$$\forall S \in]0,B[, \qquad C(t=T,S) = (S-K)^+, \quad (14.29)$$
$$\forall S \geq B, \quad \forall t \in [0,T] \qquad C(t,S) = 0. \quad (14.30)$$

Such partial integro-differential equations can then be solved numerically using the numerical methods presented in Chapter 12, in particular the finite difference method described in Section 12.4.

14.2.2 Calibration to option prices

In view of Proposition 14.1, the problem of calibrating spot characteristics of an additive process *for one maturity* to prices of European options for this maturity is equivalent to that of calibrating the characteristics of a Lévy process to the same options. However, the distribution of a Lévy process is completely determined by its marginal distribution at one fixed time. This means that if we know sufficiently many option prices for one maturity, we can calibrate an exponential-Lévy model. For an exponential additive model this is no longer the case: the distribution of the process is no longer determined by the marginal distribution at one fixed time. This gives us the necessary freedom to calibrate several maturities simultaneously, which was hardly possible in exponential-Lévy models.

Given a set of option prices with maturities $T_i, i = 1\dots m$ and an existing calibration algorithm for exponential-Lévy models (parametric or nonparametric), the simplest way to calibrate an exp-additive model is to calibrate spot characteristics separately for each maturity using the algorithm for exponential-Lévy models (see Chapter 13). If the calibrated spot characteristics satisfy the positiveness conditions of Proposition 14.1 — the forward characteristics must be positive — then we are done, the spot characteristics for different maturities that we have calibrated can correspond to a single additive process. If not, the following algorithm could be used.

ALGORITHM 14.1 Calibration of exp-additive models to option prices

Calibration of spot characteristics $(A_{T_i}, \nu_{T_i}, \gamma_{T_i})_{i=1\dots m}$ to option data with maturities $T_1, \dots, T_m$.

- *Calibrate the spot characteristics for the first maturity $(A_{T_1}, \mu_{T_1}, \Gamma_{T_1})$ using the existing calibration method for exp-Lévy models.*

- *For maturities $T_2, \ldots, T_m$ calibrate the n-th triplet using the calibration algorithm for exp-Lévy models* under the positiveness constraints*: $A_{T_n} \geq A_{T_{n-1}}$ and $\mu_{T_n}(B) \geq \mu_{T_{n-1}}(B)$ for all $B \in \mathcal{B}(\mathbb{R})$.*

These constraints are linear and typically simple to implement, both in parametric and in nonparametric models. If the Lévy measures have a density $\mu_t(x)$ then the inequalities become $\mu_{T_n}(x) \geq \mu_{T_{n-1}}(x)$. In a nonparametric calibration algorithm as the one describes in Chapter 13, where μ_{T_n} is discretized on a grid, the positiveness constraints amount to a finite number of linear inequalities, one for each grid point.

This algorithm gives a set of spot characteristics on the discrete tenor (T_i) that respects the positiveness constraints. To determine the process completely, we must somehow interpolate the spot characteristics between maturity dates and extrapolate them outside the interval $[T_1, T_m]$. Making the additional assumption that the local characteristics are piecewise constant, we obtain the following result:

$$\sigma^2(t) = \frac{A_{T_1}}{T_1}, \; t \leq T_1; \qquad \sigma^2(t) = \frac{A_{T_{i+1}} - A_{T_i}}{T_{i+1} - T_i}, \; t \in]T_i, T_{i+1}]. \tag{14.31}$$

$$\nu_t = \frac{\mu_{T_1}}{T_1}, \; t \leq T_1; \qquad \nu_t = \frac{\mu_{T_{i+1}} - \mu_{T_i}}{T_{i+1} - T_i}, \; t \in (T_i, T_{i+1}].$$

Local characteristics for $t > T_m$ can be taken equal to those on the interval $]T_{m-1}, T_m]$.

Figure 14.1 shows the results of calibrating a nonparametric exponential additive model to prices of DAX options for three different maturities. The intensity of spot measures changes roughly proportionally to time to maturity, which would be the case in an exponential-Lévy model. However, the shape of Lévy density does not remain constant: as time to maturity increases, the Lévy densities spread out, reflecting the fear of larger jumps for options with longer times to maturity. This effect clearly rejects the Lévy hypothesis (because the shape of Lévy density is maturity-dependent). In these graphs, the volatility parameter was allowed to change in order to obtain maximum similarity between Lévy measures. Despite this, the calibrated densities are very different. This shows that time dependence in volatility alone is not sufficient to explain the smile, the Lévy measure must also vary. The calibration method above allows to extract a set of implied forward Lévy triplets and can be used to test the adequacy of additive processes for modelling risk-neutral dynamics of an asset, based on option prices. We have applied here this method to DAX option prices; observations based on this data are summarized up in Table 14.1.

TABLE 14.1: Testing time homogeneity of risk-neutral dynamics using option prices

Hypothesis	Compatible with observed option prices?
(X_t) is an additive process	Yes
$A_t = \sigma^2 t$ (local volatility does not depend on time)	Yes
(X_t) is a Lévy process	No
$\nu_t = t\nu$ (local Lévy measure does not depend on time)	No
$\nu_t = \lambda(t)\nu$ (local Lévy measure depends on time only through intensity)	No

14.3 Exp-additive models vs. local volatility models

Additive processes allow, almost by definition, to reproduce option prices over a wide range of strikes and maturities. This happens because distributions of an additive process at different times, unlike those of a Lévy process, are only weakly linked to each other; knowledge of option prices for one maturity imposes almost no constraints on other maturities. This is a very nice and interesting property: indeed, additive processes provide a mathematically tractable framework to construct models with jumps that can reproduce the entire two-dimensional surface of implied volatility smile. In this respect they are comparable to local volatility models [122, 112] where the risk-neutral dynamics of the asset is given by a Markov diffusion:

$$\frac{dS_t}{S_t} = rdt + \sigma(t, S_t)dW_t. \tag{14.32}$$

Local volatility models are also capable to reproducing an arbitrary profile of implied volatilities across strikes and maturities. However, by contrast with local volatility models where computing forward smiles or pricing forward start options can be quite difficult, additive models are sufficiently simple to allow many analytical computations and where analytical computations are not possible, numerical methods (such as fast Fourier transforms) developed for exponential-Lévy models can be used without any modification. As in the case of local volatility models, this simplicity comes at the price of a strong nonstationarity of exp-additive models. The high precision of calibration to

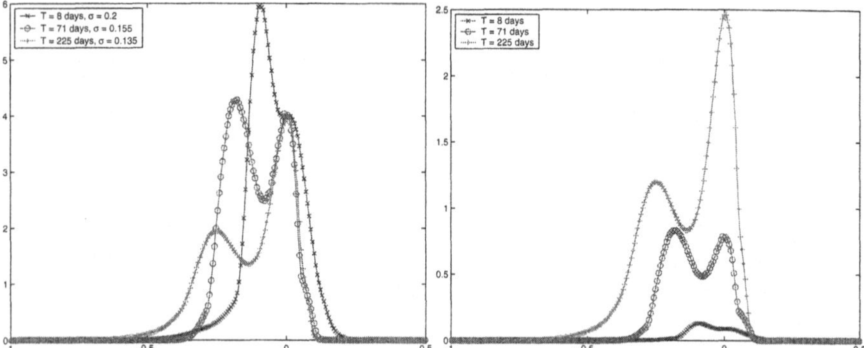

FIGURE 14.1: Left: Lévy measures, normalized by time to maturity, calibrated to DAX options of maturities $T = 8$, 71, 225 days with σ allowed to vary for maximum similarity between Lévy measures. Right: Spot Lévy measures for the same maturities.

option prices at a given date is achieved by introducing explicit time dependence into the characteristic triplet and especially the jumps. This may lead to an unrealistic forms for forward and future smiles. On one hand, due to independence of increments of an additive process, all future smiles are locked in by today's option prices: all forward smiles can be computed from today's (spot) implied volatility surface. On the other hand, due to non-stationarity of the model, forward smiles may not have the right shape.

More precisely, in an exponential additive model, the price at time T_1 of an option expiring at time T_2 will be equal to today's price of a forward start option starting at T_1 and expiring at T_2. This means that today's forward smile will become spot smile in the future. Since today's forward smile can be computed from today's spot smile, this means that all future smiles are completely determined by option prices (for all maturities) on a given date. Thus, these models do not allow for the arrival of new information between today and tomorrow: the future smiles are completely predicted. No wonder that this prediction does not seem to work too well on calibrated measures. Spot measures for a given time to maturity are more or less stationary (see Figure 13.7 in Chapter 13) and they are always different from corresponding forward measures: the latter have more mass in the tails and less mass in the center (see Figure 14.2).

To summarize, exponential additive models have enough degrees of freedom to reproduce implied volatility patterns across strike and maturity. The calibration algorithm described above translates implied volatility skews and term structures into forward Lévy triplets which can then be interpolated to give implied diffusion coefficients and implied Lévy measures and used to price path-dependent options. However the performance of exponential additive models across time is not as good as their cross-sectional performance,

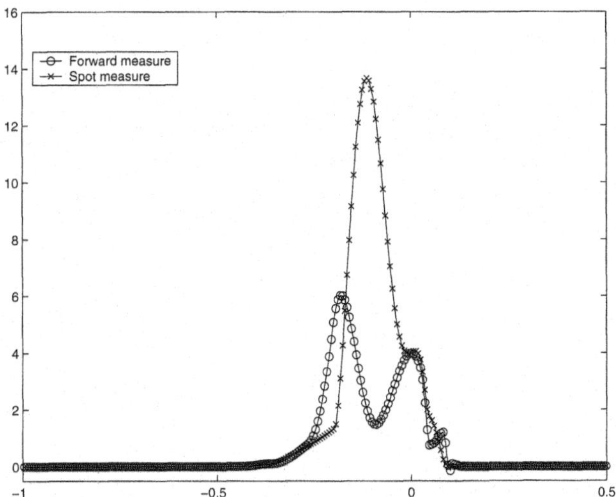

FIGURE 14.2: Comparison of a forward Lévy measure on June 13, 2001 (options starting in one month with maturity two months) to the spot Lévy measure on July 13, 2001 for options with the same maturity date.

because the predicted forward Lévy measures are too different from spot Lévy measures and because the evolution of implied volatility smile is deterministic. But additive processes go as far as one can go with independent increments. To construct models with a more stable performance across time one should relax the hypothesis of independence of increments to improve smile dynamics and gain more stationarity: this provides a motivation for the stochastic volatility models discussed in the next chapter.

Further reading

A thorough treatment of Poisson processes with generalizations to time inhomogeneous case and to Cox processes is given in Kingman [235]. Additive processes were introduced by Paul Lévy [252, 260]. Their properties are studied in the monograph by Sato [345] and their relation to semimartingales is discussed in [215].

Poisson processes with stochastic intensity, called Cox processes, have many applications especially in insurance and credit risk modelling [246] which are beyond the scope of this book. Cox processes have also been used for modelling high-frequency transaction data, in particular to accommodate the empirical features of durations between consecutive trades, see [135, 342, 343].

Chapter 15

Stochastic volatility models with jumps

The exponential-Lévy models and exponential additive models studied in Chapters 11 and 14 generalize the Black-Scholes model by introducing jumps but conserving independence of log-returns. At the level of time series of returns, these generalizations allow flexible modelling of tail behavior at various time scales. At the level of risk-neutral modelling, they allow to generate implied volatility smiles and skews similar to the ones observed in market prices.

But, as observed in Chapter 7, independence of increments is not a property observed in historical time series of returns: properties such as *volatility clustering* (see Section 7.1) suggest that the amplitude of returns is positively autocorrelated in time.

At the level of risk-neutral modelling, it is less obvious that the independence of increments can be rejected. In fact, we saw in Chapter 14 that additive processes with independent increments allow to calibrate implied volatility patterns across strike *and* maturity in a flexible way. But, as observed in Chapter 14, additive processes have the same drawback as local volatility models: they require time dependent parameters to accommodate the observed term structure of implied volatility.[1] More importantly, when the risk-neutral dynamics of the log-price is given by a Lévy process or an additive process, the implied volatility surface follows a deterministic evolution: there is no "vega" risk in such models.

Stochastic volatility models tackle these difficulties by introducing a second random process (σ_t), interpreted as the instantaneous volatility of the underlying:

$$\frac{dS_t}{S_t} = \mu dt + \sigma_t dW_t. \tag{15.1}$$

$(\sigma_t)_{t\geq 0}$ is a positive stochastic process, often chosen to be mean-reverting. Different stochastic volatility models correspond to different specifications for the volatility process σ_t.

[1] Time dependence of parameters is clearly an inconvenient feature and goes against the parsimony of the model but is not in itself an argument against a class of models: this time dependence may very well reflect an empirical reality.

In diffusion-based stochastic volatility models, reviewed in Section 15.1, σ_t is also driven by a Brownian motion, which can be correlated with W_t. Such models do account for volatility clustering, dependence in increments and long term smiles and skews, but cannot generate jumps nor can they give rise to realistic short-term implied volatility patterns. Furthermore, the positiveness of σ_t requires to use a nonlinear diffusion: an example is the CIR/square root process, discussed in Section 15.1.2, which drives the Heston model [196].

These shortcomings can be overcome by introducing jumps in these models. Jumps can be added in stochastic volatility models in two ways: in the returns and in the evolution of volatility. Adding an independent jump component to the returns of a diffusion-based stochastic volatility model improves the short-maturity behavior of implied volatility without deteriorating long-term smiles: an example of this approach is illustrated by the Bates model, discussed in Section 15.2.

Another possibility is to add jumps in the volatility process. Using a (positive) Lévy process to drive the volatility σ_t allows to build positive, mean-reverting volatility processes with realistic dynamics without resorting to non-linear models: positive Ornstein-Uhlenbeck processes, proposed as models for volatility by Barndorff-Nielsen and Shephard, are discussed in Section 15.4. These models are analytically tractable but computations become quite involved as soon as "leverage" effects are included.

A different way to build models with dependence in increments is to time change a Lévy process by a positive increasing process with dependent increments. Using this method, treated in Section 15.5, though one does not start from a "stochastic volatility" representation such as (15.1) similar implied volatility patterns can be obtained.

Sometimes the introduction of stochastic volatility models is justified by empirically observing that indicators of market volatility — such as the quadratic variation of returns or moving average estimators of the variance of returns — behave in a highly erratic manner, suggesting that "market volatility" is stochastic. As discussed in Section 7.5, erratic behavior of sample quadratic variation or sample variance does not imply in any way the existence of an additional "stochastic volatility" factor. High variability of realized variance may be due to the presence of jumps and not to the presence of stochastic volatility: random behavior of quadratic variation is in fact a generic property of all exponential-Lévy models. The main difference between exp-Lévy/exp-additive and stochastic volatility models is not their variability and marginal properties but the introduction of *dependence* in the increments which enables in turn flexible modelling of the term structure of various quantities.

This flexibility is illustrated by the following example. Let $(X_t)_{t\geq 0}$ be the log-price process. If X is a Lévy process then its fourth cumulant $c_4[X_t]$ is *always* proportional to t (see Proposition 3.13) and the kurtosis of X_t always decays as $1/t$, whatever model is chosen. In X is a "stochastic volatility" process $X_t = \int_0^t \sigma_t dW_t$, assuming that the volatility process is independent from the Brownian motion, we find that $c_4(X_t) = 3\,\mathrm{Var}(\int_0^t \sigma_s^2 ds)$, that is, many different types of dependence on t are possible.

The price for this flexibility is an increase in dimension. In a stochastic volatility models the price S_t is no longer a Markov process: the evolution of the price process is determined not only by its value but also by the level of volatility. To regain a Markov process one must consider the two-dimensional process (S_t, σ_t). This is why in practice users have privileged parametric models where efficient Fourier-based computational procedures are available.

As we saw in Chapter 9, one must distinguish statistical models for time series behavior from pricing models, i.e., models for risk-neutral dynamics. As a result of the introduction of an extra factor of randomness, the stochastic volatility models considered here lead to incomplete market models[2] and time series behavior of returns does not pinpoint option prices in a unique way. Also, since autocorrelation functions or the Markov property are not invariant under change of measure, observing volatility clustering or mean-reversion in the level of historical volatility does not necessarily imply similar behavior in the pricing models, unless we make ad hoc assumptions on the premium associated to "volatility risk." Since the main driving force behind continuous time stochastic volatility models has been the development of option pricing models, instead of making arbitrary assumptions on risk premia (as is frequently seen in the literature), we will limit our discussion to the performance of stochastic volatility models as pricing models, i.e., models for risk-neutral dynamics which, as in Chapter 13, are calibrated to option prices, not time series of underlying. References on econometric applications of stochastic volatility models are given at the end of the chapter.

15.1 Stochastic volatility models without jumps

A widely studied class of stochastic volatility models is the class of bivariate diffusion models in which the (risk-neutral dynamic) of the asset price $(S_t)_{t\geq 0}$ satisfies the stochastic differential equation

$$dS_t = \mu S_t dt + \sigma_t S_t dW_t, \tag{15.2}$$

where $(\sigma_t)_{t\geq 0}$ is called the instantaneous volatility process. In such models, the price S_t is not a Markov process anymore but (S_t, σ_t) is a Markovian diffusion. This means we have to deal with an additional dimension when using numerical methods.

[2]It is possible to introduce "stochastic volatility" models where the volatility is not an extra random factor but driven by past returns, leading to a complete market model. The models considered in the sequel (and used in practice) are not of this type.

Using bivariate diffusion models has some advantages. First, within this framework PDEs and finite difference methods are available for the computation of option prices. Second, it allows a simple description of the correlation between volatility movements and the returns, which is parameterized by the correlation coefficient of the two Wiener processes. Important features for the instantaneous volatility process are positiveness and mean-reversion. Positiveness can be obtained by setting $\sigma_t = f(y_t)$ where f is a positive function and y_t a random driving process. Mean-reversion means that the process does not wander off to infinity but oscillates around a well-defined long-term average value and is usually modelled by introducing a mean-reverting drift in the dynamics of (y_t):

$$dy_t = \lambda(\eta - y_t)dt + \ldots d\hat{Z}_t, \tag{15.3}$$

where $(\hat{Z}_t)_{t\geq 0}$ is a Wiener process correlated with (W_t). λ is usually called the *rate of mean-reversion* and η is the long-run average level of y_t. The drift term pulls y_t towards η and consequently we would expect that σ_t is pulled towards the long-run mean value of $f(y_t)$. Moreover, it will be shown later in this chapter that in models of type (15.3) the autocorrelation function, if it exists, has the form $\text{corr}(y_t, y_{t+s}) = acf(s) = e^{-\lambda s}$. This means that the rate of mean reversion is also the *memory time* of the process. This parameters governs the term structure of most quantities of interest in the model.

The Wiener process $(\hat{Z}_t)_{t\geq 0}$ can be correlated with the Brownian motion W driving the asset price. We denote by $\rho \in [-1, 1]$ the instantaneous correlation coefficient one can write

$$\hat{Z}_t = \rho W_t + \sqrt{1-\rho^2} Z_t, \tag{15.4}$$

where (Z_t) is a standard Brownian motion independent of (W_t). In most practical situations, ρ is taken to be a constant.

Some common driving processes (y_t) are

- Geometric Brownian motion [203]

$$dy_t = c_1 y_t dt + c_2 y_t d\hat{Z}_t, \tag{15.5}$$

- Gaussian Ornstein-Uhlenbeck [368]

$$dy_t = \alpha(\eta - y_t)dt + \beta d\hat{Z}_t, \tag{15.6}$$

- Cox-Ingersoll-Ross (CIR) [20, 196]

$$dy_t = \kappa(\eta - y_t)dt + v\sqrt{y_t} d\hat{Z}_t. \tag{15.7}$$

Note that the log-normal model is *not* mean-reverting: mean reversion is not an a priori constraint but it leads to realistic properties for the term structure of implied volatilities. The Gaussian OU process is mean-reverting but takes negative values, which means that one will have to apply a nonlinear transformation to it, in order to obtain a positive volatility process. Due to this reason, the model will become less mathematically tractable and option prices will be more difficult to compute. The most suitable choice from the three listed above is the CIR process because it is both mean-reverting and positive. A viable alternative to CIR is to take an Ornstein-Uhlenbeck process but drive it with a positive noise, e.g., a positive Lévy process. This approach will be discussed in Section 15.3.

When there is no correlation between the Brownian motions driving the volatility and the price process, Equation (15.2) can be rewritten using time changed Brownian motion representation:

$$\ln S_t = \tilde{\mu} t + W(v_t), \tag{15.8}$$

where $v_t = \int_0^t \sigma_s^2 ds$ is the integrated variance process. If the distribution of integrated variance is known, option prices may be computed as an average of Black-Scholes prices over this distribution. Therefore, the stochastic volatility model is mathematically tractable when the square of volatility is a linear function of the driving process (that is, $f(y) = \sqrt{y}$) because in this case the integrated variance is easy to compute. See [20] for a discussion of the role of integrated variance in option pricing.

In all cases, stochastic volatility models lead to distributions of increments with tails which are fatter normal distributions, even in absence of jumps, although precise estimates on tail decay may not be easy to obtain in specific models [150].

15.1.1 Implied volatility smiles

Stochastic volatility models generically lead to implied volatility patterns which exhibit smiles and skews: as shown by Renault and Touzi [332] when $(\sigma_t)_{t\geq 0}$ and $(W_t)_{t\geq 0}$ are independent, then the implied volatility profile $K \mapsto \Sigma_t(T, K)$ for fixed t, S_t, T is a "smile" that is, locally convex with a minimum around the forward price of the stock $K = S_t e^{r(T-t)}$. When the Wiener process driving σ_t is correlated with W_t the general analysis is more difficult, but numerous examples show that a negative correlation $\rho < 0$ leads to an implied volatility skew (downward sloping at the money) while $\rho > 0$ leads to an upward sloping implied volatility curve.

It is important to realize that, due to the symmetry of Brownian increments, the asymmetry of the (risk-neutral) distribution — therefore the implied volatility skew — is completely determined by this correlation coefficient ρ which plays a fundamental role for generating smiles and skews. A negative correlation $\rho < 0$ is often interpreted in terms of the so-called leverage effect, i.e., the empirical observation that large downward moves in the price are associated with upward moves in volatility and more generally with periods

TABLE 15.1: Stochastic volatility models without jumps

Model	Correlation	$f(y)$	y_t
Hull-White [203]	$\rho = 0$	$f(y) = \sqrt{y}$	Lognormal
Scott [358]	$\rho = 0$	$f(y) = e^y$	Gaussian OU
Stein-Stein [368]	$\rho = 0$	$f(y) = \|y\|$	Mean-reverting OU
Ball-Roma [20]	$\rho = 0$	$f(y) = \sqrt{y}$	CIR
Heston [196]	$\rho \neq 0$	$f(y) = \sqrt{y}$	CIR

of high volatility. The leverage effect is offered as an "explanation" of the downward sloping implied volatility skews. However as argued in Section 1.2, this does not explain why in some markets we observe symmetric smiles or even upward sloping smirks or why the smile effects were enhanced after the 1987 crash: does the leverage effect depend on markets or periods so negative correlation is more a "smile-fitting" constraint than an structural explanation for the smile.

But the performance of stochastic volatility models at short maturities is not very different from that of the Black-Scholes model, the effect of stochastic volatility becoming visible only at longer time scales: short-term skews cannot match empirically observed ones.

Table 15.1 gives some examples of bivariate diffusion models used for option pricing. It should be noted that, as long as $\rho = 0$, the model cannot generate an implied volatility skew and will always produce a symmetric implied volatility smile. Among these models, the Heston model is quite popular in practice. The choice of the square root (CIR) process as the instantaneous volatility leads to closed form expressions for characteristic functions of various quantities of interest in this model, enabling the use of Fourier transform methods such as the ones described in Section 11.1 for option pricing. We now discuss in more detail the CIR process, which represents the instantaneous volatility in the Heston model.

15.1.2 The square root process

The square root process, also known as the CIR process after Cox, Ingersoll and Ross. It is defined as the solution of the following stochastic differential equation:

$$y_t = y_0 + \lambda \int_0^t (\eta - y_s)ds + \theta \int_0^t \sqrt{y_s} dW_s. \tag{15.9}$$

The parameters λ, η and θ are positive constants. This process is continuous and positive because if it ever touches zero, the diffusion term disappears and the drift pushes the process in the positive direction. The precise behavior of the process near zero depends on the values of parameters: if $\theta^2 \leq 2\lambda\eta$, the process never touches zero; the drift term always pushes it away before it comes too close. In the opposite case the process will occasionally touch zero and reflect but this situation is less interesting for stochastic volatility modelling. The CIR process, like the Ornstein-Uhlenbeck process, is mean-reverting and possesses a stationary version.

Moments and correlation structure The mean of square root process can be computed by taking the expectation of (15.9):

$$E\{y_t\} = y_0 + \lambda \int_0^t (\eta - E\{y_s\})ds. \tag{15.10}$$

The solution of this equation is given by $E\{y_t\} = \eta + (y_0 - \eta)e^{-\lambda t}$, that is, η represents the long-term mean of the process and λ is responsible for the rate of mean-reversion. Other moments can be computed in a similar fashion. In particular,

$$\operatorname{Var} y_t = \frac{\theta^2\eta}{2\lambda} + \frac{\theta^2(y_0 - \eta)}{\lambda}e^{-\lambda t} + \frac{\theta^2(\eta - 2y_0)}{2\lambda}e^{-2\lambda t}.$$

In the limit $t \to \infty$ (or in the stationary case) the last two terms die out and the variance becomes equal to $\frac{\theta^2\eta}{2\lambda}$. This explains the role of θ, known as the "volatility of volatility" in this setting. In the stationary regime the autocorrelation function is given by $acf(s) = e^{-\lambda s}$. Therefore, λ can again be identified with the inverse correlation length or the inverse characteristic length of volatility clusters in this model.

Stationary distribution The infinitesimal generator of the CIR process is

$$Lf(x) = \frac{\theta^2 x}{2}\frac{\partial^2 f}{\partial x^2} + \lambda(\eta - x)\frac{\partial f}{\partial x}.$$

Denoting the distribution of y_t by μ_t, we can write, for a bounded C^2 function f, the Fokker-Planck equation:

$$\frac{d}{dt}\mu_t(f) = \mu_t(Lf),$$

where $\mu_t(f) \equiv \int f(x)d\mu_t$. In particular, if μ is the stationary distribution, it satisfies $\mu(Lf) = 0$. Choosing $f(x) = e^{iux}$, we come to the following equation for the characteristic function of μ:

$$\left(\frac{i\theta^2 u}{2} - \lambda\right)\hat{\mu}'(u) + i\lambda\eta\hat{\mu}(u) = 0.$$

Its solution satisfying $\mu(0) = 1$ has the form

$$\hat{\mu}(u) = \left(1 - \frac{i\theta^2 u}{2\lambda}\right)^{\frac{2\lambda\eta}{\theta^2}},$$

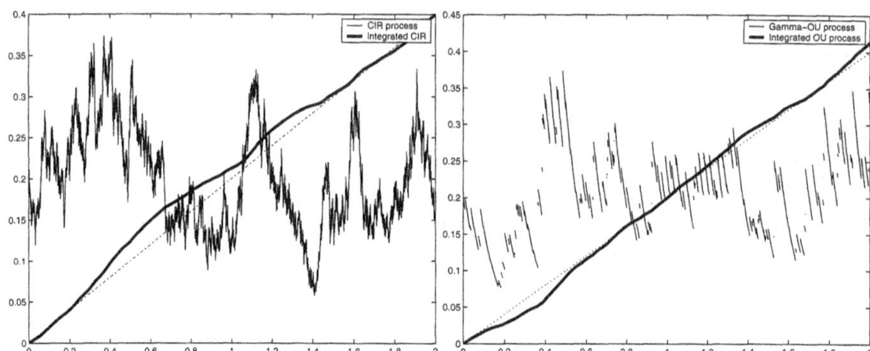

FIGURE 15.1: Left: A typical trajectory of CIR process (oscillating graph), the corresponding integrated CIR process (thick solid line) and the straight line corresponding to the mean of the integrated process (thin dotted line). Right: Positive Ornstein-Uhlenbeck process with the same correlation structure and the same stationary distribution.

where we recognize the characteristic function of a gamma distribution with parameters $c = \frac{2\lambda\eta}{\theta^2}$ and $\alpha = \frac{2\lambda}{\theta^2}$ (see Table 2.1).

Integrated CIR process In stochastic volatility models we are also interested in the integrated volatility process, given here by

$$Y_t = \int_0^t y_s ds.$$

This is an increasing process (because y_t is positive) and we can compute its mean by integrating that of y_t:

$$E\{Y_t\} = \int_0^t E\{y_s\}ds = \eta t + \frac{(y_0 - \eta)(1 - e^{-\lambda t})}{\lambda}.$$

The Laplace transform of Y_t is known in closed form [102]:

$$E\{e^{-uY_t}\} = \frac{\exp\left(\frac{\lambda^2\eta t}{\theta^2}\right)}{\left(\cosh\frac{\gamma t}{2} + \frac{\lambda}{\gamma}\sinh\frac{\gamma t}{2}\right)^{\frac{2\lambda\eta}{\theta^2}}} \exp\left(-\frac{2y_0 u}{\lambda + \gamma\coth\frac{\gamma t}{2}}\right),$$

where $\gamma = \sqrt{\lambda^2 + 2\theta^2 u}$.

Figure 15.1 (left graph) depicts a trajectory of CIR process with $y_0 = \eta = 0.2$, $\lambda = 10$ and $\theta = 0.7$ and the corresponding integrated CIR process (thick solid line). Notice how the latter line is smooth and how little it deviates from the straight line $y = \eta t$ (thin dotted line) despite the fact that instantaneous volatility varies between 5% and 40%.

15.2 A stochastic volatility model with jumps: the Bates model

As noted above, diffusion based stochastic volatility models cannot generate sufficient variability and asymmetry in short-term returns to match implied volatility skews for short maturities. The jump-diffusion stochastic volatility model introduced by Bates [41] copes with this problem by adding proportional log-normal jumps to the Heston stochastic volatility model. In the original formulation the model has the following form:

$$\frac{dS_t}{S_t} = \mu dt + \sqrt{V_t} dW_t^S + dZ_t, \tag{15.11}$$
$$dV_t = \xi(\eta - V_t)dt + \theta\sqrt{V_t} dW_t^V,$$

where (W_t^S) and (W_t^V) are Brownian motions with correlation ρ, driving price and volatility, and Z_t is a compound Poisson process with intensity λ and log-normal distribution of jump sizes such that if k is its jump size then $\ln(1+k) \sim N(\ln(1+\bar{k}) - \frac{1}{2}\delta^2, \delta^2)$. The no-arbitrage condition fixes the drift of the risk neutral process: under the risk-neutral probability $\mu = r - \lambda\bar{k}$. Applying Itô's lemma to Equation (15.11) we obtain the equation for the log-price $X_t = \ln S_t$:

$$dX_t = (r - \lambda\bar{k} - \frac{1}{2}V_t)dt + \sqrt{V_t} dW_t^S + d\tilde{Z}_t,$$

where $(\tilde{Z}_t)$ is a compound Poisson process with intensity λ and Gaussian distribution of jump sizes. This model can also be viewed as a generalization of the Merton jump-diffusion model (see Chapter 4) allowing for stochastic volatility. Although the no arbitrage condition fixes the drift of the price process, the risk-neutral measure is not unique, because other parameters of the model (for example, intensity of jumps and parameters of jump size distribution) can be changed without leaving the class of equivalent probability measures. Jumps in the log-price do not have to be Gaussian in this model. One can replace the Gaussian distribution by any other convenient distribution for the jump size without any loss of tractability, provided that the characteristic function is computable.

Option pricing In this stochastic volatility model as well as in other models of this chapter the characteristic function of the log-price is known in closed form (see below). Therefore, European options can be priced using one of the Fourier transform methods of Chapter 11. For path-dependent options closed-form expressions are not available and one must turn to numerical methods (e.g., Monte Carlo).

Characteristic function of the log-price Let us first compute the characteristic function of the continuous component X_t^c of X_t, following [196]. Let

$$f(x,v,t) = E\{e^{iuX_T^c} | X_t^c = x, V_t = v\}.$$

Applying Itô's formula to $M_t = f(X_t^c, V_t, t)$ yields

$$dM_t = \Big(\frac{1}{2}V_t\frac{\partial^2 f}{\partial x^2} + \rho\theta V_t\frac{\partial^2 f}{\partial x\partial v} + \frac{1}{2}\theta^2 V_t\frac{\partial^2 f}{\partial v^2} + (r - \lambda\bar{k} - \frac{V_t}{2})\frac{\partial f}{\partial x} \\ + \xi(\eta - V_t)\frac{\partial f}{\partial v} + \frac{\partial f}{\partial t}\Big)dt + \sqrt{v}\frac{\partial f}{\partial x}dW^S + \theta\sqrt{v}\frac{\partial f}{\partial v}dW^V.$$

Since $f(X_t^c, V_t, t)$ is a martingale we obtain, by setting the drift term to zero:

$$\frac{1}{2}v\frac{\partial^2 f}{\partial x^2} + \rho\theta v\frac{\partial^2 f}{\partial x\partial v} + \frac{1}{2}\theta^2 v\frac{\partial^2 f}{\partial v^2} + (r - \lambda\bar{k} - \frac{1}{2}v)\frac{\partial f}{\partial x} \\ + \xi(\eta - v)\frac{\partial f}{\partial v} + \frac{\partial f}{\partial t} = 0. \quad (15.12)$$

Together with the terminal condition $f(x,u,T) = e^{iux}$ this equation allows to compute the characteristic function of log-price. To solve it, we guess the functional form of f:

$$f(x,u,t) = \exp\{C(T-t) + vD(T-t) + iux\}, \quad (15.13)$$

where C and D are functions of one variable only. Substituting this into Equation (15.12), we obtain ordinary differential equations for C and D:

$$D'(s) = \frac{1}{2}\theta^2 D^2(s) + (i\rho\theta u - \xi)D(s) - \frac{u^2 + iu}{2}, \\ C'(s) = \xi\eta D(s) + iu(r - \lambda\bar{k})$$

with initial conditions $D(0) = C(0) = 0$. These equations can be solved explicitly:

$$D(s) = -\frac{u^2 + iu}{\gamma\coth\frac{\gamma s}{2} + \xi - i\rho\theta u}, \\ C(s) = ius(r - \lambda\bar{k}) + \frac{\xi\eta s(\xi - i\rho\theta u)}{\theta^2} \\ - \frac{2\xi\eta}{\theta^2}\ln\left(\cosh\frac{\gamma s}{2} + \frac{\xi - i\rho\theta u}{\gamma}\sinh\frac{\gamma s}{2}\right),$$

where $\gamma = \sqrt{\theta^2(u^2 + iu) + (\xi - i\rho\theta u)^2}$. The characteristic function without the jump term can now be found from Equation (15.13). To incorporate the jump term, since jumps are homogeneous and independent from the continuous part, we need only to multiply the characteristic function that we have obtained by the characteristic function of the jumps which in this case is simply

$$\phi_t^J(u) = \exp\{t\lambda(e^{-\delta^2 u^2/2 + i(\ln(1+\bar{k}) - \frac{1}{2}\delta^2)u} - 1)\}.$$

Finally, the characteristic function of the price process in the model of Bates is

$$\phi_t(u) = \phi_t^J(u) \frac{\exp\left(\frac{\xi\eta t(\xi - i\rho\theta u)}{\theta^2} + iut(r - \lambda\bar{k}) + iux_0\right)}{\left(\cosh\frac{\gamma t}{2} + \frac{\xi - i\rho\theta u}{\gamma}\sinh\frac{\gamma t}{2}\right)^{\frac{2\xi\eta}{\theta^2}}} \times \exp\left(-\frac{(u^2 + iu)v_0}{\gamma\coth\frac{\gamma t}{2} + \xi - i\rho\theta u}\right) \quad (15.14)$$

with γ defined above.

Implied volatility patterns in the Bates model Some implied volatility smiles attainable in the Bates model are shown in Figure 15.2. In this model there are two ways to generate an implied volatility skew. The first way is to introduce a (negative) correlation between returns and volatility movements, as in diffusion-based stochastic volatility models. Alternatively, an implied volatility skew for short-term options can be generated by asymmetric jumps (as in exp-Lévy models) even if the noise sources driving volatility and returns are independent.

Therefore the "intuition" — originating from bivariate diffusion models — that the implied volatility skew is systematically linked to a "leverage effect" is groundless: it is simply due to the symmetry (stemming from normality) of Brownian increments and gives yet another example of a property specific to the Brownian universe.

We have seen that correlation and jumps have similar effect on the implied volatility smile; is there any feature which allows to distinguish them? The answer is yes: jumps and correlation both induce an implied volatility skew but they influence the term structure of volatility differently. It is clear from Figure 15.2 that the smiles that are due to jumps are stronger at short maturities and flatten out much faster as the time to maturity increases. In contrast, smiles due to correlation can be used to obtain a skew at longer maturities but are not sufficient to explain the prices of short-maturity options. Also, introducing jumps increases the overall level of implied volatility while correlation has little effect on it. Finally, at-the-money volatility stays roughly the same in absence of jumps and tends to increase for longer maturities when the jumps are present.

These remarks shed light on a very nice feature of the Bates model: here the implied volatility patterns for long-term and short-term options can be adjusted separately. One can start by calibrating the jumps on one or two shortest maturities and then fix the jump parameters and calibrate the other ones on longer maturities. This approach will not give the best possible calibration quality but yields reasonable results. If instead of using this procedure all parameters are fitted at the same time with least-squares method, the cost function will not be convex and one may end up in a local minimum with strange parameter values [97].

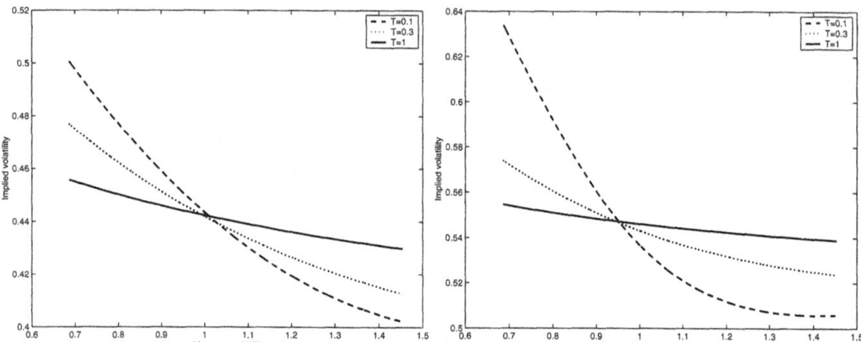

FIGURE 15.2: Implied volatility patterns in the Bates model. Left graph shows the smiles for various maturities when the jumps are absent but there is a correlation between the volatility process and the price process ($\rho = -0.5$). On the right graph the correlation is absent and the skew is due to jumps (with intensity $\lambda = 5$). The other parameters are the same for both graphs: $S_0 = 1$, $r = 0$, $V_0 = 0.2$, $\theta = 0.7$ (volatility of volatility), $\xi = 10$ (inverse correlation length), $\eta = 0.2$ (mean square volatility), $\delta = 0.1$ (jump dispersion) and $\bar{k} = -0.1$ (average relative jump size).

15.3 Non-Gaussian Ornstein-Uhlenbeck processes

Using Lévy processes as driving noise, one can construct a large family of mean-reverting jump processes with linear dynamics on which various properties such as positiveness or the choice of a marginal distribution can be imposed. We will call these processes, which are non-Gaussian generalizations of the Gaussian Ornstein-Uhlenbeck process, non-Gaussian Ornstein-Uhlenbeck processes or simply OU processes. These processes are not only convenient to model "volatility" but also have an independent interest for modelling stationary time series of various kinds such as commodity prices or interest rates [36, 307].

15.3.1 Definition and properties

Given a Lévy process (Z_t), define $(y_t)_{t\geq 0}$ as

$$y_t = y_0 e^{-\lambda t} + \int_0^t e^{\lambda(s-t)} dZ_s, \tag{15.15}$$

where the last term is the stochastic integral, defined as in Section 8.1.4, of the (nonrandom) function $e^{\lambda(s-t)}$: if (Z_t) has characteristic triplet (σ^2, ν, γ) then

$$\int_0^t e^{\lambda(s-t)} dZ_s = \gamma \int_0^t e^{\lambda(s-t)} ds + \sigma \int_0^t e^{\lambda(s-t)} dW_s$$
$$+ \int_0^t \int_{|x|\leq 1} e^{\lambda(s-t)} \tilde{J}_Z(ds \times dx) + \int_0^t \int_{|x|>1} e^{\lambda(s-t)} J_Z(ds \times dx).$$

If Z is a Brownian motion, y is the Gaussian Ornstein-Uhlenbeck process. If Z is any other Lévy process, y is a jump process which is a non-Gaussian analogue of Ornstein-Uhlenbeck process. y will be called an OU process driven by Z; Z is called the background driving Lévy process (BDLP) of y.

The process (y_t) verifies the linear "stochastic differential equation":

$$y_t = y_0 - \lambda \int_0^t y_s ds + Z_t. \tag{15.16}$$

Its local behavior is described as the sum of a linear damping term and a random term described by the increments of Z_t:

$$dy_t = -\lambda y_t + dZ_t.$$

Thus, the jumps of y_t are the same as the jumps of Z_t but "between" two jumps of Z_t, y_t decays exponentially due to the linear damping term[3]. A trajectory of a Lévy-driven OU process is shown in Figure 15.1 (right graph). The BDLP is in this case a compound Poisson subordinator (with positive jumps), so the trajectory consists of upward jumps with periods of downward exponential decay between them. This results in a more realistic *asymmetric* behavior for "volatility" than in diffusion models: volatility jumps up suddenly but simmers down gradually.

Characteristic function and characteristic triplet To compute the characteristic function of process (15.16) we need the following simple lemma.

[3] Of course this description is only correct if jumps occur in finite number, not for an infinite activity Lévy process Z_t.

LEMMA 15.1
Let $f : [0,T] \to \mathbb{R}$ be left-continuous and $(Z_t)_{t\geq 0}$ be a Lévy process. Then

$$E\left\{\exp\left(i\int_0^T f(t)dZ_t\right)\right\} = \exp\left\{\int_0^T \psi(f(t))dt\right\}, \qquad (15.17)$$

where $\psi(u)$ is the characteristic exponent of Z.

PROOF First consider a piecewise constant left-continuous function $f(t) = \sum_{i=1}^N f_i 1_{]t_{i-1},t_i]}(x)$. For this function $\int f(t)dZ_t = \sum_{i=1}^N f_i(Z_i - Z_{i-1})$ and

$$E\left\{e^{i\int f_t dZ_t}\right\} = \prod_{i=1}^N E\left\{e^{if_i(Z_{t_i}-Z_{t_{i-1}})}\right\} = \prod_{i=1}^N e^{(t_i-t_{i-1})\psi(f_i)} = e^{\int \psi(f(t))dt}.$$

This equality can now be extended by approximation to an arbitrary left-continuous function f. □

Applying the lemma to Equation (15.15), we find the characteristic function of y_t:

$$E\{e^{iuy_t}\} = \exp\{iuy_0e^{-\lambda t} + \int_0^t \psi(ue^{\lambda(s-t)})ds\}. \qquad (15.18)$$

In particular, if Z is a Brownian motion with drift μ and volatility σ, its characteristic exponent is $\psi(u) = i\mu u - \sigma^2u^2/2$ and the characteristic function of y_t becomes

$$E\{e^{iuy_t}\} = \exp\left\{iu[y_0e^{-\lambda t} + \frac{\mu}{\lambda}(1-e^{-\lambda t})] - \frac{\sigma^2u^2}{4\lambda}(1-e^{-2\lambda t})\right\},$$

which shows that in this case y_t is a Gaussian random variable. Its stationary distribution is also Gaussian, with mean μ/λ and variance $\frac{\sigma^2u^2}{4\lambda}$.

In the general case formula (15.18) allows to show that the distribution of y_t is infinitely divisible and to compute its characteristic triplet.

PROPOSITION 15.1
Let $(Z_t)_{t\geq 0}$ be a Lévy process with characteristic triplet (A,ν,γ). The distribution of y_t, defined by Equation (15.16), is infinitely divisible for every t and has characteristic triplet $(A_t^y, \nu_t^y, \gamma_t^y)$ with

$$A_t^y = \frac{A}{2\lambda}\{1-e^{-2\lambda t}\},$$

$$\gamma_t^y = \frac{\gamma}{\lambda}\{1-e^{-\lambda t}\} + y_0e^{-\lambda t},$$

$$\nu_t^y = \int_1^{e^{\lambda t}} \nu(\xi B)\frac{d\xi}{\lambda\xi}, \quad \forall B \in \mathcal{B}(\mathbb{R}),$$

where ξB is a shorthand notation for $\{\xi x, x \in B\}$.

PROOF First, let us show that ν_t^y is a Lévy measure. To see this observe that first, it is a positive measure, second

$$\nu_t^y([1,\infty)) = \int_1^{e^{\lambda t}} \nu([\xi,\infty))\frac{d\xi}{\lambda\xi} \leq \frac{e^{\lambda t}-1}{\lambda}\nu([1,\infty)) < \infty,$$

third, $\nu_t^y((-\infty,-1]) < \infty$ by the same reasoning, and finally

$$\int_{-1}^{1} z^2\nu_t^y(dz) = \int_{-1}^{1} z^2 \int_1^{e^{\lambda t}} \nu(\xi dz)\frac{d\xi}{\lambda\xi} = \int_1^{e^{\lambda t}} \frac{d\xi}{\lambda\xi^3}\int_{-\xi}^{\xi} z^2\nu(dz) < \infty.$$

Substituting ψ from the Lévy-Khinchin formula into (15.18) and after some simple transformations, we obtain:

$$E\{e^{iuy_t}\} = \exp\{iuy_0e^{-\lambda t} + \frac{i\gamma u}{\lambda}\{1-e^{-\lambda t}\} - \frac{Au^2}{4\lambda}\{1-e^{-2\lambda t}\} + \int_1^{e^{\lambda t}}\frac{d\xi}{\lambda\xi}\int_{-\infty}^{\infty}\nu(\xi dz)(e^{iuz}-1-iuz1_{|z|\leq 1})\}. \quad (15.19)$$

Since ν_t^y is a Lévy measure,

$$\int_{-\infty}^{\infty}\left|e^{iuz}-1-iuz1_{|z|\leq 1}\right|\int_1^{e^{\lambda t}}\nu(\xi dz)\frac{d\xi}{\lambda\xi} < \infty,$$

which justifies the interchange of integrals in the last term of (15.19). Performing this operation, we see that the last term of (15.19) becomes

$$\int_{-\infty}^{\infty}(e^{iuz}-1-iuz1_{|z|\leq 1})\int_1^{e^{\lambda t}}\nu(\xi dz)\frac{d\xi}{\lambda\xi}.$$

This completes the proof of proposition. □

In particular, this proposition shows that ν_t^y has a density

$$\nu_t^y(x) = \frac{\nu([x, xe^{\lambda t}))}{\lambda x}, \text{ if } x>0 \quad\text{and}\quad \nu_t^y(x) = \frac{\nu((xe^{\lambda t}, x])}{\lambda|x|}, \text{ if } x<0.$$

The finiteness of moments and the tail index of y_t are related to the moments/tail index of the driving Lévy process Z:

PROPOSITION 15.2 Moments and correlation structure

The n-th absolute moment of the Ornstein-Uhlenbeck process $(y_t)_{t\geq 0}$ exists for every $t>0$ or, equivalently, for $t>0$ if the n-th absolute moment of the driving Lévy process $(Z_t)_{t\geq 0}$ exists.

The reader can easily prove this proposition by comparing the Lévy measure of y_t (given in Proposition 15.1) to the Lévy measure of Z_t and applying Proposition 3.13.

In particular, if Z_t is square integrable, the variance of y_t is well defined, and we can, placing ourselves in the stationary case, compute the autocorrelation function $acf(s) = \frac{\text{Cov}(y_t, y_{t+s})}{\sqrt{\text{Var}\, y_t}\sqrt{\text{Var}\, y_{t+s}}}$. Equation (15.15) implies that

$$E\{y_t | y_0\} = y_0 e^{-\lambda t} + \frac{E\{Z_1\}}{\lambda}(1 - e^{-\lambda t}).$$

In the stationary case $E\{y_t\} = \lambda^{-1} E\{Z_1\}$ for all t and we can write

$$\begin{aligned}\text{Cov}(y_t, y_{t+s}) &= E\{y_t y_{s+t}\} - E\{y_t\}^2 = E\{y_t E[y_{s+t} | y_t]\} - \lambda^{-2} E\{Z_1\}^2 \\ &= e^{-\lambda s} E\{y_t^2\} - \lambda^{-1} E\{Z_1\} E\{y_t\} e^{-\lambda s} = \text{Var}\, y_t e^{-\lambda s}.\end{aligned}$$

Therefore $acf(s) = e^{-\lambda s}$. This shows that λ^{-1} plays the role of a correlation time for OU processes.

By adding together independent OU processes, one can obtain richer correlation structures, including long-range dependence, but at the price of increasing model dimension. Let $(y_t)_{t \geq 0}$ and $(\tilde{y}_t)_{t \geq 0}$ be independent OU processes. The correlation function of $y + \tilde{y}$ is a convex combination of the correlation functions of the two processes:

$$acf(s) = \frac{e^{-\lambda s}\, \text{Var}\, y_t + e^{-\tilde{\lambda} s}\, \text{Var}\, \tilde{y}_t}{\text{Var}\, y_t + \text{Var}\, \tilde{y}_t}.$$

Adding together a large, possibly infinite, number of OU processes, one can obtain every correlation function that is in the convex envelope of $\{e^{-\lambda t}, \lambda > 0\}$, that is, every decreasing completely monotonous function f satisfying $f(0) = 1$. This allows to obtain arbitrary term structures for implied volatility patterns but at the price of adding more factors. Superpositions of OU processes are discussed in [36].

Simulation Lévy driven Ornstein-Uhlenbeck processes can be simulated using Equation (15.15). For example, if a series representation for Z_t is available: $Z_t = \sum_{i=1}^{\infty} A_i 1_{V_i \leq t}$ then y_t can be represented as

$$y_t = y_0 e^{-\lambda t} + \sum_{i=1}^{\infty} A_i e^{\lambda(V_i - t)} 1_{V_i \leq t}. \tag{15.20}$$

15.3.2 Stationary distributions of OU processes (*)

Ornstein-Uhlenbeck processes are closely linked to a class of infinitely divisible distributions called *self-decomposable* distributions. The class of self-decomposable distributions (also called class L) contains all distributions μ

such that for all $b > 1$ there exists a distribution ρ_b with $\hat{\mu}(z) = \hat{\mu}(z/b)\hat{\rho}_b(z)$, where $\hat{\mu}$ denotes the characteristic function of μ. In other words, X is self-decomposable if for every $b > 0$ there exists a random variable ε_b independent from X such that $b^{-1}X + \varepsilon_b$ and X have the same law. The class of self-decomposable distributions can therefore be seen as the class of stationary distributions of AR(1) (autoregressive of order 1) processes.

Self-decomposability can be characterized in terms of Lévy measure as follows (we only give the statement in one-dimensional case; for a multidimensional version and a proof of this theorem see [345, theorem 15.10]).

PROPOSITION 15.3 Self-decomposable distributions

An infinitely divisible distribution on $\mathbb{R}$ is self-decomposable if and only if its Lévy measure has a density $\nu(x) = \frac{k(x)}{|x|}$ with k positive, increasing on $(-\infty, 0)$ and decreasing on $(0, +\infty)$.

The main property of self-decomposable distributions that makes them interesting in the context of stochastic volatility modelling is that the class of self-decomposable distributions coincides with the class of stationary distributions of Lévy-driven OU processes. This makes it possible to construct an OU process with prescribed stationary distribution.

PROPOSITION 15.4

Let $(Z_t)_{t\geq 0}$ be a Lévy process with characteristic triplet (A, ν, γ). If

$$\int_{|x|\geq 1} \ln|x|\nu(dx) < \infty \tag{15.21}$$

then the OU process $(y_t)_{t\geq 0}$ defined by Equation (15.16) has a stationary distribution μ which is self-decomposable, has characteristic exponent

$$\psi^y(u) = \lim_{t\to+\infty} \psi^y_t(u) = \lim_{t\to+\infty} \int_0^t \psi(ue^{\lambda(s-t)})ds \tag{15.22}$$

and Lévy triplet (A^y, ν^y, γ^y) with $A^y = \frac{A}{2\lambda}$, $\gamma^y = \frac{\gamma}{\lambda}$ and

$$\nu^y = \int_1^\infty \nu(\xi B)\frac{d\xi}{\lambda\xi}, \quad \forall B \in \mathcal{B}(\mathbb{R}). \tag{15.23}$$

Conversely, for every self-decomposable distribution μ there exists a Lévy process $(Z_t)_{t\geq 0}$ such that μ is the stationary distribution of the OU process driven by Z.

PROOF The direct statement. The first step is to show that (15.23) defines a Lévy measure. Observe that on one hand

$$\nu^y([1,\infty)) = \int_1^\infty \nu([\xi,\infty))\frac{d\xi}{\lambda\xi} = \frac{1}{\lambda}\lim_{\xi\to\infty} \ln(\xi)\nu([\xi,\infty)) + \frac{1}{\lambda}\int_1^\infty \ln(\xi)\nu(d\xi),$$

which is finite by theorem condition, and on the other hand

$$\int_0^1 z^2\nu^y(dz) = \int_0^1 z^2 \int_1^\infty \nu(\xi dz)\frac{d\xi}{\lambda\xi}$$
$$= \int_1^\infty \frac{d\xi}{\lambda\xi^3}\int_0^\xi z^2\nu(dz) \le C + \int_1^\infty \frac{\nu([1,\xi))}{\lambda\xi}d\xi < \infty$$

for some finite constant C by the theorem's condition.

The second step is to prove that ν^y is a Lévy measure of a self-decomposable distribution. For every $x > 0$ we have $\nu^y([x,\infty)) = \int_x^\infty \nu([z,\infty))\frac{dz}{\lambda z}$. This means that ν^y has a density $\nu^y(x) = \frac{\nu([x,\infty)}{\lambda x}$ which is a Lévy density of a self-decomposable distribution by Proposition 15.3.

The third step is to show that ψ^y is a characteristic exponent of an infinitely divisible distribution with prescribed triplet. From the proof of Proposition 15.1, we know that

$$\psi_t^y(u) = iuy_0e^{-\lambda t} + \frac{i\gamma u}{\lambda}\{1-e^{-\lambda t}\} - \frac{Au^2}{4\lambda}\{1-e^{-2\lambda t}\}$$
$$+ \int_{-\infty}^\infty (e^{iuz} - 1 - iuz1_{|z|\le 1})\int_1^{e^{\lambda t}} \nu(\xi dz)\frac{d\xi}{\lambda\xi}.$$

Since we have already shown that ν^y is a Lévy measure we can compute the limit and, using Lebesgue's dominated convergence theorem, obtain the desired result.

Finally let us prove that μ is the stationary distribution of (y_t). Let y_0 have the distribution μ. Then

$$E\{e^{iuy_t}\} = \int E\{e^{iuy_t}|y_0\}\mu(dy_0)$$
$$= \exp\left\{\int_0^t \psi(ue^{\lambda(s-t)})ds\right\}\int e^{iuy_0e^{-\lambda t}}\mu(dy_0)$$
$$= \exp\left\{\int_0^t \psi(ue^{\lambda(s-t)})ds + \lim_{r\to+\infty}\int_0^r \psi(ue^{\lambda(s-r-t)})ds\right\} = \hat{\mu}(u).$$

The converse statement. Equation (15.23) shows that ν^y has a density

$$\nu^y(x) = \frac{\nu([x,\infty))}{\lambda x}, \text{ if } x > 0, \tag{15.24}$$
$$\nu^y(x) = \frac{\nu((-\infty,x])}{\lambda|x|}, \text{ if } x < 0. \tag{15.25}$$

If the stationary distribution is known, these formulae allow to back out ν. It is easy to check that it will be a Lévy measure and that condition (15.21) will be satisfied. □

15.3.3 Positive Ornstein-Uhlenbeck processes

Positive OU processes present a particular interest in the context of stochastic volatility modelling. One can, of course, construct a positive process from every OU process, by exponentiating or squaring it, but if the OU process itself is positive, the model is linear and therefore more mathematically tractable. Positive OU processes were used for stochastic volatility modelling by Barndorff-Nielsen and Shephard [36, 38].

From Proposition 15.1 it is clear that a process of OU type is almost surely positive for every t if and only if $y_0 \geq 0$ and the driving Lévy process Z_t is a subordinator. In this case Barndorff-Nielsen and Shephard suggest to use the positive OU process to represent directly the square of volatility process:

$$d\sigma^2(t) = -\lambda\sigma^2(t)dt + dZ_t. \tag{15.26}$$

For various computations in stochastic volatility models it is important to characterize the law of integrated variance process $v_t = \int_0^t \sigma^2(s)ds$. The characteristic function of integrated OU process can be easily evaluated owing to the linear structure of the process. From Equations (15.16) and (15.15) we find:

$$v_t \equiv \int_0^t y_s ds = \frac{y_0}{\lambda}(1 - e^{-\lambda t}) + \frac{1}{\lambda}\int_0^t (1 - e^{\lambda(s-t)})dZ_s. \tag{15.27}$$

Therefore, the Laplace transform (we are using Laplace transform now because the process is positive) of v_t can be computed in the same way as the characteristic function of y_t and we obtain for $u < 0$:

$$E\{e^{uv_t}\} = \exp\left\{\frac{uy_0}{\lambda}(1 - e^{-\lambda t}) + \int_0^t l\left(\frac{u}{\lambda}(1 - e^{\lambda(s-t)})\right) ds\right\}, \tag{15.28}$$

where $l(u) = E[e^{uZ_1}]$. Sometimes we will also need the joint distribution of v_t and Z_t. Its Laplace transform is also easy to compute:

$$E[e^{uv_t + vZ_t}] = \exp\left\{\frac{uy_0}{\lambda}(1 - e^{-\lambda t}) + \int_0^t l(v + \frac{u}{\lambda}(1 - e^{\lambda(s-t)}))ds\right\}. \tag{15.29}$$

Example 15.1 OU process with gamma stationary distribution

The stationary measure of the square root process, discussed in the next section, is the gamma distribution. In this respect it is interesting to consider an OU process which admits a gamma distribution as its stationary measure and compare the corresponding stochastic volatility models based on the two processes. In particular, this will allow us to see, whether, in addition to marginal properties of σ_t, jumps in the volatility are important.

Suppose that the stationary measure μ is the gamma distribution with density $\mu(x) = \frac{\alpha^c}{\Gamma(c)}x^{c-1}e^{-\alpha x}1_{x\geq 0}$. This means that the Lévy measure ν^y

has density $\nu^y(x) = \frac{ce^{-\alpha x}}{x} 1_{x>0}$ and the drift is zero. Using Proposition 15.4 and formulae (15.24) and (15.25), we find that in this case (Z_t) has zero drift and Lévy measure with density $\nu(x) = \alpha\lambda c e^{-\alpha x}$, that is, Z is in this case a compound Poisson subordinator with exponential jump size distribution. The Laplace exponent of Z_t is $l(u) = \frac{\lambda c u}{\alpha - u}$.

In this case, the Laplace transform of the integrated variance process, given by formula (15.28), may be explicitly computed:

$$E\{e^{uv_t}\} = \exp\left\{\frac{uy_0}{\lambda}(1-e^{-\lambda t}) + \frac{ut\lambda c}{\alpha\lambda - u} + \frac{\lambda\alpha c}{\alpha\lambda - u}\ln\left(1 - \frac{u}{\alpha\lambda}(1-e^{-\lambda t})\right)\right\}.$$

□

15.4 Ornstein-Uhlenbeck stochastic volatility models

In this section we discuss the stochastic volatility models was proposed by Barndorff-Nielsen and Shephard [36, 38] where the squared volatility process is a Lévy-driven positive Ornstein-Uhlenbeck process described by Equation (15.26). The key points of this approach are the presence of jumps in the volatility, flexibility in the choice of marginal distributions for volatility and the availability of closed form expression for characteristic functions of integrated volatility and returns.

This model can be used both for statistical analysis of price series at scales ranging from several minutes to several days and for option pricing purposes. Under the historical probability the statistical model has the form

$$\begin{aligned} S_t &= S_0 \exp(X_t), \\ dX_t &= (\mu + \beta\sigma_t^2)dt + \sigma_t dW_t + \rho dZ_t, \\ d\sigma_t^2 &= -\lambda\sigma_t^2 dt + dZ_t, \quad \sigma_0^2 > 0, \end{aligned} \tag{15.30}$$

where $\rho \leq 0$, $\lambda > 0$, $(W_t)_{t\geq 0}$ is a standard Brownian motion and $(Z_t)_{t\geq 0}$ is the background driving Lévy process, a subordinator without drift, that has Laplace exponent $l(u) = E\{e^{uZ_1}\}$ and is independent from W. The term $\beta\sigma^2$ in Equation (15.30) corresponds to the volatility risk premium and the last term ρdZ_t accounts for the leverage effect. In the sequel we will call such models BNS models.

In the BNS model (15.30), the "volatility" σ_t increases by jumps and decays exponentially between two jumps. When $\rho \neq 0$, each jump in the volatility is associated to a jump in the price process, whose size is proportional to the size of volatility jump. The jumps can be interpreted as the arrival new information on the market, triggering an increase in volatility and a simultaneous fall of asset price. When $\rho = 0$, the volatility still moves by jumps but the price process has continuous sample paths. Note that, when $\rho \neq 0$ it is not correct to interpret σ_t as "volatility" since the returns are also affected by the increments of Z_t which therefore contributes to the volatility of returns.

To compute the characteristic function of the log-price process X_t, we first condition on the trajectory of Z and then use the formula (15.29) that gives the joint Laplace transform of integrated variance and driving Lévy process Z_t.

$$\begin{aligned}
E\{e^{iuX_t}\} &= E\left\{\exp\left(iu(\mu t + \beta\int_0^t \sigma_r^2 dr + \int_0^t \sigma_r dW_r + \rho Z_t)\right)\right\} \\
&= E\left\{\exp\left(iu(\mu t + \beta\int_0^t \sigma_r^2 dr + \rho Z_t) - \frac{1}{2}u^2\int_0^t \sigma_r^2 dr\right)\right\} \\
&= \exp\Big\{iu\mu t + i\sigma_0^2(iu\beta - \frac{u^2}{2})\varepsilon(\lambda, t) \\
&\quad + \int_0^t l\left(i\rho u + (iu\beta - \frac{u^2}{2})\varepsilon(\lambda, t-s)\right) ds\Big\}, \qquad (15.31)
\end{aligned}$$

where $\varepsilon(\lambda, t) \equiv \frac{1-e^{-\lambda t}}{\lambda}$.

Option pricing and calibration Barndorff-Nielsen and Shephard [37] have shown that the model (15.30) is arbitrage-free, that is, there exists an equivalent probability measure, under which the discounted price process is a martingale. Moreover, there exist many structure preserving equivalent martingale measures, that is, measures under which the discounted price process is a martingale and has the form (15.30) with some other values of parameters. The issues of measure change for models of type (15.30) are addressed in detail in [308]. Under one such structure preserving martingale probability the model (15.30) has the form

$$\begin{aligned}
S_t &= \exp(X_t), \\
dX_t &= (r - l(\rho) - \frac{1}{2}\sigma_t^2)dt + \sigma_t dW_t^Q + \rho dZ_t, \qquad (15.32) \\
d\sigma_t^2 &= -\lambda\sigma_t^2 dt + dZ_t^Q, \quad \sigma_0^2 > 0,
\end{aligned}$$

where r is the interest rate and the superscript Q refers to the fact that W_t^Q and Z_t^Q are, respectively, standard Brownian motion and subordinator with Laplace exponent $l(u)$ under the risk-neutral probability. This specification may be regarded as a risk-neutral specification of the BNS model and be used as a starting point for option pricing and calibration.

European option prices can be computed as discounted expectations of payoff under the risk-neutral probability, using one of the Fourier transform methods of Chapter 11.

Example 15.2 BNS model with gamma stationary distribution for volatility
As in Example 15.1, suppose that the volatility process has gamma stationary distribution. In this case the integral in (15.31) can be computed explicitly and we obtain (in the risk-neural case with $\mu = r - l(\rho)$ and $\beta = -\frac{1}{2}$)

$$\ln E\{e^{iuX_t}\} = iut\left\{r - \frac{\lambda c\rho}{\lambda - \rho}\right\} - \sigma_0^2 \varepsilon(\lambda, t)\frac{u^2 + iu}{2}$$
$$+ \frac{\alpha\lambda c}{\lambda(\alpha - iu\rho) + \frac{u^2+iu}{2}} \ln\left\{1 + \frac{u^2 + iu}{2}\frac{\varepsilon(\lambda, t)}{\alpha - iu\rho}\right\}$$
$$+ \lambda ct\frac{2i\lambda u\rho - u^2 + iu}{2\lambda(\alpha - iu\rho) + u^2 + iu}, \tag{15.33}$$

where $\varepsilon(\lambda, t) = \frac{1-e^{-\lambda t}}{\lambda}$. Figure 15.3 depicts the implied volatility smiles computed in this model with $\rho = 0$ (left graph) and $\rho = -0.5$ (right graph). When $\rho = 0$, the smiles are not very pronounced, they are symmetric for all maturities and quickly flatten out. A nonzero value of ρ has the effect of making the smiles stronger and asymmetric. Note however that in this model a single parameter (ρ) determines both the size of jumps in the price process (which governs the intensity of the smile effect) and the correlation between price and volatility, that is responsible for the asymmetry of the smile for all maturities. In other words, if ρ is zero, the smile is both symmetric and flat, and if ρ is nonzero, the smile is both skewed and strong. This strong relationship between jumps and volatility reduces the flexibility of calibration and makes it difficult to calibrate the jump component and volatility separately. For this reason the calibration quality in this framework may be worse than the quality achieved using the Bates model or the time changed Lévy models of the next section.

□

15.5 Time changed Lévy processes

In the Bates model the leverage effect and skew for long maturities were achieved using correlated sources of randomness in the price process and the instantaneous volatility. The sources of randomness are thus required to be Brownian motions. In the BNS model the leverage effect and skew were generated using the same jumps in the price and volatility. This does not require the sources of randomness to be Brownian motions but imposes some restrictions on attainable smiles and term structures. A third way to achieve leverage and long-term skew is to make the volatility govern the time scale of the Lévy process driving jumps in the price. This idea leads to the following

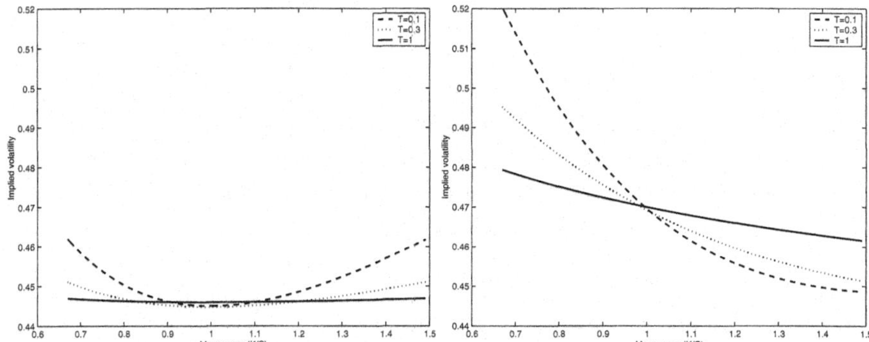

FIGURE 15.3: Implied volatility patterns in the BNS model with OU-driven stochastic volatility process with gamma stationary distribution. Prices were computed with the following parameter values: $\lambda = 10$, $\sigma_0^2 = 0.2$, $c = 8.16$, $\alpha = 40.8$ and zero interest rate. Left graph corresponds to $\rho = 0$ and right graph to $\rho = -0.5$

generic model:

$$S_t = \exp(X_t) = \exp(Y(v_t)), \tag{15.34}$$
$$v_t = \int_0^t \sigma_s^2 ds,$$

where $(Y_t)_{t\geq 0}$ is a Lévy process (with or without Brownian part). The leverage effect is preserved because in the regions of high volatility 'business time' flows faster and price jumps occur at a higher rate.

Introducing stochastic volatility into an exponential-Lévy model via time change was suggested by Carr et al. [81]. These authors argue that one should use Lévy processes without Brownian component. However, when jumps are considered to be rare events, processes with Brownian component may offer a more realistic vision of price moves (compare with discussions in Chapter 4). In the rest of this section we partially follow the treatment of Carr et al. [162].

The formulation (15.34) implies that the time change is continuous. This rules out the subordination procedure, discussed in Chapter 4, but has important implications and advantages. Winkel [384] shows that if the price process is continuously observable, Y is not a compound Poisson process and the time change does not have jumps, then the volatility can be reconstructed from the path of price process. In other words, the filtration generated by the price process alone coincides with the filtration generated by the price and volatility.

The representation 15.34 allows to write the characteristic function of the log-price process.

$$E\{e^{iuX_t}\} = e^{\psi_v(t,\psi_Y(u))},$$

where $\psi_Y(u)$ is the characteristic exponent of the Lévy process $(Y_t)_{t\geq 0}$ and $\psi_v(t,u)$ is the Laplace exponent of integrated variance process $(v_t)_{t\geq 0}$, that is, $\psi_v(t,u) = \ln E\{e^{uv_t}\}$.

To complete the description of the model we must address three questions: which Lévy processes are to be used, which volatility processes are best suited for this model and finally, how can one construct martingale versions of models of this type for arbitrage-free option pricing.

The first question is the easiest one: this model can be used to improve the performance of any of the models of Chapter 4 because for all of them characteristic exponent is known and the characteristic function of time changed process can be computed. Among infinite-activity models, the normal inverse Gaussian process and the tempered stable process are good choices: the variance gamma process has a nonsmooth density at short maturities and the generalized hyperbolic Lévy motion leads to complicated formulae involving special functions. In finite-activity setup Kou's double exponential model may be even better because it allows to control the relative sizes of positive and negative jumps.

The volatility process should, as in the diffusion setting, be positive and mean-reverting. If one wishes to compute option prices via Fourier transform its Laplace/Fourier exponent must be known in closed form. Both positive OU process and CIR process can be used; the question is whether it makes any difference. At first sight, the trajectories of the two processes are very different, even if they have the same correlation structure and same stationary distribution (see Figure 15.1). However, the price process in time changed Lévy models depends on "volatility" only through the integrated variance. The left graph in Figure 15.4 shows the distributions of integrated variance at two-year horizon for OU-driven stochastic volatility with gamma stationary distribution and for CIR stochastic volatility. The parameters of the two processes are the same as in Figure 15.1. The two curves are almost indistinguishable which means that prices processes with these two volatility patterns will not differ much. However unlike the CIR process which is a nonlinear diffusion model, the OU process is described by a *linear* equation: it can be simulated exactly and Monte Carlo methods may be easier to implement.

Martingale aspects The following lemma shows how to construct martingale versions of time changed Lévy models.

LEMMA 15.2

Let $(M_t)_{t\geq 0}$ be a positive martingale (in its own filtration) and $(v_t)_{t\geq 0}$ be an independent (increasing) time change process. Then the time changed process $M(v_t)$ is again a martingale (in the filtration generated by the time changed process $M(v_t)$ and the time change v_t).

PROOF The positiveness condition is needed to show that the time changed process is integrable. Indeed, since M_t is positive, $E\{|M(v_t)|\} = E\{M(v_t)\} = E\{M_0\} < \infty$. Now it is easy to obtain the martingale property by conditioning. Denoting by $\mathcal{F}_t^M$ the filtration of (M_t), by $\mathcal{F}_t^v$ the filtration of v_t and by $\mathcal{F}_t = \sigma(M(v_s), 0 \le s \le t) \vee \sigma(v_s, 0 \le s \le t)$ the filtration generated by $M(v_t)$ and v_t, we have:

$$
\begin{aligned}
E\{M(v_t)|\mathcal{F}_s\} &= E\{E[M(v_t)|\mathcal{F}_s \vee \mathcal{F}_t^v]|\mathcal{F}_s\} \\
&= E\{E[M(v_t)|\mathcal{F}_{v_s}^M \vee \mathcal{F}_t^v]|\mathcal{F}_s\} = E\{M(v_s)|\mathcal{F}_s\} = M(v_s).
\end{aligned}
$$

□

The positiveness condition of the lemma cannot be omitted without imposing additional integrability properties either on the martingale or on the time change. Indeed, the Cauchy process, although it is subordinate to Brownian motion (by a $\frac{1}{2}$-stable subordinator), is not a martingale because it is not integrable.

This lemma shows that martingale versions of stochastic volatility models of type (15.34) may be constructed simply by applying the time change to martingale versions of Lévy processes.

Example 15.3

The normal inverse Gaussian Lévy process (see Chapter 4) has characteristic function

$$
\phi_t(u) = \exp t\left(iu\mu + \frac{1}{\kappa} - \frac{1}{\kappa}\sqrt{1 + \kappa\sigma^2 u^2 - 2i\theta\kappa u}\right).
$$

Its martingale version has the same form with $\mu = \frac{1}{\kappa}\{\sqrt{1 - \sigma^2\kappa - 2\theta\kappa} - 1\}$. As in Example 15.1, suppose that the volatility follows an Ornstein-Uhlenbeck process with gamma stationary distribution. Then the characteristic function of log-price process is

$$
\phi_t^X(u) = \exp\left\{iurt + \psi_v\left(\frac{iu}{\kappa}\{\sqrt{1 - \sigma^2\kappa - 2\theta\kappa} - 1\} + \frac{1}{\kappa} - \frac{1}{\kappa}\sqrt{1 + \kappa\sigma^2 u^2 - 2i\theta\kappa u}\right)\right\} \quad (15.35)
$$

with $\psi_v(u) = \dfrac{uy_0}{\lambda}(1 - e^{-\lambda t}) + \dfrac{ut\lambda c}{\alpha\lambda - u} + \dfrac{\lambda\alpha c}{\alpha\lambda - u}\ln\left(1 - \dfrac{u}{\alpha\lambda}(1 - e^{-\lambda t})\right).$

The right graph in Figure 15.4 depicts the implied volatility smiles in this model for three different values of parameter κ. It shows that another intuition, stemming from continuous stochastic volatility models, is not relevant in the generalized setting combining jumps and stochastic volatility. We have seen earlier in this chapter (Section 15.1) that in continuous stochastic volatility models, when there is no instantaneous correlation between volatility and returns, the implied volatility smile is locally convex with a minimum at the money forward ($K = Se^{+r(T-t)}$). One may think that this is also true in Lévy models with stochastic volatility, at least when the jumps are symmetric. However, from Figure 15.4, we see that even if the jumps are symmetric but the jump component is sufficiently strong, the minimum of the smile does not occur at the money (on the graph it is shifted to the left). Taking smaller values of κ, we come back to the Brownian stochastic volatility setting and the minimum shifts towards the money. □

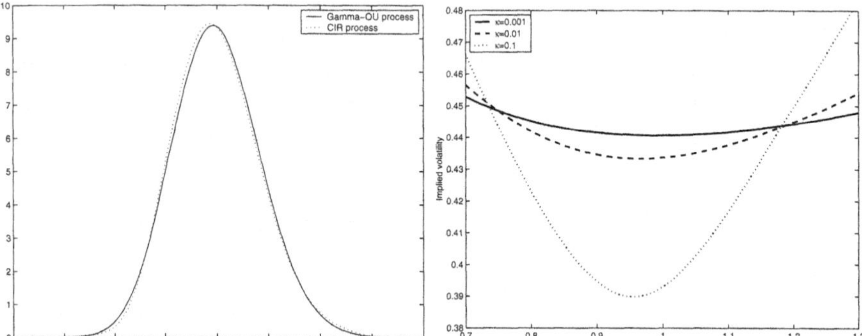

FIGURE 15.4: Left: Distributions of integrated volatility distributions for a two-year horizon for the CIR process and a positive OU process (with the same parameters as in Figure 15.1). Right: Implied volatility smiles in the NIG model with OU-driven stochastic volatility. The parameters are $T = 0.3$, $r = 0$, $\sigma = 1$, $\theta = 0$, $y_0 = 0.2$, $\lambda = 2$, $c = 2.04$ and $\alpha = 10.2$. The varying parameter κ measures the proximity of the NIG process to Brownian motion (when κ is small, they behave similarly).

15.6 Do we need stochastic volatility and jumps?

We have seen in Chapters 11 and 13 that exponential-Lévy models can generate a variety of shapes for implied volatility surfaces, including smiles and skews for a given maturity but perform poorly across maturities. In Chapter 14, we saw that it is possible to calibrate also the term structure of implied volatility using a one-factor additive process with independent log-returns. However, this led to a non-stationary risk-neutral process which implied forward implied volatilities which behave quite differently from spot implied volatilities. As discussed in the beginning of this chapter, introducing "stochastic volatility" takes the pressure off the time-dependence and allows to reproduce a realistic implied volatility pattern across maturities without having to introduce strong time variation of parameters but fails to give realistic behavior for short maturities.

From the point of view of calibration to option prices, jumps allow to reproduce strong skews and smiles at short maturities while stochastic volatility makes the calibration of the term structure possible, especially for long-term smiles. Combining jumps in returns *and* stochastic volatility allows to calibrate the implied volatility surface across strikes and maturities without introducing explicit time dependence in parameters. In addition, in a stochastic volatility model, increments of the price process are not independent, which implied that, unlike the exponential additive models of Chapter 14, knowledge of today's smile still leaves a degree of freedom for modelling forward smiles. This leads to a greater stability of calibrated parameters across time.

On the other hand, once these two ingredients — jumps in returns and "stochastic volatility" — are included, the models have a fairly good empirical performance. The results given above show that there is not much justification for introducing complex dynamics for the volatility: once the main requirements of positiveness and mean-reverting behavior have been respected, various choices for the driving process σ_t give similar results. When it comes to calibrating market option prices, the time changed Lévy model often outperforms the BNS model [352]: the OU model restricts the possible implied volatility patterns by requiring the same parameter ρ to characterize the jumps in the returns, i.e., their skewness both for short and long maturities. On the contrary, the performance of the Bates model is similar to that of time changed Lévy models. In fact, the Bates model has an additional degree of freedom: since the short-term skew is essentially due to the jump term and the long-term skew is due to the correlation parameter ρ, short and long maturities can be calibrated separately. Thus the Bates model appears to be at the same time the simplest and the most flexible of the models. Table 15.2 further compares the three models and shows their advantages compared to standard exponential-Lévy models and diffusion-based stochastic volatility models.

Further reading

Given our brief presentation of stochastic volatility models, there is a lot of room for further reading on this topic! The Bates model was introduced in [41]. A general discussion of diffusion-based stochastic volatility models can

TABLE 15.2: Stochastic volatility models with jumps

Characteristics	Diffusion-based SV	exp-Lévy	Bates	SV-OU	SV-Lévy
Historical					
Jumps in price		✓	✓	if $\rho \neq 0$	✓
Volatility clusters	✓		✓	✓	✓
Leverage effect	if $\rho \neq 0$		if $\rho \neq 0$	if $\rho \neq 0$	
Risk-neutral					
Symmetric smiles	if $\rho = 0$			if $\rho = 0$	
Skew for short maturities		✓	✓	if $\rho \neq 0$	✓
Skew for long maturities	if $\rho \neq 0$		if $\rho \neq 0$	if $\rho \neq 0$	
Flexible term structure	✓		✓	✓	✓
Pricing European options	Fourier pricing methods can be used for all models				

be found in [150, Chapters 1-2] or [261]; hedging in such models is discussed in [195]. Stochastic volatility models based on positive Ornstein Uhlenbeck processes were introduced by Barndorff-Nielsen and Shephard with an econometric motivation in [32, 36]. Option pricing under these models was studied in [307, 308]. An empirical comparison between the BNS model with OU stochastic volatility and exponential-Lévy models is given by Tompkins [379]. Stochastic volatility models constructed from time changed Lévy processes can be found in [81]. Examples of calibration and comparison between different Lévy-based stochastic volatility models from the point of view of option pricing are given in [352] and [81].

Discrete time stochastic volatility models for econometric modelling are discussed in [361, 343].

In all the models discussed in this chapter, it is possible to obtain the Fourier transform of option prices in closed form due to the special analytic form of the characteristic functions. A general class of models with this property is given by affine jump-diffusions [119, 118]: for all models in this class option prices can be computed by Fourier transform in a similar way.

Epilogue

As we have been writing this volume, the literature on jump processes in finance has been constantly growing and the material presented here can only partly reflect current research topics. Among important topics which have been left out are applications of jump processes in interest rate and credit risk modelling.

Credit events, viewed as sudden changes in the credit quality of a firm, are naturally modelled as jump processes and many of the concepts and tools used in this volume have been successfully employed for modelling default and credit risk. Credit risk modelling is a vast subject and a single chapter would have been insufficient to review current research on this topic. We refer the reader to the recent monograph by Schönbucher [351].

The development of arbitrage-free models compatible with empirical observations for pricing interest rate derivatives has also led in the recent years to the development of models based on jump-diffusions and Lévy processes for instantaneous forward rates [363, 59, 127] and LIBOR rates [166, 126]. Björk et al. [59] provide a general analysis of arbitrage restrictions in such interest rate models. Several extensions of the LIBOR market model with jumps have been proposed, based on marked point processes [166, 168, 167] and Lévy processes [126]. While the complexity of such models has prevented their widespread use in the interest rate derivatives market, they present potentially interesting aspects and merit further study, especially regarding their calibration, implementation and empirical performance. An interesting case study which illustrates the advantages of LIBOR models with jumps for pricing and hedging short term interest rate derivatives is given by Nahum and Samuelides [303].

Finally, let us note that there are other fields of application where modelling with jump processes is more natural than with diffusion processes. Examples are insurance claims and losses due to operational risk. We hope that the present volume will encourage more researchers and practitioners to contribute to this topic and improve our understanding of theoretical, numerical and practical issues related to financial modelling with jump processes.

Appendix A

Modified Bessel functions

Modified Bessel functions are also called Bessel functions of imaginary argument. They are simple special functions in the sense that they are very well studied, tabulated and fast algorithms for their computation are implemented in most computational toolboxes. Consider the equation

$$z^2 \frac{d^2 w}{dz^2} + z \frac{dw}{dz} - (z^2 + \nu^2) w = 0. \tag{A.1}$$

The modified Bessel function of the first kind $I_\nu(z)$ for $z \geq 0$ and $\nu \geq 0$ is equal to the solution of Equation (A.1) that is bounded when $z \to 0$. The modified Bessel function of the second kind $K_\nu(z)$ for for $z \geq 0$ and $\nu \geq 0$ is equal to the solution of Equation (A.1) that is bounded when $z \to \infty$. These functions are always linearly independent and admit the following series expansions:

$$\begin{aligned}
I_\nu(z) &= \sum_{k=0}^{\infty} \frac{(z/2)^{\nu+k}}{k!\Gamma(k+\nu+1)}, && \nu \text{ is not a negative integer,} \\
I_{-n}(z) &= I_n(z), && n = 0, 1, \ldots, \\
K_\nu(z) &= \frac{\pi}{2} \frac{I_{-\nu}(z) - I_\nu(z)}{\sin(\pi\nu)}.
\end{aligned}$$

For integer values of ν the last expression should be interpreted in the limiting sense. For all orders ν we have $K_{-\nu}(z) = K_\nu(z)$.

Asymptotic behavior When $z \to +\infty$ we have

$$\begin{aligned}
I_\nu(z) &= \frac{e^z}{\sqrt{2\pi z}} \left[1 + O\left(\frac{1}{z}\right) \right], \\
K_\nu(z) &= e^{-z} \sqrt{\frac{\pi}{2z}} \left[1 + O\left(\frac{1}{z}\right) \right].
\end{aligned}$$

When $z \to 0$,

$$\begin{aligned}
I_\nu(z) &\sim \frac{(z/2)^\nu}{\Gamma(\nu+1)}, \quad \nu \text{ is not a negative integer,} \\
K_0(z) &\sim -\log z, \\
K_\nu(z) &\sim \frac{1}{2}\Gamma(\nu)(z/2)^{-\nu}, \quad \nu > 0.
\end{aligned}$$

Closed form expressions for modified Bessel functions of half-integer order

$$I_{n-\frac{1}{2}}(z) = \sqrt{\frac{2}{\pi z}} z^n \left(\frac{1}{z}\frac{d}{dz} \right)^n \cosh z, \quad n = 0, 1, \ldots,$$
$$K_{n-\frac{1}{2}}(z) = \sqrt{\frac{\pi}{2z}} z^n \left(-\frac{1}{z}\frac{d}{dz} \right)^n e^{-z}, \quad n = 0, 1, \ldots$$

Integral representations Poisson integral representations:

$$I_\nu(z) = \frac{(z/2)^\nu}{\sqrt{\pi}\Gamma(\nu+\frac{1}{2})} \int_{-1}^{1} (1-t^2)^{\nu-1/2} \cosh(zt)dt,$$
$$K_\nu(z) = \frac{e^{-z}}{\Gamma(\nu+\frac{1}{2})} \sqrt{\frac{\pi}{2z}} \int_0^\infty e^{-t} t^{\nu-1/2} (1+\frac{t}{2z})^{\nu-1/2} dt.$$

Sommerfeld integral representations for K_ν:

$$K_\nu(z) = \frac{1}{2} \int_{-\infty}^{\infty} \exp(-z\cosh\psi + \nu\psi)d\psi = \frac{1}{2}\left(\frac{z}{2}\right)^\nu \int_0^\infty e^{-t-\frac{z^2}{4t}} t^{-\nu-1} dt.$$

This representation leads to the following useful formula for integral evaluation:

$$\int_0^\infty e^{-\frac{\alpha^2 t}{2} - \frac{\beta^2}{2t}} \frac{dt}{t^{1+\nu}} = 2\left(\frac{\alpha}{\beta}\right)^\nu K_\nu(\beta\alpha). \tag{A.2}$$

Relation with Bessel functions of real argument The modified Bessel function of the first kind can be expressed via the Bessel function of the first kind with imaginary argument:

$$I_\nu(z) = e^{-i\pi\nu/2} J_\nu(iz),$$

and the modified Bessel function of the second kind can be expressed via Hankel functions of the first kind:

$$K_\nu(z) = \frac{\pi}{2} e^{i\pi(\nu+1)/2} H_\nu^{(1)}(iz).$$

References

[1] Abramowitz, M. and Stegun, I., eds., *Handbook of Mathematical Functions*, Dover: New York, 1968.

[2] Ait-Sahalia, Y. and Lo, A., *Nonparametric estimation of state-price densities implicit in financial asset prices*, J. Finance, 53 (1998), pp. 499–547.

[3] Akgiray, V. and Booth, G., *The stable law model of stock returns*, J. Business Econom. Statis., 6 (1988), pp. 51–57.

[4] Allen, S. and Padovani, O., *Risk management using quasi-static hedging*, Economic Notes, 31 (2002), pp. 277–236.

[5] Alvarez, O. and Tourin, A., *Viscosity solutions of non-linear integro-differential equations*, Annales de l'Institut Henri Poincaré, 13 (1996), pp. 293–317.

[6] Amin, K., *Jump-diffusion option valuation in discrete time*, J. Finance, 48 (1993), pp. 1833–1863.

[7] Andersen, L. and Andreasen, J., *Jump-diffusion models: Volatility smile fitting and numerical methods for pricing*, Rev. Derivatives Research, 4 (2000), pp. 231–262.

[8] Andersen, T., Bollerslev, T., and Diebold, F., *Parametric and non-parametric measurement of volatility*, in Handbook of Financial Econometrics, Ait-Sahalia, Y. and Hansen, L., eds., North Holland: Amsterdam, 2003.

[9] Andersen, T., Bollerslev, T., Diebold, F., and Lays, P., *The distribution of exchange rate volatility*, J. Amer. Stat. Assoc., 96 (2001), pp. 42–55.

[10] Arneodo, A., Bacry, E., Muzy, J., and Delour, X., *What can we learn from wavelet analysis?*, in Scale Invariance and Beyond, Dubrulle, B., Graner, F., and Sornette, D., eds., Springer: Berlin, 1997.

[11] Asmussen, S. and Rosiński, J., *Approximations of small jumps of Lévy processes with a view towards simulation*, J. Appl. Probab., 38 (2001), pp. 482–493.

[12] Avellaneda, M., *The minimum-entropy algorithm and related methods for calibrating asset-pricing models*, in Proceedings of the International

Congress of Mathematicians, (Berlin, 1998), Vol. III, Documenta Mathematica, 1998, pp. 545–563.

[13] Avellaneda, M., *Minimum entropy calibration of asset pricing models*, Int. J. Theor. Appl. Finance, 1 (1998), pp. 447–472.

[14] Avellaneda, M., Buff, R., Friedman, C., Grandchamp, N., Kruk, L., and Newman, J., *Weighted Monte Carlo: a new technique for calibrating asset-pricing models*, Int. J. Theor. Appl. Finance, 4 (2001), pp. 91–119.

[15] Avellaneda, M. and Paras, A., *Managing the volatility of risk of portfolios of derivative securities: the Lagrangian uncertain volatility model*, Applied Mathematical Finance, 3 (1996), pp. 23–51.

[16] Avram, F., Kyprianou, A., and Pistorius, M., *Exit problems for spectrally negative Lévy processes and applications to Russian options*, Ann. Appl. Probab., forthcoming (2003).

[17] Bachelier, L., *Théorie de la spéculation*, Annales de l'Ecole Normale Supérieure, 17 (1900), pp. 21–86. English translation in [100].

[18] Bacry, E. and Muzy, J., *Log infinitely divisible multifractal processes*, Communications in Mathematical Physics, 236 (2003), pp. 449–475.

[19] Bacry, E., Muzy, J., and Arneodo, A., *Singularity spectrum of fractal signals from wavelet analysis: Exact results*, J. Statist. Phys., 70 (1993), pp. 635–674.

[20] Ball, C. and Roma, A., *Stochastic volatility option pricing*, J. Fin. and Quant. Anal., 29 (1994), pp. 589–607.

[21] Balland, P., *Deterministic implied volatility models*, Quant. Finance, 2 (2002), pp. 31–44.

[22] Bardhan, I. and Chao, X., *Pricing options on securities with discontinuous returns*, Stochastic Process. Appl., 48 (1993), pp. 123–137.

[23] Bardhan, I. and Chao, X., *Pricing options on securities with discontinuous returns*, Stochastic Process. Appl., 48 (1993), pp. 123–137.

[24] Bardhan, I. and Chao, X., *Martingale analysis for assets with discontinuous returns*, Math. Oper. Res., 20 (1995), pp. 243–256.

[25] Bardhan, I. and Chao, X., *On martingale measures when asset returns have unpredictable jumps*, Stochastic Process. Appl., 63 (1996), pp. 35–54.

[26] Bardou, F., Bouchaud, J., Aspect, A., and Cohen-Tannoudji, C., *Lévy Statistics and Laser Cooling*, Cambridge University Press: Cambridge, UK, 2002.

[27] Barles, G., *Convergence of numerical schemes for degenerate parabolic equations arising in finance theory*, in Numerical Methods in Finance,

Rogers, L. and Talay, D., eds., Cambridge University Press: Cambridge, 1997, pp. 1–21.

[28] Barles, G., Buckdahn, R., and Pardoux, E., *Backward stochastic differential equations and integral-partial differential equations*, Stochastics and Stochastic Reports, 60 (1997), pp. 57–83.

[29] Barles, G. and Souganidis, P., *Convergence of approximation schemes for fully nonlinear second order equations*, Asymptotic Anal., 4 (1991), pp. 271–283.

[30] Barndorff-Nielsen, O., *Processes of normal inverse Gaussian type*, Finance Stoch., (1998), pp. 41–68.

[31] Barndorff-Nielsen, O. E., *Exponentially decreasing distributions for the logarithm of particle size*, Proceedings of the Royal Society of London, A353 (1977), pp. 401–419.

[32] Barndorff-Nielsen, O. E., *Normal inverse Gaussian distributions and stochastic volatility modelling*, Scand. J. Statist., 24 (1997), pp. 1–13.

[33] Barndorff-Nielsen, O. E., *Probability densities and Lévy densities*, Maphysto Research Report 2000-18, Maphysto, May 2000.

[34] Barndorff-Nielsen, O. E. and Blaesild, P., *Hyperbolic distributions and ramifications: Contributions to theory and application*, in Statistical Distributions in Scientific Work, Taillie, C., Patil, G., and Baldessari, B., eds., Vol. 4, Reidel: Dordrecht, 1981, pp. 19–44.

[35] Barndorff-Nielsen, O. E. and Prause, K., *Apparent scaling*, Finance Stoch., 5 (2001), pp. 103–114.

[36] Barndorff-Nielsen, O. E. and Shephard, N., *Modelling by Lévy processes for financial econometrics*, in Lévy processes—Theory and Applications, Barndorff-Nielsen, O., Mikosch, T., and Resnick, S., eds., Birkhäuser: Boston, 2001, pp. 283–318.

[37] Barndorff-Nielsen, O. E. and Shephard, N., *Non-Gaussian Ornstein-Uhlenbeck based models and some of their uses in financial econometrics*, J. R. Statistic. Soc. B, 63 (2001), pp. 167–241.

[38] Barndorff-Nielsen, O. E. and Shephard, N., *Econometric analysis of realized volatility and its use in estimating stochastic volatility models*, J. R. Statistic. Soc. B, 64 (2002), pp. 253–280.

[39] Bates, D., *Estimating the stable index α in order to measure tail thickness: A critique*, Annals of statistics, 11 (1983), pp. 1019–1031.

[40] Bates, D., *The Crash of '87: Was it expected? The evidence from options markets*, J. Finance, 46 (1991), pp. 1009–1044.

[41] Bates, D., *Jumps and stochastic volatility: the exchange rate processes implicit in Deutschemark options*, Rev. Fin. Studies, 9 (1996), pp. 69–107.

[42] Bates, D., *Post-87 crash fears in the S&P 500 futures option market*, J. Econometrics, 94 (2000), pp. 181–238.

[43] Bates, D. S., *Testing option pricing models*, in Statistical Methods in Finance, Vol. 14 of Handbook of Statistics, North-Holland: Amsterdam, 1996, pp. 567–611.

[44] Becherer, D., *Rational valuation and hedging with utility based preferences*, PhD thesis, Technical University of Berlin: Berlin, 2001.

[45] Beckers, S., *A note on estimating parameters of a jump-diffusion process of stock returns*, J. Fin. and Quant. Anal., 16 (1981), pp. 127–140.

[46] Bellamy, N. and Jeanblanc, M., *Incompleteness of markets driven by mixed diffusion*, Finance Stoch., 4 (1999), pp. 209–222.

[47] Bensoussan, A., *On the theory of option pricing*, Acta Appl. Math., 2 (1984), pp. 139–158.

[48] Bensoussan, A. and Lions, J.-L., *Contrôle Impulsionnel et Inéquations Quasi-Variationnelles*, Dunod: Paris, 1982.

[49] Bertoin, J., *Lévy Processes*, Cambridge University Press: Cambridge, 1996.

[50] Bertoin, J., *Subordinators: Examples and applications*, in Ecole d'Ete de Probabilités de Saint-Flour XXVII, Vol. 1727 of Lecture Notes in Maths., Springer: Heidelberg, 1999, pp. 1–91.

[51] Bertoin, J., *Some elements on Lévy processes*, in Stochastic Processes: Theory and Methods, Vol. 20 of Handbook of Statistics, North Holland: Amsterdam, 2002.

[52] Beylkin, G., Coifman, R., and Rokhlin, V., *Fast wavelet transforms and numerical algorithms*, Communications on Pure and Applied Mathematics, 44 (1991), pp. 141–183.

[53] Biagini, F. and Guasoni, P., *Mean-variance hedging with random volatility jumps*, Stochastic Analysis and Applications, 20 (2002).

[54] Bibby, B. and Sørensen, M., *A hyperbolic diffusion model for stock prices*, Finance Stoch., 1 (1997).

[55] Bichteler, K., *Stochastic Integration with Jumps*, Vol. 39 of Encyclopedia of Mathematics and its Applications, Cambridge University Press: Cambridge, 2002.

[56] Bichteler, K. and Jacod, J., *Calcul de Malliavin pour les diffusions avec sauts: existence d'une densité dans le cas unidimensionnel*, in Seminar

on probability, XVII, Vol. 986 of Lecture Notes in Math., Springer: Berlin, 1983, pp. 132–157.

[57] Bingham, N. and Kiesel, R., *Modelling asset returns with hyperbolic distributions*, in Knight and Satchell [236], Ch. 1, pp. 1–20.

[58] Bismut, J.-M., *Calcul des variations stochastique et processus de sauts*, Z. Wahrsch. Verw. Gebiete, 63 (1983), pp. 147–235.

[59] Björk, T., Kabanov, Y., and Runggaldier, W., *Bond market structure in the presence of marked point processes*, Finance Stoch., 7 (1997), pp. 211–239.

[60] Black, F. and Scholes, M., *The pricing of options and corporate liabilities*, Journal of Political Economy, 3 (1973).

[61] Blaesild, P. and Sørensen, M., ***hyp**: A computer program for analyzing data by means of the hyperbolic distribution*, Research Report 248, Department of Theoretical Statistics, University of Aarhus, 1992.

[62] Blattberg, R. and Gonedes, N., *A comparison of the stable Paretian and Student t distributions as statistial models for prices*, Journal of Business, 47 (1974), pp. 244–280.

[63] Bollerslev, T., *Financial econometrics: Past developments and future challenges*, J. Econometrics, 100 (2001), pp. 41–51.

[64] Bondesson, L., *On simulation from infinitely divisible distributions*, Adv. Appl. Probab., 14 (1982), pp. 855–869.

[65] Bony, J., Courrège, P., and Priouret, P., *Semi-groupes de Feller sur une variété compacte et problèmes aux limites integro-différentiels du second ordre donnant un principe de maximum*, Annales de L'Institut Fourier, 18 (1968), pp. 369–521.

[66] Borodin, A. N. and Salminen, P., *Handbook of Brownian Motion—Facts and Formulae*, Birkhäuser Verlag: Basel, 1996.

[67] Bouchaud, J. and Potters, M., *Théorie des Risques Financiers*, Aléa: Saclay, 1997.

[68] Bouleau, N., *Processus Stochastiques et Applications*, Hermann: Paris, 1988.

[69] Bouleau, N. and Lamberton, D., *Residual risks and hedging strategies in Markovian markets*, Stochastic Process. Appl., 33 (1989), pp. 131–150.

[70] Boyarchenko, S. and Levendorskiĭ, S., *Barrier options and touch-and-out options under regular Lévy processes of exponential type*, Ann. Appl. Probab., 12 (2002), pp. 1261–1298.

[71] Boyarchenko, S. and Levendorskiĭ, S., *Non-Gaussian Merton-Black-Scholes Theory*, World Scientific: River Edge, NJ, 2002.

[72] Boyarchenko, S. and Levendorskiĭ, S., *Perpetual American options under Lévy processes*, SIAM J. Control Optim., 40 (2002), pp. 1663–1696 (electronic).

[73] Bremaud, P., *Point Processes and Queues: Martingale Dynamics*, Springer: Berlin, 1981.

[74] Bretagnolle, J., *Notes de lecture sur l'œuvre de Paul Lévy*, Ann. Inst. H. Poincaré Probab. Statist., 23 (1987), pp. 239–243.

[75] Byrd, R., Lu, P., and Nocedal, J., *A limited memory algorithm for bound constrained optimization*, SIAM Journal on Scientific and Statistical Computing, 16 (1995), pp. 1190–1208.

[76] Campbell, J., Lo, A., and McKinlay, C., *The econometrics of financial markets*, Princeton University Press: Princeton, 1996.

[77] Cancelier, C., *Problèmes aux limites pseudodifferentiels donnant lieu au principe du maximum*, Comm. Partial Differential Equations, 11 (1986), pp. 1677–1726.

[78] Carr, P., Ellis, K., and Gupta, V., *Static hedging of exotic options*, Journal of finance, 53 (1998), pp. 1165–1190.

[79] Carr, P. and Faguet, D., *Fast accurate valuation of American options*, working paper, Cornell University, 1994.

[80] Carr, P., Geman, H., Madan, D., and Yor, M., *The fine structure of asset returns: An empirical investigation*, Journal of Business, 75 (2002).

[81] Carr, P., Geman, H., Madan, D., and Yor, M., *Stochastic volatility for Lévy processes*, Math. Finance, 13 (2003), pp. 345–382.

[82] Carr, P. and Hirsa, A., *Why be backward? Forward equations for American options*, RISK, (2003).

[83] Carr, P. and Madan, D., *Option valuation using the fast Fourier transform*, J. Comput. Finance, 2 (1998), pp. 61–73.

[84] Carr, P. and Wu, L., *The finite moment logstable process and option pricing*, Journal of Finance, 58 (2003), pp. 753–778.

[85] Carr, P. and Wu, L., *What type of process underlies options? A simple robust test*, J. Finance, 58 (2003).

[86] Carrière, J., *Valuation of the early-exercise price for derivative securities using simulations and splines*, Insurance: Mathematics and Economics, 19 (1996), pp. 19–30.

[87] Chambers, J., Mallows, C., and Stuck, B., *A method for simulating stable random variables*, J. Amer. Stat. Assoc., 71 (1976), pp. 340–344.

[88] Chan, T., *Pricing contingent claims on stocks driven by Lévy processes*, Ann. Appl. Probab., 9 (1999), pp. 504–528.

[89] Chatelain, M. and Stricker, C., *On componentwise and vector stochastic integration*, Math. Finance, 4 (1994), pp. 57–65.

[90] Cherny, A. and Shiryaev, A., *Vector stochastic integrals and the fundamental theorems of asset pricing*, Proceedings of the Steklov Institute of mathematics, 237 (2002), pp. 6–49.

[91] Chou, C. and Meyer, P., *Sur la representation des martingales comme intégrales stochastiques de processus ponctuels*, in Séminaire de Probabilités, IX, Vol. 465 of Lecture Notes in Math., Springer: Berlin, 1975, pp. 226–236.

[92] Cont, R., *Empirical properties of asset returns: Stylized facts and statistical issues*, Quant. Finance, 1 (2001), pp. 1–14.

[93] Cont, R. and Ben Hamida, S., *Model calibration by evolutionary optimization*, working paper, CMAP, Ecole Polytechnique, February 2003.

[94] Cont, R., Bouchaud, J.-P., and Potters, M., *Scaling in financial data: Stable laws and beyond*, in Scale Invariance and Beyond, Dubrulle, B., Graner, F., and Sornette, D., eds., Springer: Berlin, 1997.

[95] Cont, R. and da Fonseca, J., *Dynamics of implied volatility surfaces*, Quant. Finance, 2 (2002), pp. 45–60.

[96] Cont, R., da Fonseca, J., and Durrleman, V., *Stochastic models of implied volatility surfaces*, Economic Notes, 31 (2002), pp. 361 – 377.

[97] Cont, R. and Tankov, P., *Calibration of jump-diffusion option pricing models: A robust non-parametric approach*, Rapport Interne 490, CMAP, Ecole Polytechnique, 2002. *Forthcoming in: Journal of Computational Finance.*

[98] Cont, R. and Voltchkova, E., *Finite difference methods for option pricing in jump-diffusion and exponential Lévy models*, Rapport Interne 513, CMAP, Ecole Polytechnique, 2003.

[99] Cooley, J. W. and Tukey, J. W., *An algorithm for the machine calculation of complex Fourier series*, Math. Comp., 19 (1965), pp. 297–301.

[100] Cootner, P., ed., *The Random Character of Stock Market Prices*, MIT Press: Cambridge, MA, 1964.

[101] Courtault, J.-M., Kabanov, Y., Bru, B., Crépel, P., Lebon, I., and Le Marchand, A., *Louis Bachelier on the centenary of "Théorie de la Spéculation"*, Math. Finance, 10 (2000), pp. 341–353.

[102] Cox, J. C., Ingersoll, J. E., Jr., and Ross, S. A., *An intertemporal general equilibrium model of asset prices*, Econometrica, 53 (1985), pp. 363–384.

[103] Crandall, M., Ishii, H., and Lions, P., *Users guide to viscosity solutions of second order partial differential equations*, Bulletin of the American Mathematical Society, 27 (1992), pp. 1–42.

[104] Dahlke, S. and Weinreich, I., *Wavelet bases adapted to pseudo-differential operators*, Appl. Comput. Harmon. Anal., 1 (1994), pp. 267–283.

[105] Danielsson, J. and de Vries, C., *Tail index and quantile estimation with very high frequency data*, Journal of Empirical Finance, 4 (1997), pp. 241–258.

[106] Daubechies, I., *Ten Lectures on Wavelets,*, SIAM: Philadelphia, 1992.

[107] Delbaen, F., Grandits, P., Rheinländer, T., Samperi, D., Schweizer, M., and Stricker, C., *Exponential hedging and entropic penalties*, Math. Finance, 12 (2002), pp. 99–123.

[108] Delbaen, F., Monat, P., Schachermayer, W., and Stricker, C., *Weighted norm inequalities and hedging in incomplete markets*, Finance Stoch., 1 (1997), pp. 181–229.

[109] Delbaen, F. and Schachermayer, W., *The fundamental theorem of asset pricing for unbounded stochastic processes*, Math. Ann., 312 (1998), pp. 215–250.

[110] Dellacherie, C. and Meyer, P., *Probabilités et Potentiel. Chapitres I à IV*, Hermann: Paris, 1975.

[111] Derman, E., *Regimes of volatility*, RISK, (1999).

[112] Derman, E. and Kani, I., *Riding on a Smile*, RISK, 7 (1994), pp. 32–39.

[113] Devroye, L., *Nonuniform random variate generation*, Springer: New York, 1986.

[114] d'Halluin, Y., Forsyth, P., and Labahn, G., *A penalty method for American options with jump-diffusion processes*, working paper, University of Waterloo, March 2003.

[115] Doléans-Dade, C., *Quelques applications de la formule de changement de variable pour les semi-martingales*, Z. Wahrsch. Verw. Gebiete, 16 (1970).

[116] Doob, J., *Stochastic processes*, Wiley: New York, 1953.

[117] Drost, T. and Nijman, F., *Temporal aggregation of GARCH processes*, Econometrica, 61 (1993), pp. 909–927.

[118] Duffie, D., Filipovic, D., and Schachermayer, W., *Affine processes and applications in finance*, Ann. Appl. Probab., 13 (2003), pp. 984–1053.

[119] Duffie, D., Pan, J., and Singleton, K., *Transform analysis and asset pricing for affine jump-diffusions*, Econometrica, 68 (2000), pp. 1343–1376.

[120] Dugac, P., Métivier, M., and Costabel, P., eds., *Siméon Denis Poisson et la Science de son Temps*, École Polytechnique: Palaiseau, France, 1981.

[121] Dumas, B., Fleming, J., and Whaley, R., *Implied volatility functions: Empirical tests*, J. Finance, 53 (1998), pp. 2059–2106.

[122] Dupire, B., *Pricing with a smile*, RISK, 7 (1994), pp. 18–20.

[123] Eberlein, E., *Applications of generalized hyperbolic Lévy motion to Finance*, in Lévy Processes—Theory and Applications, Barndorff-Nielsen, O., Mikosch, T., and Resnick, S., eds., Birkhäuser: Boston, 2001, pp. 319–336.

[124] Eberlein, E. and Jacod, J., *On the range of option prices*, Finance Stoch., 1 (1997).

[125] Eberlein, E., Keller, U., and Prause, K., *New insights into smile, mispricing and Value at Risk: The hyperbolic model*, Journal of Business, 71 (1998), pp. 371–405.

[126] Eberlein, E. and Ozkan, F., *The Lévy LIBOR model*, FDM Preprint 82, University of Freiburg, 2002.

[127] Eberlein, E. and Raible, S., *Term structure models driven by general Lévy processes*, Math. Finance, 9 (1999), pp. 31–53.

[128] Eberlein, E. and Raible, S., *Some analytic facts on the generalized hyperbolic model*, in European Congress of Mathematics, Vol. II (Barcelona, 2000), Vol. 202 of Progr. Math., Birkhäuser: Basel, 2001, pp. 367–378.

[129] El Karoui, N., Jeanblanc, M., and Shreve, S., *Robustness of the Black Scholes formula*, Math. Finance, 8 (1998), pp. 93–126.

[130] El Karoui, N. and Rouge, R., *Pricing via utility maximization and entropy*, Math. Finance, 10 (2000), pp. 259–276.

[131] Embrechts, P., Klüppelberg, C., and Mikosch, T., *Modelling Extremal Events*, Vol. 33 of Applications of Mathematics, Springer: Berlin, 1997.

[132] Emery, M. and Yor, M., eds., *Séminaire de probabilités 1967-1980: A selection in Martingale Theory*, Vol. 1771 of Lecture Notes in Math., Springer: Berlin, 2002.

[133] Engl, H. W., *Inverse problems and their regularization*, in Computational Mathematics Driven by Industrial Problems (Martina Franca,

1999), Vol. 1739 of Lecture Notes in Math., Springer: Berlin, 2000, pp. 127–150.

[134] Engl, H. W., Hanke, M., and Neubauer, A., *Regularization of Inverse Problems*, Vol. 375, Kluwer Academic Publishers Group: Dordrecht, 1996.

[135] Engle, R. and Russell, J., *Autoregressive conditional duration: A new model for irregularly spaced transaction data*, Econometrica, 66 (1998), pp. 1127–1162.

[136] Esche, F. and Schweizer, M., *Minimal entropy preserves the Lévy property: How and why*, working paper, University of Munich, 2003.

[137] Eskin, G., *Boundary Value Problems for Elliptic Pseudofifferential Equations*, American Mathematical Society: Providence, RI, 1981.

[138] Falconer, K., *Fractal geometry*, John Wiley & Sons: Chichester, 1990.

[139] Fama, E., *Mandelbrot and the stable Paretian hypothesis*, Journal of Business, 36 (1963), pp. 420–429.

[140] Fama, E., *The behavior of stock market prices*, Journal of Business, 38 (1965), pp. 34–105.

[141] Feller, W., *An Introduction to Probability Theory and its Applications*, Vol. II, John Wiley & Sons: New York, 1971.

[142] Fisher, R. and Tippett, L., *Limiting forms of the frequency distribution of the largest and smallest members of a sample*, Proc. Cambridge Phil. Soc., 24 (1928), pp. 180–190.

[143] Fleming, W. and Soner, H., *Controlled Markov Processes and Viscosity Solutions*, Springer: New York, 1993.

[144] Föllmer, H. and Elliott, R., *Orthogonal martingale representation*, in Stochastic Analysis, Mayer-Wolf, E., Merzbach, E., and Schwartz, A., eds., Academic Press: Boston, 1991, pp. 139–152.

[145] Föllmer, H. and Kramkov, D., *Optional decompositions under constraints*, Probab. Theor. Relat. Fields, 109 (1997), pp. 1–25.

[146] Föllmer, H. and Leukert, P., *Efficient hedging: cost vs. shortfall risk*, Finance Stoch., 4 (2000), pp. 117–146.

[147] Föllmer, H. and Schied, A., *Stochastic Finance*, De Gruyter: Berlin, 2002.

[148] Föllmer, H. and Schweizer, M., *Hedging of contingent claims under incomplete information*, in Applied Stochastic Analysis, Davis, M. and Elliott, R., eds., Vol. 5, Gordon and Breach: London, 1991, pp. 389–414.

[149] Föllmer, H. and Sondermann, D., *Hedging of non-redundant contingent claims*, in Contributions to Mathematical Economics, Hildenbrand, W. and Mas-Colell, A., eds., North Holland: Amsterdam, 1986, pp. 205–223.

[150] Fouque, J.-P., Papanicolaou, G., and Sircar, K. R., *Derivatives in financial markets with stochastic volatility*, Cambridge University Press: Cambridge, 2000.

[151] Friedman, A. and Robin, M., *The free boundary for variational inequalities and non-local operators*, SIAM Journal of Control and Optimization, 16 (1978), pp. 347–372.

[152] Frisch, U. and Parisi, G., *Fully developed turbulence and intermittency*, in Turbulence and Predictability in Geophysical Fluid Dynamics and Climate Dynamics (Proc. Intl. Summer School Phys. Enrico Fermi), Ghil, M., ed., North Holland: Amsterdam, 1985, pp. 84–88.

[153] Fristedt, B. and Gray, L., *A Modern Approach to Probability Theory*, Birkhäuser: Boston, 1997.

[154] Frittelli, M., *The minimal entropy martingale measure and the valuation problem in incomplete markets*, Math. Finance, 10 (2000), pp. 39–52.

[155] Gallant, R. and Tauchen, G., *Which moments to match?*, Econometric Theory, 12 (1996), pp. 657–681.

[156] Galtchouk, L., *Représentation des martingales engendrées par un processus à accroissements indépendants (cas des martingales de carré intégrable)*, Ann. Inst. H. Poincaré Sect. B (N.S.), 12 (1976), pp. 199–211.

[157] Garroni, M. and Menaldi, J., *Second Order Elliptic Integro-Differential Problems*, CRC Press: Boca Raton, FL, 2001.

[158] Garroni, M. and Menaldi, J.-L., *Green Functions for Second Order Parabolic Integro-Differential Problems*, Vol. 275 of Pitman Research Notes in Mathematics Series, Longman Scientific & Technical: Harlow, 1992.

[159] Garroni, M. G. and Menaldi, J.-L., *Maximum principles for integro-differential parabolic operators*, Differential Integral Equations, 8 (1995), pp. 161–182.

[160] Geman, H., *Pure jump Lévy processes for asset price modeling*, Journal of Banking and Finance, 26 (2002), pp. 1297–1316.

[161] Geman, H., Madan, D., and Yor, M., *Asset prices are Brownian motion: Only in business time*, in Quantitative Analysis in Financial Markets, Avellaneda, M., ed., World Scientific: River Edge, NJ, 2001, pp. 103–146.

[162] Geman, H., Madan, D., and Yor, M., *Time changes for Lévy processes*, Math. Finance, 11 (2001), pp. 79–96.

[163] Gerber, H. and Shiu, E., *Pricing perptual options for jump processes*, North American actuarial journal, 2 (1998), pp. 101–112.

[164] Gikhman, I. and Skorokhod, A., *Introduction to the Theory of Random Processes*, Dover Publications Inc.: Mineola, NY, 1996. Translated from the 1965 Russian original.

[165] Glasserman, P., *Monte Carlo Methods in Financial Engineering*, Springer: New York, 2003.

[166] Glasserman, P. and Kou, S., *The term structure of forward rates with jump risk*, Math. Finance, 13 (2003), pp. 383–410.

[167] Glasserman, P. and Merener, N., *Cap and swaption approximations in LIBOR market models with jumps*, J. Comput. Finance, 7 (2003), pp. 1–36.

[168] Glasserman, P. and Merener, N., *Numerical solution of jump-diffusion LIBOR market models*, Finance Stoch., 7 (2003), pp. 1–27.

[169] Glowinski, R., Lawton, W., Ravachol, M., and Tenenbaum, E., *Wavelet solution of linear and nonlinear elliptic, parabolic and hyperbolic problems in one space dimension*, in Proceedings of the 9th International Conference on Numerical Methods in Applied Sciences and Engineering, SIAM: Philadelphia, 1990.

[170] Goll, T. and Kallsen, J., *Optimal portfolios for logarithmic utility*, Stochastic Process. Appl., 89 (2000), pp. 31–48.

[171] Goll, T. and Rüschendorf, L., *Minimax and minimal distance martingale measures and their relationship to portfolio optimization*, Finance Stoch., 5 (2001), pp. 557–581.

[172] Goll, T. and Rüschendorf, L., *Minimal distance martingale measures and optimal portfolios consistent with observed market prices*, in Stochastic Processes and Related Topics (Siegmundsburg, 2000), Taylor & Francis: London, 2002, pp. 141–154.

[173] Good, I., *Some statistical applications of Poisson's work*, Statist. Sci., 1 (1986), pp. 157–180.

[174] Gopikrishnan, P., Meyer, M., Amaral, L., and Stanley, H., *Inverse cubic law for the distribution of stock price variations*, European Physical Journal B, 3 (1998), pp. 139–140.

[175] Gourieroux, C. and Monfort, A., *Simulation Based Econometric Methods*, Oxford University Press: Oxford, UK, 1996.

[176] Grandits, P., *On martingale measure for stochastic processes with independent increments*, Teor. Veroyatnost. i Primenen., 44 (1999), pp. 87–100.

[177] Grosswald, , *The student t distribution of any degree of freedom is infinitely divisible*, Z. Wahrsch. Verw. Gebiete, 36 (1976), pp. 103–109.

[178] Grunewald, B., *Absicherungsstrategien für Optionen bei Kurssprüngen.*, Deutscher Universitäts-Verlag: Wiesbaden, 1998.

[179] Grunewald, B. and Trautmann, S., *Varianzminimierende hedgingstrategien für optionen bei möglichen kurssprüngen*, Zeitschrift für Betriebswirtschaftliche Forschung, Sonderheft 38 (1997), pp. 43–87.

[180] Guillaume, D., Dacorogna, M., Davé, R., Müller, U., Olsen, R., and Pictet, O., *From the birds eye view to the microscope: A survey of new stylized facts of the intraday foreign exchange markets*, Finance Stoch., 1 (1997), pp. 95–131.

[181] Guillaume, D., Dacorogna, M., Davé, R., Müller, U., Olsen, R., and Pictet, O., *Volatilities at different time resolutions: analyzing the dynamics of market components*, Journal of Empirical Finance, 4 (1997), pp. 213–239.

[182] Gukhal, C., *Analytical valuation of American options on jump-diffusion processes*, Math. Finance, 11 (2001), pp. 97–115.

[183] Halgreen, C., *Self-decomposability of the generalized inverse Gaussian and hyperbolic distributions*, Z. Wahrsch. Verw. Gebiete, 47 (1979), pp. 13–18.

[184] Hall, J., Brorsen, B., and Irwin, S., *The distribution of future prices: A test of the stable Paretian and mixture of normals hypothesis*, J. Fin. and Quant. Anal., 24 (1989), pp. 105–116.

[185] Halsey, T., Jensen, M., Kadanoff, L., Procaccia, I., and Shraiman, B., *Fractal measures and their singularities: The characterization of strange sets.*, Physical Review A, 33 (1986), pp. 1141–1151.

[186] Hamilton, J., *Time Series*, Princeton University Press: Princeton, 1994.

[187] Hansen, L., *Large sample properties of generalized method of moments estimators*, Econometrica, 50 (1982), pp. 1029–1054.

[188] Härdle, W., *Applied Nonparametric Regression*, Cambridge University Press: Cambridge, UK, 1985.

[189] Härdle, W. and Linton, O., *Applied non-parametric methods*, in Handbook of Econometrics, Engle, R. and McFadden, J., eds., Vol. IV, Elsevier Science: Amsterdam, 1994, Ch. 38, pp. 567–611.

[190] Harrison, J. and Kreps, D., *Martingales and arbitrage in multiperiod security markets*, J. Economic Theory, 2 (1979), pp. 381–408.

[191] Harrison, J. M. and Pliska, S. R., *Martingales and stochastic integrals in the theory of continuous trading*, Stochastic Process. Appl., 11 (1981), pp. 215–260.

[192] Harrison, J. M. and Pliska, S. R., *A stochastic calculus model of continuous trading: Complete markets*, Stochastic Process. Appl., 15 (1983), pp. 313–316.

[193] He, H., *Convergence from discrete time to continuous time contingent claims prices*, Rev. Fin. Studies, 3 (1990), pp. 523–546.

[194] He, S.-W., Wang, J.-G., and Yan, J.-A., eds., *Semimartingale Theory and Stochastic Calculus*, CRC Press: Boca Raton, 1992.

[195] Heath, D., Platen, E., and Schweizer, M., *Numerical comparison of local risk minimization and mean-variance hedging*, in Jouini and Pliska [225], pp. 509–538.

[196] Heston, S., *A closed-form solution for options with stochastic volatility with applications to bond and currency options*, Rev. Fin. Studies, 6 (1993), pp. 327–343.

[197] Hodges, S. and Neuberger, A., *Optimal replication of contingent claims under transaction costs*, Review of futures markets, 8 (1989), pp. 222–239.

[198] Hols, M. and De Vries, C., *The limiting distribution of extremal exchange rate returns*, Journal of Applied Econometrics, 6 (1991), pp. 287–302.

[199] Honoré, P., *Pitfalls in maximum likelihood estimation of jump-diffusion models*, tech. rep., University of Aarhus Graduate School of Business, 1995.

[200] Hubalek, F. and Krawczyk, L., *Variance optimal hedging and Markowitz efficient portfolios for processes with stationary independent increments*, working paper, Vienna Technical University, April 2003.

[201] Hugonnier, J. and Kramkov, D., *Utility maximization in incomplete markets with random endowments.* Forthcoming in Ann. App. Probab., 2003.

[202] Hull, J., *Options, Futures and Other Derivative Securities*, Prentice Hall: Upper Saddle River, NJ, 1997.

[203] Hull, J. and White, A., *The pricing of options on assets with stochastic volatilities*, J. Finance, XLII (1987), pp. 281–300.

[204] Hurd, T. and Choulli, T., *The role of Hellinger processes in mathematical finance*, Entropy, 3 (2001), pp. 141–152.

[205] Ikeda, N. and Watanabe, S., *Stochastic Differential Equations and Diffusion Processes*, Kodansha: Tokyo, 1981.

[206] Itô, K., *On stochastic processes 1 (infinitely divisible laws of probability)*, Japan J. Math., 18 (1942). Reprinted in *Kiyosi Itô Selected Papers*, Springer: New York, 1987.

[207] Itô, K., *Stochastic differentials*, Appl. Math. Optim., 1 (1974/75), pp. 374–381.

[208] Jacob, N., *Pseudo-Differential Operators and Markov Processes, Volume I: Fourier Analysis and Semi-Groups*, World Scientific: Singapore, 2001.

[209] Jacob, N. and Schilling, R., *Pseudodifferential operators and Lévy processes*, in Lévy Processes—Theory and Applications, Barndorff-Nielsen, O., Mikosch, T., and Resnick, S., eds., Birkhäuser: Boston, 2001, pp. 139–168.

[210] Jacod, J., *Calcul Stochastique et Problèmes de Martingales*, Vol. 714 of Lecture Notes in Math., Springer: Berlin, 1979.

[211] Jacod, J., *Sharp estimates for the Euler scheme for Lévy driven stochastic differential equations*, Prépublication 656, Laboratoire de Probabilités, Université de Paris VI, 2001.

[212] Jacod, J., *The Euler scheme for Lévy driven stochastic differential equations: Limit theorems*, Prépublication 711, Laboratoire de Probabilités, Université Paris VI, 2002.

[213] Jacod, J., Méléard, S., and Protter, P., *Explicit form and robustness of martingale representations*, Annals of Probability, 28 (2000), pp. 1747–1780.

[214] Jacod, J. and Protter, P., *Probability Essentials*, Springer: Berlin, 2000.

[215] Jacod, J. and Shiryaev, A. N., *Limit Theorems for Stochastic Processes*, Springer: Berlin, 2nd ed., 2002.

[216] Jaffard, S., *Multifractal formalism for functions. Part I: Results valid for all functions*, SIAM Journal on Mathematical Analysis, 28 (1997), pp. 944–970.

[217] Jaffard, S., *Multifractal formalism for functions. Part II: Self-similar functions*, SIAM Journal on Mathematical Analysis, 28 (1997), pp. 971–998.

[218] Jaffard, S., *The multifractal nature of Lévy processes*, Probab. Theor. Relat. Fields, 114 (1999), pp. 207–227.

[219] Jaillet, P., Lamberton, D., and Lapeyre, B., *Variational inequalities and the pricing of American options*, Acta Appl. Math., 21 (1990), pp. 263–289.

[220] Jakobsen, E. and Karlsen, K., *A maximum principle for semicontinuous functions applicable to integro-partial differential equations*, working paper, Dept. of Mathematics, University of Oslo, 2003.

[221] Jansen, D. and De Vries, C., *On the frequency of large stock returns*, Rev. Econ. Stat., 73 (1991), pp. 18–24.

[222] Jiang, G., *Implementing asset pricing models when assets are predictable and discontinuous*, in Knight and Satchell [236], Ch. 7, pp. 167–224.

[223] Joe, H., *Multivariate models and dependence concepts*, Chapman & Hall: London, 1997.

[224] Jorion, P., *On jump processes in the foreign exchange and stock markets*, Rev. Fin. Studies, 1 (1988), pp. 427–445.

[225] Jouini, E. and Pliska, S., eds., *Option Pricing, Interest Rates and Risk Management*, Cambridge University Press: Cambridge, 2001.

[226] Kabanov, Y., *Arbitrage theory*, in Jouini and Pliska [225], pp. 3–42.

[227] Kallenberg, O., *Random Measures*, Akademie-Verlag: Berlin, 1976.

[228] Kallenberg, O., *Foundations of Modern Probability*, Springer: New York, 1997.

[229] Kallsen, J. and Shiryaev, A. N., *The cumulant process and Esscher's change of measure*, Finance Stoch., 6 (2002), pp. 397–428.

[230] Karatzas, I. and Shreve, S., *Methods of Mathematical Finance*, Vol. 39 of Applications of Mathematics, Springer: Berlin, 2000.

[231] Karoui, N. E. and Quenez, M., *Dynamic programming and pricing of contingent claims in an incomplete market*, SIAM J. Control and Optimization, 33 (1995), pp. 29–66.

[232] Kearns, P. and Pagan, A., *Estimating the density tail index for financial time series*, Review of Economics and Statistics, 79 (1997), pp. 171–175.

[233] Kendall, D., *Obituary: Paul Lévy*, J. R. Statistic. Soc. A, 137 (1974), pp. 259–260.

[234] Kesten, H., *Hitting probabilities of single points for processes with independent increments*, Mem. Amer. Math. Soc., 178 (1969), pp. 1473–1523.

[235] Kingman, J., *Poisson Processes*, Vol. 3 of Oxford Studies in Probability, Oxford University Press: New York, 1993.

[236] Knight, J. and Satchell, S., eds., *Return Distributions in Finance*, Butterworth Heinemann: Oxford, 2001.

[237] Koponen, I., *Analytic approach to the problem of convergence of truncated Lévy flights towards the Gaussian stochastic process.*, Physical Review E, 52 (1995), pp. 1197–1199.

[238] Kou, S., *A jump-diffusion model for option pricing*, Management Science, 48 (2002), pp. 1086–1101.

[239] Kou, S. and Wang, H., *Option pricing under a jump-diffusion model.* Working Paper, 2001.

[240] Kramkov, D., *Optional decomposition of supermartingales and hedging contingent claims in incomplete security markets*, Probab. Theor. Relat. Fields, 105 (1996), pp. 459–479.

[241] Kunita, H., *Representation of martingales with jumps and applications to mathematical finance.* 2003.

[242] Kunita, H. and Watanabe, S., *On square integrable martingales*, Nagoya Math. J., 30 (1967), pp. 209–245.

[243] Kushner, H. J. and Dupuis, P., *Numerical Methods for Stochastic Control Problems in Continuous Time*, Vol. 24 of Applications of Mathematics, Springer: New York, 2nd ed., 2001.

[244] Kyprianou, A. and Palmowski, Z., *A martingale review of some fluctuation theory for spectrally negative Lévy processes.* 2003.

[245] Lamberton, D. and Lapeyre, B., *Introduction au Calcul Stochastique Appliqué à la Finance*, Ellipses: Paris, 1997. English translation: *Introduction to Stochastic Calculus Applied to Finance*, Chapman & Hall: London, 1996.

[246] Lando, D., *On Cox processes and credit risky securities*, Rev. Derivatives Research, 2 (1998), pp. 99–120.

[247] Le Cam, L., *Paul Lévy, 1886–1971*, in Proceedings of the Sixth Berkeley Symposium on Mathematical Statistics and Probability (Univ. California, Berkeley, Calif., 1970/1971), Vol. III: Probability theory, University of California Press: Berkeley, CA, 1972, pp. xiv–xx.

[248] Léandre, R., *Estimation dans $L^p(\mathbf{R}^n)$ de la loi de certains processus à accroissements indépendants*, in Séminaire de probabilités, XIX, 1983/84, Vol. 1123 of Lecture Notes in Math., Springer: Berlin, 1985, pp. 263–270.

[249] LePage, R., *Multidimensional infinitely divisible variables and processes II*, in Lecture Notes in Math., Vol. 860, Springer: Berlin, 1980, pp. 279–284.

[250] Levendorskii, S., *Pricing of the American Put under Lévy processes*, Research Report 2002-44, Maphysto, 2002.

[251] Lévy, P., *Sur les intégrales dont les éléments sont des variables aléatoires indépendantes*, Ann. Scuola Norm. Sup. Pisa, 3,4 (1934). Reprinted in [257].

[252] Lévy, P., *Théorie de l'Addition des Variables Aléatoires*, Gauthier Villars: Paris, 1937.

[253] Lévy, P., *Quelques aspects de la pensée d'un mathématicien. Introduction. Première partie: Souvenirs mathématiques. Deuxième partie: Considérations philosophiques*, Librairie Scientifique et Technique Albert Blanchard: Paris, 1970.

[254] Lévy, P., *Œuvres de Paul Lévy. Vol. I*, Gauthier-Villars.: Paris, 1973. Analyse, Publiées sous sa direction par Daniel Dugué avec la collaboration de Paul Deheuvels et Michel Ibéro.

[255] Lévy, P., *Œuvres de Paul Lévy. Vol. II*, Gauthier-Villars: Paris, 1974. Analyse-Géométrie. Physique théorique.

[256] Lévy, P., *Œuvres de Paul Lévy. Vol. III*, Gauthier-Villars: Paris, 1976. Éléments aléatoires.

[257] Lévy, P., *Œuvres de Paul Lévy. Vol. IV*, Gauthier-Villars: Paris, 1980. Processus stochastiques.

[258] Lévy, P., *Œuvres de Paul Lévy. Vol. V*, Gauthier-Villars: Paris, 1980. Mouvement brownien.

[259] Lévy, P., *Œuvres de Paul Lévy. Vol. VI*, Gauthier-Villars: Paris, 1980. Théorie des jeux.

[260] Lévy, P., *Processus Stochastiques et Mouvement Brownien*, Editions Jacques Gabay: Sceaux, 1992.

[261] Lewis, A., *Option Valuation under Stochastic Volatility*, Finance Press, 2000.

[262] Lewis, A., *A simple option formula for general jump-diffusion and other exponential Lévy processes.* available from `http://www.optioncity.net`, 2001.

[263] Lindskog, F. and McNeil, A., *Common Poisson shock models: Applications to insurance and credit risk modelling.* Available from `www.risklab.ch`, September 2001.

[264] Loève, M., *Paul Lévy, 1886–1971*, Ann. Probability, 1 (1971), pp. 1–18.

[265] Longin, F., *The asymptotic distribution of extreme stock market returns*, Journal of Business, 69 (1996), pp. 383–408.

[266] Longstaff, F. and Schwartz, E., *Valuing American options by simulation: a simple least-squares approach*, Rev. Fin. Studies, 14 (2001), pp. 113–147.

[267] Loretan, M. and Phillips, P., *Testing the covariance stationarity of heavy-tailed time series*, Journal of empirical finance, 1 (1994), pp. 211–248.

[268] Lux, T., *On moment condition failure in German stock returns*, Empirical economics, 25 (2000), pp. 641–652.

[269] Ma, J., Protter, P., and Zhang, J., *Explicit representations and path regularity for martingale representations*, in Lévy Processes—Theory and Applications, Barndorff-Nielsen, O., Mikosch, T., and Resnick, S., eds., Birkhäuser: Boston, 2001.

[270] Madan, D., *Financial modeling with discontinuous price processes*, in Lévy Processes—Theory and Applications, Barndorff-Nielsen, O., Mikosch, T., and Resnick, S., eds., Birkhäuser: Boston, 2001.

[271] Madan, D., Carr, P., and Chang, E., *The variance gamma process and option pricing*, European Finance Review, 2 (1998), pp. 79–105.

[272] Madan, D. and Milne, F., *Option pricing with variance gamma martingale components*, Math. Finance, 1 (1991), pp. 39–55.

[273] Madan, D. and Seneta, E., *Simulation of estimates using the empirical characteristic function*, Internat. Statist. Rev., 55 (1987), pp. 153–161.

[274] Maddala, G. and Rao, C., eds., *Statistical Methods in Finance*, Vol. 14 of Handbook of Statistics, Elsevier, 1996.

[275] Malevergne, Y., Pisarenko, V., and Sornette, D., *Empirical distributions of log-returns: Between the stretched exponential and the power law?*, Arxiv Preprint 0305089, UCLA, 2003.

[276] Mancini, C., *Disentangling the jumps from the diffusion in a geometric jumping Brownian motion*, Giornale dell'Istituto Italiano degli Attuari, LXIV (2001), pp. 19–47.

[277] Mandelbrot, B., *The fractal geometry of nature*, Freeman: San Francisco, 1982.

[278] Mandelbrot, B. B., *The variation of certain speculative prices*, Journal of Business, XXXVI (1963), pp. 392–417.

[279] Mandelbrot, B. B., *Fractals and Scaling in Finance: Discontinuity, Concentration, Risk.*, Springer: New York, 1997.

[280] Mandelbrot, B. B., *Scaling in financial prices. I. Tails and dependence*, Quant. Finance, 1 (2001), pp. 113–123.

[281] Mandelbrot, B. B., *Scaling in financial prices. II. Multifractals and the star equation*, Quant. Finance, 1 (2001), pp. 124–130.

[282] Mandelbrot, B. B., Calvet, L., and Fisher, A., *The multifractality of the Deutschmark/US Dollar exchange rate*, Discussion paper 1166, Cowles Foundation for Economics: Yale University, 1997.

[283] Mandelbrot, B. B. and Van Ness, J., *Fractional Brownian motion, fractional noises and applications*, SIAM Review, 10 (1968), pp. 422–437.

[284] Mantegna, R. and Stanley, H., *Stochastic process with ultraslow convergence to a Gaussian: the truncated Lévy flight*, Physical Review Letters, 73 (1994), pp. 2946–2949.

[285] Mantegna, R. and Stanley, H., *Scaling behavior of an economic index*, Nature, 376 (1995), pp. 46–49.

[286] Matache, A. and Schwab, C., *Wavelet Galerkin methods for American options on Lévy-driven assets*, working paper, ETHZ: Zurich, 2003.

[287] Matache, A., von Petersdorff, T., and Schwab, C., *Fast deterministic pricing of options on Lévy driven assets*, ESAIM: M2AN, (Forthcoming).

[288] Matacz, A., *Financial modeling and option theory with the truncated Lévy process*, Int. J. Theor. Appl. Finance, 3 (2000), pp. 143–160.

[289] McCulloch, J., *Financial applications of stable distributions*, in Maddala and Rao [274], Ch. 14, pp. 393–425.

[290] Merton, R., *Theory of rational option pricing*, Bell Journal of Economics, 4 (1973), pp. 141–183.

[291] Merton, R., *Option pricing when underlying stock returns are discontinuous*, J. Financial Economics, 3 (1976), pp. 125–144.

[292] Metwally, S. and Atiya, A., *Using Brownian bridge for fast simulation of jump-diffusion processes and barrier options*, Journal of Derivatives, 10 (2002), pp. 43–54.

[293] Meyer, G., *The numerical valuation of options with underlying jumps*, Acta Math. Univ. Comenian., 67 (1998), pp. 69–82.

[294] Meyer, G. and Van Der Hoek, J., *The evaluation of American options with the method of lines*, Advances in Futures and Options Research, 9 (1997), pp. 265–285.

[295] Meyer, P., *Processus de Poisson ponctuels, d'après K. Itô*, in Séminaire de Probabilités, V (Univ. Strasbourg, année universitaire 1969–1970), Vol. 191 of Lecture Notes in Math., Springer: Berlin, 1971, pp. 177–190.

[296] Meyer, P. A., *Un cours sur les intégrales stochastiques*, in Séminaire de Probabilités, Vol. 511 of Lecture Notes in Math., Springer: Berlin, 1976, pp. 245–398.

[297] Mikulyavichyus, R. and Pragarauskas, G., *On the uniqueness of solutions of the martingale problem that is associated with degenerate Lévy operators*, Lithuanian Mathematics Journal, 33 (1994), pp. 352–367.

[298] Miyahara, Y., *Minimal entropy martingale measures of jump type price processes in incomplete assets markets*, Asia-Pacific Financial Markets, 6 (1999), pp. 97–113.

[299] Miyahara, Y. and Fujiwara, T., *The minimal entropy martingale measures for geometric Lévy processes*, Finance Stoch., 7 (2003), pp. 509–531.

[300] Morozov, V., *On the solution of functional equations by the method of regularization*, Soviet Math. Doklady, 7 (1966), pp. 414–417.

[301] Müller, U., Dacorogna, M., and Pictet, O., *Heavy tails in high-frequency financial data*, in A Practical Guide to Heavy Tails: Statistical Techniques for Analysing Heavy Tailed Distributions, Adler, R., Feldman, R., and Taqqu, M., eds., Birkhäuser: Boston, 1998, pp. 55–77.

[302] Muzy, J., Delour, J., and Bacry, E., *Modeling fluctuations of financial time series: From cascade processes to stochastic volatility models*, Eur. J. Phys. B, 17 (2000), pp. 537–548.

[303] Nahum, E. and Samuelides, Y., *A tractable market model with jumps for pricing short-term interest rate derivatives*, Quant. Finance, 1 (2001), pp. 270–283.

[304] Nakano, Y., *Minimization of shortfall risk in a jump-diffusion model*, working paper, Hokkaido University, 2002.

[305] Nelsen, R., *An Introduction to Copulas*, Springer: New York, 1999.

[306] Neveu, J., *Processus ponctuels*, in École d'Été de Probabilités de Saint-Flour VI—1976, Vol. 598 of Lecture Notes in Math., Springer: Berlin, 1977, pp. 249–445.

[307] Nicolato, E., *Stochastic volatility models of Ornstein-Uhlenbeck type*, PhD thesis, Aarhus University: Aarhus, 2000.

[308] Nicolato, E. and Venardos, E., *Option pricing in stochastic volatility models of Ornstein-Uhlenbeck type*, Math. Finance, 13 (2003), pp. 445–466.

[309] Norris, J., *Integration by parts for jump processes*, in Séminaire de Probabilités, XXII, Vol. 1321 of Lecture Notes in Math., Springer: Berlin, 1988, pp. 271–315.

[310] Nualart, D. and Schoutens, W., *Chaotic and predictable representations for Lévy processes*, Stochastic Process. Appl., 90 (2000), pp. 109–122.

[311] Nualart, D. and Schoutens, W., *Backward stochastic differential equations and Feynman-Kac formula for Lévy processes, with applications in finance*, Bernoulli, 7 (2001), pp. 761–776.

[312] Orey, S., *On continuity properties of infinitely divisible distributions*, Annals of mathematical statistics, 39 (1968), pp. 936–937.

[313] Pagan, A., *The econometrics of financial markets*, Journal of Empirical Finance, 3 (1996), pp. 15–102.

[314] Pan, J., *The jump-risk premia implicit in options: evidence from an integrated time-series study*, Journal of financial economics, 63 (2002), pp. 3–50.

[315] Petrov, V., *Sums of Independent Random Variables*, Springer: Berlin, 1975.

[316] Pham, H., *Optimal stopping, free boundary, and American option in a jump-diffusion model*, Appl. Math. Optim., 35 (1997), pp. 145–164.

[317] Pham, H., *Optimal stopping of controlled jump-diffusion processes: A viscosity solution approach*, Journal of Mathematical Systems Estimation and Control, 8 (1998), pp. 1–27.

[318] Pham, H., *Hedging and optimization problems in continuous financial models*, in Mathematical Finance: Theory and Practice, Cont, R. and Yong, J., eds., Vol. 1 of Series in Contemporary Applied Mathematics, Higher Education Press: Beijing, 2000, pp. 353–381.

[319] Prause, K., *The Generalized Hyperbolic Model: Estimation, Financial Derivatives, and Risk Measures*, PhD thesis, Universität Freiburg i. Br., 1999.

[320] Press, S., *A compound events model for security prices*, Journal of Business, 40 (1967), pp. 317–335.

[321] Press, W. H., Teukolsky, S. A., Vetterling, W. T., and Flannery, B. P., *Numerical Recipes in C: the Art of Scientific Computing*, Cambridge University Press: Cambridge, 1992.

[322] Prigent, J., *Weak Convergence of Financial Markets*, Springer: New York, 2002.

[323] Protter, P., *Stochastic integration without tears*, Stochastics, 16 (1986), pp. 295–325.

[324] Protter, P., *Stochastic integration and differential equations*, Springer: Berlin, 1990.

[325] Protter, P., *A partial introduction to financial asset pricing theory*, Stochastic Process. Appl., 91 (2001), pp. 169–203.

[326] Protter, P. and Dritschel, M., *Complete markets with discontinuous security price*, Finance Stoch., 3 (1999), pp. 203–214.

[327] Rachev, S., ed., *Handbook of heavy-tailed distributions in finance*, Elsevier: Amsterdam, 2003.

[328] Rachev, S. and Mittnik, S., *Stable paretian models in finance*, Wiley: New York, 2000.

[329] Raible, S., *Lévy processes in finance: theory, numerics and empirical facts*, PhD thesis, Freiburg University, 1998.

[330] Rebonato, R., *Volatility and Correlation in the Pricing of Equity, FX and Interest Rate Options*, Wiley: Chichester, 1999.

[331] Reiss, R. and Thomas, M., *Statistical Analyis of Extreme Values*, Birkhäuser: Basel, 2nd ed., 2001.

[332] Renault, E. and Touzi, N., *Option hedging and implied volatilities in a stochastic volatility model*, Math. Finance, 6 (1996), pp. 279–302.

[333] Resnick, S., *Extreme values, regular variation and point processes*, Springer: Berlin, 1987.

[334] Resnick, S., *Adventures in stochastic processes*, Birkhäuser: Boston, 1992.

[335] Revuz, D. and Yor, M., *Continuous Martingales and Brownian Motion*, Springer: Berlin, 1999.

[336] Rheinländer, T. and Schweizer, M., *On L^2 projections on a space of semimartingales*, Annals of probability, 25 (1997), pp. 1810–1831.

[337] Rong, S., *On solutions of backward stochastic differential equations with jumps and applications*, Stochastic Process. Appl., 66 (1997), pp. 209–236.

[338] Rosiński, J., *Series representations of Lévy processes from the perspective of point processes*, in Lévy Processes—Theory and Applications, Barndorff-Nielsen, O., Mikosch, T., and Resnick, S., eds., Birkhäuser: Boston, 2001.

[339] Ross, S., *Options and efficiency*, Quarterly Journal of Economics, 90 (1976), pp. 75–89.

[340] Rubenthaler, S., *Numerical simulation of the solution of a stochastic differential equation driven by a Lévy process*, Stochastic Process. Appl., 103 (2003), pp. 311–349.

[341] Rüschendorf, L. and Woerner, J. H., *Expansion of transition distributions of Lévy processes in small time*, Bernoulli, 8 (2002), pp. 81–96.

[342] Rydberg, T. H., *The normal inverse Gaussian Lévy process: simulation and approximation*, Comm. Statist. Stochastic Models, 13 (1997), pp. 887–910.

[343] Rydberg, T. H. and Shephard, N., *Dynamics of trade-by-trade mouvements: decomposition and models*, Journal of Financial Econometrics, 1 (2003), pp. 2–25.

[344] Samorodnitsky, G. and Taqqu, M., *Stable Non-Gaussian Random Processes*, Chapman & Hall: New York, 1994.

[345] Sato, K., *Lévy Processes and Infinitely Divisible Distributions*, Cambridge University Press: Cambridge, UK, 1999.

[346] Sato, K.-I., *Self-similar processes with stationary increments*, Probab. Theor. Relat. Fields, 89 (1991), pp. 285–300.

[347] Savage, L. J., *The Foundations of Statistics*, John Wiley & Sons Inc.: New York, 1954.

[348] Sayah, A., *Equations d'hamilton jacobi du premier order avec termes integro-differenitels: Parties i et ii*, Comm. Partial Differential Equations, 16 (1991), pp. 1057–1093.

[349] Schachermayer, W., *No arbitrage: on the work of David Kreps*, Positivity, 6 (2002), pp. 359–368.

[350] Schachermayer, W., *A super-martingale property of the optimal portfolio process.*, Finance Stoch., 7 (2003), pp. 433–456.

[351] Schonbucher, P., *Credit derivatives pricing models*, Wiley: Chichester, 2002.

[352] Schoutens, W., *Lévy Processes in Finance: Pricing Financial Derivatives*, Wiley: New York, 2003.

[353] Schwab, C. and Schotzau, D., *Time discretization of parabolic problems by the hp-version of the discontinuous Galerkin finite element method*, SIAM J. Numer. Analysis, 83 (2000), pp. 837–875.

[354] Schwartz, L., *Quelques réflexions et souvenirs sur Paul Lévy*, Astérisque, (1988), pp. 13–28. Colloque Paul Lévy sur les Processus Stochastiques (Palaiseau, 1987).

[355] Schweizer, M., *Risk-minimality and orthogonality of martingales*, Stochastics Stochastics Rep., 30 (1990), pp. 123–131.

[356] Schweizer, M., *On the minimal martingale measure and the Föllmer-Schweizer decomposition*, Stochastic Anal. Appl., 13 (1995), pp. 573–599.

[357] Schweizer, M., *A guided tour through quadratic hedging approaches*, in Jouini and Pliska [225], pp. 538–574.

[358] Scott, L. O., *Pricing stock options in a jump-diffusion model with stochastic volatility and interest rates: applications of Fourier inversion methods*, Math. Finance, 7 (1997), pp. 413–426.

[359] Sepp, A., *Analytical pricing of lookback options under a double-exponential jump diffusion process.* Available from the author's Web site, 2003.

[360] Sepp, A., *Pricing double-barrier options under a double-exponential jump-diffusion process: Applications of Laplace transform.* Available from the author's Web site, 2003.

[361] Shephard, N., *Statistical aspects of ARCH and stochastic volatility*, in Time Series Models, Chapman and Hall: London, 1980, pp. 1–67.

[362] Sheynin, O., *S. D. Poisson's work in probability*, Arch. History Exact Sci., 18 (1977/78), pp. 245–300.

[363] Shirakawa, H., *Interest rate option pricing with Poisson-Gaussian forward rate dynamics*, Math. Finance, 1 (1991).

[364] Shiryaev, A., ed., *Probability theory. III. Stochastic Calculus*, Vol. 45 of Encyclopaedia of Mathematical Sciences, Springer: Berlin, 1998. A translation of *Current problems in mathematics. Vol. 45* (Russian), Akad. Nauk SSSR, Vsesoyuz. Inst. Nauchn. i Tekhn. Inform.: Moscow, 1989.

[365] Sklar, A., *Random variables, distribution functions, and copulas—a personal look backward and forward*, in Distributions with Fixed Marginals and Related Topics, Rüschendorf, L., Schweizer, B., and Taylor, M. D., eds., Institute of Mathematical Statistics: Hayward, CA, 1996.

[366] Soner, H., *Jump Markov Processes and Viscosity Solutions*, Vol. 10 of IMA Volumes in mathematics and applications, Springer Verlag: New York, 1986, pp. 501–511.

[367] Sørensen, M., Bibby, B., and Skovgaard, I., *Diffusion-type models with given marginal distribution and autocorrelation function*, Preprint 2003-5, University of Copenhagen, 2003.

[368] Stein, E. and Stein, J., *Stock price distributions with stochastic volatility: An analytic approach*, Rev. Fin. Studies, 4 (1991), pp. 727–752.

[369] Stricker, C., *On the utility indifference price*, Prépublication 30, Laboratoire de Mathématiques de Besançon: Besançon, 2002.

[370] Stroock, D. W., *Markov processes from K. Itô's perspective*, Princeton University Press: Princeton, 2003.

[371] Stutzer, M., *A simple nonparametric approach to derivative security valuation*, Journal of Finance, 101 (1997), pp. 1633–1652.

[372] Taleb, N., *Dynamic Hedging: Managing Vanilla and Exotic Options*, John Wiley & Sons: New York, 1997.

[373] Tankov, P., *Dependence structure of Lévy processes with applications in risk management*, Rapport Interne 502, CMAP, Ecole Polytechnique, 2003.

[374] Taqqu, M., *A bibliographical guide to self-similar processes and long range dependence*, in Dependence in Probability and Statistics, Eberlein, E. and Taqqu, M., eds., Birkhäuser: Boston, 1986, pp. 137–162.

[375] Taqqu, M., *Bachelier and his time: a conversation with bernard bru*, Finance Stoch., 5 (2001), pp. 3–32.

[376] Tavella, D. and Randall, C., *Pricing Financial Instruments: the Finite Difference Method*, Wiley: New York, 2000.

[377] Taylor, S., *Paul Lévy*, Bull. London Math. Soc., 7 (1975), pp. 300–320.

[378] Thomée, V., *Galerkin Finite Element Methods for Parabolic Problems*, Vol. 25 of Series in Computational Mathematics, Springer: Berlin, 1997.

[379] Tompkins, R., *Stock index futures markets: Volatility models and smiles*, Journal of Futures Markets, 21 (2001), pp. 43–78.

[380] Von Neumann, J. and Morgenstern, O., *Theory of Games and Economic Behavior*, Princeton University Press: Princeton, New Jersey, 1944.

[381] Wiktorsson, M., *Improved convergence rate for the simulation of Lévy processes of type G*. Available from the author's Web site, 2001.

[382] Willinger, W. and Taqqu, M., *Pathwise stochastic integration and applications to the theory of continuous trading*, Stochastic Process. Appl., 32 (1989), pp. 253–280.

[383] Wilmott, P., *Derivatives*, Wiley: New York, 2000.

[384] Winkel, M., *The recovery problem for time-changed Lévy processes*, Research Report 2001-37, Maphysto, October 2001.

[385] Yor, M. and de Sam Lazaro, J., *Sous-espaces denses de l^1 ou h^1 et représentation de martingales*, in Séminaire de Probabilités, Vol. 649 of Lecture Notes in Math., Springer: Berlin, 1978, pp. 265–309.

[386] Yor, M. and Nguyen-Ngoc, L., *Wiener-Hopf factorization and pricing of barrier and lookback options under general Lévy processes*. 2002.

[387] Yu, J., *Testing for finite variance in stock returns distributions*, in Knight and Satchell [236], Ch. 6, pp. 143–164.

[388] Zajdenweber, D., *Propriétés autosimilaires du CAC40*, Revue d'Economie Politique, 104 (1994), pp. 408–434.

[389] Zhang, X., *Options américaines et modèles de diffusion avec sauts*, C. R. Acad. Sci. Paris Sér. I Math., 317 (1993), pp. 857–862.

[390] Zhang, X., *Analyse Numérique des Options Américaines dans un Modèle de Diffusion avec Sauts*, PhD thesis, Ecole Nationale des Ponts et Chaussées, 1994.

[391] Zhang, X., *Numerical analysis of American option pricing in a jump-diffusion model*, Math. Oper. Res., 22 (1997), pp. 668–690.

[392] Zhang, X., *Valuation of American options in a jump-diffusion model*, in Numerical methods in finance, Cambridge University Press: Cambridge, 1997, pp. 93–114.

[393] Zhu, J., *Modular Pricing of Options: An Application of Fourier Analysis*, Springer: Berlin, 2000.

[394] Zolotarev, V., *One Dimensional Stable Distributions*, American Mathematical Society: Providence, RI, 1986.

[395] Zvan, R., Vetzal, K., and Forsyth, P., *Swing low, swing high*, RISK, 11 (1998), pp. 71–75.

Symbol Description

Symbol	Description
$\sigma(\mathcal{A})$	The smallest σ-algebra, with respect to which $\mathcal{A}$ is measurable. $\mathcal{A}$ may be a collection of subsets or a collection of functions
$\#A$	Number of elements in A
$\mathbb{P}(A)$	Probability of event A
$E[X]$	Expectation of random variable X
Φ_X	Characteristic function of a random variable X
Ψ_X	Cumulant generating function) of a random variable X
$\phi_t(.)$	Characteristic function of a Lévy process
$\psi(.)$	Characteristic exponent of a Lévy process
M_X	Moment generating function of X
J_X	Jump measure of a cadlag process X
$X \stackrel{d}{=} Y$	X and Y have the same distribution
$X_n \stackrel{d}{\to} X$	(X_n) converges to X in distribution
$\mu_n \Rightarrow \mu$	(μ_n) converges weakly to μ
$X_n \stackrel{\mathbb{P}}{\to} X$	(X_n) converges to X in probability
i.i.d.	Independent and identically distributed
a.s.	Almost surely
a.e.	Almost everywhere
$\mathcal{B}(E)$	Borel σ-algebra of E
$\omega \in \Omega$	scenario of randomness
$\mathbf{a}.\mathbf{b}$	Scalar product of vectors $\mathbf{a}$ and $\mathbf{b}$

Subject index

Underlined numbers show the pages where definitions are given.

Printed in the United States
by Baker & Taylor Publisher Services